97-IRM-2

METALS REMOVAL TECHNOLOGIES FOR URBAN STORMWATER

by:

Pauline D. Johnson
Robert Pitt
S. Rocky Durrans
Matilda Urrutia
The University of Alabama, Tuscaloosa

Shirley Clark
The University of Alabama, Birmingham

2003

Water Environment Research Foundation

The Water Environment Research Foundation, a not-for-profit organization, funds and manages water quality research for its subscribers through a diverse public-private partnership between municipal utilities, corporations, academia, industry, and the federal government. WERF subscribers include municipal and regional water and wastewater utilities, industrial corporations, environmental engineering firms, and others that share a commitment to cost-effective water quality solutions. WERF is dedicated to advancing science and technology addressing water quality issues as they impact water resources, the atmosphere, the lands, and quality of life.

For more information, contact:
Water Environment Research Foundation
635 Slaters Lane, Suite 300
Alexandria, VA 22314-1177
Tel: (703) 684-2470
Fax: (703) 299-0742
www.werf.org
werf@werf.org

This report was co-published by the following organizations. For non-subscriber sales information, contact:

Water Environment Federation
601 Wythe Street
Alexandria, VA 22314-1994
Tel: (800) 666-0206
Tel: (703) 684-2452
Fax: (703) 684-2492
www.wef.org
pubs@wef.org

IWA Publishing
Alliance House, 12 Caxton Street
London SW1H 0QS, United Kingdom
Tel: +44 (0) 20 7654 5500
Fax: +44 (0) 20 7654 5555
www.iwapublishing.com
publications@iwap.co.uk

Library of Congress Catalog Card Number: 2003105658

WERF ISBN: 978-1-893664-93-7 IWAP ISBN: 978-1-84339-682-6

The University of Alabama, Tuscaloosa
The University of Alabama, Birmingham

ACKNOWLEDGMENTS

The project team would like to acknowledge and thank USI Infrastructure and Dr. Stephen Allen of the Chemical Engineering Department at Queens University, Belfast for their support on the project. Special thanks are due Elizabeth Y. Graham, analytical technician at the University of Alabama, Tuscaloosa.

Report Preparation

Principal Investigator:
Pauline Johnson, Ph.D.
The University of Alabama, Tuscaloosa

Project Team:
Robert Pitt, Ph.D., P.E
Rocky Durrans, Ph.D., P.E
Matilda Urrutia, Ph.D.
University of Alabama, Tuscaloosa

Shirley Clark, Ph.D., P.E.
University of Alabama, Birmingham

Student Team:
Jason Thomas Kirby
Sarah Gill
Renee Morquecho
Sumandeep Shergill
Laith Al-Faqih
Joe Brown
University of Alabama, Tuscaloosa

Project Subcommittee

C. Mel Lytle Ph.D., Chair
San Joaquin County Dept. of Public Works, Water Resource Division

Gregory A. Brodie
Tennessee Valley Authority

Terry Cooke
URS Corporation

Ronald W. Crites, P.E.
Brown and Caldwell

David R. Yonge, Ph.D.
Washington State University

Water Environment Research Foundation Staff

Director of Research: Daniel M. Woltering, Ph.D.
Research Program Director: Jeff C. Moeller, P.E.

ABSTRACT AND BENEFITS

Abstract:

This study of innovative methods for capturing metals from stormwater mainly focused on two major areas of investigation, media filters and swales.

Twelve media were chosen for initial evaluation. The three best performing media, peat-sand mix, compost, and zeolite, were then selected for in-depth study. This investigation demonstrated the importance of characterizing the stormwater before selecting a treatment medium because the type and quantity of metals, pH, and other runoff characteristics can vary a great deal among sites. For example, media that were effective at high metals concentrations were outperformed by some media at the low metals concentrations typically found in stormwater.

In the swale study, the hydraulic characteristics of grass swales appear to be more important than grass species for removing heavy metals from stormwater during single storm events. The outdoor swale tests showed highly variable results. Many of the concentration reductions were quite large, but some "negative removals," possibly associated with scour of previously deposited materials, were also noted. Because of the potentials for both sediment deposition and scouring, swales can either improve or deteriorate water quality during storm events. Long-term performance considering infiltration has shown significant heavy metal retention in swale systems. Data from the phytoremediation study suggest that the difference grass species perform similarly. In other words, any grass added to the surface of a swale system would represent a positive influence on metal uptake. Centipedegrass is possibly the best choice for its resilience to drought, its nutritional frugality, and its greater ability to accumulate important contaminant metals such as copper, zinc, and lead.

Benefits:

The regulation of metals in urban runoff may be a reality in the near future. Results from this work have important implications for the development of low-cost, low-maintenance systems for the control of metals in urban runoff in that they:

- Provide results of a broad study of metals removal from stormwater;
- Further the understanding of the effectiveness of conventional and novel media and filter flow configuration on the capture of metals from urban runoff through laboratory and field-scale studies;
- Identify the most suitable media for different site conditions and relative benefits of upflow columns;
- Provide design information for typical low volume flows in stormwater drainage swales integrating water quality data;
- Identify and preliminarily quantify the role of common grass varieties in phytoremediation of metals in runoff, particularly in the low metal concentration milieu of stormwater runoff;

- Answer critical questions about the stormwater particulate size-metal fraction associations in regard to treatability and the significance of filterable forms of metals;
- Determine that disposal issues are a significant consideration in media selection; and
- Establish guidelines for the development of working systems for the capture of metals from urban runoff.

Keywords: Heavy metals, runoff, swales, phytoremediation, filters

TABLE OF CONTENTS

LIST OF TABLES

LIST OF FIGURES

LIST OF ACRONYMS

ANOVA	Analysis of variance
ASV	Anodic stripping voltammetry
ATP	Adenosine triphosphate
BTC	Breakthrough curve
BDST	Bed depth service time
BET	Brunauer, Emmett, and Teller
BOD_5	Biochemical oxygen demand
BMP	Best management practice
BVRT	Bed volume residence time
CEC	Cation exchange capacity
CFMF	Cross-flow microfiltration
CHN	Carbon, hydrogen, and nitrogen [elemental analysis]
COD	Chemical oxygen demand
CV	Coefficient of variation
CSF	Compost Stormwater Filter System
CSO	Combined sewer overflow
CSTR	Continuously Stirred Tank Reactor
DDT	Dichlorodiphenyl trichloroethane
DMSO	Dimethylsulfoxide
DO	Dissolved oxygen
DOC	Dissolved organic carbon
EDTA	Ethylenediamine tetraacetic acid
ESS	Experimental sewer system
FBM	Flow balancing method
GAC	Granulated activated carbon
HDPE	High-density polyethylene
HDTMA	Hexadecyltrimethylammonium bromide
HRT	Hydraulic Residence Time
ICP-OES	Inductively coupled plasma/optical emission spectrophotometer
ID	Inside diameter
IJC	International Joint Commission
IOCS	Iron oxide coated sand
IR	Infrared radiation
LCM	Low concentration metal
LEACHM	Leaching Estimation and Chemistry Model
LOD	Limits of detection
LOQ	Limits of quantification
MCTT	Multi-chambered treatment train
MDL	Method detection limit
MMHg	Methylmercury
NALMS	North American Lakes Management Society
NPDES	Nonpoint Discharge Elimination System

NSS	Not statistically significant
NTU	Nephelometric turbidity unit
NURP	Nationwide Urban Runoff Program
ORP	Oxidation-reduction potential
PA	Phosphoric acid
PAH	Polyaromatic hydrocarbons
PCB	Polychlorinated biphenyls
PEL	Probable effect level
PVC	Polyvinyl chloride
PZC	Point of zero charge
QA/QC	Quality control/quality assurance
QAPP	Quality assurance project plan
RCRA	Resource Conservation and Recovery Act
SOP	Standard operating procedure
SPLITT	Split-flow thin-cell
SRB	Sulfate-reducing bacteria
SS	Suspended solids
SSO	Sanitary sewer overflow
STDS	Solids Transport and Deposition Study
SWDA	Solid Waste Disposal Act
SYSTAT	System for Statistics
TCLP	Toxicity characteristic leaching procedure
TDS	Total dissolved solids
TKN	Total Kjeldahl nitrogen
TMDL	Total maximum daily load
TOC	Total organic carbon
TSS	Total suspended solids
UAB	University of Alabama at Birmingham
U.S. EPA	U.S. Environmental Protection Agency
USGS	United States Geological Survey
WDNR	Wisconsin Department of Natural Resources
WERF	Water Environment Research Foundation
WET	Effluent toxicity test
WinSLAMM	Windows Version of Source Loading and Management Model
XTT	Sodium 3′-[1-(phenylamino)-carbonyl]-3,4-tetrazolium]-bis[4-methoxy-6-nitro]benzenesulfonic acid hydrate]

Executive Summary

The United States Environmental Protection Agency (U.S. EPA) has identified stormwater runoff as a leading cause of water quality impairment nationwide. Heavy metals in urban stormwater primarily originate from automobile-related activities and the exposure of building materials to rain.

Heavy metals may occur as dissolved, colloidal, or particulate-bound species; however, most metals are predominantly associated with particulates. Particle associations and speciation critically affect the toxicity and bioavailability of metals and are greatly dependent upon chemical and physical parameters. Short-term, or acute, toxicity is rare for stormwater, but longer exposures to contaminated stormwater and to contaminated sediment can cause longer-term, chronic toxicity as illustrated by major changes in benthic organisms. A successful control strategy for the reduction of heavy metals from stormwater must, therefore, be effective in capturing a variety of metals that are in the particle-bound, colloidal, and dissolved states.

Opportunities for the control of metal-bearing urban runoff can be grouped into three basic strategies: 1) source avoidance or reduction; 2) passive systems; and 3) installed treatment technologies. Although the concept of pollution avoidance and minimization at critical sources is the preferred option for the control of metals in urban runoff, it is unlikely that any one method will provide either a realistic or effective solution to the problem.

The scope of this project was to address emerging stormwater control technologies for the capture of heavy metals from urban runoff. The processes selected for investigation were media filtration systems and grass swales. The goal of this study was to gain further understanding of the role of these technologies in the removal of metals from urban runoff and to provide guidelines for system design.

ES.1 Filter Studies

Twelve media were initially evaluated through kinetic and equilibrium tests. The results emphasized the importance of characterizing the stormwater before selecting a treatment media since the type and quantity of metals, pH, and other runoff characteristics can vary a great deal between sites. For example, determining the range of metal concentrations to be treated is crucial to selecting the best media, since the removal efficiencies of the media relative to each other changed with varying metal concentration. Media that were effective at high metals concentrations were outperformed by some media at the low metals concentrations typically found in stormwater. In addition, some media that offer potential as sorbents of metals may create water quality problems of their own. In equilibrium tests for all media, correlations were evident between the metals sorbed and the Ca, Mg, K, and Na ions desorbed. The ions desorbing and their quantity varied depending on the media and exchange metals present. The three best performing media (peat-sand mix, compost, and zeolite) were then selected for in-depth study.

Based on the unsteady state tests representing unsaturated media, the order of preference for removal on a mass basis was Pb>Cr, Cu>Cd>Zn, Fe for the peat-sand, Pb,Cr>Cu,Fe>Zn,Cd for the St. Cloud zeolite, and Cd>Zn>Pb>Cu>Fe>Cr for the compost. The order of preference for removal based equilibrium conditions on the low concentration metal uptake tests, was Cd, Pb>Zn,Cu>Cr>Fe for peat-sand, Zn,Cd>Pb>Cu>Cr>Fe for St. Cloud zeolite, and Cd>Zn>Pb> Cu>Cr>Fe for compost. In extensive comparisons of the three media, the peat-sand mix performed best at removing dissolved metals.

All three media removed total suspended solids well over all runs, but only peat showed any removal of total dissolved solids. Compost increased the total dissolved solids concentrations. As expected, the behavior of the total suspended solids and particulate-bound metals paralleled each other. Copper and chromium typically had lower removal efficiencies than the other metals; one possible explanation is that these metals had a larger fraction of their particulate-bound metal concentrations associated with smaller particulates which passed more easily through the columns. Peat had the best removal efficiencies for particulate-bound metals. Removal efficiencies of compost and zeolite were approximately the same.

Peat-sand had the greatest headloss and the greatest change in headloss over the course of each run. Zeolite and compost beds offered considerably less headloss. The degree of change in headloss increased as the total suspended solids concentration of the influent increased for all media. The advantages and disadvantages of the three media are summarized in the table below:

Table ES-1.

Media	Main Advantages	Main Drawbacks
Peat-Sand Mix	Best metal capture capability	Most detrimental impact on pH, the greatest headloss, and showed the most potential for clogging.
Compost	Second best metal capture capability	Added color to the effluent. Less impact on the pH of the effluent, less headloss, and exhibited less potential for clogging.
Zeolite	Lowest metal capture capability	Less impact on the pH of the effluent, less headloss, and exhibited less potential for clogging.

Upflow columns proved more effective than downflow columns in the control of detention time and a reduction in clogging of the media by solids and associated head loss in the columns. At residence times of 3–10 minutes, most of the suspended solids settled out in the sump area of the columns. Compost and zeolite columns showed little sign of increased headloss during these runs, even when influent suspended solids concentrations were around 400 mg/L.

Samples of spent media from column studies underwent toxic characteristic leaching procedure (TCLP) analysis for those metals regulated by the Resource Conservation and Recovery Act (RCRA). Under the Land Disposal Restrictions program, a restricted waste may be land disposed only if a TCLP extract of the waste, or a TCLP extract of the treatment residue of the waste, does not exceed the regulatory levels for hazardous constituents listed for that waste. The regulatory levels are 1 mg/L for cadmium and 5 mg/L for chromium and lead. Chromium and lead did not present a problem, however, the cadmium concentration in the leachate was above the acceptable limit for peat-sand, and was close to the limit for zeolite and compost. Compost performed best in retaining all three RCRA metals.

Pilot-scale studies confirmed the modeling equations for a few pollutants for some media and indicated that the modeling approach adopted has the potential to provide an estimate of the life of the filter, i.e., the time until media replacement is needed for applications where the influent concentration is not quite as low as it was in the two detention ponds used in this pilot scale study. The best application for these filters may then be to further treat the effluent from a detention pond in critical source areas, such as scrap metal recyclers, rather than as a polisher for effluent from a detention pond in a residential area. During these field pilot-scale and related full-scale media filtration tests, the minimum attainable effluent concentrations for these metals appears to be approximately 5–10 μg/L for copper and lead, and approximately 10–20 μg/L for zinc, irrespective of media.

ES.2 Swale Hydrologic and Phytoremediation Studies

Grass swales reduce heavy metal discharges through a combination of sedimentation, infiltration through the soil, and biological uptake. Grass swales can incorporate many positive features to reduce heavy metal discharges to surface and groundwater. Sedimentation occurs due to increased channel lining roughness, infiltration is common, and amended soils can easily be incorporated in swale liners. Critical source area controls can be used incorporating special sorption/filtration media. Finally, sedimentation in wet detention ponds can be improved to better remove and contain finer particulates, and post-treatment with wetlands can be used for increased biological control processes.

The outdoor phytoremediation study on the role of common grass varieties in the capture of metals from runoff indicated that both centipedegrass and zoysiagrass varieties, common in many lawns in the southeastern United States, exhibit a capability to uptake metals from soil that receives stormwater. Experiments in indoor swales with spiked stormwater, however, suggested that centipedegrass had a greater maximum ability to accumulate Cu and Zn than zoysiagrass and Kentucky bluegrass, commonly used in northern areas of the country. Centipede seemed to also be able to accumulate more Pb. All three grasses accumulated Cd and Cr at similar rates of uptake, but the extent of uptake was small (~30 mg/Kg and 10 mg/Kg, respectively). Zoysia accumulated the least Pb, Cu and Zn of all three varieties in these studies. However, the overall results suggest that there is relatively similar behavior among the different grass species. In other words, any grass added to the surface of a swale system would represent a positive influence on metal uptake from runoff. However, if a recommendation should be made for systems designed to have a long life as filters for metal accumulation from stormwater, centipede could possibly be the best choice for its resilience to drought, its nutritional frugality, and its greater ability to accumulate key contaminant metals, such as Cu, Zn, and Pb.

Indoor laboratory swale experiments produced retardance curves for centipede, Kentucky bluegrass, and zoysia species for flow depths that will optimize heavy metal removal. The hydraulic characteristics of grass swales may be more important than the grass species for removing heavy metals from stormwater during single storm events. Because of the potentials for both sediment deposition and scouring, swales can improve or deteriorate water quality during storm events. The outdoor swale tests showed highly variable results. Many of the concentration reductions were quite large, but some "negative removals" possibly associated with scour of previously deposited materials, were also noted. However, mass metal removals would generally occur due to infiltration in the swale.

ES.3 Treatability Testing

Treatability tests were performed to assess the effectiveness of different treatment trains and processes quantifying improvements in stormwater toxicity and metals capture. The treatability tests included intensive analyses of samples from twelve sampling locations in the Birmingham, AL, area that all had elevated toxicant concentrations, compared to the other urban source areas initially examined. The treatability tests conducted were: settling column, flotation, screening and filtering, photodegradation, aeration, combined photodegradation and aeration and an undisturbed control sample. More than 900 toxicity tests were performed using the Microtox procedure. Turbidity tests were also conducted on all samples. Results indicated a reduction in toxicity as the level of treatment increased. All samples, with one exception, showed dramatic reductions in toxicity with increasing settling times. Even though the data are separated into three source groups, as expected, there were greater apparent differences between the treatment methods than between the sample groupings.

ES.4 Metals Associations

Metal-particulate association tests using Celex-100 resin revealed that more than 90% of the filterable forms of calcium, magnesium, potassium, iron, and zinc were in ionic forms, with very little colloidal, or other bound forms. Also, more than 80% of the filterable chromium and lead were also ionic, while only about 50% of the filterable copper and 30% of the filterable cadmium were ionic. This data can be used to estimate the level of control that may be associated with different designs of particle trapping devices. Some pollutants can be significantly reduced by a reduction in particulates, such as suspended solids, total phosphorus and most heavy metals. Other pollutants, such as nitrates, are reduced much less, even after filtration down to 0.45 μm.

Experiments were also conducted to examine the likelihood of the metals disassociating from the particulates under pH conditions ranging from about 4–11 with both weak and strong acids. These tests indicated that the heavy metals of concern remain strongly bound to the particulates during long exposures at the extreme pH conditions likely to occur in receiving water sediments. They will also likely remain strongly bound to the particulates in stormwater control device sumps or detention pond sediments where particulate-bound metals are captured.

Related tests were conducted as part of the filter media evaluation task of this research to measure the disassociation potential of heavy metals and nutrients under aerobic and anaerobic conditions having extreme Eh values. Studies on the effect of anaerobiosis on metal retention by filter systems indicated that heavy metals were not mobilized from filter systems under anaerobic conditions. It was found that metal retention within the filters was not different from what was observed in oxygenated environments. However, it is plausible that under certain specific environmental conditions, co-precipitation of metals by iron- and sulfate-reducing bacteria may take place in stormwater treatment systems.

ES.5 Design Guidelines

Much information has been collected during this research project that can be directly used for the design of stormwater controls for the reduction of heavy metals. This report includes overviews of the associations of metals with different stormwater fractions. These associations are also useful in predicting the performance of sedimentation controls in removing heavy metals. This information can be used to predict the heavy metal control associated with a wet detention pond that is designed to remove particles of a specific minimum size.

Additionally, material on the selection of filtration media, the clogging rates under different conditions, and the metal retention capacity of the different media are summarized. A simple approach for selecting filtration media and determining the sizing of stormwater filters is outlined. Investigations of several important issues associated with the upflow filters are discussed. These filters hold promise in being much more cost effective than the more commonly used traditional slow filters. Additional tests were also conducted that investigated the clogging behavior of the different media under upflow conditions. Both the hydraulic and the water quality objectives of a swale design are examined and information is included on the selection of the grass species and other channel parameters. This subsection also shows how the important project information on the determination of channel roughness is used in the design of channels.

CHAPTER 1.0

OBJECTIVES AND ORGANIZATION OF THE REPORT

1.1 The overall goals of the project are outlined below

1. Identify which attributes significantly affect filter design.
2. Enable users to select the best media, required pre-treatment, and filter size for their specific application.
3. Develop a model to predict performance.
4. Develop a method by which the hydraulic characteristics of shallow open channel flow in a swale can be predicted on the basis of flow rate, channel cross-sectional geometry and slope, and grass type and condition.
5. Determine the role of various grasses in runoff metals capture.

1.2 Organization and overview of the Report

The report begins with a summary of the work to date on the role heavy metals in urban receiving waters and watersheds. Chapter 2 is a discussion of prior investigations that have examined the development and testing of innovative controls of heavy metals in urban stormwater. This is a discussion of the treatment options for stormwater heavy metals, while the associated appendix A is a summary of heavy metal effects and sources in urban receiving waters.

Subsequent sections, chapters 3-6, in the report describe current work on specific areas of the experimental program for the control of heavy metals in urban areas.

.

Filter studies are covered in chapters 3 and 4 and the organization of the work is as follows.

Filter Study Experimental Work: Part One: Chapter 4

1. The first stage in the examination of stormwater filters was a series of batch experiments and downflow lab-scale filter column studies to determine the parameters for a series of design equations to use in modeling the removal ability of the filters.
2. Pilot-scale downflow filters were then built using the media examined in stage 1.1. The results of the pilot scale studies were compared with the data from the stage one lab-scale columns to examine how well the design equations predicted removal.
3. Preliminary lab-scale tests to examine upflow filters were performed.

Filter Study Experimental Work: Part Two: Chapter 3

1. The media examined in part one along with several additional media were further examined in batch tests to compare metal removal ability and determine which media had the best potential to remove metals in runoff.
2. Additional lab-scale work was performed to examine the potential of upflow filtration. Upflow filters were constructed using the three most effective media from stage 2.1 and metal removal efficiency was examined for different rates of flow.
3. The roll of biomass in metal uptake was examined.

Swale studies are covered in chapter 5. The chapter is divided into two sections: 5.1 hydraulic studies and 5.2 phytoremediation studies. Treatability Studies and Design Guidelines are covered in chapters 6 and 7.

Chapter 6 is an investigation of associations of heavy metals (along with some major constituents and nutrients) with different-sized particulates in stormwater. The binding strengths of these metals to the particulates were also examined by using a sequential extraction procedure using different acids and bases under several pH values. Also, experiments examining the characteristics of the filterable (<0.45 μm) portion of the heavy metals were also conducted. Appendix H summarizes much historical information concerning the characteristics of stormwater that affects its treatability. That appendix is an excerpt from an on-going EPA research project that is examining advanced filtration techniques for stormwater treatment.

Each of the chapters 3-6 are self contained units with sections on background information, experimental design and setup, results, discussion, conclusion.. References for all chapters appear at the end of the report.

CHAPTER 2.0

GENERAL LITERATURE REVIEW OF STORMWATER HEAVY METAL CONTROLS

2.1 Introduction

The development and testing of innovative controls of heavy metals in urban stormwater should begin with prior investigations that have examined metals in urban areas. Therefore, the discussion presented here and in Appendix A is a summary of the work to date on the role heavy metals in urban receiving waters and watersheds. This is a discussion of the treatment options for stormwater heavy metals, while the associated appendix is a summary of heavy metal effects and sources in urban receiving waters.

Some material in this discussion was excerpted from the following sources:

Pitt, R. *Stormwater Quality Management.* CRC/Lewis, New York, publication pending.

Pitt, R., M. Lilburn, S. Nix, S.R. Durrans, S. Burian, J. Voorhees, and J. Martinson *Guidance Manual for Integrated Wet Weather Flow (WWF) Collection and Treatment Systems for Newly Urbanized Areas (New WWF Systems).* U.S. Environmental Protection Agency. 612 pgs. Cooperative agreement #CX 824933-01-0. publication pending.

These materials have been updated based on the annual wet weather flow literature reviews prepared for Water Environmental Research for the past six years by project team members (Clark, *et al.* 2002; Clark, *et al.* 2001; Fan, *et al.* 2000; O'Conner, *et al.* 1999; Field, *et al.* 1998; and Field, *et al.* 1997).

2.2 Stormwater Controls for Heavy Metals

During a recent EPA research project, Pitt, *et al.* (1999) using many other current literature sources, found that it is beneficial to develop stormwater management design guidelines based on local rain categories. Small events, making up the majority of the number of rain events (but little of the annual runoff), but commonly exceed metal criteria. The runoff from these smaller events is relatively easy to control through simple infiltration or on-site reuse. Moderate-sized rains are responsible for the majority of the annual runoff volume and pollutant discharges. The runoff from these events can be significantly reduced, but usually not totally eliminated, through infiltration. The larger flows exceeding the infiltration capacity will have to be treated to reduce pollutant concentrations and excessive discharge rates. Large rains that approach (and may exceed) the capacities of the drainage system produce little of the annual flows and are rare. Significant pollutant concentration reductions during these large events would be difficult and very expensive because of the very large flows involved. However, runoff flow rates (and energy) should be reduced to produce instream flowrate distributions less than critical values needed to protect in-stream habitat. In general, needed benefits can best be realized with significant runoff volume reductions. These are much easier to implement at the time of development. Implementing appropriate stormwater controls at the time of development is much more cost effective and will provide much more effective levels of control than if only retrofitting is used. Extensive retro-fitting may only result in low to moderate levels of stormwater control (generally about 25% reductions over the whole watershed), while implementation of a broad set of appropriate controls (using both sedimentation and infiltration) at the time of development could result in control levels approaching 75%, or more.

Stormwater treatment at critical source areas is usually quite different from treatment at other areas. The control of small critical area contributions to urban runoff may be the most cost-effective approach for treatment/reduction of stormwater toxicants. The general features of the critical source areas appear to be large paved areas, heavy vehicular traffic (especially frequent and large numbers of vehicle starts, such as at convenience stores) and outdoor use or storage of problem pollutants. Most of these areas are quite small, being small commercial establishments. However, large areas of continuous pavement (such as at shopping malls) are also critical sources. Runoff from these larger areas can usually be controlled through the use of preferred wet detention ponds, but runoff control for small areas usually requires special applications.

Sedimentation is the most common fate and control mechanism for particulate related pollutants. This would be common for most stormwater pollutants, as noted above. Particulate removal can occur in many conventional stormwater control processes, including catchbasins, screens, drainage systems, and detention ponds. Sorption of pollutants onto solids and metal precipitation increase the sedimentation potential of these pollutants and also encourages more efficient bonding of the pollutants in soils, preventing their leaching to groundwaters. Detention ponds are probably the most common management practice for the control of stormwater runoff. If properly designed, constructed, and maintained, wet detention ponds can be very effective in controlling a wide range of pollutants. The performance of wet detention ponds can be more than 90 percent removal of suspended solids, 70 percent for BOD_5 and COD, nutrient removals of about 60 to 70 percent, and heavy metal removals of about 60 to 95 percent. Catchbasins are very small sedimentation devices. Adequate cleaning can help to reduce the total solids and lead urban runoff yields by between 10 and 25 percent, and COD, total Kjeldahl nitrogen, total phosphorus, and zinc by between 5 and 10 percent. Other important fate mechanisms available in wet detention ponds, but which are probably not as important in small sump devices such as catchbasins, include volatilization and photolysis. Biodegradation, biotransformation, and bioaccumulation (into plants and animals) may also occur in ponds.

Upland infiltration devices (such as infiltration trenches, porous pavements, percolation ponds, and grass roadside drainage swales) are located at urban source areas. Infiltration (percolation) ponds are usually located at stormwater outfalls, or at large paved areas. These basins, along with perforated storm sewers, can infiltrate flows and pollutants from all upland sources combined. Infiltration devices can safely deliver large fractions of the surface flows to groundwater, if carefully designed and located. Local conditions that can make stormwater infiltration inappropriate include steep slopes, slowly percolating soils, shallow groundwater, and nearby groundwater uses. Grass filter strips may be quite effective in removing particulate pollutants from overland flows. The filtering effects of grasses, along with increased infiltration/recharge, reduce the particulate sediment load from urban landscaped areas. Grass swales are another type of infiltration device and can be used in place of concrete curb and gutters in most land uses, except possibly strip commercial and high density residential areas. Grass swales allow the recharge of significant amounts of surface flows. Swales can also reduce the concentration of pollutants, due to filtration. Soluble and particulate heavy metal (Cu, Pb, Zn, and Cd) concentrations can be reduced by at least 50 percent, COD, nitrate nitrogen, and ammonia nitrogen concentrations can be reduced by about 25 percent, but only very small concentration reductions can be expected for organic nitrogen, phosphorus, and bacteria.

Sorption of pollutants to soils is probably the most significant fate mechanism of toxicants in biofiltration devices. Many of the devices also use sedimentation and filtration to remove the particulate forms of the pollutants from the water. Incorporation of the pollutants onto soil with subsequent biodegradation, with minimal resultant leaching to the groundwater, is desired. Volatilization, photolysis, biotransformation, and bioconcentration may also be significant in grass filter strips and grass swales. Underground French drains and porous pavements offer little biological activity to reduce toxicants.

The following summary of the performance of stormwater controls for heavy metal removal is divided into sections for public works activities, critical source controls, infiltration and bioretention, sedimentation and wetland, CSO controls, and chemical treatment options.

2.2.1 Public Works Activities

2.2.1.1 Catchbasins

Storm drainage system inlet structures can be separated into three general categories. The first category is a simple inlet that is comprised of a grating at the curb and a box, with the discharge located at the bottom of the box which connects directly to the main storm drainage or combined sewerage.

This inlet simply directs the runoff to the drainage system and contains no attributes that would improve water quality. However, large debris (several cm in size) may accumulate (if present in the stormwater, which is unlikely). The second type of inlet is similar to the simple inlet, but it contains a sump that typically extends 0.5 to 1 m below the bottom of the outlet. This is termed a catchbasin in the U.S., or a gully pot in the U.K., and has been shown to trap appreciable portions of the course sediment. The third category is also similar to the simple inlet, but contains some type of screening to trap debris. These include small cast iron perforated buckets placed under the street grating, as used in Germany, large perforated and lipped stainless steel plates placed under the street grating, as used in Austin, Texas, and a number of proprietary devices incorporating filter fabric or other types of screening placed to intercept the stormwater flow.

Catchbasin performance has been investigated for some time in the U.S. Sartor and Boyd (1972) conducted controlled field tests of a catchbasin in San Francisco, using simulated sediment in fire hydrant water flows. They sampled water flowing into and out of a catchbasin for sediment and basic pollutant analyses. Lager, *et al.* (1977) was the first EPA funded research effort that included a theoretical laboratory investigation to evaluate sedimentation in catchbasins and to develop effective designs.

Catchbasins, simple inlets, man-holes, and sewerage sediment accumulations were monitored at more than 200 locations in Bellevue, Washington, in two mixed residential and commercial study areas (Pitt 1985). These locations were studied over three years to monitor accumulation of sediment and sediment quality. The sediment in the catchbasins and the sewerage was found to be the largest particles that were washed from the streets. The sewerage and catchbasin sediments had a much smaller median particle size than the street dirt and were therefore more potentially polluting than the particulates that can be removed by street cleaning. Cleaning catchbasins twice a year was found to allow the catchbasins to capture particulates most effectively. This cleaning schedule was found to reduce the total residue and lead urban runoff yields by between 10 and 25 percent, and COD, total Kjeldahl nitrogen, total phosphorus, and zinc by between 5 and 10 percent (Pitt and Shawley 1982).

Earlier EPA research (Lager, *et al.* 1977) found that an optimal catchbasin design should have the following dimensions: if the outlet pipe is D in diameter, its bottom should be located about 2.5D below the street level and 4D from the bottom of the catchbasin sump. The overall height of the catchbasin should therefore be 6.5D, with a diameter of 4D.

Three storm drain inlet devices were evaluated in Stafford Township, New Jersey, by Pitt and Field (1998) as part of an EPA-funded study. An optimally designed catchbasin with a sump and two representative designs that used filter material. A broad list of constituents were analyzed during 12 monitored events, including heavy metals. The catchbasin with the sump was the only device that showed important and significant removals for several pollutants (suspended solids averaged 32%).

The performance of Drain Inlet Inserts (Fossil Filter and StreamGuard) and an oil/water separator in treating runoff from four California Department of Transportation (Caltrans) maintenance stations was evaluated (Othmer, *et al.* 2001). Drain Inlet Insert results to date showed that reductions in metals, hydrocarbons, and solids were consistent with expectations; however, frequent flow bypass required more maintenance than anticipated. Oil/water separator results showed no discernable difference between influent and effluent hydrocarbon concentrations at the low levels measured.

2.2.1.2 Grit Traps

Several proprietary stormwater treatment devices are being marketed throughout North America. These devices can be located underground. Unfortunately, comprehensive testing with actual stormwater is not available for most of these devices. The designs and demonstrations are mostly based on reduction of relatively large particles that rarely occur in stormwater. The suspended solids in stormwater is mostly in the range of 1 to 100 μm, with only a small fraction of the mass (usually <10%) associated with particles greater than 100 μm. These devices are designed to capture particle sizes that have typically been found on streets, not in the runoff water (Pitt 1987). These devices are excellent grit chambers (and can probably capture floating oils) and can be used to prevent sand-sized particles from accumulating in sewerage. Very little scour of the captured grit material is also likely with these devices. However, they are not likely to provide important reductions of most stormwater pollutants, especially heavy metals and other toxicants. If a site is grossly contaminated with oils or grit, then a proprietary oil/water separator or grit chamber is needed, but further treatment will also likely be necessary, such as with a media filter.

Washbusch (1999) reported on an extensive evaluation of a proprietary urban stormwater treatment unit (the Stormceptor) in Madison, Wisconsin, conducted by the United States Geological Survey (USGS) and the Wisconsin Department of Natural Resources (WDNR). The evaluation was conducted on a 4.3 acre city maintenance yard site over 9 months and for 45 runoff events having rainfall depths ranging from 0.02 to 1.31 in., plus some snowmelt influence from imported snow to the maintenance yard. About 90% of the runoff was treated by the unit, and the remainder flows bypassed the unit during periods of high flows. The overall effectiveness of the unit was 33% for suspended solids, 17% for total phosphorus, 34% for total PAH, and from 20 to 30% for most heavy metals. In general, dissolved constituents were unaffected by the unit. Performance during the early spring events was degraded by high saline snowmelt water, which had elevated conductivity from on-site deicer storage, that significantly slowed particle settling.

2.2.1.3 Street Cleaning

Street cleaning was extensively studied during early EPA-funded research projects. It was thought to be an effective runoff water quality control practice because of the large quantities of pollutants found on streets during early stormwater research in the U.S. (Sartor and Boyd 1972). Because streets were assumed to contribute most of the urban runoff flows and pollutants, street cleaning was assumed to be a potentially effective practice. Unfortunately, few data have shown street cleaning to be effective because of the different sized particles that street cleaners remove compared to the particles that are mostly removed by rains. Furthermore, in many areas, rains are relatively frequent and keep the streets cleaner than typical threshold values that most street cleaners can remove. However, in the arid west of the U.S., rains are very infrequent, allowing streets to become quite dirty during the late summer and fall. Extensive street cleaning during this time has been shown to result in small to moderate suspended solids and heavy metal reductions in runoff (Pitt 1979, Pitt and Shawley 1982). In other areas of the U.S., especially in the wet southeast where large and frequent rains occur, street cleaning is likely to have much less direct water quality benefits, beyond possible important litter and floatable control.

Conventional street cleaning does not have a significant effect on stormwater quality because conventional street cleaners preferentially remove the large particles, and the smaller particles from the street that are most effectively removed during rains. Valiron (1992) confirmed the many earlier U.S. studies by showing that street cleaners only remove about 15% of the finest particles (less than 40 μm), while close to 80% of the largest particles (>2,000 μm) are removed. Sutherland (1996, and with Jelen 1996) conducted tests using a new style street cleaner that shows promise in removing large fractions of most of the street dirt particulates, even the small particles that are most heavily contaminated. The Enviro Whirl I, from Enviro Whirl Technologies, Inc. is capable of much improved removal of fine particles from the streets compared to any other street cleaner ever tested. This machine was also able to remove large fractions of the fine particles even in the presence of heavy loadings of large particles. This is a built-in tandem machine, incorporating rotating sweeper brooms within a powerful vacuum head. Model analyses for Portland, OR, indicate that monthly cleaning in a residential area may reduce the suspended solids discharges by about 50%, compared to only about 15% when using the older mechanical street cleaners that were tested during the early 1980s. This equipment is currently being evaluated in large-scale tests by the Wisconsin Department of Natural Resources and WI Dept. of Transportation (Bannerman, personal communication).

The pollutant loads conveyed by street cleaning waters, by street runoff and the maximum pollutant load removed by street washing were measured for three streets in central Paris, France. For suspended solids and organic matter, the pollutant load removed on a daily basis from street surfaces by street cleaning waters was found to be similar to that removed during one rainfall event. However, it was five times lower for heavy metals. It was also shown that the total mass of pollutants stored on the street surface is significant, and that the effects of street cleaning may be limited (Gromaire, *et al.* 2000).

Normal street cleaning operations for aesthetics and traffic safety purposes are not very satisfactory from a stormwater quality perspective. These objectives are different and the removal efficiency for fine and highly polluted particles is very low. Unless the street cleaning operations can remove the fine particles, they will always be limited in their pollutant removal effectiveness. Some efficient machines are now available to clean porous pavements and infiltration structures, and new tandem machines that incorporate both brooms and vacuums have recently been shown to be very efficient, even for the smaller particles. Conventional street cleaning operations preferentially remove the largest particles, while rain preferentially remove the smallest particles. In addition, street cleaners are very inefficient when the street dirt loadings are low, when the street texture is course, and when parked cars interfere. However, it should also be noted that streets are not the major source of stormwater pollutants for all rains in all areas. Pavement is the major source of pollutants for the smallest rains, but other areas contribute significant pollutants for moderate and large rains. Therefore, the ability of street cleaning to improve runoff quality is dependent on many issues, including the local rain patterns and other sources of runoff pollutants.

2.2.2 Critical Source Area Controls

There are numerous controls that have been used at critical source areas, such as catchbasin inserts, grit chambers, and many other proprietary devices. Unfortunately, the performance of many of these devices has been disappointingly low, and/or their service life very short, requiring frequent maintenance. However, filtration has been effectively used in many critical source areas. Detailed discussions of stormwater filtration are included in other sections of this report. In most situations at critical source areas, combination controls using a treatment train approach is likely needed to provide high levels of heavy metal control, especially when filterable fractions also need control. The following discussion is a description of an example of this type of control.

Example of combination practice using grit removal, sedimentation, and filtration/sorption: The Multi-Chambered Treatment Train (MCTT). The multi-chambered treatment train is an example of a stormwater device that utilizes a combination of processes. It was developed based on EPA sponsored research on treatability of stormwater at critical source areas (Pitt, *et al.* 1995 and 1999). The MCTT contains aeration, sedimentation, sorption, and sand/peat filtration. The MCTT is most suitable for use at relatively small and isolated paved critical source areas, from about 0.1 to 1 ha (0.25 to 2.5 acre) in area, where surface land is not available for stormwater controls. Typical locations include gas stations, junk yards, bus barns, public works yards, car washes, fast food restaurants, convenience stores, etc., and other areas where the stormwater has a high probability of containing high concentrations of oils and filterable toxic pollutants that are difficult to treat by other means. A typical MCTT requires between 0.5 and 1.5 percent of the paved drainage area, which is about 1/3 of the area required for a well-designed wet detention pond, and is generally installed below ground. A pilot-scale MCTT was constructed in Birmingham, AL, at a large parking area at the University of Alabama at Birmingham campus, and tested over a six month monitoring period. Two additional full-scale MCTT units have also been constructed and were monitored as part of Wisconsin's 319 grant from the U.S. EPA. Complete organic and metallic toxicant analyses, in addition to conventional pollutants, are included in the evaluation of these units. Currently, Caltrans is also monitoring the effectiveness of 3 MCTT units in Los Angeles county.

The MCTT includes a special catchbasin (based on Lager, *et al.*'s 1977 design) followed by two tank chambers that is intended to reduce a broad range of suspended solids and stormwater toxicants (volatile, particulate, and dissolved). The runoff enters the catchbasin chamber by passing over a flash aerator (small column packing balls with counter-current air flow) to remove any highly volatile components present in the runoff (unlikely). This catchbasin also serves as a grit chamber to remove the largest (fastest settling) particles. The second chamber serves as an enhanced settling chamber to remove smaller particles and has inclined tube settlers to enhance sedimentation. The settling time in this main settling chamber usually ranges from 20 to 70 hours. This chamber also contains fine bubble diffusers and sorbent pads to further enhance the removal of floatable hydrocarbons and additional volatile compounds. The water is then pumped to the final chamber at a slow rate to maximize pollutant reductions. The final chamber contains a mixed media (sand and peat) slow filter, with a filter fabric layer. The MCTT is typically sized to totally contain all of the runoff from a 6 to 20 mm (0.25 to 0.8 in) rain, depending on treatment objectives, inter-event time, typical rain size, and rain intensity for an area.

During monitoring of 13 storms at the Birmingham pilot-scale MCTT facility (designed for 90% toxicity reductions), the following overall median removal rates were observed: 96% for total toxicity (as measured using the Microtox™ screening test), 98% for filtered toxicity, 100% for lead, and 91% for zinc. It had very effective removal rates for both filtered and particulate stormwater toxicants and suspended solids. Increased filterable toxicant removals were obtained in the peat/sand mixed media filter/ion exchange chamber, at the expense of increased color, lowered pH, and depressed COD and nitrate removal rates. The main settling chamber resulted in substantial reductions in total and dissolved toxicity, lead, and zinc. The filter/ion exchange unit was also responsible for additional filterable toxicant reductions. However, the catchbasin/grit chamber did not indicate any significant improvements in water quality, although it is an important element in reducing maintenance problems by trapping bulk material. Results from the full-scale tests of the MCTT in Wisconsin (Corsi, *et al.* 1999) collaborate the high levels of treatment observed during the Birmingham pilot-scale tests. Table 2-1 shows the treatment levels that have been observed during seven tests in Minocqua (during one year of operation) and 15 tests in Milwaukee (also during one year of operation), compared to the pilot-scale Birmingham test results (13 events). These data indicate high reductions for lead (93 to 96%) and zinc (90 to 91%). The reductions of dissolved heavy metals (filtered through 0.45 μm filters) were also all greater than 65% during the full-scale tests. The Wisconsin MCTT effluent concentrations were very low for all of the constituents monitored: <5 μg/L for cadmium and lead, and <20 μg/L for copper and zinc. Caltrans data indicate lower percentage reductions (due to cleaner influent), but produced similar excellent effluent.

Table 2-1. Performance Data for Full-Scale MCTT Tests, Compared to Birmingham Pilot-Scale MCTT Results (median percentage reductions and median effluent quality)

	Milwaukee MCTT (15 events)	Minocqua MCTT (7 events)	Birmingham MCTT (13 events)
Microtox® toxicity (total)	na[a]	na	100 (0%)
Microtox® toxicity (filtered)	na	na	87 (3%)
Cadmium (total)	91 (0.1 μg/L)	na	18 (0.6 μg/L)
Cadmium (filtered)	66 (0.05 μg/L)	na	16 (0.5 μg/L)
Copper (total)	90 (3 μg/L)	65 (15 μg/L)	15 (15 μg/L)
Copper (filtered)	73 (1.4 μg/L)	na	17 (21 μg/L)
Lead (total)	96 (1.8 μg/L)	nd (<3 μg/L)	93 (<2 μg/L)
Lead (filtered)	78 (<0.4 μg/L)	na	42 (<2 μg/L)
Zinc (total)	91 (<20 μg/L)	90 (15 μg/L)	91 (18 μg/L)
Zinc (filtered)	68 (<8 μg/L)	na	54 (6 μg/L)

na[a] : not analyzed

Source: Pitt, *et al.* 1999 and Corsi, *et al.* 1999

2.2.3 Infiltration/Bioretention Stormwater Controls for Heavy Metal Reductions

Numerous papers describe the successful use of stormwater infiltration throughout the world. Musiake, *et al.* (1990) described the use of shallow infiltration facilities is Tokyo, and Stenmark (1990) described the use of infiltration facilities in cold climates. Other stormwater infiltration experience has been described by Wada and Miura (1990), Harada and Ichikawa (1993),Yamada (1993), and Duchene, *et al.* (1993).

The radical alteration of the local hydrologic cycle due to urbanization has prompted the use of infiltration of stormwater. As an example, Krijci, *et al.* (1993) described the mandatory use of stormwater infiltration in Switzerland to decrease the burden on combined and separate sewerage systems. The 1992 Swiss Water Pollution Control Law requires that unpolluted wastewater must be infiltrated. If local conditions prevent infiltration, then special authorization is required and detention is used. A simple system is used to determine the suitability of stormwater for infiltration, depending on the area drained and the use of the groundwater. As an example, runoff from roofs, bike lanes, and walking paths must be infiltrated in all areas, even if the groundwater has high importance as a drinking water source. Surface infiltration is required (and subsurface infiltration is prohibited) for this runoff in most drinking water protection zones. The infiltration of roadway and parking area runoff is more restricted, where only surface infiltration is allowed for all areas. Any infiltration of highway and freeway runoff is only allowed in exceptional situations. In all cases, "clean" water (runoff from yard drainage, spring water, groundwater, and cooling water) is forbidden to be discharged into the storm drainage system.

Conradin (1995) describes how Zurich is complying with the Swiss Water Pollution Control Law. The city has 50 to 100 year old sewerage, about 80% being combined sewerage and 20% being separate sewerage. Clean flows (fountain water, spring water, yard drainage, cooling system water, and possibly roof runoff) are required to be diverted from the sewerage. All other stormwater will be directed to the combined sewerage and newly renovated treatment plants. The city is converting its system to a partially separate system that collects the clean water and directly diverts it to the Limmat River. Zurich is building open brooks along streets and walkways to collect these waters. The open brooks provide natural water channels and aesthetically revitalizes the urban area. About 12 km of brooks have been built as of 1995, and as much as 30 km total are planned. The current brooks divert about 150 L/sec from the sewerage. The brooks are designed to carry about two to five times the dry weather flows, with excess diverted to the sewerage and the treatment plants.

A recurrent theme in the literature is concern for lack of appropriate design guidelines for infiltration practices (Petersen, *et al.* 1993). Very little design guidance for specific stormwater infiltration practices existed for Europe before 1991. Somewhat more guidance had been available in the U.S. However, much of the U.S. guidance had been transferred from other areas of the country having greatly different rainfall, topographic, soil, and frost conditions, with little modification. In addition, long-term performance information on infiltration practices is also limited. This makes predictions of useful life very difficult. The high failure rate of many types of infiltration practices, mostly associated with lack of any maintenance, is also of great concern, along with concerns of groundwater contamination. However, the extensive and successful use of stormwater infiltration in Tokyo, and elsewhere, plus the absence of groundwater contamination problems from stormwater infiltration for most areas, indicates that stormwater infiltration is feasible in many situations. Newer guidelines (such as described by Pratt and Powell 1993) also offer a uniform and reasonably conservative approach for the design of infiltration devices. The goal is to identify those conditions where stormwater infiltration is most likely to be successful, stress the use of robust infiltration practices (such as grass swales), be conservative in useful life estimates, provide appropriate pretreatment, and ensure adequate maintenance. It is important to use alternative stormwater controls (such as detention and biofiltration) in areas and situations that are marginal for infiltration.

Bioretention uses plants and soil to filter runoff from developed areas. Laboratory and field tests showed good removal of metals, P, and NH_3 with little or negative removal of nitrate. A significant mulch layer was found to be important for metals removal (Davis, *et al.* 1998b and Pitt, *et al.* 1999b).

2.2.3.1 General Infiltration Practices

The Long Island and metropolitan Washington D.C. NURP projects (EPA 1983) examined the performance of several types of infiltration devices. The Long Island project studied a series of interconnected percolating catchbasins which were found to reduce stormwater discharges by more than 99 percent. The Washington D.C. study found that porous pavement reduced the pavement pollutant runoff loadings by 85 to 95 percent, while an infiltration trench reduced urban runoff flows by about 50 percent. The EPA concluded that, with a reasonable degree of site specific design considerations to compensate for soil characteristics, infiltration devices can be very effective in controlling urban runoff. Local conditions that can make recharge inappropriate include steep slopes, slowly percolating soils, shallow groundwater, and close-by important uses of the groundwater.

Prince George's County, Maryland first introduced bioretention devices (such as the "rain garden" to infiltrate roof runoff) in 1990. Utilizing physical, chemical, and biological treatment processes within an aerobic soil media/vegetated filter system, bioretention has been shown to be highly effective in removing pollutants such as heavy metals and nutrients from urban runoff. By capturing, infiltrating or filtering stormwater runoff close to the source, the use of bioretention treatment can also restore hydrologic functions (Winogradoff and Coffman, 1999).

The behavior of heavy metals (Cd, Cu, Zn, Cr, Pb), nutrients (organic C, P, and N parameters), and major ions were investigated during percolation of roof runoff water through an artificial infiltration site. The concentrations of various components were determined in rainwater, roof runoff, and infiltrating water at various depths in the soil (Mason, *et al.* 1999). Runoff from highways contains significant loads of heavy metals and hydrocarbons, according to German regulations it should be infiltrated over embankments to support groundwater recharge. To investigate the decontaminating effect of greened embankments, soil-monoliths from highways with high traffic densities were taken. Soils were analyzed to characterize the contamination in relation to distance and depth for lead, zinc, copper, cadmium, and PAH (Dierkes and Geiger 1999). They found that the highest concentrations were found in the top 5 cm of soil and within two meters of the street.

Dupre, *et al.* (1999) used ultrafiltration at different pHs to investigate the affinity of certain metal ions to form humate complexes in natural waters with a high dissolved organic carbon concentration. The log K constants for sorption to organic colloids (metal-humate complexes) were as follows: Al, Ga, Fe, Th, U, Y, Re (more than 7) >> Cr (5.5) >> Co (3) > Rb, Ba, Sr, Mn, Mg (approximately 2). The tendency of these metals to bind to organic colloids in water should be similar to their ability to bind to organic soils in the vadose zone during infiltration or to organic filtration media in a stormwater filter. Pitt, *et al.* (1999b) described adding compost to the natural soil in an infiltration system to improve both the flow and pollutant removal characteristics of the system. When compared to a natural soil infiltration area, the compost-amended soil system significantly increased the removal of both the quantity of surface runoff and the concentrations of many toxicants typically found in urban runoff. The drawback to the compost amendments was the increased nutrient loads in surface runoff.

Barbosa and Hvitved-Jacobsen (2001) suggested a method for design and evaluation of the design of infiltration ponds for treating highway runoff in semiarid climates. The design was based on capture and infiltration of the most polluted runoff and accounted for the rainfall and soil hydraulic characteristics in the determination of the design volume. Seasonal variations in rainfall and evaporation were considered. Soil characteristics (hydraulic conductivity, texture, pH, and cation exchange capacity), the infiltrated volume, and the infiltrated area were used to calculate the movement of the most mobile heavy metal, Zn, in the soil below the basin. In Dubai, United Arab Emirates, high groundwater levels restrict the amount of infiltration that is permissible and influence the design of the ponds (wet versus dry) (Darnell, *et al.* 2001b). Design criteria specific to Dubai were established to maximize the efficiency of systems that incorporate the use of detention ponds. The criteria included minimum storage pond volumes, clear times, pond geometry, water depth, the use of linings, and alternative land uses. The first pond constructed using the design criteria is now in operation collecting runoff and excess groundwater flows.

2.2.3.2 Grass Swales, Grass Filter Strips, and Bioretention Devices

Grass swale drainages can be used in place of concrete curb and gutter drainages in most land uses, except most strip commercial, manufacturing industrial, and high density residential areas. Grass swales reduce urban runoff problems by a combination of mechanisms. Infiltration of the runoff and associated pollutants is probably the most important process. Filtering of particulate pollutants in grass waterways may also occur, but the flows are usually too large (and deep) to permit effective filtering by the grass. Groundwater contamination concerns are frequently raised whenever stormwater infiltration is proposed. Pitt (1996) reported that groundwater contamination is not a major concern for most stormwaters, if using surface spreading (such as occurs in grass swales). Lind and Karro (1995) also reported on the accumulation of stormwater pollutants in the surface soils of swales, minimizing groundwater contamination problems.

Several large-scale urban runoff monitoring programs have included test sites that were drained by grass swales. Bannerman, *et al.* (1979), as part of the International Joint Commission (IJC) monitoring program to characterize urban runoff inputs to the Great Lakes, monitored a residential area served by swales and a similar residential area served by concrete curb and gutters in the Menomonee River watershed in the Milwaukee area. This monitoring program included extensive flow and pollutant concentration measurements during a variety of rains. They found that the swale drained area, even though it had soils characterized as poorly drained, had significantly less flows and pollutant yields (up to 95% less) as compared to the curb and gutter area.

The ability of grass swales to reduce source area sheetflow pollutant concentrations was also monitored by the Durham, New Hampshire NURP project (EPA 1983). A special grass swale was constructed to treat runoff from a commercial parking lot. Flow measurements were not available to measure pollutant yield reductions, but pollutant concentration reductions were found. Soluble and particulate heavy metal (Cu, Pb, Zn, and Cd) concentrations were reduced by about 50 percent.

Wang, *et al.* (1980) monitored the effectiveness of grass swales at several freeway sites in Washington. They found that 55 to 75 m of grass swale removed most of the heavy metals in the runoff. Lead was more consistently and effectively removed than the other metals, possibly because of its greater association with particulates in the runoff. Lead concentration reductions, with 55 m grass swales, were typically 80 percent, or more, while copper was reduced by about 60 percent, and zinc was reduced by about 70 percent. They concluded that it may be necessary to remove the contaminated sediments and replant the grass periodically to prevent the dislodgment of the deposited polluted sediment.

Yu, *et al.* (1993) constructed and monitored a grass filter in Charlottesville, VA. A 4 ha paved commercial area drained to the 3,800 m^2 grass filter. Stormwater was directed to the grass filter via an infiltration trench and a level spreader. The level spreader system cost about $15,000 (1986). The filter had moderate removals for suspended solids (54 to 84%), total phosphorus (25 to 40%), and zinc (47 to 55%), but only poor removals for nitrate nitrogen (-27 to 20%) and lead (-16 to 55%). The effectiveness of grass strips in controlling highway-runoff contaminants was studied by Newberry and Yonge (1996). The largest portion of metals was found to be retained within the first 1 m of grass strip and 10 mm of depth, with 84% of the applied Zn, 93% of Pb, and >99% of Cd and Cu being retained by the grass strip.

2.2.3.3 Porous Pavements

Porous pavement is a "hard" surface that can support a certain amount of activity, while still allowing water to pass through. Porous pavement is generally used in areas of low traffic, such as service roads, storage areas, and parking lots. Several different types of porous pavement exist. Open mixes of asphalt appear to be similar to regular asphalt, but only use a specific size range of rocks in the hot mix. The porosity of the finished asphalt is much higher than regular asphalt, if properly designed and constructed. Concrete grids have open holes up to several cm wide, possibly containing sand or gravel. It is possible to plant grass in the holes, if traffic is very light and if light and moisture conditions are adequate. Tests have found few problems with porous pavement in areas having severe winters. They can be designed to eliminate all of the runoff from paved areas.

Numerous papers have described successful applications of porous pavements throughout the world. Niemczynowicz and Hogland (1987) describe tests of porous pavements in Sweden. Hogland (1990) gave an overview of porous pavement use in the U.S. and in Sweden. Pratt (1990) described design and maintenance issues, Nawang and Saad (1993) and Sztruhar and Wheater (1993) presented results of experimental field tests of porous pavements, and Fujita (1993) described the extensive use of porous pavements as part of the Experimental Sewer System in Tokyo.

Research at the University of Guelph in Ontario (Thompson and James 1995) has found that porous pavement systems can also be effective filters to remove particulate pollutants from the runoff, even with an underdrain that captures the runoff after pavement percolation. Runoff from typical pavement also had greater masses of pollutants than runoff from the porous pavements. Porous pavement research at the University of Essen in Germany (personal communication, Wolfgang Geiger 1995) also found significant water quality benefits from using porous pavement systems.

French experiments in Nantes, Bordeaux and Paris, have shown that porous pavements (with substantial subsurface reservoir capacity) were very efficient in reducing the pollutant loads discharged into the receiving water (Baladès, *et al.* 1995a and 1995b). These French studies have shown that the pollutant removal efficiencies of suspended solids can be between 50 and 70%, between 54 and 89% for COD, and between 78 and 93% for lead. These reductions were associated with the high amount of infiltration of water, and associated pollutants, through the pavements, away from the surface drainage. These experiments confirm results from previous studies in other countries (Hogland, *et al.* 1987, Pratt, *et al.* 1989 and 1995).

The retention ability of a 50-year old porous pavement structure and the soil below was investigated by Legret, *et al.* (1996 and 1999) and was modeled by LEACHM. Lead, copper, and zinc were not found in significant quantities in the soil below the pavement, i.e., surface retention of those metals, while cadmium was found to have migrated to a depth of 30 cm below the pavement. Porous pavement was found to reduce storm-runoff-pollutant loadings significantly at an experimental site that monitored about 30-rainfall events over four years. Influent-pollutant concentrations of SS were reduced by 64% and 79% for Pb.

2.2.4 Sedimentation/Wetland Controls for Heavy Metal Reductions

There is probably more information concerning the design and performance of detention ponds in the literature than for any other stormwater control device. Wet detention ponds are also a very robust method for reducing stormwater pollutants. They typically show significant pollutant reductions as long as a few design-related attributes are met (most important being size). Many details are available to enhance performance and safety that should be followed. Many processes are responsible for the pollutant removals observed in wet detention ponds. Physical sedimentation is the most significant removal mechanism. However, biological and chemical processes can also contribute important pollutant reductions. The extensive use of aquatic plants, in a controlled manner, can provide additional pollutant removals. Magmedov, *et al.* (1996), for example, report on the use of wetlands for treatment of stormwater runoff in the UK and in the Ukraine, including design guidelines. Wet detention ponds also are suitable for enhancement with chemical and advanced physical processes. Lamella separators, air floatation, filtration, and UV disinfection are examples of treatment enhancements being investigated in France (Bernard, *et al.* 1996; Delporte 1996).

2.2.4.1 Wet Detention Ponds

Wet pond stormwater quality benefits have been commonly reported in the literature since the 1970s. The Nationwide Urban Runoff Program included full-scale monitoring of nine wet detention ponds (EPA 1983). The Lansing project included two up-sized pipes, plus a larger detention pond. The NURP project located in Glen Ellyn (west of Chicago) monitored a small lake, the largest pond monitored during the NURP program. Ann Arbor, Michigan, monitoring included three detention ponds, Long Island, New York, studied one pond, while the Washington D.C. project included one pond. About 150 storms were completely monitored at these ponds, and the performances ranged from negative removals for the smallest up-sized pipe installation, to more than 90% removal of suspended solids at the largest wet ponds. The best wet detention ponds also reported BOD_5 and COD removals of about 70%, nutrient removals of about 60 to 70%, and heavy metal removals of about 60 to 95%.

Hvitved-Jacobsen, *et al.* (1994) examined the most effective treatment systems for treating urban and highway runoff in Denmark. They concluded that wet detention ponds were the most efficient and suitable solution for the removal of most pollutants of concern from both highway and urban runoff. Denmark does not have any effluent standards and the acceptable pollutant discharges are therefore determined based on specific receiving water requirements. They found that both urban and highway runoff were mostly causing accumulative (chronic) effects (associated with suspended solids, toxicants, and nutrient discharges) and treatment designs therefore need to be based on long-term pollutant mass discharge reductions. It was evident that relatively low concentrations of pollutants must be reduced, and that large volumes of water must be treated in a short time period. For these reasons, and for the specific pollutants of concern, they concluded that wet detention ponds were the most effective option, even though the first wet detention pond was only constructed in Denmark in 1989. Their recommended design was based on: detention pond volume (about 250 m^3 per effective hectare of drainage area), water depth, pond shape, use of plants (covering at least 30% of the water surface), and the use of a grit removal forebay. The resulting mass removals using this design were excellent for suspended solids (80 to 90%) phosphorus (60 to 70%) and heavy metals (40 to 90%).

Crunkilton and Kron (1999) measured the toxicity of stormwater runoff before and after it had been allowed to flow through a pilot-scale wet-detention basin. Selected heavy metals and PAH compounds were measured in incoming and outgoing-settling basin water. *Daphnia magna* and *Pimephales promelas* (fathead minnow) were exposed to pre and post-settling basin treated stormwater runoff for three test periods of 14 days each in 1996 and 1997. Petterson (1999) examined the partitioning of heavy metals in particulate-bound and dissolved phases in a stormwater pond in Goteborg, Sweden. The results showed a clear variation in lead partitioning affected by specific conductivity.

Caltrans initiated a study in two Districts (Los Angeles and San Diego, California) to examine the benefits, technical feasibility, costs and operation and maintenance requirements of retrofitting extended detention facilities into existing highway and related infrastructure (Taylor, *et al.* 2001). Sampling results showed average suspended solids removal was 73%, total metals removal varied between 61% and 75%, while dissolved metals removal varied between 16% and 44%. Concrete lined dry ponds showed lower removal rates. Major removal of sediment is estimated to be required every 10 years.

Potential Sediment Contamination Problems. It is necessary to plan for required maintenance during the design and construction of detention ponds. Ease of access of heavy equipment and the possible paving of a sediment trap near the inlet would ease maintenance problems. Frequently, concern arises about the safety of disposing sediments from wet detention ponds. There have been several studies that have addressed this issue. Deposited sediment can be heavily polluted and may require special disposal practices. Sediment concentrations of up to 100,000 mg organic carbon, several thousand mg lead, several hundred mg zinc, and more than ten mg arsenic per kg dry sediment are not uncommon for lakes receiving urban runoff (Pitt and Bozeman 1979). Dredged sediment is usually placed directly onto trucks, or is placed on the pond banks for dewatering before hauling to the disposal location. One common practice is to keep an area adjacent to the detention pond available for on-site sediment disposal. Small mounds can be created of the dried sediment and covered with top soil and planted.

Dewberry and Davis (1990) analyzed sediments from 21 ponds in northern Virginia. They found trace metals in many of the sediments, but the available forms of the metals were significantly less than applicable toxic thresholds. They concluded that the dredged materials could be safely disposed either on-site or at sanitary landfills without danger of health problems. However, they recommend that sediment samples from specific ponds be analyzed before dredging.

Yousef and Lin (1990) conducted extensive pond water quality and sediment quality analyses in six wet detention ponds in Florida as part of a Florida Dept. of Transportation study to develop pond maintenance procedures. The sediment accumulation rates were found to be between 0.25 and 0.72 cm per year and correlated with pond age, size of drainage basin and size of pond. The bottom material was found to be poorly graded sand. Appreciable amounts of heavy metals (Cu: 7 to 73 μg/g, Ni: 12 to 82 μg/g, Pb: 84 to 1025 μg/g, and Zn: 13 to 538 μg/g), and nutrients (N: 1.1 to 5.2 mg/g, and P: 0.1 to 1.2 mg/g) were found in the surface layers of the sediments. However, the concentrations of the pollutants decreased rapidly with depth, generally being less than 10% of the surface sediment concentrations below 20 cm beneath the water-sediment interface. The bottom sediments were also analyzed to determine the TCLP extractable portions of the metals. These were found to be significantly less than the whole sediment metal concentrations (Cu: 0.13, Ni: 0.31, Pb: 0.27, and Zn: 0.33). They determined that the TCLP extractable fraction was lowest for sediments having higher clay and organic material. They concluded that the sediments could be removed during normal maintenance operations and disposed of on non-agricultural land.

MacDonald, *et al.* (1999) investigated the removal of heavy metals in a detention pond in Scotland. They found that the sediment in the pond may reach unacceptable concentrations of heavy metals within twenty years. Karouna-Renier and Sparling (2001) investigated the accumulation of Cu, Zn, and Pb by macroinvertebrates collected in Maryland stormwater treatment ponds serving commercial, highway, residential and open-space watersheds to determine if watershed land-use classification influenced metal concentrations in macroinvertebrates, sediments, and water. Composite Zn concentrations in odonates from ponds with commercial development (mean = 113.82 $\mu g\ g^{-1}$) were significantly higher than concentrations in the other land-use categories. Similarly, Cu levels in odonates from commercial ponds (mean = 27.12 $\mu g\ g^{-1}$) were significantly higher than from highway (mean = 20.23 $\mu g\ g^{-1}$) and open space (mean = 17.79 $\mu g\ g^{-1}$) ponds. However, metal concentrations in sediments and water did not differ significantly among land-uses. The levels of Cu, Zn, and Pb in invertebrates from all ponds were less than dietary concentrations considered toxic to fish.

2.2.4.2 Dry Detention Ponds

Dry detention ponds provide less, and highly variable, water quality improvements than wet ponds, mostly due to increased scour of previously settled sediment.

Stanley (1996) examined the pollution removal performance at a dry detention pond in Greenville, NC, during eight storms. The observed reductions were low to moderate for suspended solids (42 to 83%), phosphate (-5 to 36%), nitrate nitrogen (-52 to 21%), ammonia nitrogen (-66 to 43%), copper (11 to 54%), lead (2 to 79%), and zinc (6 to 38%). Stanley also summarized the median concentration reductions at dry detention ponds studied by others. In all cases, the removals of the stormwater pollutants is substantially less than would occur at well designed and operated wet detention ponds. The resuspension of previously deposited sediment during subsequent rains was typically noted as the likely cause of these low removals.

Maintenance problems in dry ponds had also been commonly noted in earlier Maryland surveys. Bartone and Uchrin (1999) compared the performance of two dry-stormwater-detention facilities, one having a concrete low-flow channel, and the other with a vegetated low-flow channel during four events. As they expected, the detention pond having the concrete channel was ineffective for stormwater quality control. However, the basin with the vegetated channel was also found essentially ineffective for water quality improvement, with flushing of previously captured pollutants being the most likely reason for the poor performance.

2.2.4.3 Wetlands

Martin (1989) monitored a 0.2 acre wet detention pond followed by a 0.7 acre wetland. The system provided moderate to high pollutant reductions of solids, lead, and zinc (between 50 and 80%) and smaller reductions for nitrogen and phosphorus (between 30 and 40%). At low discharges and with large storage volumes, the pond was found to be moderately well mixed with residence times not much less than the maximum expected if operating under ideal mixing conditions, with little short-circuiting apparent. At higher discharges and with less storage volume, significant short-circuiting occurred.

The hydrology and water quality of an urban wetland receiving stormwater runoff from a municipal maintenance garage were measured during the summer of 1993 to evaluate the wetland's water quality enhancement function. According to Thurston (1999), hydrologic and analytical data together suggest that sedimentation was the primary mechanism actively reducing water column concentrations of lead and petroleum hydrocarbons introduced to the wetland via stormwater runoff.

Shutes, *et al.* (1997a and 1997b) reported on the results of two case studies of the pollution-removal performance of constructed wetlands for urban-stormwater-runoff treatment. The design of constructed wetlands included primary-treatment components, substrate composition, and flowrate regulation. Two subsurface-flow-constructed wetlands were tested for pollutant-removal performance in conjunction with an online-stormwater-detention pond. The wetlands were able to maintain removal of SS (46%), dissolved metals (Cu 50%), and organic C (10%) (Rochfort, *et al.* 1997). The retention of heavy metals in a wet reed-bed wetland were compared to retention in a dry-detention pond near London, United Kingdom, by Hares and Ward (1999). A higher percentage removal occurred in the monitored wetland facility than in the dry pond.

Scholes, *et al.* (1999) described the performance of two wetlands located near London, United Kingdom. Removal efficiencies were greater than 65% for Zn, Cd, Pb, and Cu, at one site, while they were only about one-half of these rates at the other site. They also investigated the role of microorganisms in removing the heavy metals and two metal tolerant strains were found to accumulate Pb and Zn. It was believed that this ability may prove to be an important year-round pollutant removal process. Lead and petroleum hydrocarbons (originating from a municipal public works yard) were monitored in an urban wetland by Thurston (1999). Levels of these compounds in the sediments were higher near the inlet than elsewhere. It was concluded that sedimentation was the primary mechanism responsible for reducing the concentrations. Dombeck, *et al.* (1999) examined the assimilation of trace metals in wetland sediments at the Sacramento (California) Regional County Sanitation District's Constructed Wetland Demonstration Project. Six metals (silver, cadmium, copper, mercury, lead and zinc) consistently exceed mass removal rates of 60%. Chromium and nickel were observed in concentrations approaching or exceeding the freshwater sediment Probable Effect Levels (PEL). It was expected that the nickel enters the wetland in a predominantly dissolved and strongly organic-complexed form.

To address the unknowns and design concerns of a stormwater wetland system, a bench-scale vegetative-treatment-cell study has been conducted for the purpose of examining nutrient and metal dynamics and removal efficiencies of three individual plant species under various stormwater-pollutant loadings and detention times. The plant species being evaluated are: cattails (*Typha latifolia*), reeds (*Phragmites sp.*), and bulrushes (*Scirpus sp.*) (O'Shea, *et al.* 1999). The research presented by Fassman and Yu (2001) examined the relative pollutant removal capability for specific wetland vegetation. Results of water quality analysis show various trends with respect to pollutant removal and time. In general, export of TP was shown in all vegetated cells. Maximum export for vegetated cells was observed usually during the first week of a treatment, especially for *Typha* spp. and *Phragmites* spp. By the end of most treatments, TP removal was observed in all cells. COD removal was seen in all cells for all weeks of each treatment (average removal ranged from 54-78%). TSS results were inconsistent. Export of Cu was noted for *Scirpus* spp. whereas all other plants showed positive removal. TOC was removed by all plants (average removal 24 – 69%), with the exception of the *Scirpus* spp. cell. Goulet, *et al.* (2001) tested a first-order removal model to predict metal retention in a young constructed wetland receiving agricultural and urban runoff. The wetland retained metals best during summer and fall. The first-order removal model predicted Fe and Mn retention in the spring and dissolved Zn retention from spring to fall in both years. However, first-order removal models failed to fit summer, fall and winter data fur almost every metal under investigation (Fe, Mn, dissolved Cu, dissolved As) suggesting that HRTs (< 1-25 days) did not affect metal retention during these seasons. Therefore, the first-order removal model is inadequate to predict metal retention on a seasonal basis.

The ability of a wetland to mitigate highway-runoff pollutants, particularly metals, was investigated by Mitchell, *et al.* (2001a). Fifty-seven rainfall-runoff events were monitored. Metal concentrations were reduced via flow through the grassy median and then the wetland system. Average heavy metals removals were Zn (67%), Fe (67%), Pb (54%) and Ni (45%). Metals were reduced some even during winter months and low temperatures. Wetland sediment analysis indicated a broad range of concentrations of metals. Rodgers, *et al.* (2001) reported on the A-01 effluent outfall, which collected both normal daily process flow and stormwater runoff from a industrial park area and did not meet its South Carolina NPDES permit limits for metals, toxicity, and total residual chlorine at the outfall sampling point. Installation of a constructed wetland system including a basin to manage stormwater surges reduced the problematic constituent concentrations to below the NPDES permit limits before the sampling point. The constructed treatment wetland system proved its ability to treat industrial wastewaters containing metals with low O&M costs.

2.2.5 Combined Sewer Overflow/Sanitary Sewer Overflow Control

In most urban areas, stormwater is directed to subsurface drainage systems. In areas having combined sewer systems, such as in most of Europe, in the large cities of Asia, and in many older cities of the U.S., this additional water causes overflows of raw or poorly treated domestic sewage during periods of moderate to heavy rainfalls. Even in areas having separate sewerage systems, the use of conventional subsurface sewerage radically alters the receiving waters. The frequent and high flows in receiving waters causes detrimental biological conditions, causes increased erosion of channels, causes flood damage, and dramatically reduces the amount of rainfall that recharges the local groundwaters. This recharge reduction causes severe low flow problems in many areas during prolonged dry periods, further worsening the biological habitat, decreasing recreation benefits, and reducing the assimilative capacity for downstream wastewater discharges.

The Technical University of Denmark has conducted numerous research projects concerning the benefits of infiltration as a source area control to reduce combined sewer overflows (Geldof, *et al.* 1994; Mikkelsen, *et al.* 1994; Rosted Petersen, *et al.* 1994; and Jacobsen and Mikkelsen 1996). Rosted Petersen, *et al.* (1994), for example, found that the optimal solution for reducing CSO volumes by 40% required infiltrating 65% of the paved areas using infiltration trenches having total storage volumes of 3.6 mm. This corresponds to a return period of 0.04 years (about 2 weeks), in contrast to the commonly applied design return periods of 2 to 10 years.

Infiltration improves the receiving water quality in areas served by either combined or separate sewers (Geldof, *et al.* 1993). Decreased amounts, frequencies, and durations of overflows from combined systems have dramatically lowered the discharges of many pollutants. The number of overflows in combined sewers in Tokyo have decreased from about 36 per year to about 7 in areas served by the experimental sewer system (ESS) that incorporates many infiltration elements. Phosphorus and heavy metals in separate sewer discharges can be substantially reduced with the widespread use of infiltration (Hvitved-Jacobsen, *et al.* 1992). The most difficult problems related to the Tokyo infiltration facilities have been clogging and groundwater contamination (Fujita 1993). A high-pressure water jet has been successfully used to restore clogged permeable pavements, along with other measures to protect the devices. Groundwater monitoring has been conducted for ten years in the ESS area, with no indication of groundwater contamination. However, efforts to improve service life and to protect groundwater quality are continuing.

Lim and Lim (1999) described a unique scheme to implement urban stormwater pond collection systems for potable water use in Singapore. Close monitoring and control of pollution through the adoption of stringent anti-pollution measures and enforcement actions have resulted in the collection of generally good quality raw water from these urbanized catchments with very low levels of heavy metals and low coliform counts as compared to raw water from largely forested catchments.

The Actiflo® method is a compact physical-chemical water- treatment system that is being used increasingly for treating wastewater and CSOs. The method uses weighted settling combined with lamella settling and typical efficiencies are: SS-85%, COD-60%, TKN-18%, total P-85% and also removes heavy metals efficiently (Plum, *et al.* 1997).

Ab Razak and Christensen (2001) investigated the benefits of the Milwaukee deep tunnel for temporary storage of storm water and sewage, on the water quality of the Milwaukee, Menomonee, and Kinnickinnic Rivers. Statistical analysis of BOD, phosphorus, suspended solids, fecal coliforms, zinc, and chloride indicated that the Menomonee River benefited the most from the deep tunnel since 1994 when the tunnel came on line. Fecal coliforms inside the CSO area, and to a certain extent BOD and zinc levels, exhibited the most significant decline.

2.2.6 Chemical Treatment of Stormwater

Chemical addition has been used for many years in water treatment, and in lake management. More recently, full-scale implementations of chemical assisted settling has been used for the treatment of stormwater in wet detention ponds or at outfalls into small urban lakes. The chemicals tested and used include alum (generally a complex of aluminum and sulfate), ferric chloride, and aluminum chloride compounds, plus various coagulant aids.

The addition of alum in Colston's (1974) tests further increased the reduction of particulate residue, COD, and turbidity to about 85 to 97 percent. Gietz (1981), in a series of laboratory tests in Ontario, found that an alum dosage of 4 to 6 mg/L was the most effective for highly polluted runoff. Over-dosages of alum and ferric chloride generally gave poor results. He found that it was difficult to add the correct dosage of coagulant because of the changing pollutant concentrations in the runoff. Low flow velocities also reduced mixing effectiveness and may require mechanical assistance. The flocs that were formed with the coagulants were easily disturbed by runoff turbulence.

Kronis (1982), in a series of Ontario bench and pilot scale tests, found that disinfection of stormwater with NaOCl at 5 mg/L available chlorine reduced fecal coliform populations to less than 10 organisms per 100 ml. He identified alum dosages of 30 mg/L as a preferred flocculant, with 10 to 30 percent increases in removals of particulate residue, BOD5, COD, and total phosphorus as compared to plain sedimentation. However, chemical assisted settling generally produced moderate and erratic reductions in bacteria populations. Disinfection in wet detention ponds may be expensive, but it may be the only feasible method of significantly reducing bacteria populations in areas with serious bacteria problems.

Heinzmann (1993) described the development of a coagulation and flocculation treatment procedure for stormwater in Berlin. He found that because the stormwater was weakly buffered and was very soft, a polyaluminum chloride, with a cationic coagulant aid (polyacrylamid), was most suitable. A constant dosage of 0.06 mmol/L (as Al) was used, resulting in pH levels always greater than 6. The constant dosage was possible because the pH and buffering capacity of the stormwater was relatively constant during storms. He found that the best enhanced stormwater treatment process used coagulation and flocculation in a pipe designed for both microfloc and macrofloc formation, and final separation by filtration. The filtration was much better than the one hour sedimentation typically used in Berlin sedimentation tanks. He did find that a six minute flocculation time was sufficient before filtration. He found significant removals of phosphorus (<0.2 mg/L), organic compounds (including PCB and PAHs), solids (<5 mg/L), lead and copper. However, very poor removal of zinc was noted, and pollution prevention (decreased use of galvanized metals) was recommended. In the one-hour sedimentation tanks, without any chemical addition, the phosphorus (about 0.5 mg/L) and solids (about 50 mg/L) effluent concentrations were not nearly as low. The costs for this enhanced treatment (7 to 10 DM/m^3 in 1990) was about 10 to 40% higher than with the ordinary one-hour sedimentation tanks alone.

Pitt and Dunkers (1992 and 1995a) described a full-scale stormwater treatment plant, using the Karl Dunkers' system for treatment of separate stormwater and lake water. This system has been operating since 1981 in Lake Rönningesjön, near Stockholm, Sweden. The treatment facility uses ferric chloride and polymer precipitation and crossflow lamella clarifiers for the removal of phosphorus. Excess flows are temporarily stored before treatment inside an in-lake flow balancing tank (the Flow Balancing Method, or FBM). The stored excess stormwater is then pumped from the flow balancing storage tanks to the treatment facility during dry weather. The overall phosphorus removal rate for the 11 years from 1981 through 1991 was about 17 kg/year. About 40% of the phosphorus removal occurred in the FBM from sedimentation processes, while the remaining occurred in the chemical treatment facility. This phosphorus removal would theoretically cause a reduction in phosphorus concentrations of about 10 μg/L per year in the lake, or a total phosphorus reduction of about 100 μg/L during the data period since the treatment system began operation. About 70% of this phosphorus removal was associated with the treatment of stormwater, while about 30% was associated with the treatment of lake water. The lake phosphorus concentration improvements averaged about 50 μg/L. This was only about one-half of the theoretical improvement, probably because of sediment-water interchange of phosphorus, or other unmeasured phosphorus sources.

The 1996 NALMS (North American Lakes Management Society) conference in Minneapolis/St. Paul included several presentations describing the use of alum for stormwater treatment. Harper and Herr (1996) describe the historical use of alum to treat stormwater entering Lake Ella in Tallahassee, FL, which began in 1986. A liquid slurry of alum is injected into the major storm drainage entering the lake, on a flow-weighted basis during rains. The alum forms precipitates with phosphorus, suspended solids, and heavy metals, which then settle in the lake. This treatment system resulted in immediate and substantial improvements to Lake Ella water quality. There are currently 23 alum stormwater treatment systems in Florida. Harper and Herr (1996) report that alum treatment of stormwater has consistently achieved 90% reductions in total phosphors, 50 to 70% reductions in total nitrogen, 50 to 90% reductions in heavy metals, and >99% reductions in fecal coliform bacteria. The precipitates of the phosphorus and heavy metals have been shown to be extremely stable over a wide range of dissolved oxygen and pH conditions.

Herr and Harper (1996) also reported on a very large alum project at Lake Maggiore in St. Petersburg, FL. This 156 ha lake receives stormwater from a 927 ha watershed. Water quality problems were noted as early as the 1950s that included fish kills, algal blooms, nuisance macrophyte algal growths, and high bacteria levels. An environmental assessment determined that an 80% reduction in the annual phosphorus discharges from the stormwater and baseflow would result in an acceptable trophic status for the lake. Five alum treatment plants were then designed and will be operational in August 1997, comprising the largest alum stormwater treatment system ever built.

An alum pilot-scale treatment system for stormwater, located in Minnesota, was described by Kloiber and Brezonik (1996). This system injected 1 mg/L (as Al) alum into a storm sewer at a pumping station just upstream of a 1.2 acre wet detention pond. The few minutes travel time between injection and the pond allowed 75 to 80% reductions in soluble reactive phosphorus. However, the pond retained only 40% of the added aluminum, increasing to 70% when a coagulant aid was used. The lowest total aluminum concentration in the pond effluent was 0.26 mg/L, still exceeding the water quality standard. They concluded that closer evaluations of the toxicity and bioavailability of the aluminum associated with alum stormwater treatment is needed. During treatability tests of stormwater from critical source areas, Pitt, *et al.* (1995) found that alum addition significantly increased the toxicity of the water (as indicated using the Microtox screening procedure).

2.3 Summary of Stormwater Heavy Metal Effects, Sources and Control

Heavy metals have been extensively studied in urban areas for many years. As shown in Appendix A, total recoverable heavy metals in stormwater frequently exceed water quality standards and criteria. However, most of the metals (with the typical exception of zinc) are predominantly associated with particulates (> 0.45 μm). Toxicologists generally discount the ecological effects of particulate bound metals. However, the fine particulates associated with stormwater have been shown to cause substantial toxicity during various controlled experiments. Short-term, acute, toxicity is rare for stormwater, but longer exposures to contaminated stormwater, and to contaminated sediment, cause longer-term, chronic toxicity, illustrated by major changes in benthic organisms. Of course, other factors will also affect these observations, such as sediment texture and interstitial water oxygen content. However, the fine sediment coating larger rocks and aquatic plants, has been shown to be especially toxic to test organisms.

Heavy metals in stormwater mostly originate from automobile-related activities, and exposure of rain to building materials (especially zinc and copper). Treated wood and tires are also common sources of metals in residential and commercial areas. Industrial areas, however, may contribute heavy metals to stormwater by improper disposal of wastewaters and the exposure of materials and products to rain. The decreased use of leaded gasoline and leaded paint in the U.S. has caused substantial reductions in stormwater lead concentrations over the past several decades. Other pollution prevention options are available to reduce other metal discharges (use of alternative building materials, for example).

The reduction of stormwater heavy metals has mostly focused on the removal of stormwater suspended solids. Wet detention ponds generally provide the largest percentage removals of heavy metals, but have much smaller effects on the dissolved metal fractions. Sorption of dissolved metals on various media has been demonstrated to effectively remove some dissolved heavy metals. Combination devices using both sedimentation and sorption may be the most effective means of removing a large portion of all the metal species. Chemical precipitation and coagulation has also been shown to be effective, especially for copper, lead, and zinc. Infiltration of stormwaters having heavy metals most likely can be safely conducted, as most metals migrate poorly through surface soils. Exceptions would be for industrial stormwaters which may have excessive dissolved metal fractions, areas having shallow groundwater (especially for sandy soils), and where subsurface disposal is used, without the benefit of infiltration through surface soils. Amended soils have been shown to substantially reduce the migration of heavy metals to groundwater and can therefore be used to further reduce the potential for groundwater contamination when infiltrating stormwater in sandy soils.

A number of conventional and emerging stormwater controls can be better designed to provide more effective control of heavy metals. Grass swales can incorporate many positive features to reduce heavy metal discharges to surface and groundwaters: sedimentation occurs due to increased channel lining roughness, infiltration is common, and amended soils can easily be incorporated. Critical source area controls can be used incorporating special sorption/filtration media. Finally, sedimentation in wet detention ponds can be improved to better remove and contain finer particulates, and post-treatment with wetlands can be used for increased biological control processes.

CHAPTER 3.0

Filter Media Evaluation and Filter Column Studies

3.1 Filter Tasks Experimental Objectives

The two main objectives of this study were to examine and compare several media to determine their advantages and disadvantages for removing metals from urban runoff and identify the factors that effect filter performance and design. An overview of all filter tasks and the allocation of experimental work is outlined below.

3.1.1 Filter Tasks

- Filter Task 1: A summary of the work to date on the treatment options for stormwater heavy metals in urban receiving waters and watersheds (Chapter Two). A summary of heavy metal effects and sources in urban receiving waters(Appendix A). Review of filter media in the treatment of stormwater metals (Chapter 3).
- Filter Task 2: Assess and quantify the ability of various media to remove metals from stormwater under lab scale and pilot scale conditions. (Chapters Three (lab scale) and Four (pilot scale))
- Filter Task 3: Characterize the physical properties of the media with particular reference to their sorption properties, endurance and stability in the packed column environment of the filter. (Chapters Three and Four)
- Filter Task 4: Determine the effect of varying process conditions on the metal capture process. Evaluate the influence of surface phenomena on the rate and extent of metal uptake (Chapters Three and Four).
- Filter Task 5: Conduct laboratory scale fixed-bed column studies to evaluate the dynamic characteristics of the metal capture process. (Chapter 4).
- Filter Task 6: Investigate the Roll of Biomass in Metals Uptake (Chapter 5).
- Filter Task 7: Full Scale Field Studies (Chapter 4).
- Filter Task 8: Investigate leaching and disposal options for the media (Chapters 3 and 6).

3.1.2 Filter Work Plan of Work

3.1.2.1 Experimental Work: Part One

1. The first stage in the examination of stormwater filters was a series of batch experiments and downflow lab-scale filter column studies to determine the parameters for a series of design equations to use in modeling the removal ability of the filters.
2. Pilot-scale downflow filters were built using the media examined in stage one. The results of the pilot scale studies were compared with the data from the stage one lab-scale columns to examine how well the design equations predicted removal.
3. Preliminary lab-scale tests to examine upflow filters were performed.

3.1.2.2 Experimental Work: Part Two

1. The media examined in part one along with several additional media were further examined in batch tests to compare metal removal ability and determine which media had the best potential to remove metals in runoff.
2. Additional lab-scale work was performed to examine the potential of upflow filtration. Upflow filters were constructed using the three most effective media from stage one of part two and metal removal efficiency was examined for different rates of flow.
3. The roll of biomass in metal uptake was examined.

3.1.2.3 Guide to Discussion of Filtration Experimental Work

The pilot scale field study performed at the University of Alabama at Birmingham is discussed in Chapter 4. The lab scale study performed at the University of Alabama Tuscaloosa is discussed in sections 3 through 5 of this chapter. Several filter tasks overlap both lab and pilot scale studies and therefore these tasks are discussed in both chapters as they relate to the filter work. A literature review on filters and filter media is covered in generally in chapter 2 and a filter-specific literature review can be found in section 2 of this chapter.

3.2 Introduction

3.2.1 Filters

Stormwater filters remove pollutants through a variety of mechanisms, primarily straining of larger particles and sorption and ion-exchange of "dissolved" pollutants. The effluent from filters are generally discharged via an outfall or runoff to a surface receiving water, but if adequate treatment levels are achieved, the effluent can be infiltrated to promote groundwater recharge.

Filters are particularly suitable to areas where the water table is too high for infiltration methods and/or groundwater needs protection. One advantage of various filter applications over other stormwater treatment methods is their reduced spatial footprint compared to other methods. Another advantage is that they can be installed below ground. Disadvantages are that filter beds can be expensive to construct and may require considerable maintenance (Urbonas and Stahre 1993).

Many stormwater treatment devices currently in use, or in development, use sedimentation and straining as the primary pollutant removal mechanisms. In filtration pollutants are trapped either on the surface of the filter or in pores and it can be very effective for solids removal: 90 percent removal can be achieved over a wide range of particle sizes (diameter between 6 and 41 μm) (Clark and Pitt 1999). Pre-sedimentation followed by filtration is preferred over sedimentation or filtration alone because although most particulate matter in urban runoff is in the coarser size range, metals in runoff are associated with particles in all size ranges(Dennison 1996). Furthermore not all contaminants are particulate-bound. Traditional filtration media such as sand that are characterized filtration/straining mechanisms are ineffective in removing soluble pollutants (Pitt, et al. 1995; Claytor and Schueler 1996). The incorporation of non-traditional media such as peat-sand into filter beds can further enhance effluent quality by removing dissolved or colloidal-bound metals through ionic adhesion and ion-exchange (Urbonas 1999).

3.2.2 Filter Media

The effectiveness of different filter media for a given pollutant load is highly variable: media that is excellent for one pollutant may provide no treatment for another. In addition, some media that offer potential as sorbents of metals may create water quality problems of their own. The application of filter media to the removal of metals from industrial wastewater is well established, however, the transfer of this technology to stormwater treatment presents unique challenges. Several media have been examined further for their potential to capture metals/pollutants from stormwater. The following discussion highlights some of the advantages and disadvantages of the media investigated. The review includes sections on the following media: sand, activated carbon, peat moss, compost, iron oxide coated sand, iron infused media, zeolite, cotton (cellulose-based fibers), and the Forest Products Research Laboratory's agrofiber.

3.2.2.1 Sand

The traditional and most common filter media used for stormwater filtration is sand. The following advantages of sand filters over both retention devices and infiltration devices have been listed (Galli, 1992 as noted in Debo and Reese, 1995):

- They have a history of successful employment in diverse applications including dry climates and areas with high water table or poor draining soils.
- They have a proven ability for small (less than 10 acre) applications.
- Their pollutant removal is moderate to high for most pollutants.
- Sand filters are more resistant to clogging and easier and cheaper to repair if clogged than infiltration trenches or basins.
- Sand filters can be easily coupled with flood control BMPs.
- On a dollar per runoff volume treated basis, sand filters are more economical to construct than infiltration devices or oil and grit chambers.
- Removal of contaminated soil and surface scum does not involve costly and elaborate measures.

Slow sand filters are characterized by slow flow rates, an extremely narrow range of particle sizes, no chemical pretreatment, relatively long filter runs between cleanings, and surface scraping and sand removal for cleaning (Collins, et al. 1992). Filtration rates may be as much as fifty times slower than rapid sand filters; consequently, slow sand filters require significantly more surface area in order to filter comparable volumes of water (Crittenden, et al. 1993). Fine sand/silt filters remove particulates by direct straining on the surface of the filter media. The combination of grain size and bed depth will determine the effectiveness of the filter (Naghavi and Malone 1986; Tobiason, et al. 1993). Smaller diameter media affects the removal rate of larger particles and the rate of head-loss development. Head-loss-causing particles are distributed more evenly through the filter in larger diameter media, and, therefore, capturing a particle with larger media results in less head loss than capturing it with smaller media. Filtration velocity, to a lesser extent than media size, also affects removal efficiency, bed depth use, and head loss. Head loss is directly proportional to velocity in new filters, but for ripened filter beds, the direct proportionality does not apply. Increased velocity pushes particles deeper into the filter bed prior to capture, thus allowing more of the filter depth to be used in particle capture (Clark, et al. 1992). Since particulate removal from the wastewater is a function of both sand depth and sand particle size, using coarse sand and a deep bed is recommended by Farooq and Al-Yousef (1993) because this type bed will require less cleaning than fine sand in a shallow bed filter.

Void formation ("wormholes," or pores larger than a sand particle's diameter) has been observed in traditional sand filters. These wormholes start with holes in the surface deposit and remain open despite the continuing flow of solids into them (Ives 1989). This development of areas of preferential flow (macropores, fingering, or funneled flow paths) means that the fluid follows the local wetting front in wormholes and bypasses the matrix pore space. Filtration efficiency for preferential flow pathways is much smaller than it is for matrix pore flow because flow through preferential pathways is more rapid with less time for straining and/or sorption (Steenhuis, et al. 1990).

Slow sand filters have been found to be extremely effective in removing suspended particles, and effluent turbidities are consistently below 1.0 NTU (Collins, et al. 1992; Eighmy, et al. 1992; Farooq and Al-Yousef 1993). Sand filters were shown to effectively remove biochemical oxygen demand (BOD) (86%), suspended solids (68%), turbidity (88%), and total coliform bacteria (99%) from sanitary wastewater (Farooq and Al-Yousef 1993). Generally, dissolved oxygen concentrations and pH decrease in sand filters. Particulate COD, particulate organic nitrogen, and particulate phosphorus typically are removed during filtration, even before the filter is ripened. However, very little of the soluble fraction of the above constituents is removed (Fujii, et al. 1987).

When looking at metal binding to sand by type of reaction site, the strength is relatively low in the ion-exchangeable fraction and increases in the non-exchangeable fraction because metals in the non-exchangeable fraction likely are incorporated within the crystalline lattice or strongly sorbed to the mineral surface. When looking at binding by sand particle size, the smallest sized grains tend to have the greatest mass concentration of metals. Lead binds more strongly to the smaller particles, while arsenic, copper, and zinc show similar affinities for all size fractions (Van Benschoten, et al. 1994).

Sand filtration removed approximately 50% of suspended solids for waters that contained mostly colloidal-sized particles and 80 to 100% for waters whose solids were larger (Biskner, et al. 1978; Dahab, et al. 1991). At a Superfund site, it was noted that solids breakthrough occurred much earlier than expected, possibly because the filter was not in continuous operation (Dahab, et al. 1991).

Sand filters are most effective at removing particulates and particulate bound contaminants. Sand filtration, without modification of the media by ripening or by adding a adsorbent surface coating such as manganese or ferric oxide, is not effective at removing dissolved constituents (Selim and Wang 1994), including dissolved metals. Ripening of the filter or the addition of an iron-oxide coating can increase dissolved contaminant removal. Adding media with ion-exchange and/or sorption properties, such as peat or compost, to sand can also improve the removal of dissolved contaminants.

Clark (1996) found that sand filters could lose previously trapped contaminants by flushing with subsequent flows until the filter is aged and a biofilm forms that will more permanently retain the contaminants. Ripening, the development of a bacterial biofilm ("schmutzdecke") on the sand filter, improves the removal ability of the filter because the captured particles aid in the collection of subsequent particles by partially blocking and restricting passage through the pores. This increased efficiency occurs for all particle sizes initially but eventually only continues for small sizes with the removal efficiency decreasing and possibly becoming negative for larger particles (Fujii, et al. 1987; Clark, et al. 1992; Tobiason, et al. 1993).

Sand filtration for stormwater treatment began on a large scale in Austin, Texas. The Austin sand filters are used both for single sites and for drainage from larger areas, generally less than fifty acres. The filters are designed to hold and treat the first one-half inch of runoff with very good pollutant removal ability. According to the City of Austin design guidelines (1988), the minimum sand depth should be 18 inches. These filters may have either gravel, a geotextile, or other fabric on top of the sand to prevent premature clogging with large particles. The assumed pollutant removal efficiencies for an Austin sand filter, which are based upon the preliminary results of the City's stormwater monitoring program, are in Table 3-1.

The Delaware sand filter has a similar design to the Austin filters, with an 18-inch sand depth and a 6-inch gravel underdrain. Each filter has a minimum of 6 to 12 inches of ponding depth/storage head available on top of the filter. These filters are intended for sites that have impervious areas that will drain directly to the filter, such as fast-food restaurants and gas stations, and they often precede an infiltration device in order to prevent or postpone clogging of the device (Shaver 1994). According to Delaware's guidelines, the sand filter can be expected to adequately remove particulates (TSS removal efficiency 75-85 percent) but not soluble compounds. Monitoring of a Delaware sand filter at a parking lot near National Airport in Alexandria, Virginia, showed that the filter had an average 72 percent removal of total phosphorus, >80% removal of total suspended solids, and >90% removal of zinc (Galli 1990).

Table 3.1. Reported Pollutant Removal Efficiencies for Sand Filters (Data derived from City of Austin 1988)

Pollutant	Removal Efficiency (%)
Fecal Coliform Bacteria	76
Total Suspended Solids (TSS)	70
Total Nitrogen	21
Total Kjeldahl Nitrogen	46
Nitrate – Nitrogen	0
Total Phosphorus	33
Biochemical Oxygen Demand (BOD)	70
Total Organic Carbon	48
Iron	45
Lead	45
Zinc	45

3.2.2.2 Iron Oxide Coated Sand

Another technique for improving the removal efficiency of a sand filter is by adding an adsorbent coating, usually an iron or manganese oxide, to the sand grains, thus providing adsorption sites for the ions in solution. The iron oxide coating on sandy soil can bind metals of all sizes very strongly, greater than 90% removal for lead, copper, and zinc (Sansalone and Hird 1999). In an industrial application, copper removal on an iron-oxide coated sand was 100% through the first 35 to 40 bed volumes of water treated. The iron-oxide coated sand was easily regenerated with no significant loss of capacity (Lai, et al. 1994).

Benjamin et al. (1996) looked at filters packed with iron oxide coated sand (IOCS). Typical removal rates for soluble metals were 80% for copper (average of 0.2 mg/l in influent), 90% for Pb (average of 0.8mg/l in influent), and 98% for Cd (average of 1.5 mg/l in influent). Typical overall (particulate and soluble) removal efficiencies were 99% or higher for all three metals (influent fluctuated between 3 and 7 mg/l of each metal). These experiments were run at a pH of 9.0. Satpathy and Chaudhuri (1995) found IOCS to be unsuitable for removing chromium from a mixed-oxidation state chromium plating waste (pH of 7.5 to 7.8). Adsorption of dilute concentrations of hexavalent chromium, which was the predominant form in the waste, decreases as pH increases. The removal of chromium that did occur was due to adsorption of trivalent chromium, which increases as pH increases. However, Satpathy and Chaudhuri (1995) also found IOCS to be suitable for removing cyanide and cadmium from a cadmium plating waste (pH of 11.5 to 11.8). Joshi and Chaudhuri (1996) examined IOCS in laboratory studies and found it to be promising as a possible low-cost small system or home treatment unit for removing arsenic(III) and arsenic(IV) (1mg/l influent concentrations) from groundwater (pH 7.5-7.8) in developing countries.

Comparing IOCS to regular sand, IOCS outperformed regular sand at removing Cu from solution at a pH of 8.0 (Benjamin et al., 1996). Edwards and Benjamin (1989) compared IOCS to free ferrihydrite and uncoated sand in batch tests. They looked at filterable (0.1 um filter) metals in solution (influent Pb=5.2mg/l, Cu=3.2mg/l, Ni=2.9mg/l, Cd=2.8mg/l, Cr(III)=2.6mg/l) as a function of the solution pH. Differences between IOCS and free ferrihydrite were relatively insignificant. When compared with uncoated sand, IOCS had better removal of Cd at pH's of between about 5 up to about 9.3, of Ni at pH's between about 6.5 to 7 up to about 9.3, Pb at pH's up to about 8.5, and was about the same for Cr(III) throughout. Uncoated sand appeared to be better at removing copper at pH's below about 6. At pH values outside those listed, uptake by the two sands did not appear significantly different. Note: Only 4 to 5 pH values were examined, so the pH values where differences were noted are approximate. Edwards and Benjamin (1989) also looked at IOCS verses uncoated sand in columns. At a pH of 8.5 soluble Cd, Pb, and Ni were removed more effectively by IOCS; there did not appear to be a significant difference in Cr(III) or Cu removals. At pH's of 10, 9.3 and 8.5, particulate metal removals (*floc or particulate bound*) were better for IOCS; at a pH of 8, Pb removal was better for IOCS, but removal efficiencies for the other metals were about the same.

IOCS was also found to be effective at removing Cu and Cd complexed with ammonia (solution pH=10), but was ineffective at removing metals complexed with EDTA (solutions of pH=10 and pH=4.5) (Benjamin et al., 1996). Edwards and Benjamin (1989) found IOCS efficient at removing Zn (as well as Cu and Cd) complexed with ammonia, while uncoated sand had virtually no metal removal.

IOCS has been found to be fairly stable. Some iron particles were seen in the backwash of IOCS after it was first prepared, but this diminished after a few backwash cycles. The solubility of the Fe is pH dependent, above about pH=4, iron dissolution was found to be undetectable (Benjamin et al., 1996). IOCS did not have a greater initial headloss than uncoated sand, but headloss increases have been found to be generally greater than uncoated sand. The headloss was also found to be pH dependent for IOCS (increasing as pH decreased), but not for uncoated sand (Edwards and Benjamin, 1989). IOCS filters can be successfully backwashed to remove particulates and regenerated to remove adsorbed soluble metals and restore adsorption capacity (Edwards and Benjamin, 1989; Joshi and Chaudhuri, 1996; Benjamin et al., 1996).

Researchers have also examined the potential of manganese oxide coated media and combination iron and manganese oxide coated media for removing dissolved metals from stormwater. Surface area and surface charge are very important parameter impacting metal sorption onto media. A larger surface area provides more sites for sorption. Since the metals to be removed are cations, having a media with a net negative charge at the pH of the stormwater will facilitate metal removal. This can be accomplished either by some form of pH control to bring the pH of the stormwater into the optimum range for the media or by using a media with a point of zero charge (PZC) below the natural pH of the stormwater. Most stormwater pH values are in the range of 6-8 (Lui et al., 2001) and one summary of reported studies gave a range of 4.5-8.7 (Makepeace et al., 1995). Liu et al. (2001) found that manganese oxide coated media began to show a net negative surface charge at a pH level of 6 or greater. Iron oxide coated media had a PZC in the range of 9.2 and therefore the surface had a net positive charge in the pH range of 6-8. But the iron oxide coated media had, depending on the coating method used, three or more times the surface area of the manganese oxide coated sand. However, two-step coated media that were first coated with iron oxide as a scratch coating and then coated with manganese oxide had both a larger surface area than the plain iron oxide coated media and began to show a net negative surface charge at a lower pH (~5) than the plain manganese oxide coated media.

3.2.2.3 Iron-Infused Media

An iron-infused media, patented by Manfred Woog at Environmental Products International and used for silver recovery, is being marketed as a filter media by Stormwater Management Inc. The iron-infused media is a stable phenolic open-cellular resin that has been impregnated with iron particles at the time of polymerization. The media has been investigated for phosphorus uptake from stormwater (Wigginton and Lenhart, 2000), however, at the time of going to press, only two lab column studies had been completed for metal uptake from stormwater using this media. The two tests showed a 20-30% removal of zinc (Wigginton – personal communication, 2001).

3.2.2.4 Compost

The use of leaf compost as a filtration media is gaining popularity. Composts made from yard waste (primarily leaves) have a very high capacity for adsorbing heavy metals, oils, greases, nutrients, and organic toxins due to the humic content of the compost. These humic compounds are stable and insoluble, and have a high molecular weight. They act like polyelectrolytes and remove toxicants from the runoff either by adsorption or ion-exchange (Clark and Pitt 1999). Compost also has a high buffering or capacity (Carnes and Lossin 1970). When the influent is between pH 6.7 and 8.3, the effluent is consistently between pH 7.0 and 8.0 (Clark and Pitt 1999).

According to Debo and Reese (1995), it is important to both physical performance and pollutant removal that compost have the following characteristics: (1) mature, (2) humic, (3) low contaminant levels, (4) high permeability, and (5) locally obtainable. Compost filter media also provides a use for a waste material. Removal rates were about 95% of TSS, 40% of TP, 88% of zinc, 87% of hydrocarbons, 67% of copper, and 61% of chromium for the studies quoted by Debo and Reese

Clark (1996) found that a mixed compost-sand filter provided moderate to excellent removal of many pollutants when the runoff was not pre-settled. When the runoff was pre-settled, compost sand did little to remove contaminants, and caused color, hardness, and COD to increase. However, heavy metal removal was good, even for pre-settled runoff. Unlike plain sand, the compost-sand began removing pollutants immediately – it was not necessary to wait for the filter to ripen. However, compost filters have also been noted to leach dissolved solids, nitrate, phosphorus, and boron (Debo and Reese, 1995; Clark and Pitt, 1998).

Coconut fiber compost has been shown to remove hexavalent chromium from wastewater, with the best removal occurring when the wastewater pH was 2.0. Removal decreased as the pH of the wastewater increased. An undesirable color leached from the compost when the removal occurred at pH 10. Removal patterns were similar to that seen for peat moss treating hexavalent chromium-containing wastewater (Sharma and Forster 1993). Urban waste-compost-amended soils have shown a high affinity for heavy metals (Pb, Cu, Zn), compared to the original soil (Guisquiani, et al. 1995).

Composts tend to act as ion-exchangers, i.e., whenever a pollutant sorbs to the media, an ion is 'leached off.' For the compost provided by CSF Systems (now Stormwater Management, Inc.), soluble phosphorus is one of the ions leached off during ion exchange (Clark and Pitt 1999). Soluble phosphorus likely is released from the captured solids through microbial action and since the compost only has a weak anion exchange capacity, the soluble phosphorus is not removed from the water once it is leached. Testing also has shown an increase in boron and nitrate in the compost effluent (CSF Systems 1994).

The first-generation composted leaf filter (proprietary deciduous leaf compost mixture) for stormwater treatment was developed by W&H Pacific (also called CSF Systems, now Stormwater Management, Inc.) for Washington County (Washington), the Unified Sewer Agency, and the Metropolitan Service District of Washington County (W&H Pacific 1992a). Early filter designs consisted of a bottom impermeable membrane with a drainage layer above it. Above the drainage layer is a geotextile fabric upon which rests the compost material. The toxicant removal occurs in the compost layer by filtration, adsorption, ion exchange, biodegradation, or by a combination of these processes. A small pre-settling area (less than 1 minute detention time) was recommended; otherwise, the larger particles and flotables could prematurely clog the filter and reduce its treatment efficiency. Filter design flow was selected as 2.25 gallons per minute (0.30 m^3/minute) for a required compost bed surface area of 200 ft^2/cfs (60,435 m^2/m^3/second). The results from a prototype Compost Storm Water Filter System (CSF) are given in Table 3-2. Currently, the compost filter is pre-packaged into cartridges that are sold by Stormwater Management, Inc. The cartridges have design flow rates up to 15 gpm (2 m^3/minute). The cartridges use horizontal flow, which encourages the pre-settling of the larger solids in the runoff outside the cartridge.

Three years of testing on the CSF has shown that the filter is excellent at removing metals and hydrocarbons from the runoff (CSF Systems 1994). Based upon the sample results, the first flush of a storm had the heaviest pollutant loadings, and the filter had the highest removal efficiencies during this first flush. The average first flush removal rates for the 3 years of operation are given in Table 3-3.

Table 3-2. Reported Compost Filter Pollutant Removal Efficiencies (Data derived from CSF Systems 1994)

Pollutant	REMOVAL RATE (%)
Turbidity	82
Total Solids	49
Suspended Solids	92
Total Volatile Suspended Solids	89
COD	70
Settleable Solids	95
Total Phosphorus	49
Ammonia	60
Total Kjeldahl Nitrogen	57
Copper	7
Zinc	83
Lead	83
Aluminum	84
Iron	91
Petroleum Hydrocarbons	84
Oil and Grease	81

Table 3-3. Reported Compost Filter Removal Efficiencies: First Flush (Data derived from W&H Pacific 1992b; CSF Systems 1994)

Pollutant	REMOVAL RATE (%)
Turbidity	86
Total Solids	63
Total Suspended Solids	94
Settleable Solids	98
Total Volatile Suspended Solids	97(*)
COD	79
Total Phosphorus	63
Ammonia	65
Total Kjeldahl Nitrogen	72
Copper	83
Zinc	86
Lead	86
Aluminum	88
Iron	93

* Results are from the first year of operation only.

Lenhart (1997) describes the application of a CSF compost filter cartridge system at a 15-acre site with a 136,000-ft^2 (12,600-m^2) building and a parking lot with a capacity of 775 vehicles.

The cartridge system was installed underground in two 2.5 m by 5.5 m precast concrete vaults, with a total capacity of 2 m^3/minute (1.2 cfs). The approximate diameter of an individual cartridge is 0.6 m giving a media surface area of approximately 0.16 m^2. The individual cartridges have a maximum flow rate of 15 gpm (0.06 m^3/minute), although orifice control of the flow rate is available as needed.

3.2.2.5 Peat

Peat is formed by the decay of plant material in a waterlogged, oxygen poor environment. The physical structure and chemical composition of peat are determined by the plants (mosses, sedges and other wetland plants) from which it is formed. Peat is highly organic, comprising mainly of humic and fulvic acids and cellulose.

Peat moss (sphagnum moss) is a fibrous ("fibric") peat, typically brown and/or yellow in color, with easily identifiable undecomposed fibrous material and with a bulk density generally less than 0.1 g/cc (Galli 1990). Its chemical and physical structure (pore volume of 80-90%; Karamanev, et al. 1994) encourages water retention, and it can contain up to approximately 90% water by weight (Leahy and Birkinshaw 1992). However, peats lose most of their hydraulic conductivity when compressed. Two different flow regimes exist in peat moss filters because of the peat's three-level, fractal-like structure. At low velocities, the liquid flows through the peat moss particles; however, at high velocities (above the critical velocity of approximately 0.1 cm/s), the liquid mainly flows between the solid aggregates with only a small amount penetrating the particles (Karamanev, et al. 1994).

Peat also exhibits a hysteresis between wetting and drying curves, likely because as the material dries out, it becomes more hydrophobic and, consequently, more difficult to rewet (da Silva, et al. 1993), with severely dried peats (≥ 35% moisture loss) being exceptionally difficult to rewet. Possible reasons for this phenomenon include macropore collapse and high micropore suction pressures. Drying also shrinks humic molecules. The peat initially will repel new water; however, continuous re-wetting eventually will lead to water penetrating all pore spaces, flushing out any accumulated color-producing organic acids (Mitchell and McDonald 1992).

Peat can also leach organic compounds, especially colored organic matter, with the amount of leaching depending upon season (for an outdoor filter) and soil moisture. One possible explanation for the correlation of peat moisture and color distribution/intensity is the change in pH and water content during filtration. Peat showed a rapid initial rise in color and acidity, followed by a gradual decline. The drying time between filtration events indicates the size of the "store" of water-soluble, color-producing organic acids. When the filter is initially wetted, this "store" is released, and the effluent becomes colored (Mitchell and McDonald 1992). As an example, the major drawback to a peat-filter system treating sanitary wastewater was the tea color of the effluent (Daigle 1993).

Peat's ability to uptake metals has been extensively studied (Brown et al, 2000a). Malterer et al. (1996) presents an overview of peat, its nature, and its uses. Some of the facts they cite include: Peat has a high adsorption capacity for transition metals and polar organic molecules. The adsorption properties of peat are greatly enhanced by the highly porous nature of peat and its high specific area. It has been shown that in addition to metals, peat can also remove nutrients, suspended solids, organic matter, oils, and odors. Peat moss is a complex material with lignin, organic acids and cellulose as major constituents. These constituents, especially lignin and humic acid, bear polar functional groups, such as alcohols, aldehydes, ketones, carboxylic acids, phenolic hydroxides and ethers that can be involved in chemical bonding (Chaney and Hundemann, 1979). Under microscopic examination, peat presents a cellular structure (Coupal and Lalancette, 1976). Because of the polar character of peat, the specific adsorption potential for dissolved solids, such as metals and polar organic molecules, is quite high. These properties have consequently led to the examination of the potential of peat as an agent for the purification of wastewaters contaminated with dissolved metals (Lalancette, 1974).

Chaney and Hunderman (1979) reported sphagnum peat to be effective in removing Cd from simulated hypochlorite-oxidized cadmium cyanide electroplating wastewaters. They found that dissolved Cd was reduced from concentrations of about 560 micrograms/l to generally <3 micrograms/l. Sharma and Forester (1993) found sphagnum peat to be effective in removing hexavalent chromium from aqueous solutions. Gardea-Torresdey et al. (1996) found that sphagnum peat moss could uptake 16.1±1.40 mg of Cu(II) per gram of peat moss at a pH of 4.0. In a study of Cu(II), Cd(II), Zn(II), and Ni(II) binding on various classifications of peat, Gossett et al. (1986) found that the maximum binding capacities in 10 mM metal cation solutions were in the range 180-200 mMol/kg dry weight. Coupal and Lalancette (1976) concluded that peat can adsorb most metals very efficiently, up to 4% of the weight of dry peat. They found that for solutions of metals less than 10ppm, 1.5lb of peat can purify 250 gal of wastewater. In column studies Shiskowski and Viraraghavan (1993) found that a 300mm peat column had a chromium removal efficiency of around 94% and a 500mm column had a chromium removal efficiency of around 96.6% (influent concentrations ranged from 0.348 to 0.619 mg/l). Gossett et al. (1986) noted that the metal removal efficiency in unbuffered solutions is significant in concentrations ranging from 0.01 to 100 mM, with maximum extraction ratios being obtained in the 0.1 - 1.0 mM range. However, the sorption capacity of peat is finite and reversible and is controlled by the pH of the solution (Galli 1990). At higher pHs, the metals are typically strongly bound, and little desorption occurs in low molarity caustic solutions. In high molarity caustic solutions, the peat itself disintegrates, releasing the captured metals (Sharma and Forster 1993).

In buffered solutions, the order of sorption for four metal ions to peat is $Ni^{2+} > Cu^{2+} > Cd^{2+} = Zn^{2+}$, independent of peat origin. Above pH 3, copper binding is similar to nickel and is dependent upon the pH of the solution; cadmium and zinc present a similar pH dependence but are less strongly bound than the copper and nickel. Only the nickel cation, however, is bound strongly enough not to be desorbed when the pH is dropped to below 1.5 (Sharma and Forster 1993).

Brown (1993) compared several types of peat and similar peats from different locations, as well as two activated carbons and bone char. The factors examined included uptake capacity for Cd and Cu, cation exchange capacity, and the quantity of Ca, Mg, and Fe present on the media. The results of these comparisons are shown in Table 3.4.

Table 3-4: Comparison of Peat and Related Filter Media

Media Type	Cu capacity, mg/g	Cd capacity, mg/g	CEC, mequiv./ 100g	Ca, mmol/ g	Mg, mmol/ g	Fe, mmol /g
Fibrous extract from blended peat	10	15	95	0.083	0.026	0.004
Blended milled peat dried to 12% moisture	16	22	139	0.107	0.053	0.025
Blended milled peat	27	30	149	0.177	0.033	0.026
Sphagnum peat	32	35	155	0.238	0.033	0.053
Milled peat	58	62	210	0.340	0.033	0.079
Milled peat	14	21	141	0.107	0.032	0.015
Milled fen peat	23	26	144	0.187	0.015	0.056
Graded crushed sod peat	26	28	156	0.255	0.033	0.026
Sphagnum peat	18	20	139	0.056	0.049	0.004
High ash milled peat	50	50	201	0.312	0.009	0.336
Peat char	2	1	20			
Bone char activated carbon	58	105	16			
Filtasorb F400 activated carbon	13	14	9			
PK 1-3 activated carbon	40	39	13			
Lignite char	3	5	46			
Commercial horticultural grade sphagnum peat	18	20	126	0.042	0.053	0.013
Lignite	23	33	135	0.240	0.102	0.078

Bone char had the highest sorption capacity of all the samples and peat and lignite chars the lowest. Among the peat samples, sorption capacity varied a great deal. The best indication of sorption capacity in the peat and lignite samples was the amount of calcium present in the samples and the cation exchange capacity determinations.

The mechanism by which metal ions are sorbed onto peat has been a matter of considerable debate (Debo and Reese, 1995). Variations in peat type, peat preparation, methodology, and metals studied made the comparison of results difficult. Based on the research conducted to date, it would appear that metal sorption by peat is not the result of one mechanism, but several. The mechanisms involved and their degree of importance seem to depend on the materials and the methods used. These mechanisms include ion-exchange, surface adsorption, chemisorption, complexation, and adsorption-complexation. Factors affecting uptake include pH, loading rates, and the presence of competing metals. The optimum sorption pH is different for each metal, but a generalized optimum pH range for most divalent metal ions can be considered to be between 3.5 and 6.5. Loading rates of metal-bearing wastewater also significantly influence sorption. Removal of metals is most efficient at low loading rates; as the loading rates increase, the efficiency decreases. Recovery of metals and regeneration of the peat is possible without an appreciable effect on performance. Although the presence of more than one metal in a solution creates competition for sorption sites and less of a particular ion may be bound, the total sorption capacity has been found to increase (Brown et al., 2000a).

Immobilization of a metal by peat depends on (1) the metal ion capture chemistry, (2) solute transport rates from the bulk solution to the adsorbent surface, and (3) the transport rates and equilibria within the adsorbent's interstices. For metal adsorption on peat, film diffusion appears to be the rate-controlling step; although at small peat-to-metal ratios, internal mass transfer also greatly influences the sorption. A three-step model has been used to describe the metal immobilization process by peat: (1) solute mass transfer from the solution to the particle surface, (2) ion-exchange reactions at fixed sites on the peat, and (3) internal diffusion of solute. At high peat concentrations, film and external mass transfers are most important, while at low peat concentrations, intra-particle diffusion controls the reaction rate (Allen, et al. 1992). The sorption curves for the metals are not linear, regardless of the peat-metal combination, indicating that the peat-metal complex stoichiometry and thermodynamics are probably dependent both on the free metal concentration and on pH (Gosset, et al. 1986).

Clark (1996) found mixed peat-sand media to have moderate to excellent pollutant removal for both non-presetteled and presettled runoff. The best average removal efficiency generally occurred for the presettled runoff. As with compost-sand, the peat-sand filter was found to start removing pollutants immediately, unlike plain sand, which had to ripen.

The release of color upon wetting is one drawback to using peat. Another potential problem is that peat may leach some nutrients, depending on the soil and water chemistry and water level. Sphagnum peat generally will release significantly more phosphorus and ammonium than Carex peat, especially in waters with a high sulfate concentration (Koerlsman, et al. 1993). Temperature also influences the amount of ammonium, potassium, and phosphate leached. Nutrient leaching has been shown to increase twofold to threefold after the peat was frozen (Koerlsman, et al. 1993). Also, in unbuffered metal solutions, the pH drops between 0.2 and 0.6 pH units during filtration (Gosset, et al. 1986), likely due to the release of humic and fulvic acids (Sharma and Foster 1993). During stormwater filtration tests, pH drops of one to two pH units were sometimes observed (Clark, 1996).

The Peat-Sand Filter System designed by the Metropolitan Washington Council of Governments (Washington, D.C.) has a good grass cover on top underlain by 12 to 18 inches of peat. The peat layer is supported by a 4-inch mixture of sand and peat that is supported by a 20- to 24-inch layer of fine to medium grain sand. The mixture layer is needed because it will provide the necessary continuous contact between the peat and the sand layers and ensure uniform water flow (Galli, 1990). Expected pollutant removal efficiencies are given in Table 3-5. Peat moss has been used as a growth medium for red maple and cranberry seedlings used to treat urban stormwater runoff containing lead and zinc (Vedagiri and Ehrenfeld, 1991).

Table 3-5. Peat-Sand Filter Pollutant Removal Efficiencies (Data derived from Galli 1990)

Pollutant	Removal Efficiency (%)
Suspended Solids	90
Total Phosphorus	70
Total Nitrogen	50
BOD	90
Trace Metals	80
Bacteria	90

3.2.2.6 Zeolite

Zeolites are a type of mineral and are defined as any of a group of hydrated silicates of aluminum with alkali metals, commonly occurring as secondary minerals in cavities in basic volcanic rocks: used for their molecular sieve properties because they undergo dehydration with little or no change in crystal structure (Webster's, 1996). Zeolites are preferred adsorbents in the chemical process industry because they are inorganic and non-flammable and can withstand very high temperatures (Vaughn 1988). They have been used in such applications as natural gas purification (chabazite), radioactive waste disposal (clinoptilolite), ammonia recovery from sewage effluents (clinoptilolite), and various petroleum and petrochemical catalyst applications (erionite, mordenite). The ability to control access to the reactive sites through zeolite selection, as well as the size and stereochemistry of the site itself, makes molecular-level control of reactions possible (Vaughn 1988).

Because micropore size is uniform in zeolites, these adsorbents have a sharp cut-off of sorption with increasing molecular size. Although the framework primarily determines the pore size, the free aperture, particularly in the smaller 8-ring sieves, may be modified by ion exchange, again tailoring the zeolite. Adsorption forces for zeolites include van der Waals forces, induced dipole interactions, and other electrostatic forces (polarization, dipole and quadrupole interactions) (Ruthven 1988).

When Al^{3+} is substituted for Si^{4+} in the zeolite framework, a net negative charge on the molecule results. This is compensated for by a nonframework cation (e.g., Na^+), which is 'held' in the pores of the structure. Because this cation is not a part of the crystalline lattice, it is relatively easy to exchane for other cations (Vaughn 1988). Ion-exchange and adsorption processes for zeolites often are complicated because the zeolite has two distinct pore structures: micropores in the crystals and macropores in the binder, both of which can participate in sorption (Robinson, et al. 1994). Zeolites have internal and external surface areas of up to several hundred meters squared per gram. They can have CECs of up to several equivalents per kilogram (Haggerty and Bowman 1994).

Because of the exchangeable cations, zeolites are polar adsorbents. Molecules such as water or ammonia (high dipole), CO_2, N_2 (quadrupolar) or aromatic hydrocarbons (π layer interaction) therefore adsorb more strongly than nonpolar compounds. This affinity generally increases with increasing charge on the exchangeable cation and decreasing cation radius, but its effect may be masked by water, which, because it is strongly bound to a zeolite, will reduce the zeolite's affinity for other less polar molecules. When removing contaminants from an aqueous solution, considerable amounts of water will be present in the intra-crystalline fluid (Ruthven 1988).

Zeolites are negatively charged and do not absorb anions well. In one experiment, anion sorption was greatest when the HDTMA satisfied the zeolite's total external cation-exchange capacity. Anion retention (4 mmol/kg for CrO_4 and >2 mmol/kg for SeO_4 compared to 1 mmol/kg for both on natural, unmodified zeolite) resulted from the formation of an HDTMA-anion precipitate on the zeolite surface (Eyde 1993; Haggerty and Bowman 1994). Some zeolites are unstable at low pH because the framwork aluminum is hydrolyzed. One approach to exchanging transition metals at low pH is to first form ammonia complexes by dissolving them in dilute aqueous ammonium hydroxide and then carrying out the exchange at high pH (Vaughn 1988). The modified surface, however, is stable at low pH, at higher ionic strength, and with organic solvents (Eyde 1993; Haggerty and Bowman 1994).

There are many types of zeolite. Numerous zeolite deposits occur throughout the Western US and clinoptilolite is the most common (Leppert, 1990). Clinoptilolite can be used for the removal of ammonium ions from drinking water (Blanchard et al., 1984), and has been used for radioactive waste disposal and ammonia recovery from sewage effluents (Clark, 1996). Clinoptilolites, as well as other zeolites, have also been investigated for removal of heavy metals from waters. (Blanchard et al., 1984; Zamzow and Murphy, 1992; Al-Haj Ali and El-Bishtawi, 1997).

For the U.S. Bureau of Mines, zeolites are an alternative to conventional precipitation removal techniques for metals such as lead (Eyde 1993). Zamzow and Murphy (1992) examined 24 zeolites, including 15 clinoptilolites from different locations, for uptake of Cu, Cd, and Zn. In column studies looking at uptake capacity the range of uptake for the different clinoptilolites was 0.40 to 1.37 meq/g sorbent for Cd, 0.33 to 1.88 meq/g for Cu, and 0.17 to 1.35 meq/g of Zn. Column tests were also run using one of the mid-range clinoptilolites and Rio Tinto copper mine wastewater. The composition of the minewater was, in mg/L, 73 Al, 550 Ca, 27 Cu, 210 Fe, 22 Mn, 1.4 Pb, and 11 Zn (pH=2.62). The research indicated that 10 kg of the clinoptilolite could effectively treat 3.33 L of the wastewater and the capacity of the zeolite for all the ions in the minewater was 0.4 meq/g.

3.2.2.7 Ion-Exchange Resins

Ion exchange is a unit process in which ions of a given species are displaced from an insoluble exchange material by ions of a different species in solution. Although both natural and synthetic ion exchange resins are available, synthetic resins are more widely used because of their durability (Metcalf and Eddy, 1991). Synthetic resins have been created that have a selective affinity for heavy metals, preventing the resin capacity from being consumed by ions such as calcium, magnesium and sodium (Totura, 1996). Synthetic ion-exchange resins are widely used in industrial applications. The main impediment to use for urban runoff is resin cost.

Brown et al. (2000b) studied the uptake of Cu, Cd, Zn, and Pb onto raw peanut hulls, peanut hull pellets (hulls were formed into capsules using a binder), and a commercial grade ion exchange resin. A simple cost analysis based on metal capture per dollar of media cost showed that the hulls and hull pellets outperformed the resin in terms of gram metal removed per media cost. Though the capacity of the resin was greater than the hulls or pellets, with resin verses pellet capacity ratios ranging from 6:1 to 9:1 (resin:pellet) depending on the metal, the reverse was true for a cost analysis: the ratio of grams metal removed per dollar of media cost for resin verses pellets ranged from 1:13 to 1:17 (resin:pellet) depending on the metal.

3.2.2.8 Activated Carbon

Activated carbon separation is commonly used in the water treatment and chemical process industries and in hazardous waste cleanup as an effective method for removing trace organics from a liquid. Activated carbon can be made from coal; wood; almond, coconut, walnut, and peanut hulls; and various other materials. Activation of carbon can be achieved by either physical or chemical activation. Physical activation of carbon involves two steps. First, the material is heated to a red heat in a retort to drive of the hydrocarbons but with an insufficient air supply to maintain combustion. This process produces a char. Then the char is activated by exposure to an oxidizing gas (steam, CO_2) at high temperature. The gas causes a porous structure to develop in the char creating a large internal surface area (Metcalf and Eddy, 1991). Chemical activation of carbon is a process in which uncarbonized carbonaceous materials are mixed with chemicals capable of dehydrating and oxidizing. The chemically treated material is then heated to create the necessary pore structure. The activating agents, such as zinc chloride, phosphoric acid, and sulfuric acid are subsequently washed out and recovered (Rice and Robson, 1982). Numerous studies have examined the ability of various types of activated carbon to remove metal ions from solution.

The removal of free heavy metal ions from solution by activated carbon is generally held to be due to surface adsorption, i.e. an electrostatic interaction between the positively charged metal ions and the negatively charged carbon surface. Chu and Hashim (2000) examined the adsorption of free Cu(II) and EDTA-chelated Cu(II) and concluded that electrostatic binding was important in the adsorption of both. They found low adsorption levels of Cu(II) at pH 3 and high adsorption levels in the pH range 4-6, while EDTA-chelated Cu(II) showed a reversed adsorption trend. They concluded that the pH of the solution impacts the surface charge of the activated carbon as well as the charge property of Cu(II) ions and EDTA-chelated Cu(II) species.

Surface complexation models have been proposed in which the carbon's functional groups can be divided into two types: acidic (carboxyl, phenolic, quinonoid, and normal lactone) groups and basic (chromene and pyrone-like) groups (Dahab, et al. 1991). Solution pH less than the pK_a does not significantly affect the adsorption capacity of the activated carbon, but when the pH is greater than the pK_a, there is a linear decrease in adsorption capacity with the increase in pH (Nelson and Yang 1995).

Adsorption is not the only way in which metals may be removed. Reed et al. (1995) concluded that Pb removal was primarily due to precipitation on the carbon surface or in the carbon pore. At pH values greater than around 5.5 (the pH at which solution precipitation begins), $Pb(OH)_2$ forms. At pH values below the precipitation pH, adsorption is the dominant removal mechanism. At higher pH values, Pb precipitation on the carbon surface or in the pore liquid becomes important.

Hexavalent chromium is effectively removed using activated carbon by pH-dependent adsorption. More than 80% of inorganic and organic mercury in a solution was removed by a commercial GAC, with even greater removals resulting when humic acid or nitrilotriacetic acid was added (Sharma and Foster 1993). Activated carbon from peanut shells was seven times more effective than commercially-available carbon at the removal and recovery of mercury from solution, possibly due to its lower ash/higher carbon content. Peanut hull carbon adsorption also was not as pH dependent as commercially-available carbon. Rice-husk and coconut-shell activated carbon also has been effective in the removal of heavy metals from aqueous solutions (Namasivayam and Periasamy 1993; Sharma and Foster 1993).

Netzer and Hughes (1984) compared 10 different activated carbons (some powdered and some granular and made from different raw materials) for uptake of copper, lead, and cobalt from aqueous solution. They found significant differences in the ability of the different types of activated carbon to uptake metals. Perez-Candela et al. (1995) also looked at several different activated carbons (powder and granular, different raw materials, different activation procedures). The study found that adsorption of Cr(IV) depended on the porous texture, material used, and preparation procedure, with the best performance being obtained from those prepared by physical activation instead of chemical activation.

Brown (1993) tested two commercially available activated carbons, one bituminous coal based (Chemviron Filtasorb F400) and the other a wood based product (Norit PK 1-3), for uptake of Cd and Cu in solution (made using stock salts and de-ionized, distilled water). The Filtasorb had a capacity of 13 mg/g for Cu and 14 mg/g for Cd, the PK 1-3 had a capacity of 40 mg/g for Cu and 39 mg/g for Cd.

A sand filtration/carbon treatment system has been shown to reduce a diversity of organophosphate, organochlorine, and pyrethroid pesticide residues down to below their detection limit. The sand filtration step removed the pesticides associated with particulate matter while the carbon adsorbed the non-particulate pesticides in the solution (Moore, et al. 1985).

Slow GAC filters achieve excellent removals (> 90%), with the removal efficiency limited by the depth of the filter. This dependence is due to 'slowness' of the transport kinetics and attachment mechanisms inherent in activated carbon sorption. One problem with activated carbon is its exponential head loss curve, thus requiring more frequent filter cleaning (Collins, et al. 1992). GAC has a very small net surface charge and is ineffective at removing free or hydrated metal ions, unless they are complexed with easily-adsorbed organics prior to filtration. However, once they are complexed with these insoluble organics, the complexed metals are readily adsorbed onto the carbon, which results in the desired high removal rates (Anderson and Rubin 1981). The primary drawback to the use of activated carbon, however, is its cost.

3.2.2.9 Bone Char

Bone char, a form of activated carbon, is derived from the carbonization of selected grades of animal bones by heating dry bones in an airtight iron retort at 600-900 degrees Celsius for about 8h. Comparing the potential for metal ion removal with activated carbon: bone char provides a porous carbon surface (9-11 wt%) for physical adsorption, but also has a hydroxyapatite surface (70-76 wt%) for ion exchange (Cheung et al., 2000). Bone char has been used for the deflouridation of water and for color removal in sugar refining.

Brown (1993) also tested bone char for uptake of Cd and Cu in solution (made using stock salts and de-ionized, distilled water). Bone char had a capacity of 58 mg/l Cu and 105 mg/l Cd, which were higher than the capacities found for several types of peat, activated carbon and lignite examined in the same study (See Table 3-4). Cheung et al. (2000) found that bone char had a maximum uptake capacity for Cd of about 0.534mmole Cd per gram of bone char. Promising methods for regenerating bone char have also been studied (Christoffersen et al., 1991).

3.2.2.10 Cotton (Cellulose-based) Fibers

Cellulose-based, carbonaceous residues, such as wheat straw, have been used to remove nitrogen from reclaimed wastewater in a nitrification/denitrification sequence in which the material is then a source of carbon for the microbial colonies that perform the nitrification and denitrification process. Significant reductions in BOD, organic carbon, chlorophyll, phosphorus, algae, and clay concentrations in the influent were also found (Lowengart, et al. 1993). If the material has a poor nutrient content (such as wheat straw), nitrogen and phosphorus are removed from the influent water by the microbial biomass (Diab, et al. 1993).

Fibrous cotton materials have been used to clean up oil spills and for demulsifying oily emulsions (Viraraghavan, et al. 1989). An example of a fibrous cotton used for oil spill cleanup is a material named Enretech I. It is a light-weight, non-toxic, 100 percent cellulose product (waste from cotton milling) that can be used to clean up oil spills, especially in areas where it is difficult to clean due to access limitations. It can also be used in areas such as tank storage sites, fueling locations, oil production fields, and oil field pipe treatment yards to collect slow leaks. This material type has also been shown to be effective at cleaning up fuel, oil, paint, or coolant spills on highways (RAM Services, Inc. 1995).

3.2.2.11 Forest Products Agrofiber

The Forest Products Research Lab agrofiber product was developed as both an economic oil adsorbent and as an economic ion-exchange medium for pollutant removal from water. Kenaf and jute fibers, along with forest wastes such as barks and pine needles, have been found to efficiently remove copper from water. Chemical treatment of the kenaf with reactive yellow-2 significantly increased the adsorption capacity of the kenaf for copper (Forest Products Research Lab 1995).

3.2.2.12 Other Media

In addition to the media discussed above a wide variety of other materials have also been examined for their metal removal efficiency. Different researchers have studied a wide range of low-cost adsorbents. Some of these include corncobs, peanut hulls, coal, tires, and straw.

Larsen and Schierup (1981) studied the ability of barley straw to remove heavy metals in both capacity and column experiments. They compared the capacity of the straw to the metal removal capacity of activated carbon, pine sawdust, and precipitation with calcium carbonate. They found that one gram of barley straw adsorbed Zn, Cu, Pb, Ni, and Cd in amounts ranging from 4.3 to 15.2 mg. By comparison, one gram of activated carbon removed 6.2 to 19.5 mg, pine sawdust removed 1.3 to 5.0 mg, and precipitation with calcium carbonate 1.6 to 19.8 mg. The total amount of heavy metals removed in competitive column experiments was 408 mg/g of straw. They also found that by using a 0.5N HCl solution to regenerate the straw, a column could be reused about five times.

Henderson et al. (1977) studied the ability of softwood sawdust, corncobs, and finely ground scrap rubber to remove Cd, Zn, Cu, Hg, and Ni from wastewater. Wood fines were not very effective in the adsorption of heavy metals. The only metals removed to an appreciable amount were Cu and Ni. Corncob fragments, although large in size, removed a good percentage of Cu and Cd in low concentration solutions. The scrap rubber removed Hg and Cu well from both high and low concentration solutions. However, Ni, Cd, and Zn were moderately adsorbed by the rubber at high concentrations, but only slightly adsorbed at low concentrations. They concluded that these materials were not as effective at removing heavy metals as peanut hulls or composted bark, which they had examined in a previous study.

Brown et al. (2000b) studied the uptake of Cu, Cd, Zn, and Pb onto raw peanut hulls (40-mesh), peanut hull pellets (hulls were formed into capsules using a binder), and a commercial grade ion exchange resin. The goal was to evaluate peanut hull pellets as a low-cost, low-maintenance, biomass filter for the removal of metals from dilute aqueous waste streams. For all media-metal studies, over 85% of the total metal-ion capture occurred within the first 10 minutes of the run, over 90% within the first 20 minutes of the run, and over 96% within 40 minutes. The results of the equilibrium capacity studies are given in Table 3.6.

Table 3-6: Equilibrium Capacity of Media for Metals in Single-Component Systems for Brown et al. (2000)

Media	Capacity for Pb mg/g	Capacity for Cu mg/g	Capacity for Cd mg/g	Capacity for Zn mg/g
Peanut Hulls	30	8	6	9
Peanut Hull Pellets	30	10	6	10
Resin	Greater than 200	85	50	90

Netzer et al. (1974) examined the use of discarded automotive tires for the removal of trace metals from wastewaters. The tires were used in conjunction with lime to remove aluminum, cadmium, chromium, cobalt, copper, iron, lead, manganese, mercury, nickel, silver and zinc. According to Netzer et al. rubber is as effective as activated carbon, but is cheaper and easier to use. The study by Henderson et al. (1977), discussed earlier, had only moderate success with ground rubber, however they did not specify the type used. It is possible there may be differences between the material studied by Henderson and that studied by Netzer. Netzer et al.'s results showed greater than 99% removal, or less than 0.1ppm residual metal concentration, achieved for all the metals in continuous bench scale studies. The percent removal for several of the metals had high pH dependence

Friedman and Weiss (1972) examined mercury adsorption by various agricultural products, including by-products and derivatives. They looked at walnut expeller meal, peanut skins, wool, rice straw, plum pit shells, peanut hulls, rice hulls, sugar cane bagasse, chicken feathers, flour, gelatin, silk, and other materials. They found evidence that mercury uptake by the various materials was roughly proportional to protein content. The keratins, wool and feathers, were very effective. The materials with high tannin contents such as walnut expeller meal and peanut skins showed evidence of possibly being even more effective. The materials that were mainly carbohydrate, such as cellulose, starch, rice straw, and bagasse, took up very little mercury.

The use of low-cost adsorbents for metals removal is a growing field that could have considerable application in remediating urban runoff. The adsorbents discussed above are plentiful, readily available, and inexpensive. In some cases the use of these adsorbents will provide a useful application for materials that are essentially waste products, such as used automotive tires and peanut hulls.

Many of these reported studies were conducted using small-scale columns with high concentrations of a single metal. While providing some indication of potential application benefits for stormwater treatment, those results should not be compared to the results from studies that have used stormwater in full-scale tests. This research program therefore examined a variety of potentially useful filtration media using stormwater and mixed metal solutions for more appropriate comparisons.

3.2.3 Microbial Uptake of Metals: Uptake Mechanisms and Potential Impact on Filter Performance

Metal-microbe interactions, generally of concern due to metal toxicity in microbes, are also of interest in corrosion, geochemical cycling of metals, and for the potential to remediate metal pollution. Microbial mediated removal of metals can occur for several reasons. Some metal uptake occurs due to a microorganism's need for certain metals, such as iron, for its metabolic processes. However, many heavy metals can be toxic to microbes. Microorganisms deal with metal toxicity using various mechanisms, including precipitation, exopolymer production, bioimmobilization/biosorption, biosequestration, volatization, efflux systems, and valence transformation (Shafikhani, 2001). Many of these mechanisms result in the removal of heavy metals from the surrounding environment either by the microbes retaining the metals or by precipitation of the metals on or near the microbes. Metals may also be retained or precipitated by microbes due to the normal metabolic processes of the cell which are not related to obtaining needed metals or dealing with metal toxicity. For example, some metabolic by-products of certain cells will react with metals to form precipitates.

Microorganisms are capable of accumulating and immobilizing trace metals by several methods. Active metal accumulation or transformation requires the participation of living, metabolically active cells. Metal accumulation that is dependent on the metabolic activity of the cell is called bioaccumulation. Passive sequestering and transformation of heavy metals by microbes is through physical-chemical reactions that do not require the participation of live cells. This passive uptake is called biosorption. Several processes by which metals are accumulated or transformed by microorganisms are discussed below.

3.2.3.1 Methylation and Demethylation (Active)

Methylation, can increase the toxicity and bioavailability of methylated metals, the most well-known example being mercury.

3.2.3.2 Oxidation and Reduction (Active)

The oxidation and reduction of metals by bacteria can impact soluble metals concentrations in solution. The most commonly studied metals appear to be Fe and Mn. Many bacteria, algae, yeast and fungi can either oxidize Mn(II) or reduce Mn(III) and Mn(IV) oxides. Nealson et al. (1989) discusses Mn oxidation and reduction by microbes in depth. The oxidation of Mn can reduce soluble Mn concentrations in two ways, first Mn(II) is the only state stable as a soluble cation, the other states form Mn oxides, which are solid phases. Second, Mn(IV) oxides are excellent chelating agents for divalent cations including Mn(II).

In the environment iron largely exists as either Fe(II) or Fe(III), which are the reduced ferrous and oxidixed ferric forms, respectively, and most of the oxidation and reduction of iron is microbially catalyzed. Fe(II) is fairly soluble in water and therefore mobile, while Fe(III) tends to form insoluble Fe(III) oxyhydroxides and is therefore fairly immobile. Fe(III) oxyhydroxides are also very efficient at adsorbing a variety of chemical species, including trace metals and phosphate ions (Chapelle, 2001).

3.2.3.3 Extracellular Precipitation (Active)

Occurs when microorganisms produce metabolic products that are excreted and result in the immobilization of metals. For example, under anaerobic conditions sulfate-reducing bacteria produce hydrogen sulfide and bicarbonate while consuming carbon and sulfate. SRB oxidize simple organic compounds, such as acetic acid and lactic acid, by utilizing sulfate as an electron acceptor, as shown in the following equations (Kim et al., 1999):

$$SO_4^{2-} + CH_3COOH = H_2S + 2HCO_3^-$$
$$3SO_4^{2-} + 2CH_3CH(OH)COOH = 3H_2S + 6HCO_3^-$$

The hydrogen sulfide reacts with soluble heavy metals to form insoluble metal sulfide minerals (Brierley et al., 1989). Sulfate-reducing bacteria have been investigated and used for the treatment of acid mine drainage. Paques Inc. markets a sulfate and metal removal technology called THIOPAQ®. This process removes metal and sulfate from mine waters in two biological steps that are complemented by solids separation. The first step uses sulfate-reducing bacteria to convert sulfate to hydrogen sulfide. The second step involves sulfur-oxidizing bacteria converting excess hydrogen sulfide to elemental sulfur (De Veght et al., 1998).

SRBs are a unique and ubiquitous group of prokaryotic microorganisms that are found in a variety of environmental niches (Singleton, 1993). Given the abundance of SRBs and their presence in many soil, sediment and aquatic environments, the importance of this process overall may be substantial.

3.2.3.4 Intracellular Accumulation and Complexation (Active)

Microorganisms can accumulate intracellularly both metabolically essential metals (Ca, K, Na, Fe, Mg) and nonmetabolic metals (Ni, Cd, Co, etc.). This process can be an energy dependant function requiring active respiration by the microbial cell. Active uptake generally requires a specific transport system. Microorganisms have well-developed transport systems capable of accumulating metals against a concentration gradient. When a metal is taken into the cell, ions of an equivalent charge are released from the cell (Brierley et al., 1989).

3.2.3.5 Extracellular Complexing and Subsequent Accumulation (Passive)

Some microorganisms generate substances that have a high binding efficiency for metals. Examples include the production of siderophore systems in bacteria and generation of metal binding polymers (Brierley et al., 1989).

Exopolymers. Bacteria can form protective exopolymer layers such as capsules and slime layers above the cell wall, which are capable of binding metals. Capsules, usually polysaccharides with a repeating sequence of two to six sugar subunits, are exopolymers tightly linked with the cell surface. Slime layers are similar exopolymers but have a more transient association with the cell surface and often slough off (Thompson and Beveridge, 1993). Metal binding is generally considered a direct consequence of the presence of negatively charged functional groups on the exopolymer. These groups include hydroxyl, phophate, pyruvate, succinate, and uronate. Binding of positively charged ions can occur rapidly, resulting in equilibrium changes in ion concentrations within the biofilm (Ford, 1993). Other surface layers which can bind metals include

S-layers and sheaths. S-layers are paracrystalline arrays of proteins or glycoproteins sometimes found on the outer surface of bacterial walls (Beveridge, 1989; Thompson and Beveridge, 1993). Sheaths are relatively rare, and can be a loose arrangement of homo- or htereo-polymers that encase chains of cells or be more ordered and exist as proteinaceous hoops which are stacked together to form long cylinders which encase a chain of cells (Thompson and Beveridge, 1993).
Siderophores. These are relatively low-molecular-weight, usually ferric specific coordination compounds excreted under low iron stress by aerobic and facultative anaerobic bacteria and by fungi for the purpose of securing iron from the environment (Neilands, 1989).

3.2.3.6 Binding to the Cell Surface (Passive)

Microorganisms can accumulate metabolic and nonmetabolic metals by precipitating or binding the metals onto cell walls or cell membranes (Bremer and Geesey, 1993; Brierley et al., 1989). Passive metal sequestration by different parts of the cell can occur via complexation, coordination, chelation of metals, ion exchange, adsorption, and inorganic microprecipitation (Volesky, 1990).

Most experiments using intact bacteria or larger eucaryotic microorganisms point to cell surfaces as being the most important mechanism for metal complexation and mineral development (Thompson and Beveridge, 1993). Sar et al. (2001), studying Ni(II) accumulation in the bacteria *Pseudomonas aeruginosa* (using living cells), found that 88% of the accumulated Ni was retained in the cell envelope region (58% by the periplasma; 30% by the membrane components). The Ni load in the cytoplasm was only 11.3% of the total nickel load. Sar et al. note that similar results have been found for other metal-bacteria interactions and that peripheral sequestration of toxic cations exterior to the cytoplasm has also been suggested as the possible microbial strategy for metal detoxification/or resistance. Davidova and Kaspora (1992a), studying biosoption by yeast cell walls (using non-living cells), found that the walls of *T. famata* cells, which are resistant to high cobalt concentrations, have the ability to sorb significantly more cobalt than the walls of *C. utilis* cells, which are less resistant to high cobalt concentrations.

Bacteria. Most cell walls of Gram-positive bacteria contain reasonably large amounts of peptidoglycan and an anionic polymer, such as a teichoic or a teichuronic acid. As far as is known, all walls of Gram-positive bacteria are negatively charged in media capable of supporting growth. Anionic sites include carboxylate (from peptidoglycan) and phosphate (from teichoic acid). Positively charged sites are exclusively ammonium, from D-alanine (teichoic acid), amino sugar (glycan), and diaminapimelic acid (peptide portion of peptidoglycan). These are the charges that mediate the interaction between cell walls and metal ions (Doyle, 1989). The peptidoglycan of Gram-negative bacteria has similar but fewer reactive sites for metal binding than Gram-positive walls. It is also shielded from the external milieu by the outer membrane (Beveridge, 1989). Gram-negative cells have been found to adsorb less metal than Gram-positive cells (Churchill et al., 1995). Archaeobacteria are bacteria capable of surviving in extreme environments. The diversity of the environments they occupy results in wide variety in the structure and chemical composition of the cell wall. There is not a great deal of information on the interactions between archaeobacteria and metals (Thompson and Beveridge, 1993)
Fungi. Fungal cell walls are protected by a rigid cell wall, the same as bacteria. But the fungal cell wall has different chemical and structural characteristics than the bacterial cell wall. Various polysaccharides comprise up to 90% of the cell wall. They are often complexed with proteins,

lipids, and other substances. The fungal cell wall is a multilaminate, microfibrillar structure. Phosphodiester and carboxyl groups present confer electrical surface potential to the cell wall (Remacle, 1990). Davidova and Kaspora (1993a) studied the untreated cell walls of two types of yeast, *C. utilis and T. famata*, and found evidence that ion-exchange was the principal mechanism of biosorption of heavy metals. In aqueous solution with respect to the metals the cell walls exhibited the properties of a weak cation-exchange resin due to the presence of negatively charged functional groups. Davidova and Kaspora (1993b) examined several components of the cell walls of *C. utilis and T. famata* and attempted to determine which components were responsible for metal sorption. They found that the capacity of the yeast cell walls to sorb metals was strongly associated with the structural organization of the wall and with the degree of dissociation and accessibility of the negative-charged functional groups. Disruption of this structure profoundly affected the adsorptive capacity of the cell walls. Not only the cell wall proteins, but also its polysaccharides, the mannans and glucans, appeared to be likely participants in metal sorption.
Algae. Algal cell walls are structurally similar to fungal cell walls. They have a multilayered, microfibrillar framework generally consisting of cellulose and interspersed with amorphous material. Like bacteria algal cells are often covered by mucilaginous layers with a significant metal sorption capacity due to the presence of uronic acids (Remacle, 1990).

3.2.3.7 Application of Microbes to Metal Contamination Treatment

Many studies have looked at using bacteria, fungi, and algae to treat metals contaminated water. Seenska-Pobell et al. (1999) studied the interaction of three Bacillus strains and their spores, recovered from a uranium wastepile, with different heavy metals present in the drain waters of the pile. It was found that these strains could selectively accumulate several heavy metals (Mn, Co, Ni, Cu, Zn, Cd, Hg, Pb, and U). Some of the strains, especially their spores were capable of irreversibly binding large amounts of metals such as Al, Cd, and U. Some microbial treatments have found their way into practical application, such as the use of sulfate-reducing bacteria discussed earlier.

The use of microbial biomass as an adsorbent media in filter applications has also been examined. Brady et al. (1994) examined the application of cross-flow microfiltration (CFMF) in yeast-based bioaccumulation processes. They found that the use of serial batteries of CFMF-based yeast bioaccumulators reduced the concentration of toxic heavy metals in waters significantly. Due to the practical difficulties of using live biomass in filter applications, the use of immobillized and/or non-living biomass has been explored (Gabriel et al., 2001; Churchill et al., 1995). Of the various processes, extracellular complexation with polymers or siderophores and extracellular binding to the cell wall are of primary interest in practical applications. The other processes require active participation of the microflora or the use of living microorganisms (Brierly et al., 1989). Churchill et al. (1995) notes that some studies have found that prepared biomass can adsorb metals with equal or greater efficiency than viable cells. Brierly et al. (1989) discusses several applications using living and non-living biomass systems.

In filters that use more common media (sand, peat, compost, etc.) instead of microbial biomass, microbial populations often take hold and grow naturally, depending on the nature of the media and the characteristics of the water (nutrients, toxicity, etc.) entering the filters. However, a review of the literature turns up few studies dealing with the impact of naturally occurring bacteria on metals uptake in filter applications.

The most common types of filters are sand filters. The effect of microorganisms in sand filters, especially the 'slime layer' that forms on sand filters, has been examined to some extent for the impact on metal removal. Eighmy et al. (1992) used linear regressions to examine the correlation between biomass and Fe and Mn oxidizing bacteria in three slow sand filters to Fe and Mn extractable (by 1N HNO_3) from used media. Eight of eleven core samples taken showed significant correlation between Mn oxidizers and extractable Mn. Six of the eleven cores showed significant correlations between biomass and extractable Mn. Extractable Fe was not as frequently correlated to heterotrophic iron precipitators or biomass.

3.3 Filter Tasks Experimental Design

3.3.1 Media

The media chosen for initial screening were selected using a range of criteria. One of the objectives was to look at renewable products as filtration media. With this in mind, the compost, cotton mill waste, peanut hulls, kudzu and an agro-fiber were examined. Compost has also become one of the more common media marketed for runoff treatment. In addition two media, peat and zeolite, used in stormwater treatment were chosen: Peat has been used for stormwater treatment for several years. Zeolite and iron-oxide coated sand are fairly new media being proposed for stormwater treatment. Media commonly used in other (industrial/municipal) water treatment applications, an activated carbon and a commercially developed ion-exchange media, were examined to see how they performed with stormwater. Sand, the most common media used for stormwater filters, was examined as a comparison against the other media and was used in mixed media systems when pressure drop was a problem.

- Compost: Leaf compost obtained from Stormwater Management Inc.
- Peat-Sand Mix: Consists of 50% sand and 50% peat moss by volume. The sand was a common construction grade and the peat was a national commercial brand (Majestic Growth Sphagnum Peat Moss). The peat-sand mix was prepared by placing peat and sand into separate five gallon buckets and lightly packing them to make sure there were no voids in the containers. The density of the sand when packed in this manner was 79.62 lb/ft^3 and the density of the peat was 9.85 lb/ft^3. This gives a weight ratio of 7.7:1 (sand:peat). The equal volumes of peat and sand were then placed in a concrete mixer to ensure uniform distribution. A sieve analysis of the amended soil showed that the mix has a D60 = 0.405mm, a D30 = 0.277 mm and a D10 = 0.185 mm.
- Activated Carbon: A long pelletized carbon called "Reef Carbon" that is available through Aquatic Ecosystems.
- Ammonia Zeolite: A clinoptilolite zeolite marketed for ammonia removal by Aquatic Ecosystems Inc.
- St. Cloud Zeolite: Another clinoptilolite zeolite, this one is mined from St. Cloud mines – one of the largest zeolite mines in North America.
- Agrofiber: The Forest Products Research Lab agro-fiber product was developed as an economic ion-exchange media for pollutant removal from water. Kenaf and Jute

fibers, along with forest wastes such as barks and pine needles, have been found to efficiently remove copper from water. Chemical treatment of the kenaf with reactive yellow-2 significantly increased the capacity of kenaf for copper (Forest Products Research Lab, 1995 – as quoted in Clark, 1996).

- Cotton Mill Waste (ENRETECH I): A cellulose waste fiber from cotton milling developed for cleaning up oil spills - has the consistency of blown-in fiberglass or mineral wool insulation.
- Iron-Oxide Coated Sand (IOCS): Prepared using the same construction grade sand used to prepare the peat-sand mix. First the sand was acid washed (pH of 1.0 for 24 hours), rinsed with distilled, deionized water, and dried at 105°C. The iron-oxide coating was then applied using a procedure originally outlined by Benjamin and Edwards (1989). In this procedure, 200 grams of dried sand, 20 grams of ferric nitrate [$Fe(NO_3)_3 \bullet 9H_2O$], and approximately 50 ml of water (enough to cover the sand) were placed in a 1L flat-bottomed glass flask. The flask was agitated gently for about 2 minutes to dissolve the ferric nitrate. Then the uncovered flask was placed in an oven at 110°C for 20 hours. The coated media was then washed with deionized water until the runoff was clear. Analysis of both the IOCS and the plain sand determined that the iron concentration was approximately 5.6 mg of iron per gram of IOCS compared to 0.6 mg of iron per gram of plain sand.
- Lewatit TP207: A specialty resin designed for the selective removal of heavy metals manufactured by Bayer. It has a crosslinked polystyrene matrix with iminodiacetic acid as its function group. Na is the ionic form for the exchange. The resin costs $320 per cubic foot.
- Kudzu: Kudzu was collected, air-dried under a fume hood, and ground. Particle size: 0.425 – 0.710 mm.
- Peanut Hulls: Ground, dried, and autoclaved peanut hulls. Particle size: 0.710 mm – 2.36 mm.
- Sand: The construction grade sand used for the peat-sand mix and the iron-oxide coated sand will be looked at for comparison.

3.3.2 General Procedures

Table 3-7 lists the standard methods used to analyze the samples collected during the experiments that are discussed in this chapter.

Table 3-7: Standard Methods Used:

Parameter	Type	Method	Instrument
Metals	Critical	Digestion - S. M. 3015 Analysis - S. M. 3500	Microwave ICP
Total Dissolved Solids	Non-critical	S. M. 2540 C	
Total Suspended Solids	Non-critical	S. M. 2540 D	
pH	Non-critical	EPA 150	pH meter

3.3.2.1 Sample Collection for Runoff Characterization and Media Studies

In order to characterize the runoff, stormwater samples were collected during a series of rainfall events between October 2001 and July 2002. The samples were collected in front of a curb inlet from parking lot runoff at the rear of H.M Comer Hall on the University of Alabama Campus in Tuscaloosa, Alabama. Two samples were collected for each event - one a 'first flush' and one a composite. For first flush samples, runoff was collected every 5 minutes for the first 20 minutes of flow. For composite samples runoff was collected every 30 minutes over the duration of stormwater flow. Samples were collected and stored using nitric acid washed bottles and tubes. Runoff from this site was also used in the lab-scale column studies and the filter media tests as detailed in the project QAPP.

3.3.2.2 Runoff Characterization

Collected samples were examined for dissolved and suspended solids and total and dissolved metals. TDS and TSS were sampled and analyzed according to standard methods. For total metals determinations, unfiltered samples were acidified with nitric acid, digested (microwave method), and then analyzed on the ICP-OES (Perkin Elmer DV4000). For dissolved metals determinations the stormwater samples were filtered through 0.45 μm metricel membrane filters, placed in test tubes, preserved with nitric acid, and analyzed on the ICP.

3.3.2.3 Preparation of Metal Ion Solutions

The metal solutions used for all the experiments were made by dissolving analytical grade metallic salts from Fisher Scientific in distilled, deionized, $\approx$18 meg-ohm resistivity water. The salts used include $ZnSO_4*H_2O$, $CdSO_4$, Cadmium Nitrate Tetrahydrate, $CuSO_4$, $Cr(NO_3)_3*9H_2O$, $FeSO_4*7H_2O$, and $Pb(NO_3)_2$. Metal solutions for experiments using lab-prepared solutions as opposed to runoff were prepared by making single-component metal ion stock solutions of 1000mg/l. Additional concentrations were prepared using the stock solution. To prepare small volumes of spiked runoff for the batch equilibrium studies (outlined below) a known volume of stock solution was added to a known volume of runoff. To allow the spiked runoff to equilibrate the spiked runoff was shaken for 8-10 hours at 150 rpm on an Innova 2100 Platform Shaker from NewBrunswick Scientific. The spiked runoff was analyzed for total and dissolved metals of concern.

3.3.3 Experimental Design for Filter Task 2: *Assess and quantify the ability of various media to remove metals from stormwater.*

Limited data is available in the literature as to the capacity and rate of metals uptake onto many of the media. These experiments were replicated during this study because the wide range in experimental design and analytical techniques of different investigators make comparison of the data virtually impossible. In addition, previous studies focused on industrial discharges rather than stormwater. The uptake of copper, cadmium, zinc, lead, chromium and iron by various media (peat, sand, peat-sand, composted leaves, cotton mill waste, activated carbon, resin and others) was investigated. Capacity data, uptake ability at low metal concentrations, and rate of uptake data were obtained with isotherm and kinetic investigations under steady state batch equilibrium conditions. Leaching of metals from the various media was also examined. The purpose of these initial tests was to compare the different media and use the collected data to decide which media would have the best potential as filtration media to remove metals from stormwater. The selected media were then examined further. The criteria used for selecting the final media were:

1. The ability of the media to remove metals at low concentrations, typical of what might be found in runoff.
2. Whether significant amounts of metals or major ions were leached from the media.
3. Rate of uptake
4. The overall capacity of the media for metals.
5. Additional factors such as impact of media on pH and color of effluent and the cost of the media.

This data allowed comparisons to be made between the media in order to select the media that had the best capacity for metals.

3.3.3.1 Equilibrium Adsorption and Desorption Experiments Procedures

Equilibrium adsorption isotherm studies were performed to determine the relationship between metal sorption/desorption from the media and equilibrium metal ion concentration in solution. Fifty milliliters of solution was placed with a known mass of media in plastic 60 ml jars then sealed with parafilm and screw top lids. These jars were shaken for 4 h at 250 rpm on an Innova 2100 Platform Shaker from NewBrunswick Scientific. Then 5-10 ml samples of solute were filtered through 0.45 μm metricel membrane filters from Gelman Sciences, preserved with 0.01-0.02 ml of ultra pure nitric acid to create a 2% nitric acid solution, and refrigerated until analysis with a Perkin Elmer Inductively Coupled Plasma – Optical Emission Spectrophotometer (ICP-OES). This procedure was used to examine the 'ultimate' uptake capacity of the media for different metals (ultimate given the listed conditions), the uptake ability of the different media for low concentrations of metals from actual stormwater, the leaching of metals from the media, and the sorption/desorption of the major ions Ca, Mg, K, and Na from the media during metal uptake. Table 3-8 lists the tests for which this procedure was used and the specifics for each test. Blanks were run for all tests and all tests were run in triplicate.

Table 3-8. Procedure specifics for equilibrium sorption and desorption experiments

Test	Matrix	Metals in Matrix	Initial Metal Concentration	Mass of Media
Uptake Capacity (Equilibrium Isotherm Curve)	Distilled Deionized Water	Zn (single-ion solutions)	10 to16 concentrations ranging from 0mg/l to 1000mg/l)	0.25 grams
Uptake Ability at Low Metal Concentrations / Metals Leached from Media	Spiked Runoff	Zn, Cu, Cr, Cd, Pb, Fe (mixed-ion solutions)	3 concentrations ranging from 0.05 mg/l to 0.5 mg/l of each metal.	1.00 grams
Sorption and Desorption of Ca, Mg, K, and Na	These ions were analyzed for in all of the above tests.			

3.3.3.2 Kinetic Experimental Procedures

The purpose of these tests was to examine the rate of uptake of metals by the media. The kinetic tests were performed using baffled batch adsorbers constructed according to standardized tank design for Continuously Stirred Tank Reactors (CSTRs) (Furasawa and Smith, 1973; 1974). Each unit consisted of a plastic vessel with a capacity of 2 liters. The vessels were fitted with eight plexiglass baffles and agitation was provided by a flat six-blade impeller driven by a variable speed motor. All components of the batch system design: impellers, baffles and vessel, were constructed from perspex and HDPE to minimize metal-ion sorption by fixtures. Ten grams of media were added to a stirred tank that contains 2 liters of solution and the timer started. At increasing time intervals aliquots of sample were withdrawn by plastic syringe over a four-hour period. Sampling consists of removing 10 ml of the solution from the tank with a plastic syringe. The solution was immediately filtered through 0.45 μm metricel membrane filters into test tubes. The samples were preserved with 0.02ml of ultra-pure nitric acid to create a 2% nitric acid solution. Acidifying the samples kept the metals in solution. The tubes were then sealed with parafilm and refrigerated. Analysis was performed using a dual view model Perkin Elmer Inductively Coupled Plasma – Optical Emission Spectrophotometer (ICP-OES) DV3000.

3.3.3.3 Influence of Media on Solution pH:

Changes in pH due to the addition of the media were examined by monitoring changes in pH throughout the experiment for all the batch kinetic studies, with pH recorded when samples were withdrawn. The pH was monitored using an Accumet™ Model 15 pH meter from Fisher Scientific with a standard glass combination electrode with a silver/silver chloride reference.

3.3.4 Experimental Design for Filter Task 3: *Characterize the physical properties of the media with particular reference to their sorption properties.*

The properties of porous solids that render them useful adsorbents in contaminant capture include high porosity and surface area as well as the physical and chemical nature of the internal adsorptive surfaces. Table 3-9 outlines the information necessary for the characterization of the sorbents.

Table 3-9. Characterization Tests for Sorbents

Test	Method	Citation(s)
particle size analysis	D2862, D421 and D422	ASTM
BET surface area	BET	Brunauer, Emmett, Teller (1938)
bulk density		ASTM
moisture content	D2867 or D2974	ASTM
pH	D3838-80	ASTM
end group titration (CEC)	MB^+ and/or $trien^+$ absorption	Meier and Kahr (1999)

3.3.4.1 Particle size analysis:

One of the principal parameters of importance in any filter application is the particle size of the media, both the range of sizes and the distribution of particles within that range. A sieve shaker test is usually performed on all granular media. The particle size distribution affects head loss through the column, density of the packed media, and other factors influencing column performance and design. The procedure for particle size analysis is given as ASTM D 421 and D 422, sample preparation and procedures, respectively.

3.3.4.2 Surface area:

Surface area determinations based on the Brunauer, Emmett, and Teller (BET) model were performed using a Sorptomatic 1990® (CE Instruments, Austin, TX). According to the BET method, the surface area of a sorbent may be calculated from the N_2-isotherm observed at the boiling point of nitrogen. By the analysis of the adsorption and desorption curves within the relative pressure ranges [p/p_0] of approximately 0.05 and 0.33, the volume is determined which corresponds to the quantity of nitrogen necessary for a monomolecular layer. The surface needed for one nitrogen molecule is 16.2Å^2; therefore the specific surface of the sample may be calculated in m^2/g (Brunauer et al., 1938). The BET method is generally used as the standard measure of surface area of an adsorbent.

3.3.4.3 Moisture content:

The moisture content of the media was determined using the ASTM method listed in Table 3-9.

3.3.4.4 pH:

The pH of the system is known to have an effect on the adsorptive properties of a given sorbent. The pH of the media can influence the chemistry of the water treated, resulting in precipitation of metals or effects on ion exchange. The pH of all media was determined using ASTM method D3838-80.

3.3.4.5 End group titration (CEC):

The cation-exchange capacity (CEC) of a material is defined as the sum of the exchangeable cations it can adsorb at a given pH. Alternatively, the CEC is a measure of the negative charge present at the sorbent surface. The CEC is generally measured to evaluate the ability of certain soils to sorb K^+ (from fertilisers), heavy metals, and various other target ions whose mobility in the soil is an issue of concern.

The CEC is a function of available surface charge per unit area of material, the pH at which exchange occurs, and the relative affinities of the ions to be exchanged for the material surface. The method developed by Meier and Kahr was used here (Meier and Kahr, 1999).

3.3.4.6 Other tests:

Carbon tetrachloride activity; iodine number; carbon, hydrogen, and nitrogen (CHN) elemental analysis; and infrared radiation (IR) surface group analysis were tests that were specified in the initial testing rubric for the media used in this study. These tests were more suited to and/or developed for comparison of activated carbons. The media being tested in this study were sufficiently different from activated carbon (both structurally and in terms of function) that the expense of performing these tests outweighed the value of whatever results were expected.

3.3.5 Experimental Design for Filter Task 4: *Determine the effect of varying process conditions on the metal capture process. Evaluate the influence of surface phenomena on the rate and extent of metal uptake.*

Using data from task 2, the media were assessed and ranked and the best sorbent media were selected for in depth investigation. This series of investigations was divided into two parallel sections namely kinetic batch contact time experiments and batch equilibrium isotherm studies. Kinetic and equilibrium data generated yield vital clues as to the mechanism of metal capture. Much of the data collected during task 2 was also used for task 4. Additional batch kinetic and equilibrium experiments were run as needed to further examine the metal uptake process. Experimental procedures were the same as those outlined for task 2.

3.3.6 Experimental Design for Filter Task 5: *Conduct laboratory scale fixed-bed column studies to evaluate the dynamic characteristics of the metal capture process.*

This investigation focused on breakthrough time and hydraulic parameters affecting performance of the column. Stormwater runoff was collected and treated using lab-scale upflow filter columns. Stormwater was used in the parametric study to investigate the effect of variables such as media and flowrate on column performance. In addition, physical and hydraulic parameters affecting head loss in the column were studied. The use of upflow columns allowed for better process control than downflow filters. These tests may also gave some indication as to whether the use of upflow columns might extend filter life in comparison to downflow columns. In tests with several media in downflow columns at the University of Alabama at Birmingham (UAB), it was found that the surface of the columns often became clogged before the sorption capacity of the media was reached (Clark, 1996).

Variable flow rate experiments were performed to examine how metal removal efficiency in the columns changes with increasing flow rate. The aim was to see how quickly runoff can be pumped through the columns before removal efficiency begins to fall off. Given the design of the column, increasing flow rate also reduces the time available for sedimentation of particulates to occur in the sump. Therefore, increasing flowrate through the filters should decrease the volume of runoff that can be pumped through the filter before clogging occurs. The goal is to optimize the balance between flow rate and clogging potential.

3.3.6.1 Filter Column Design

All column materials, pump tubing, and stormwater holding containers were made of plastic and PVC. The columns were approximately 26 inches tall and have an inside diameter of 1.5 inches. A sump area was provided below the media to allow space for particulates to settle. The depth of media in the columns was approximately 12 inches. The stormwater was pumped from a 32 gallon plastic holding container using Masterflex® Model 7518-10 peristaltic pumps. The stormwater entered at the side of the column below the media and above the sump and flowed upward through the media. The effluent exited the column at the top through a section of vinyl tubing.

3.3.6.2 Variable Flow Rate Experiments

The first step in these experiments was to determine the bed volume for the column. To do this the column was packed with the media to be examined and assembled with the sump area pre-filled with water. Water was then pumped into the column from a graduated cylinder until the water began to emerge from the top of the packed media. The volume of water pumped was the dry bed volume of the column. However, for the purpose of these tests, it was necessary to determine the wet bed volume. To determine the wet bed volume, the wetted column was drained using the pump. After refilling the sump area, water was again pumped from a graduated cylinder into the column until water began to emerge from the top of the packed media. The volume pumped equaled the wet bed volume. This procedure was repeated three times and the results averaged to determine the wet bed volume for each column.

The second step was to determine the upper limit of the flow rate through the column. Water was pumped at an increasing rate through the column until (1) blowout of the media occured or (2) the column setup began to show signs of excessive stress (i.e. leaks at couplings, excessive ballooning of the plastic couplings, excessive vibration of the pump tubing and/or column). The only media that showed problems with stress on the columns (over the range of flow rates of interest in these tests) was the peat-sand media. This is discussed further in section 3.4.3. No blowout occurred for any of the media.

Finally five flow rates were selected to give a range of hydraulic residence times. Each column was run with the following hydraulic residence times: 10 minutes, 5 minutes, 3 minutes, 1 minute and 30 seconds. Three series were run for each column - a series consists of running each flow rate once - for a total of 15 runs for each media. The order in which the different flow rates were run was randomized within each series. Media was not changed between series.

Each column run began by calibrating the pumps running to the various columns to the correct flow rate. Then 20 bed volumes were run through the columns to insure that uptake had stabilized for the column at that speed, then a composite sample was collected over another 10 bed volumes. Since exact calibration of the pumps was difficult, a record was kept of the time over which the composites were collected and the final volume of the composites. From these measurements, the actual residence time was computed for each run. The composite sample was analyzed for dissolved metals, total metals (dissolved plus particulate bound), total dissolved solids, total suspended solids, color and pH. Influent samples were pulled for each run to compare with the effluent composite samples. Influent samples were drawn once during the shorter runs (1 minute and 30 second BVRT), but were pulled several times at set intervals during the longer runs and mixed to account for variations over time. When pulling influent samples the lag time between influent entering and exiting the column was taken into account. All media were run at the same time and influent was pumped from the same source so that comparisons could be made between media as well as between speeds. An impeller that was attached to a variable speed motor set at 180 rpm continuously mixed the influent. Headloss was monitored for all runs using a pressure gauge mounted at the base of the column. Initial headloss and final headloss (after 30 bed volumes were run) were noted for each run.

Figures 3-1 and 3-2 show the column design and experimental set up for the variable flow column experiments.

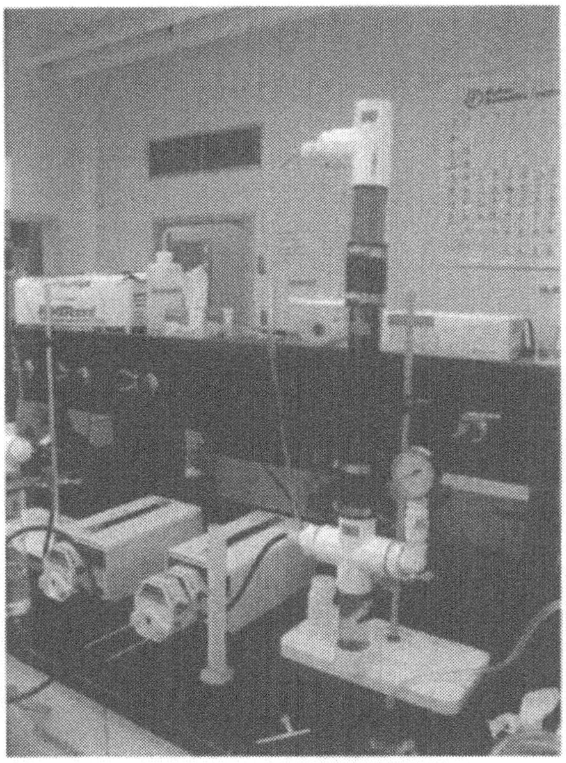

Figure 3-1. Experimental Column Design

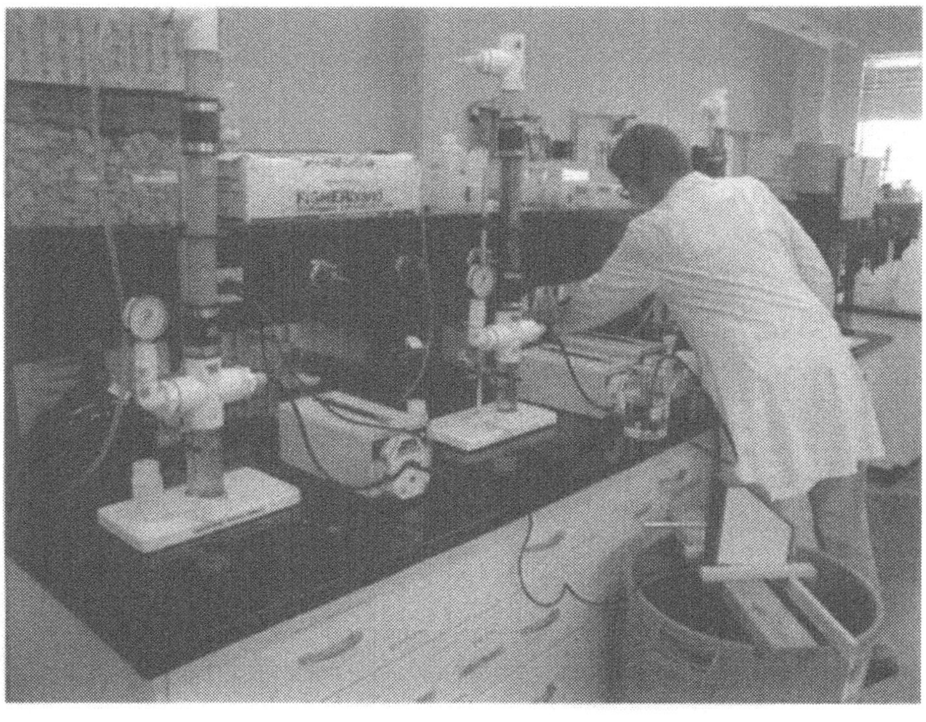

Figure 3-2. Variable Flow Experiments Column Set-Up

3.3.7 Experimental Design for Filter Task 6: *The Roll of Biomass in Metals Uptake*

Engineered filter systems developed to treat metal contaminated waters in field situations frequently develop microbial populations. The effect of microbial growth within filter systems on stormwater metal retention was studied in laboratory column experiments. To determine the effect of microbial growth on metals sorption by the media, a series of experiments were performed using identical columns – one column a control and the other designed to inhibit microbial growth. The inhibited growth and control columns were run in parallel and samples were taken at regular intervals to test for microbial activity and effluent metal concentration.

3.3.7.1 Column Sterilization

Filter media, columns, tubing, and water containers were sterilized by autoclaving. The media were isolated from light by covering the columns with aluminum foil to prevent growth of autotrophic organisms.

3.3.7.2 Determination of Microbial Presence in Samples

The degree of the microbial presence in the samples of filtered water was determined using the test of reduction of XTT by active microorganisms (procedure after Roslev and King, 1993). The procedure was:

1. Dissolve XTT in medium heat up to 50 degrees Celsius in a water bath to ensure solution. Make up a 10mM stock (MW 764.5). Store at 40 degrees Celsius in the dark until used.
2. Mix 1:1 with samples in sterile 10 ml serum vials. Final XTT concentration is 5mM.
3. Incubate at 35 degrees Celsius and 120 rpm for 8 hours.
4. Remove 0.5 ml from serum vials and mix with 0.5 ml of autoclaved, deionized water. Centrifuge at 8,000 *x* g for 8 minutes. If color development is too small try undiluted supernatant.
5. Measure color in supernatant at 470 nm with medium without XTT as reference.

3.3.7.3 Determination of XTT Standard Curve

Although the above procedure provides a measure of relative microbial concentration, it does not provide a quantitative measure of microbial concentration. To correct this a XTT standard curve was determined for each media.

1. For each of the media examined, one of the non-sterile columns was started a few days earlier than the rest of the filters to allow a microbial population to begin to develop.
2. Samples of water from the filters were taken and the microbial population grown.
3. The absorbance at 470 nm was determined for several known quantities of microbes, and a standard curve for mg protein/L vs. absorbance generated.

3.3.8 Experimental Design for Filter Task 8: *Investigate disposal options for the media.*

Toxicity Characteristic Leaching Procedure (TCLP) was performed on the three filter column media (peat-sand mix, St. Cloud zeolite, and compost) to establish the regulatory classification of saturated media with a range of media metal concentrations under RCRA or SWDA. This categorization can be significant in determining the cost and options for disposal.

The TCLP is a laboratory test that is designed to simulate leaching in a municipal landfill. It is used to determine the mobility of organic and inorganic analytes in liquid, solid and multi-phased wastes. The general procedure for waste containing greater than or equal to 0.5% solids is (Method 1311 in EPA publication SW-846, *Test Methods for Evaluating Solid Waste, Physical/Chemical Methods*):

1. Any liquid present in the waste is separated from the solid phase and stored for later analysis.
2. If the solid media is too large to fit through a 9.5 mm (0.375 inch) standard sieve then the particle size of the solid media is reduced.
3. A minimum of 100g of solid media and a volume of extraction fluid equal to 20 times the weight of the solid media is placed in a container and rotated end over end at 30±2 rpm for 18±2 hours. The extraction fluid used depends on the alkalinity of the solid media.
4. After extraction the liquid is separated from the solid media by filtration through a 0.6 to 0.8 μm glass fiber filter.
5. If the original liquid waste from step 1, if there was any, is compatible with the liquid extract from step 4 then the two are combined and analyzed for leached analytes together. If the two liquids are not compatible then they are analyzed separately and the results are mathematically combined.

Before TCLP analysis the selected media will be exposed to different concentrations of multi- and/or single-metal solution such that each media will have samples with varying metals concentrations. A blank will also be included for each media. The purpose of this is to determine the amount of metals that each media can hold before it will be considered a RCRA hazardous waste based on the TCLP.

The metal solutions used for these tests were pre-settled runoff spiked with Cr, Cd, and Pb since these were the three metals examined by this study that are regulated by RCRA. The selection of the metal solution concentrations began by examining the maximum metal concentrations allowed in the TCLP extract. The regulatory levels are 1 mg/L for cadmium and 5 mg/L for chromium and lead. Given the ~20 mL of extraction fluid per 1 gram of solid media ratio the minimum mg/g metal concentration that might result in reaching the regulatory levels (given complete leaching of metals) is 0.02 mg/g for cadmium, and 0.1 mg/g for chromium and lead. This became the minimum degree of saturation wanted on the media. Concentrations below this did not have the potential to exceed regulatory levels. Then it was necessary to select the different metal concentrations for the spiked runoff to be used such that the mg/g metal concentration on the media exceeded the minimum but remained in a range reasonable for media from a stormwater filter. Therefore, before selecting the metal concentrations for the experiment it was necessary to examine adsorption of Cr, Cd, and Pb from spiked runoff by the media.

To simplify calculations the 1g media to 20 ml of solution ratio used in TCLP extraction was also used in the equilibrium curve determinations and in saturating the media samples to be analyzed by TCLP. During the preliminary equilibrium tests 2.5 g of each media was mixed with 50 ml of runoff, spiked to different concentrations, in 60 ml plastic powder jars and shaken at 250 rpm for 4 hours. Multi-metal solutions with all three metals were used. Concentrations started at 5 mg/l Cr, Pb and 1 mg/L Cd, (the lowest concentration which could theoretically result in exceeding leaching regulatory levels – if all metal was adsorbed and then leached during TCLP) and ranged upward. After shaking, samples were withdrawn, filtered, and analyzed for dissolved metals. Equilibrium adsorption curves (C_e vs. mg/g uptake) were plotted and from these the final solution concentrations to be used to saturate the media were selected.

Once the solution concentrations were selected the samples for TCLP analysis were generated. 100 grams of each media was mixed with 2 liters of multi- and/or single metal solutions at different concentrations in 2-L HDPE bottles and shaken at 225 rpm on a platform shaker for 4 hours. Samples of the metal solution were taken before and after contact with the media and used to calculate the quantity of dissolved metals adsorbed by the media. The media was separated from the metal solution by filtering. Filtering was performed using vacuum filtration. The media-metal solution mix was first filtered through a plastic fine mesh screen to collect the larger, coarser material. The solution that passed through the fine mesh was then filtered through Whatman 40 filter paper to collect the finer media particles. While most of the liquid was removed the media was not completely dried but left slightly moist since media in a field filter column rarely dries completely and it is unlikely that the media would be dried completely before disposal. The TCLP analysis was then performed on the saturated media samples.

3.4 Results and Discussion

3.4.1 Runoff Characterization

The type and quantity of metals in runoff can vary widely depending on location and conditions. Makepeace et al. (1995) reviewed a wide sampling of papers that had analyzed stormwater runoff samples. The overall range and the range of the means from these papers for the metals of concern in this study are shown in Table 3-10. One thing to note is that the data in Table 3-10 does not distinguish between dissolved and total metals.

Table 3-10. Metal concentration ranges (overall and mean) and pH reported by various studies for stormwater samples. From Makepeace et al. (1995).

Metal	Concentration Range (mg/l)	Concentration Range of Means (mg/L)
Cadmium	0.00005-13.73	0.003-0.011
Chromium	0.001-2.30	0.010-0.23
Copper	0.00006-1.41	0.0065-0.15
Iron	0.08-440.0	0.988-12.0
Lead	0.00057-26	0.0209-1.558
Zinc	0.0007-22	0.0166-0.58
pH	4.5-8.7	

Runoff was used for many of the lab scale tests conducted in this study. The runoff used for these tests was collected parking lot runoff. A series of runoff samples was collected from this parking lot in order to characterize the runoff. The results of these samplings are given in Tables 3-11 and 3-12. Iron, zinc and copper were the most prevalent metals detected in the collected runoff along with small amounts of particulate bound lead.

In the various experiments in this study both spiked distilled, de-ionized water and spiked runoff were used depending on the experiment. The spiked de-ionized, distilled water was used in the 'ultimate' uptake capacity and kinetic rate of uptake tests. The other metal uptake tests used runoff as the matrix. The runoff used for the low concentration metal (LCM) uptake tests was spiked to different levels to examine uptake by the media at different concentrations. The spiking level for the column tests was then selected such that there would be sufficient metals present to quantify concentration changes accurately in the analysis but such that the metal concentrations would still be reasonable for stormwater runoff. Table 3-13 summarizes the water matrix and metal spiking data for all of the experiments.

Table 3-11. Results of Runoff Characterization Tests on 'First Flush' Samples from Parking Lot in Rear of H.M. Comer Building on the University of Alabama Campus.

Date		10/5/2001	11/20/2001	11/24/2001	11/27/2001	6/22/2002	6/27/2002
Cu mg/l	total	0.036	0.051	0.018	0.030	0.181	0.075
	dissolved	0.017	0.030	0.012	0.030	0.038	0.013
Cr mg/l	total	0.015	< 0.010	< 0.010	*0.013*	0.016	0.011
	dissolved	< 0.003	< 0.010	< 0.010	< 0.010	0.002	< 0.002
Pb mg/l	total	0.024	0.014	*0.004*	*0.009*	*0.038*	0.037
	dissolved	< 0.004	< 0.003	< 0.003	< 0.003	< 0.008	< 0.008
Zn mg/l	total	0.203	0.136	0.034	0.079	0.256	0.197
	dissolved	*0.009*	0.051	*0.001*	0.006	0.028	< 0.003
Cd mg/l	total	< 0.001	< 0.001	< 0.001	< 0.001	< 0.0004	< 0.002
	dissolved	< 0.001	< 0.001	< 0.001	< 0.001	*0.001*	< 0.002
Fe mg/l	total	5.872	4.469	1.373	2.823	5.976	6.924
	dissolved	0.020	0.033	0.016	0.035	0.027	0.013
Na mg/l	total	0.855	2.996	0.755	1.618	2.651	0.337
	dissolved	0.878	2.929	0.828	2.396	0.906	0.241
Mg mg/l	total	8.618	8.673	1.970	3.309	24.934	8.665
	dissolved	2.048	5.422	1.130	1.203	5.680	0.501
Ca mg/l	total	35.760	46.060	10.189	15.702	139.323	20.949
	dissolved	18.959	38.692	8.956	10.054	41.593	4.454
K mg/l	total	2.784	7.224	1.450	2.653	10.119	2.194
	dissolved	1.567	5.881	1.227	2.004	3.203	0.684
pH		7.42			7.84	7.65	7.85
Color						285.00	55.00
TSS mg/l		271.33	260.00	33.00	82.00	104.00	348.67
TDS mg/l		106.70	309.33	59.80	75.40	396.55	34.10

Note: Italicized concentrations were above ICP-OES LOD (limits of detection) but below LOQ (limits of quantification). Concentrations below LOD are reported as less than the LOD value for the ICP run in which that sample was tested.

Table 3-12. Results of Runoff Characterization Tests on Composite Samples from Parking Lot in Rear of H.M. Comer Building on the University of Alabama Campus.

Date		10/5/2001	11/20/2001	11/24/2001	11/27/2001	6/22/2002	6/27/2002
Cu mg/l	total	0.059	0.034	0.027	0.046	0.072	0.064
	dissolved	0.041	0.022	0.021	0.011	0.031	0.015
Cr mg/l	total	*0.005*	< 0.010	< 0.010	< 0.010	0.007	*0.005*
	dissolved	< 0.003	< 0.010	< 0.010	< 0.010	*0.001*	< 0.002
Pb mg/l	total	*0.009*	*0.007*	< 0.003	*0.004*	*0.014*	0.027
	dissolved	< 0.004	< 0.003	< 0.003	< 0.003	< 0.008	< 0.008
Zn mg/l	total	0.089	0.080	0.026	0.045	0.116	0.137
	dissolved	0.019	0.039	0.009	*0.001*	0.022	*0.003*
Cd mg/l	total	< 0.001	< 0.001	< 0.001	< 0.001	< 0.0004	< 0.002
	dissolved	< 0.001	< 0.001	< 0.001	< 0.001	< 0.0004	< 0.002
Fe mg/l	total	1.898	1.967	0.641	1.274	2.755	4.006
	dissolved	0.025	0.023	0.019	0.058	0.024	0.012
Na mg/l	total	0.545	1.817	0.605	2.447	1.230	0.367
	dissolved	0.532	1.806	0.620	1.517	0.862	0.286
Mg mg/l	total	4.049	5.662	1.395	2.181	10.946	4.992
	dissolved	1.617	4.359	1.001	1.182	5.333	0.831
Ca mg/l	total	20.089	34.712	8.705	12.447	61.485	16.197
	dissolved	13.781	31.146	7.690	10.557	39.950	7.302
K mg/l	total	2.461	4.623	1.508	2.264	4.166	1.837
	dissolved	2.032	4.187	1.370	1.820	2.677	1.025
pH		7.33			7.46	7.71	7.92
Color						260.00	45.00
TSS mg/l		70.00	91.00	11.00	38.00	103.33	204.00
TDS mg/l		74.80	216.20	46.80	65.00	348.70	41.80

Note: Italicized numbers were above ICP-OES LOD (limits of detection) but below LOQ (limits of quantification). Concentrations below LOD are reported as less than the LOD value for the ICP run in which that sample was tested.

Table 3-13. Summary of Metal Concentration Ranges for all Metal Adsorption Experiments.

Test		Metal Concentration Ranges for Individual Tests (mg/L)			
		'Ultimate' uptake capacity	Kinetic-rate of uptake	LCM uptake	Variable flow column studies
Matrix		Distilled, deionized water	Distilled, deionized water	Stormwater runoff	Stormwater runoff
Cu mg/l	total			0.089-0.511	0.168-0.498
	dissolved	0	0.47,0.78-0.87	0.074-0.398	0.127-0.168
Cr mg/l	total			0.043-0.461	0.136-0.339
	dissolved	0	0.49,0.84-0.95	0.029-0.346	0.092-0.129
Pb mg/l	total			0.045-0.461	0.126-0.530
	dissolved	0	0.54,0.60-0.99	0.028-0.350	0.050-0.122
Zn mg/l	total			0.071-0.452	0.329-0.807
	dissolved	0-1000	1.01,1.61-1.76	0.065-0.426	0.254-0.361
Cd mg/l	total			0.070-0.415	0.206-0.412
	dissolved	0	0.57,0.93-1.00	0.074-0.461	0.198-0.265
Fe mg/l	total			0.099-0.526	0.871-9.313
	dissolved	0	0.82,1.33-1.48	0.031-0.368	0.132-0.323
pH		<3.5* - 6	4.43-4.53	6.64-7.01	6.9-7.7
TSS mg/l		0	0	3.7-6.0	29-460
TDS mg/l		0	0	62-74	115-173

*The lower pH limit for the zinc test was hard to quantify, as the Zn concentration increased the pH dropped until Zn concentrations reached between 200 and 300ppm then the pH readings began to increase. Unsure of the reason for this, perhaps there was interference caused by the high Zn concentrations which resulted in inaccurate readings.

3.4.2 Filter Task Two Results and Discussion

3.4.2.1 Batch Equilibrium Capacity Tests Results

To examine the relationship between metal uptake by the media and equilibrium metal concentration in the solution and compare the maximum uptake capacity of the media under similar conditions batch equilibrium capacity tests were performed. Zn was chosen because it is one of the most common metals found in runoff. Also, while most metals in runoff have the largest fraction of the metal concentration bound to particulates in the runoff instead of being dissolved up to 50% of the Zn in runoff is often in the dissolved phase. Metals in the dissolved phase are more available to aquatic life and therefore present more potential for harm. The procedure outlined in section 3.3.3.1 and Table 3-8 was used for these tests.

The results of the equilibrium capacity tests were used to generate equilibrium uptake capacity graphs (Figures 3-3 through 3-14) plotting metal concentration at equilibrium vs. mg of metal removed from solution per gram of media. Each concentration was run in triplicate and the error bars are shown on the graph. From these graphs, an estimate of the ultimate uptake capacity of the media under the test conditions were made.

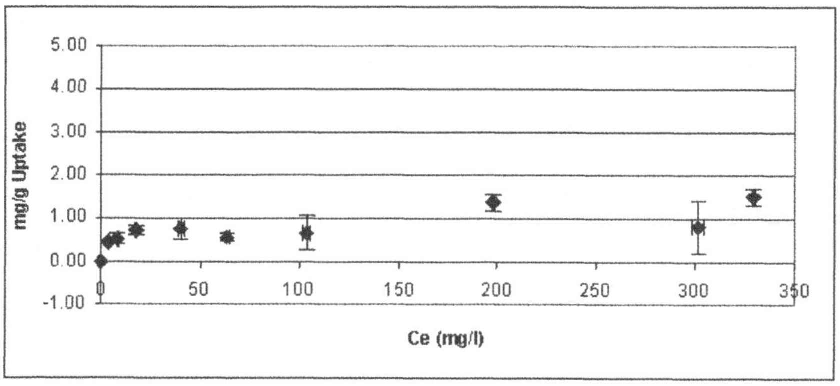

Figure 3-3. Batch Equilibrium for Zinc onto Peat-Sand. Average of 3 Runs with Error Bars.

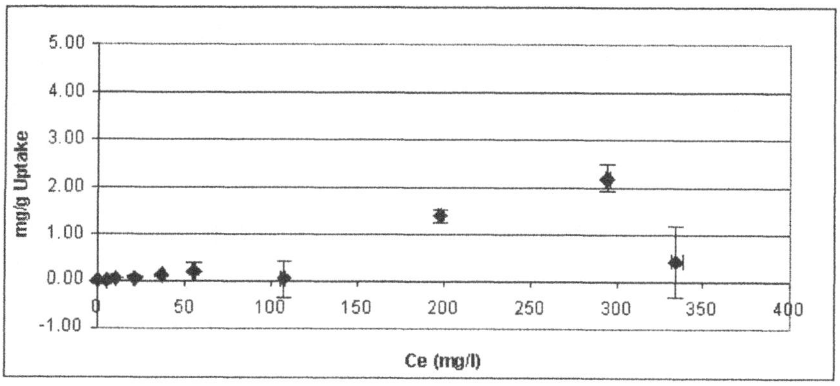

Figure 3-4. Batch Equilibrium for Zinc onto Iron-Oxide Coated Sand. Average of 3 Runs with Error Bars.

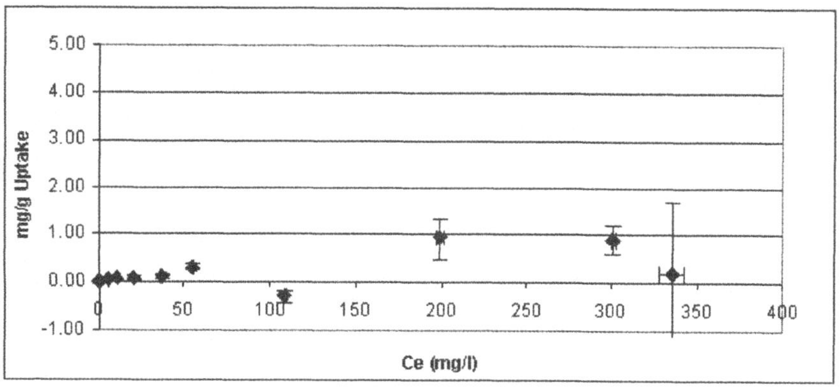

Figure 3-5. Batch Equilibrium for Zinc onto Sand. Average of 3 Runs with Error Bars.

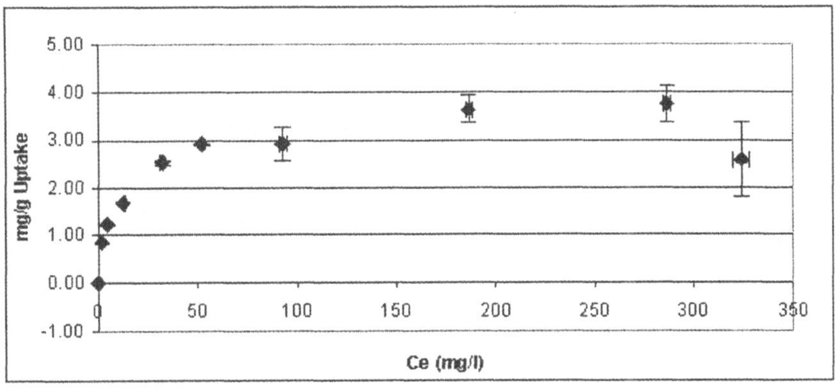

Figure 3-6. Batch Equilibrium for Zinc onto St. Cloud Zeolite. Average of 3 Runs with Error Bars.

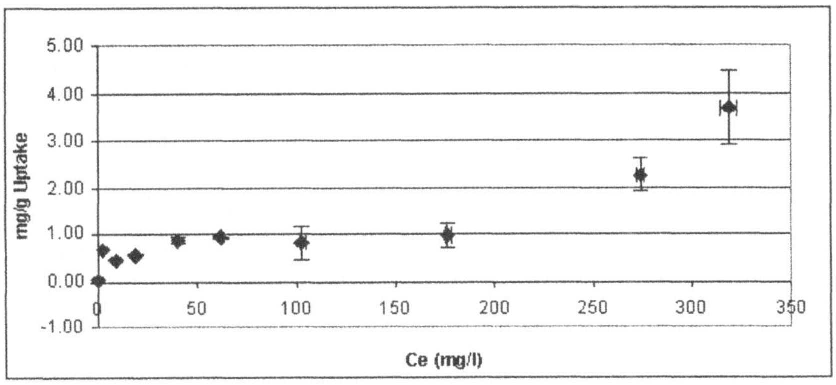

Figure 3-7. Batch Equilibrium for Zinc onto Activated Carbon. Average of 3 Runs with Error Bars.

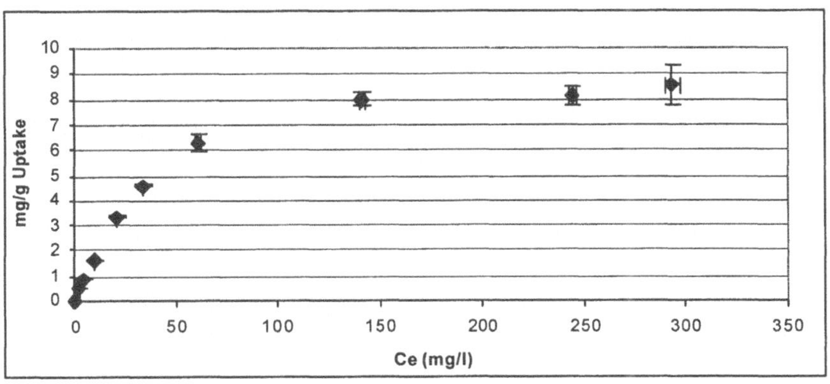

Figure 3-8. Batch Equilibrium for Zinc onto Kudzu. Average of 3 Runs with Error Bars.

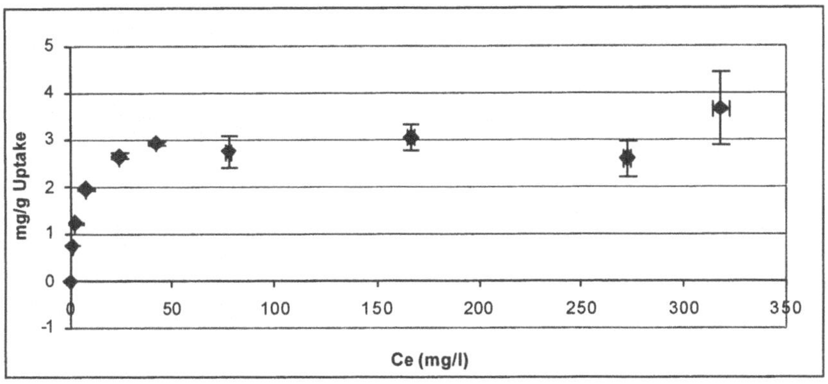

Figure 3-9. Batch Equilibrium for Zinc onto Peanut Hulls. Average of 3 Runs with Error Bars.

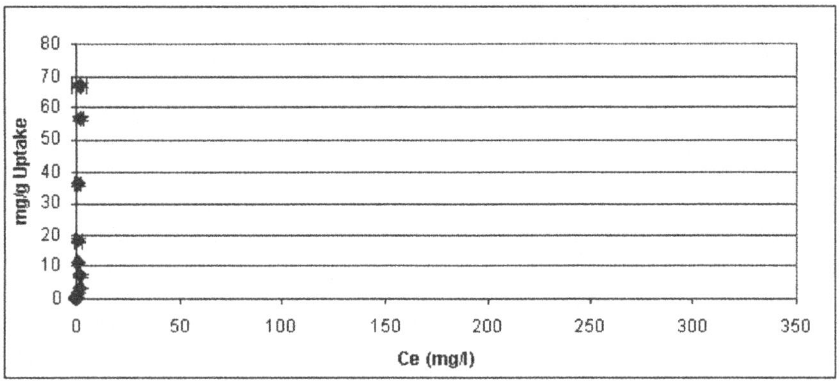

Figure 3-10. Batch Equilibrium for Zinc onto TP 207 Resin. Average of 3 Runs with Error Bars.

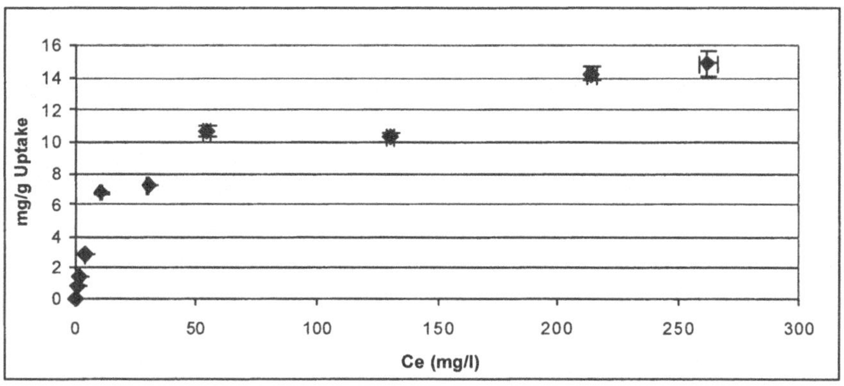

Figure 3-11. Batch Equilibrium for Zinc onto Compost. Average of 3 Runs with Error Bars.

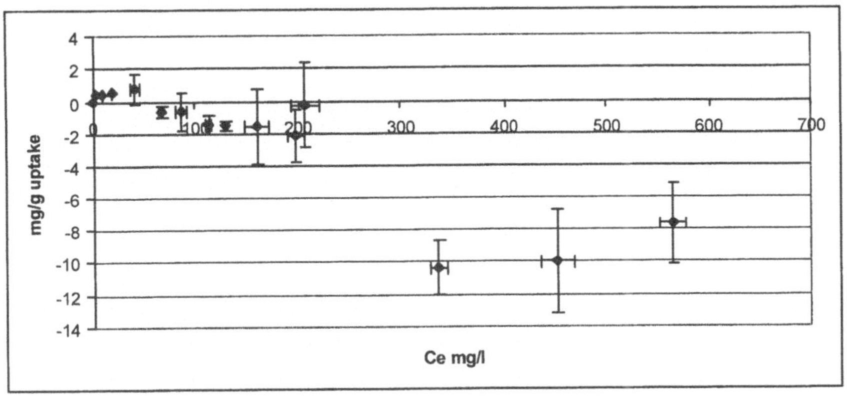

Figure 3-12. Batch Equilibrium for Zinc onto Cotton Mill Waste. Average of 3 Runs with Error Bars.

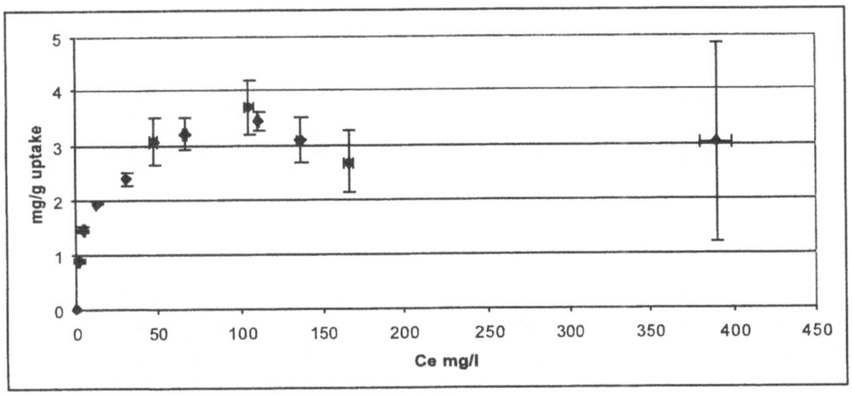

Figure 3-13. Batch Equilibrium for Zinc onto Ammonia Zeolite. Average of 3 Runs with Error Bars.

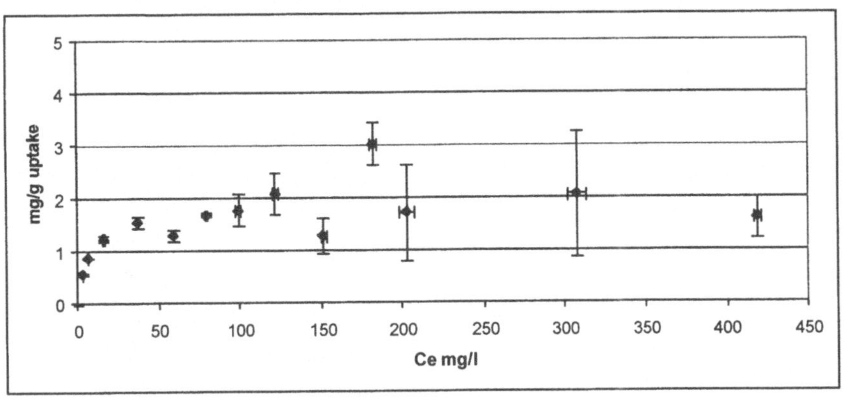

Figure 3-14. Batch Equilibrium for Zinc onto Agrofiber. Average of 3 Runs with Error Bars.

Table 3-14 lists the approximate maximum zinc uptake capacity for the given conditions for each of the media.

Table 3-14. Approximate Uptake Capacity of Media for Zinc

Media	Maximum Uptake of Zinc (mg/g)
TP 207*	90
Compost	15
Kudzu	8.5
Peanut Hulls	3.5
St. Cloud Zeolite	3.5
Ammonia Zeolite	3.5
Activated Carbon	3.0
Agrofiber	2.0
Iron-Oxide Coated Sand	2.0
Peat-Sand Mix	1.25
Cotton Waste	< 1
Sand	< 1

*Metals concentrations necessary to achieve saturation on the resin were not reached in the current study but a study by Parrish (1998) using an identical procedure found a maximum uptake of 90 mg/g.

Standard equilibrium curves were observed for zinc uptake onto all media except the cotton mill waste. There was significant leaching of zinc from the cotton mill waste at higher equilibrium concentrations A slight decrease in metals uptake at higher equilibrium concentrations was observed for iron oxide coated sand and plain sand: one possible explanation for this shift is the accompanying increasing acidity of the solution with increasing metals concentration. As metal concentration increased, the initial pH of the solutions mixed with the media decreased. The decreased pH may have inhibited uptake, or in the case of the cotton mill waste, caused metals to leach from the media. The commercial resin was outstanding in metals capture however there was an accompanying dramatic increase in pH of the solution that was cause for concern.

3.4.2.2 Low Metal Concentration Uptake Tests

Understanding the ultimate capacity of the media is important. However, for applications to stormwater filters, it is more important to examine how the media will perform at the low metal concentrations actually found in stormwater.

The behavior of the media with actual runoff was examined in batch equilibrium tests under the conditions listed in Table 3-8. The two things that were examined in these low metal concentration uptake tests were whether metals or major ions were being leached and how well each media removed metals at low concentrations.

For the low metal concentration equilibrium tests, actual runoff collected from the parking lot in the rear of H.M Comer hall on the University of Alabama at Tuscaloosa campus was used. Three sets of runoff, spiked to different levels, were run in triplicate. An additional set consisting of blanks with distilled, deionized water was also run. Table 3-15 lists the metal concentrations, pH, total suspended solids (TSS) concentration, and total dissolved solids (TDS) concentration of the runoff used for each set.

Table 3-15. Characteristics of Spiked Runoff used for Low Metal Concentration Uptake Tests – Metal concentrations in micrograms per liter and solids concentrations in milligrams per liter.

Metal		Set 1	Set 2	Set 3
Cu (µg/L)	Dissolved	74	202	398
	Total	89	271	511
Cr (µg/L)	Dissolved	29	173	346
	Total	43	244	461
Pb (µg/L)	Dissolved	28	165	350
	Total	45	228	461
Zn (µg/L)	Dissolved	65	228	426
	Total	71	241	452
Cd (µg/L)	Dissolved	74	234	461
	Total	70	211	415
Fe (µg/L)	Dissolved	31	192	368
	Total	99	534	526
pH		7.01	6.8	6.64
TSS (mg/L)		3.7	6	5.7
TDS (mg/L)		62	67	74

Figures 3-15 through 3-20 show the percent removal of metals for the media and Table 3-16 summarizes the results by listing the media from best to worst for each metal in each set.

In Figures 3-15 through 3-20 and Table 3-16 the names of some of the media are abbreviated as follows: Iron-Oxide Coated Sand = IOCS; Peat-Sand Mix = Peat; TP 207 Commercial Resin = Resin; Cotton Mill Waste = Cotton; St.Cloud Zeolite = S. Zeolite; Ammonia Zeolite = A. Zeolite; Peanut Hulls = Peanuts. The (-) symbol in Table 3-16 indicates percent removal was negative i.e. metals were leached from the media instead of being removed from solution.

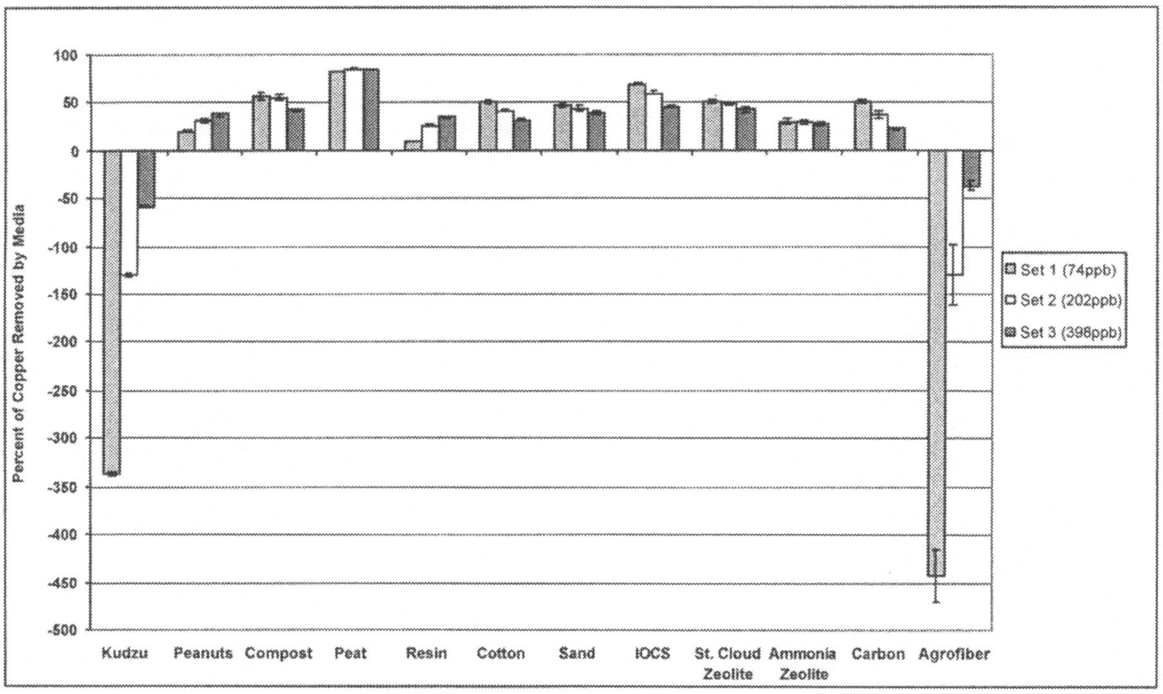

Figure 3-15. Percent of Dissolved Copper Removed from Runoff

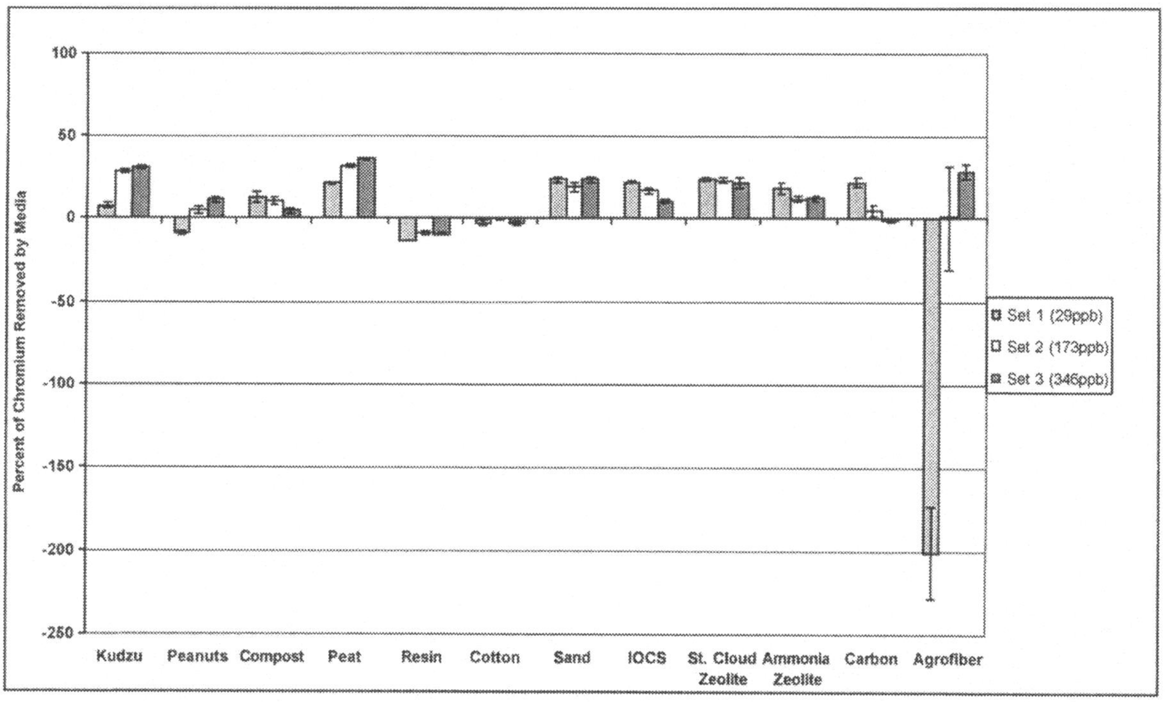

Figure 3-16. Percent of Dissolved Chromium Removed from Runoff

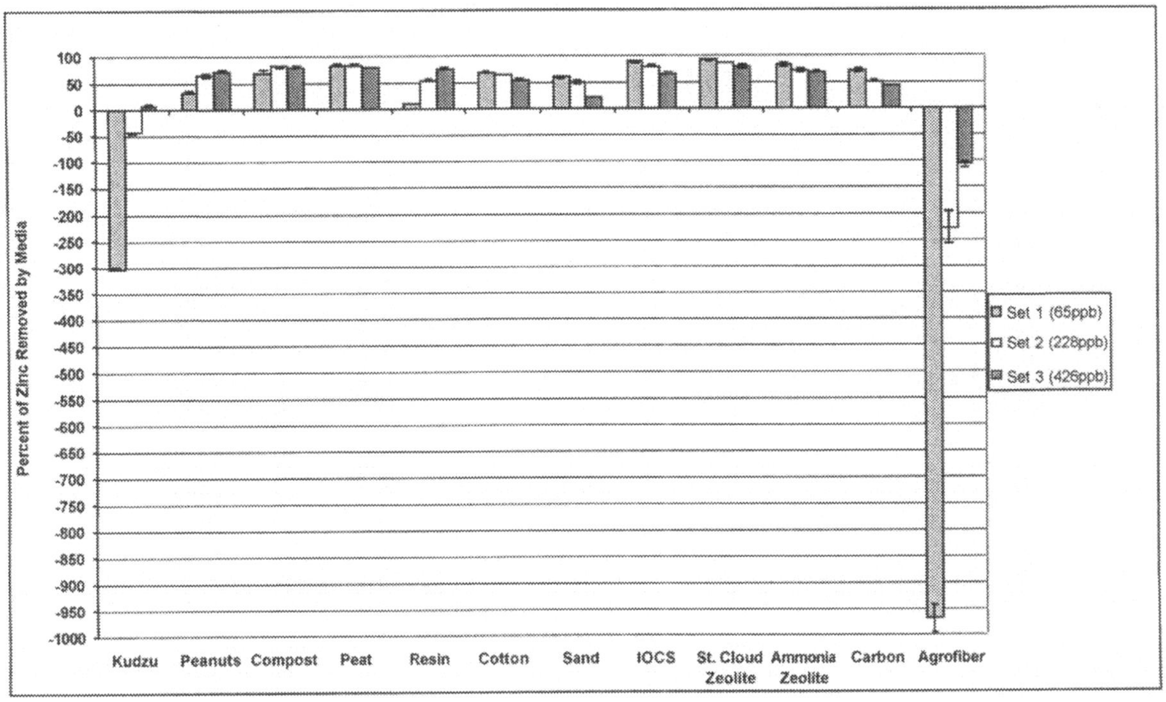

Figure 3-17. Percent of Dissolved Zinc Removed from Runoff

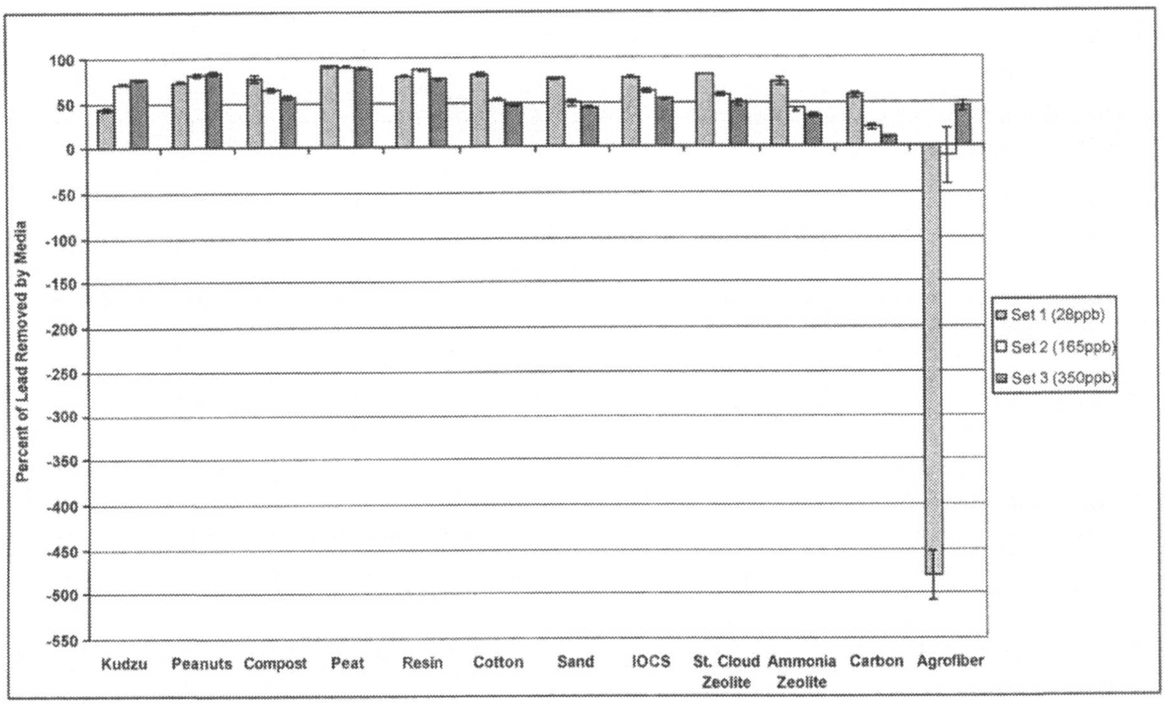

Figure 3-18. Percent of Dissolved Lead Removed from Runoff

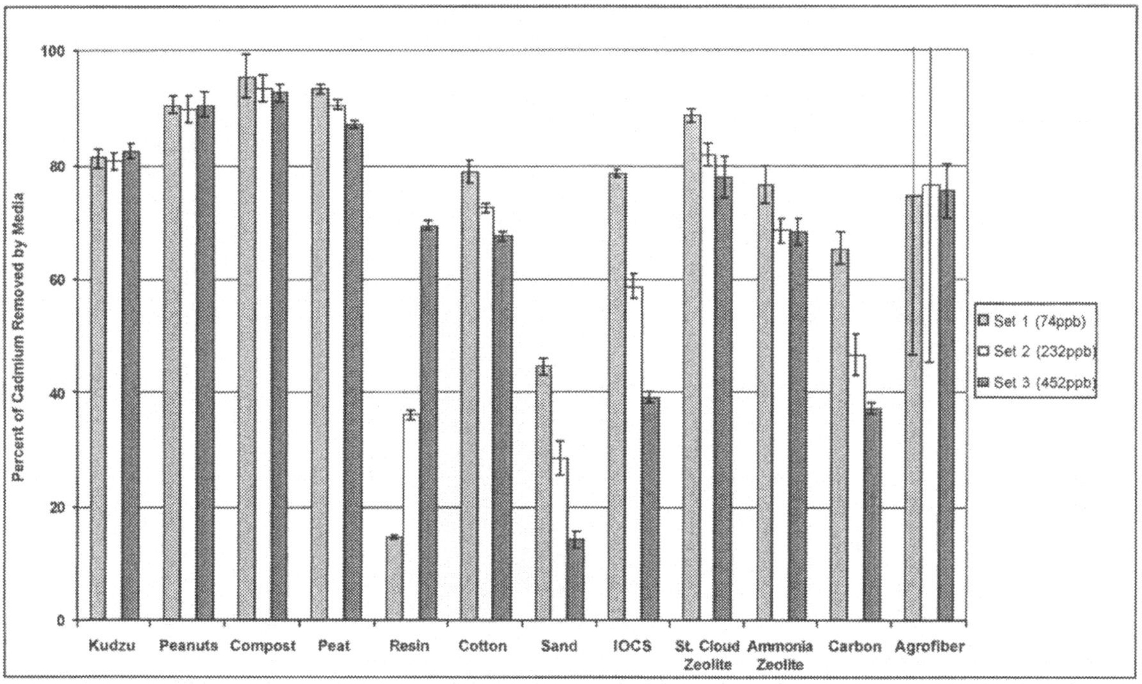

Figure 3-19. Percent of Dissolved Cadmium Removed from Runoff

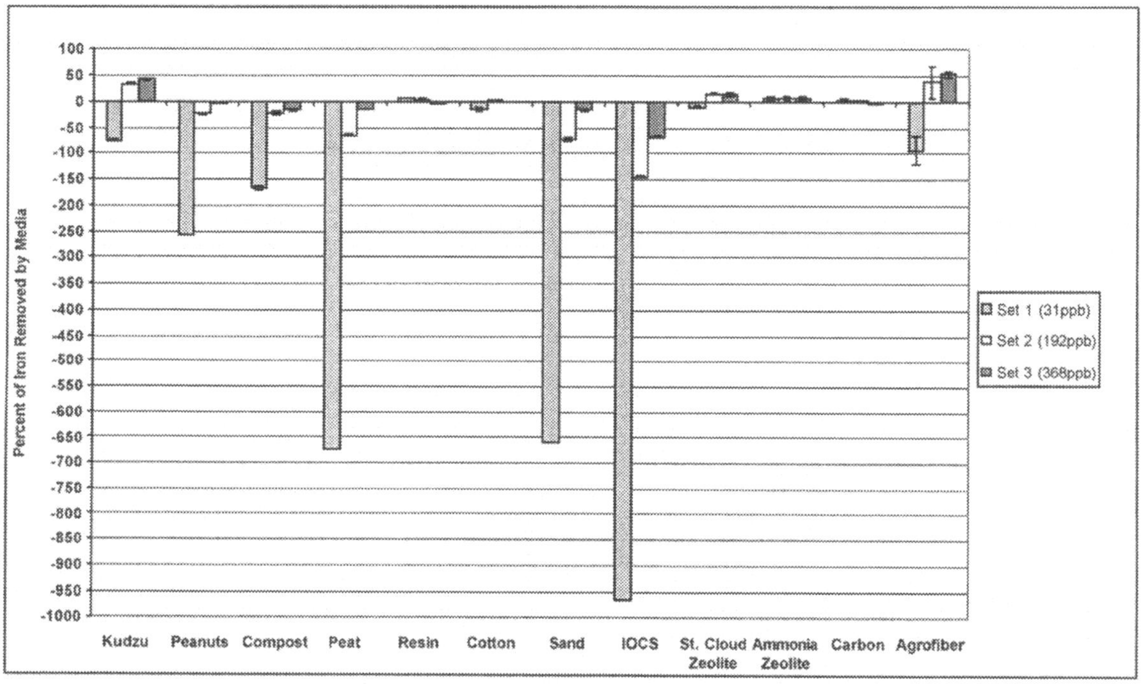

Figure 3-20. Percent of Dissolved Iron Removed from Runoff

Table 3-16. Summary of Results from the Low Metal Concentration Uptake Tests - Media from Best to Worst (based on percent removal) for Each Metal in Each Set.

Metal	Cu			Cr			Pb		
Set	Set 1	Set 2	Set 3	Set 1	Set 2	Set 3	Set 1	Set 2	Set 3
	Media (% removal)	Media (% removal)	Media (% removal)	Media (% removal)	Media (% removal)	Media (% removal)	Media (% removal)	Media (% removal)	Media (% removal)
Best	Peat (82.3%)	Peat (84.3%)	Peat (83.9%)	S. Zeolite (24.0%)	Peat (32.2%)	Peat (36.3%)	Peat (90.2%)	Peat (89.6%)	Peat (87.9%)
	Iron (69.0%)	Iron (59.5%)	Iron (45.1%)	Sand (23.5%)	Kudzu (28.6%)	Kudzu (30.8%)	Cotton (80.7%)	Resin (86.3%)	Peanuts (82.9%)
	Compost (56.3%)	Compost (55.4%)	Compost (41.8%)	Carbon (22.1%)	S. Zeolite (22.9%)	Agrofiber (28.4%)	S. Zeolite (80.7%)	Peanuts (80.9%)	Kudzu (75.9%)
	S. Zeolite (50.7%)	S. Zeolite (47.8%)	S. Zeolite (41.6%)	Iron (22.0%)	Sand (19.2%)	Sand (23.8%)	Resin (78.9%)	Kudzu (70.5%)	Resin (75.1%)
	Carbon (50.5%)	Sand (43.3%)	Sand (38.5%)	Peat (21.4%)	Iron (16.8%)	S. Zeolite (21.8%)	Iron (77.2%)	Compost (64.0%)	Compost (55.2%)
	Cotton (49.8%)	Cotton (41.2%)	Peanuts (37.5%)	A. Zeolite (18.0%)	A. Zeolite (11.9%)	A. Zeolite (12.0%)	Compost (75.6%)	Iron (61.4%)	Iron (52.4%)
	Sand (46.3%)	Carbon (36.6%)	Resin (35.0%)	Compost (12.8%)	Compost (10.4%)	Peanuts (11.1%)	Sand (75.1%)	S. Zeolite (57.6%)	S. Zeolite (47.8%)
	A. Zeolite (29.9%)	Peanuts (31.3%)	Cotton (31.4%)	Kudzu (7.3%)	Peanuts (4.9%)	Iron (10.0%)	Peanuts (73.0%)	Cotton (51.7%)	Cotton (46.6%)
	Peanuts (20.0%)	A. Zeolite (28.8%)	A. Zeolite (27.3%)	Cotton (-)	Carbon (4.3%)	Compost (4.4%)	A. Zeolite (71.7%)	Sand (47.9%)	Agrofiber (44.3%)
	Resin (9.9%)	Resin (26.0%)	Carbon (22.9%)	Peanuts (-)	Agrofiber (0.9%)	Carbon (-)	Carbon (56.8%)	A. Zeolite (41.4%)	Sand (43.9%)
	Kudzu (-)	Kudzu (-)	Agrofiber (-)	Resin (-)	Cotton (-)	Cotton (-)	Kudzu (42.5%)	Carbon (20.5%)	A. Zeolite (34.3%)
Worst	Agrofiber (-)	Agrofiber (-)	Kudzu (-)	Agrofiber (-)	Resin (-)	Resin (-)	Agrofiber (-)	Agrofiber (-)	Carbon (10.4%)
Metal	Zn			Cd			Fe		
Set	Set 1	Set 2	Set 3	Set 1	Set 2	Set 3	Set 1	Set 2	Set 3
	Media (% removal)	Media (% removal)	Media (% removal)	Media (% removal)	Media (% removal)	Media (% removal)	Media (% removal)	Media (% removal)	Media (% removal)
Best	S. Zeolite (90.8%)	S. Zeolite (85.4%)	Compost (78.6%)	Compost (95.6%)	Compost (93.5%)	Compost (92.7%)	A. Zeolite (7.6%)	Agrofiber (38.0%)	Agrofiber (52.8%)
	Iron (87.7%)	Peat (82.1%)	S. Zeolite (78.6%)	Peat (93.3%)	Peat (90.6%)	Peanuts (90.6%)	Resin (6.4%)	Kudzu (33.5%)	Kudzu (41.4%)
	Peat (82.3%)	Compost (80.5%)	Peat (77.2%)	Peanuts (90.6%)	Peanuts (89.8%)	Peat (87.1%)	Carbon (2.8%)	S. Zeolite (14.4%)	S. Zeolite (12.2%)
	A. Zeolite (79.9%)	Iron (79.1%)	Resin (74.8%)	S. Zeolite (88.7%)	S. Zeolite (81.9%)	Kudzu (82.4%)	S. Zeolite (-)	A. Zeolite (7.0%)	A. Zeolite (6.7%)
	Carbon (70.2%)	A. Zeolite (70.8%)	Peanuts (71.8%)	Kudzu (81.4%)	Kudzu (80.8%)	S. Zeolite (77.8%)	Cotton (-)	Resin (4.7%)	Cotton (-)
	Compost (69.1%)	Peanuts (63.8%)	A. Zeolite (69.2%)	Cotton (78.9%)	Agrofiber (76.6%)	Agrofiber (75.5%)	Kudzu (-)	Carbon (1.6%)	Carbon (-)
	Cotton (67.6%)	Cotton (62.9%)	Iron (64.4%)	Iron (78.6%)	Cotton (72.4%)	Resin (69.3%)	Agrofiber (-)	Cotton (1.1%)	Resin (-)
	Sand (58.3%)	Carbon (50.8%)	Cotton (54.4%)	A. Zeolite (76.6%)	A. Zeolite (68.5%)	A. Zeolite (68.3%)	Compost (-)	Compost (-)	Peanuts (-)
	Peanuts (32.0%)	Sand (48.8%)	Carbon (42.9%)	Agrofiber (74.7%)	Iron (58.7%)	Cotton (67.5%)	Peanuts (-)	Peanuts (-)	Peat (-)
	Resin (8.9%)	Resin (54.1%)	Sand (20.7%)	Carbon (65.5%)	Carbon (46.9%)	Iron (39.1%)	Sand (-)	Peat (-)	Compost (-)
	Kudzu (-)	Kudzu (-)	Kudzu (7.8%)	Sand (44.7%)	Resin (36.0%)	Carbon (37.2%)	Peat (-)	Sand (-)	Sand (-)
Worst	Agrofiber (-)	Agrofiber (-)	Agrofiber (-)	Resin (14.7%)	Sand (28.6%)	Sand (14.2%)	Iron (-)	Iron (-)	Iron (-)

The ability of the media to remove metals from spiked runoff varied depending on the metal being examined and its concentration. The peat-sand mix, the St. Cloud zeolite, compost, peanut hulls, and iron-oxide coated sand showed the best overall performance, even though the other media sometimes out-performed them for certain metals at different concentrations. Kudzu and agrofiber caused an increase in the dissolved concentrations of several metals. Kudzu also released approximately 1.5 - 2mg/g of Mg, 3.5 mg/g of Ca and 18 mg/g of K. Other media that released noticeable levels of Na, Mg, Ca, or K included TP 207 resin (7.5 mg/g Na), peanut hulls (5.5 mg/g K) and compost (0.9 mg/g of K).

3.4.2.3 Batch Kinetic Studies

The procedure used was outlined previously in section 3.3.3.2. A lab prepared mixed metal solution (Cu, Cr, Cd, Pb, Zn, and Fe at concentrations of approximately 1.0 mg/l of each metal) was used as the test solution. The most promising media from the low concentration metal tests, and two others, selected for comparison, were examined. The media were then compared by plotting the normalized fraction of metal removed verses time. The normalized fraction of metal removal was calculated as $(C_t-C_f)/(C_o-C_f)$ where C_t is the metal concentration in mg/l at time t, C_o is the initial metal concentration in mg/l and C_f is the final metal concentration in mg/l. The plots show how quickly each metal concentration reached equilibrium. Figures 3-21 to 3-26 compare the normalized fraction of metal removal vs. time between the five media for each of the metals. Table 3-17 shows $(C_t-C_f)/(C_o-C_f)$ at 10, 30, and 120 minutes for each media.

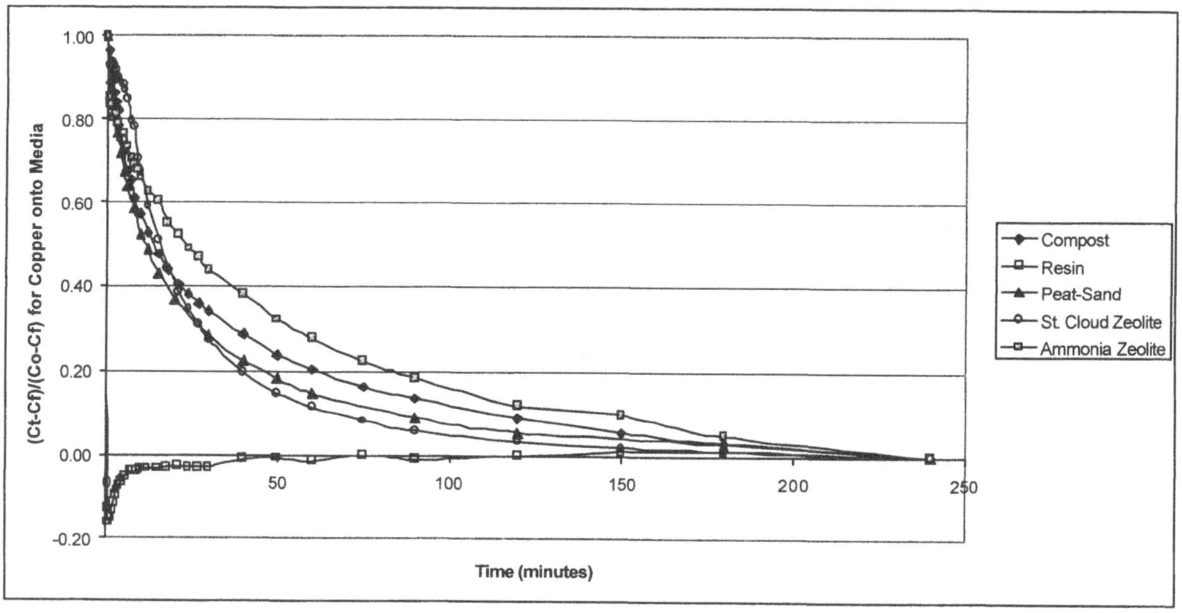

Figure 3-21. Normalized Fraction of Metal Concentration Remaining verses Time for Copper onto Media for a Mixed Metal Solution.

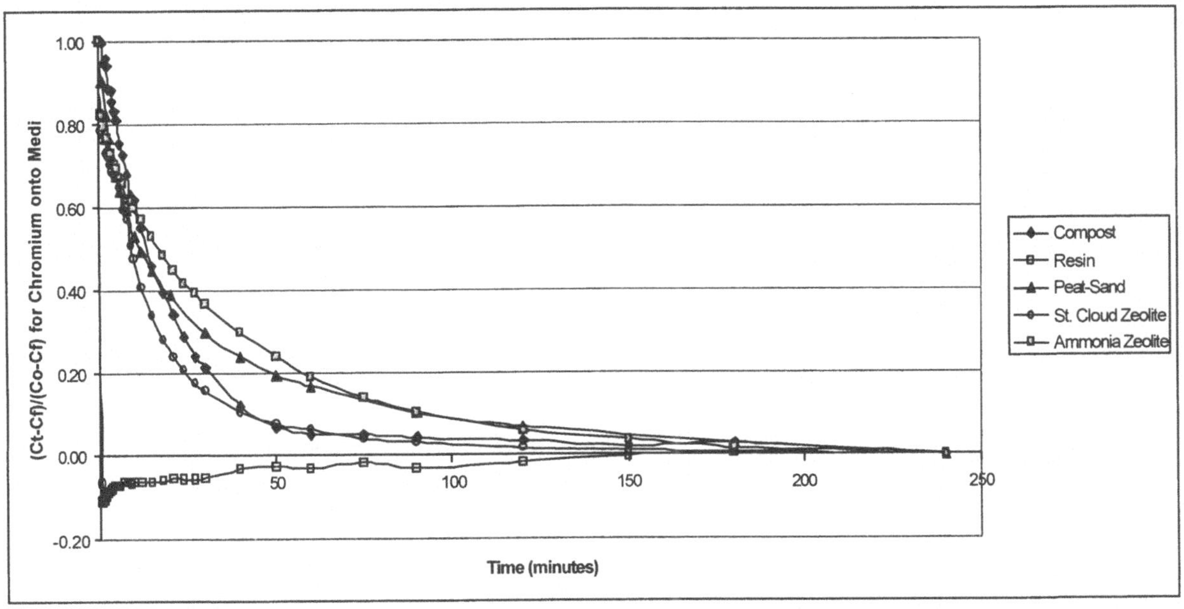

Figure 3-22. Normalized Fraction of Metal Concentration Remaining verses Time for Chromium onto Media for a Mixed Metal Solution.

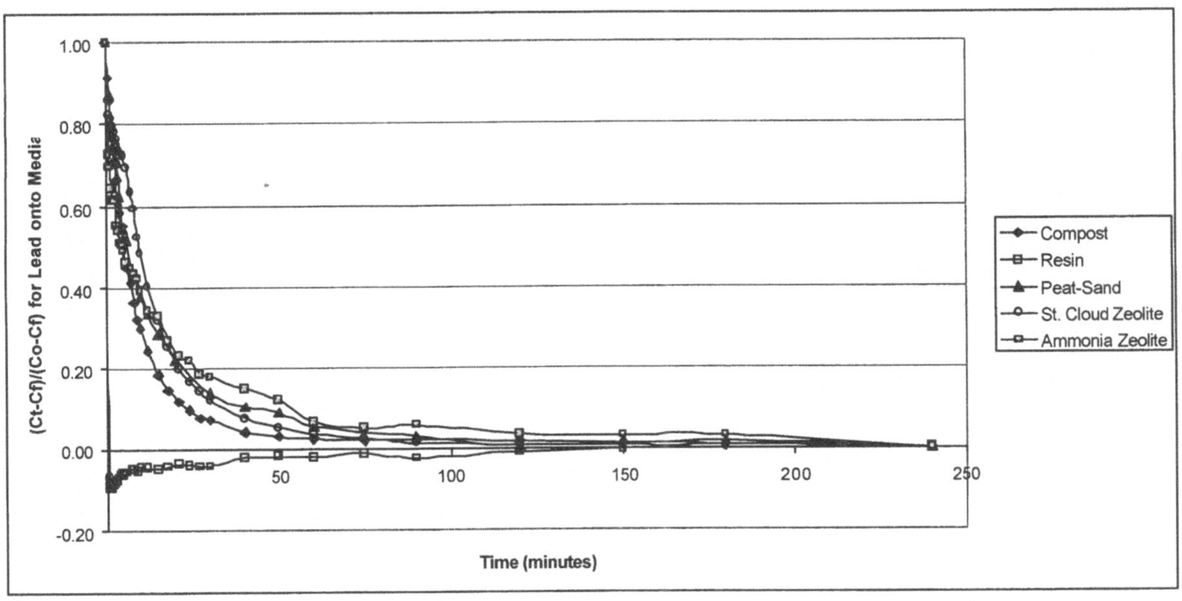

Figure 3-23. Normalized Fraction of Metal Concentration Remaining verses Time for Lead onto Media for a Mixed Metal Solution.

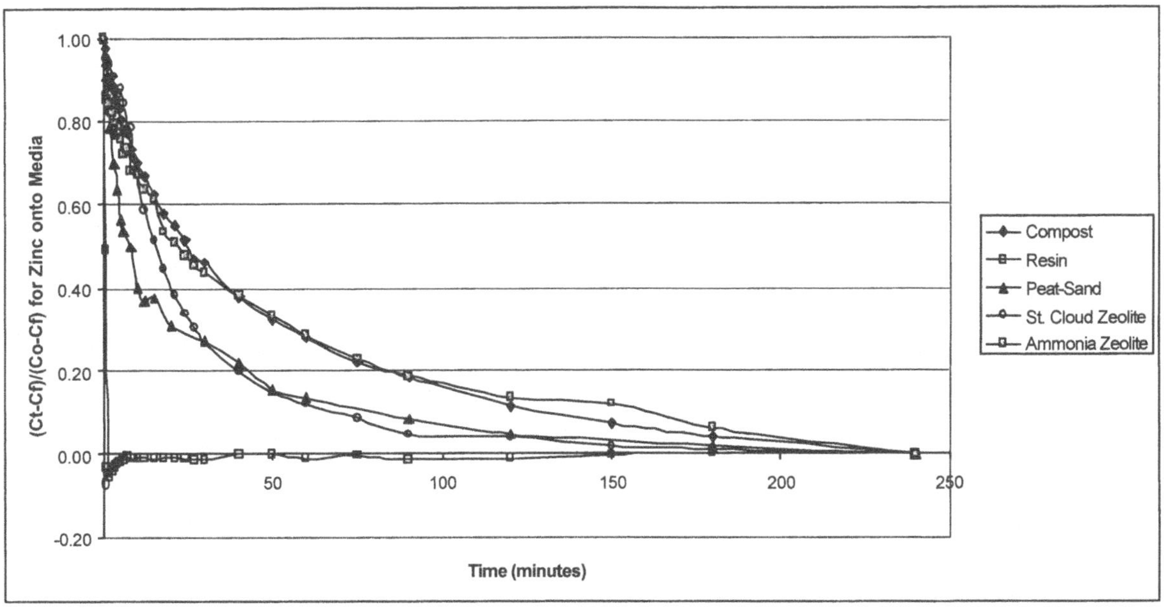

Figure 3-24. Normalized Fraction of Metal Concentration Remaining verses Time for Zinc onto Media for a Mixed Metal Solution.

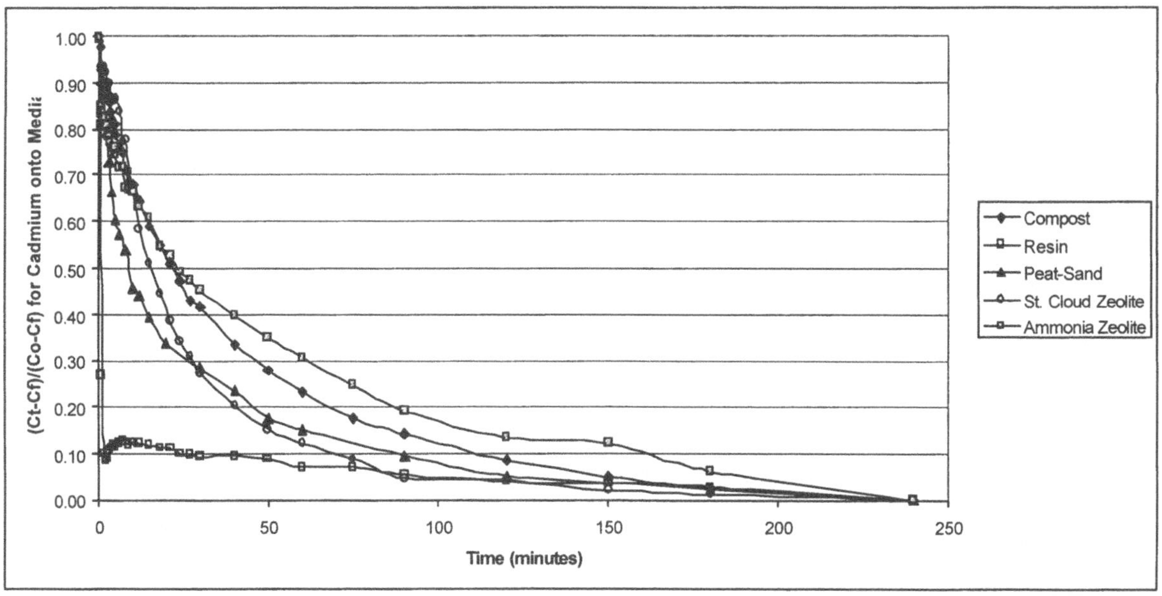

Figure 3-25. Normalized Fraction of Metal Concentration Remaining verses Time for Cadmium onto Media for a Mixed Metal Solution.

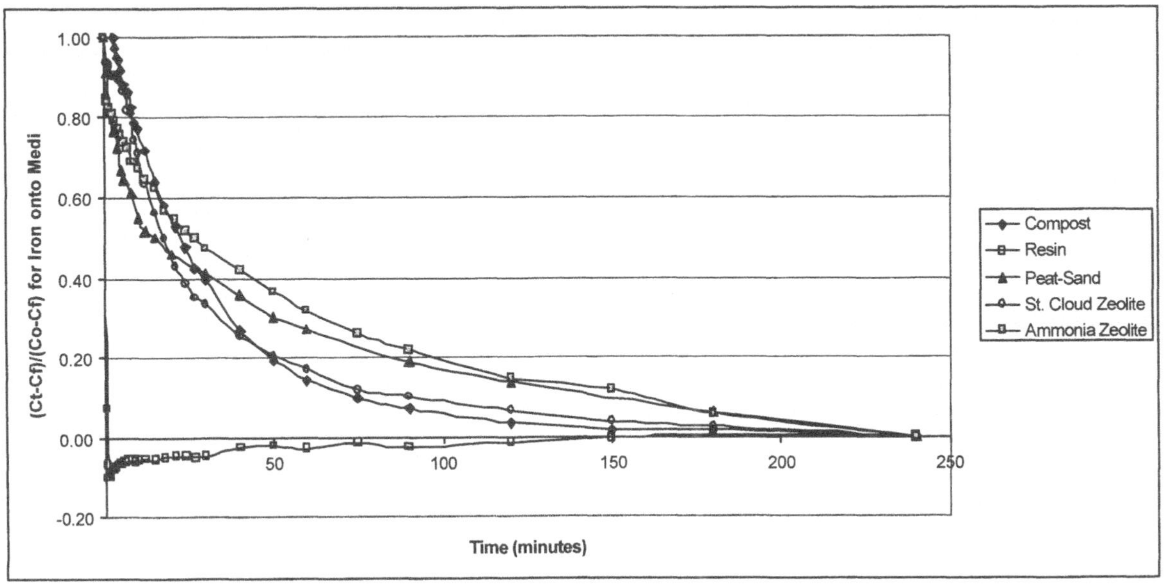

Figure 3-26. Normalized Fraction of Metal Concentration Remaining verses Time for Iron onto Media for a Mixed Metal Solution.

Table 3-17. $(C_t$-$C_f)/(C_o$-$C_f)$ at 10, 30, and 120 minutes for each media.

	Cu $(C_t$-$C_f)/(C_o$-$C_f)$			**Cr** $(C_t$-$C_f)/(C_o$-$C_f)$			**Pb** $(C_t$-$C_f)/(Co$-$C_f)$		
	10 min	**30 min**	**120min**	**10 min**	**30 min**	**120min**	**10 min**	**30 min**	**120min**
Compost	0.57	0.34	0.09	0.62	0.22	0.03	0.30	0.07	0.02
TP 207 Resin	-0.035	-0.030	0.000	-0.064	-0.056	-0.017	-0.046	-0.043	-0.011
Peat-Sand	0.52	0.28	0.06	0.53	0.30	0.07	0.38	0.14	0.01
St. Cloud Zeolite	0.67	0.27	0.04	0.47	0.16	0.02	0.49	0.12	0.01
Ammonia Zeolite	0.66	0.44	0.12	0.60	0.36	0.06	0.39	0.18	0.03
	Zn $(C_t$-$C_f)/(C_o$-$C_f)$			**Cd** $(C_t$-$C_f)/(C_o$-$C_f)$			**Fe** $(C_t$-$C_f)/(C_o$-$C_f)$		
	10 min	**30 min**	**120min**	**10 min**	**30 min**	**120min**	**10 min**	**30 min**	**120min**
Compost	0.70	0.46	0.12	0.68	0.42	0.09	0.77	0.39	0.04
TP 207 Resin	-0.008	-0.014	-0.010	0.123	0.093	0.040	-0.055	-0.048	-0.015
Peat-Sand	0.40	0.27	0.05	0.46	0.28	0.05	0.55	0.41	0.14
St. Cloud Zeolite	0.67	0.27	0.04	0.66	0.27	0.04	0.71	0.33	0.07
Ammonia Zeolite	0.67	0.44	0.13	0.66	0.45	0.13	0.67	0.47	0.15

As far as the kinetics of uptake, the resin far exceeded the other media, with essentially all uptake occurring in the first 2 minutes, while the other media took much longer to reach the same levels. One thing noted with the resin is that there was some re-release of the metals to solution after the initial uptake period. At this time the cause of this re-release is unknown. When compared with the resin the rate of metal uptake of the other media were fairly similar to each other. The rate of uptake of the other media relative to each other varied depending on the metal and the time, but overall peat seemed to have slightly faster uptake than the other media during the first minutes of the test.

The pH was monitored during the kinetic studies– each time a sample was withdrawn the pH was recorded. The initial pH was approximately 4.5 for all of the kinetic studies. Resin had a dramatic impact on the pH, increasing it to about 10.5 (Figure 3-27). In comparison, compost increased the pH to about 6.5, peat-sand decreased the pH to 4.1, and the zeolites increased the pH to about 5.

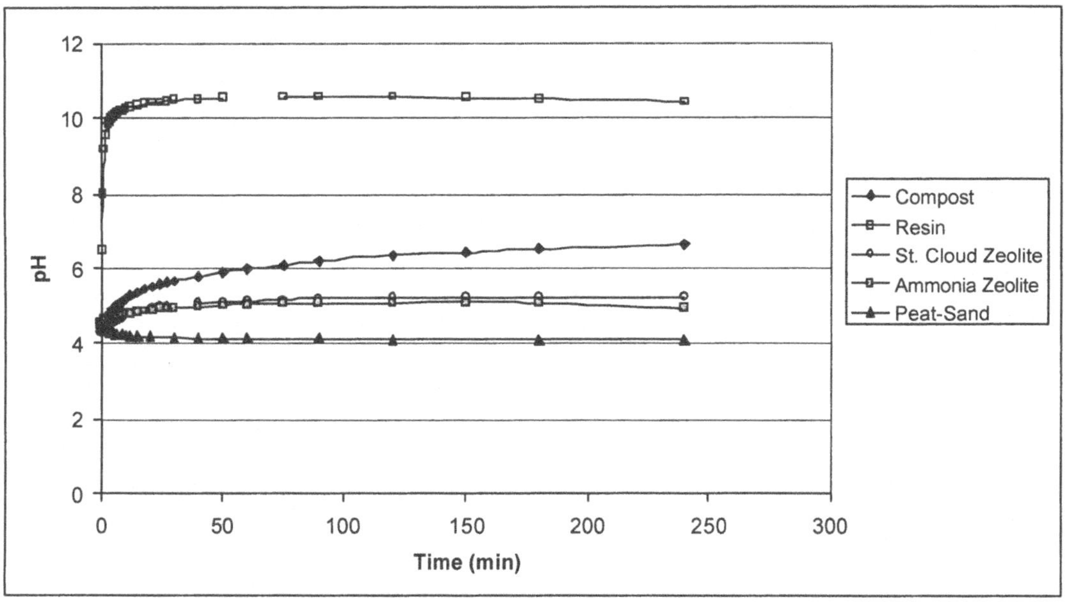

Figure 3-27. pH vs. Time for Various Media during Kinetic Tests

3.4.2.4 Filter Task 2 Discussion

One of the original project objectives was to include a 'renewable source' product or by-product as one of the media. With this in mind, the compost, cotton mill waste, peanut hulls, kudzu and agrofiber were examined. However, kudzu and agrofiber leached metals when in contact with the runoff. Changes in water chemistry due to contact with the media may have caused more of the metals to enter the dissolved phase. The kudzu also released large quantities of K. Although peanut hulls and cotton mill waste did have good uptake for some metals the compost had equivalent or better uptake (except for lead where peanuts had slightly better uptake). The peanuts also released larger amounts of K than compost. Compost was also one of the better performing media overall in the low concentration tests and had the second highest ultimate uptake capacity after the resin. For these reasons, it was decided to continue looking at compost, but drop the agrofiber, kudzu, peanuts, and cotton from further investigation.

The resin was found to increase the dissolved concentration lead. Given the nature of the resin, this was most likely due to changes in the water chemistry causing changes in the speciation of the metals as opposed to leaching from the media. The results of the kinetic tests support this conclusion. Monitoring of pH during the kinetic tests (Figure 3-27) shows the dramatic effect of resin on the pH of the effluent – increasing it from 4.5 to 10. Since the resin did not perform as well as many of the other media in the low concentration equilibrium studies, released large quantities of sodium, and had a detrimental impact on the pH in the kinetic studies, as well as being the most costly of the media, it was also decided to eliminate the resin from further investigation.

Among the seven remaining media, the best performing media were peat-sand mix, St. Cloud zeolite, compost, and iron-oxide coated sand. The iron-oxide coated sand performed well, but was less effective than the peat-sand mix. For these tests, it was decided to focus on the peat-sand mix, St. Cloud zeolite and compost for further examination.

3.4.3 Filter Task Three Results

Tables 3-18 through 3-21 and Figures 3-28 through 3-30 give the results of the media characterization studies on the media selected for the column tests. The surface area data was also performed on the reef carbon, cotton mill waste and agrofiber.

Table 3-18. Moisture content and bulk density of media

Medium	*bulk density (g/cm^3)*	*moisture content*
St Cloud Zeolite	1.91	3%
peat/sand mix	1.99	1%
compost	2.00	3%

Table 3-19. pH of media

Medium	*pH*
St Cloud Zeolite	6.8
peat/sand mix	4.3
compost	3.7

Table 3-20. Media cation exchange capacity

Medium	*CEC (mmol/100g)*
St Cloud Zeolite	7.03
peat/sand mix	1.44
compost	9.38

Table 3-21. Surface area data

Medium	*Specific surface area, macroporous (m2/g)*
Compost	<0.5
Zeolite	16.93
Reef Carbon	853.40
Cotton	<0.5
Agrofiber	<0.5
*Peat	26.6

*The peat-sand mix was not analyzed during the surface area tests conducted in this study. The data quoted for peat (without sand) in the above table is from Allen (1981).

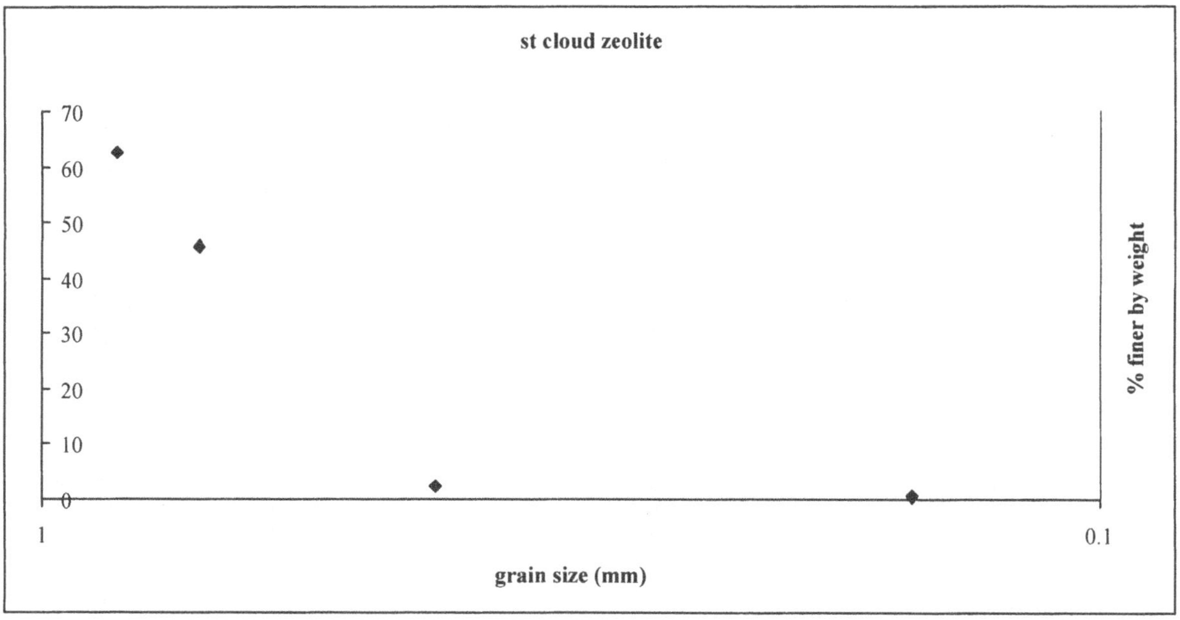

Figure 3-28. Particle size analysis of St. Cloud Zeolite

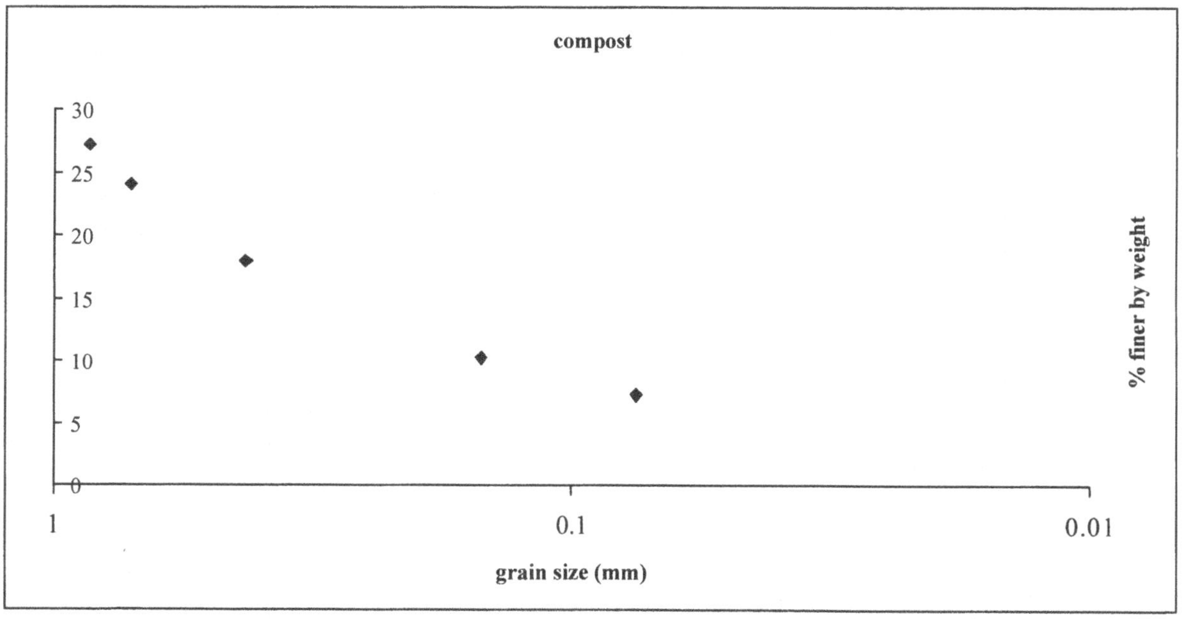

Figure 3-29. Particle size analysis of Compost

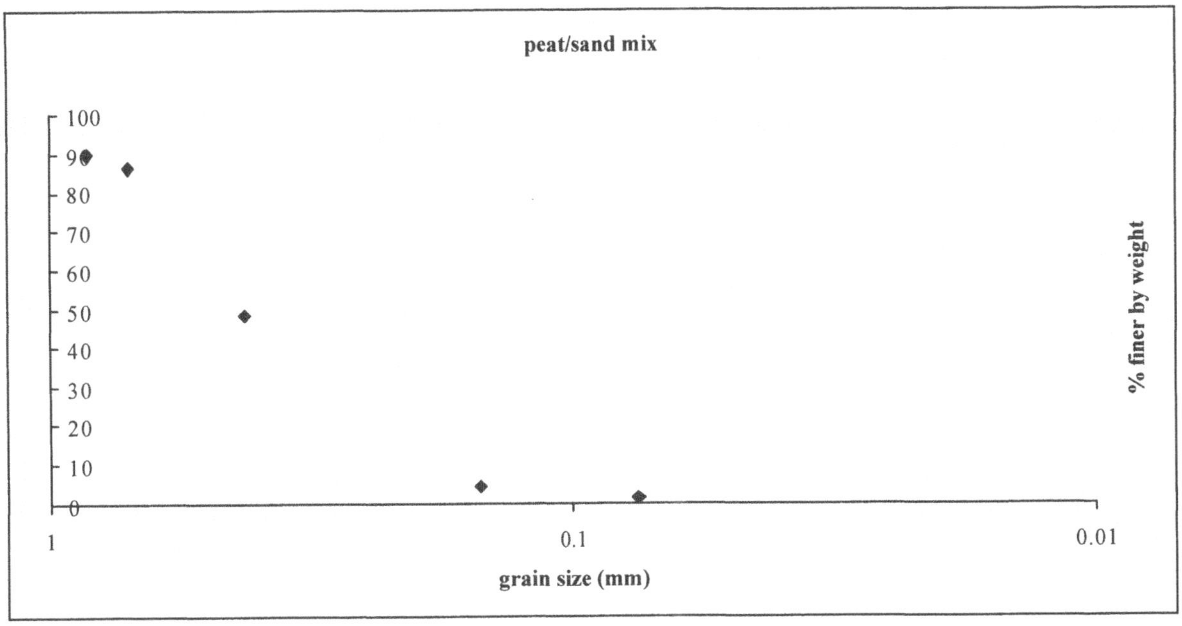

Figure 3-30. Particle size analysis of peat-sand mix

3.4.4 Filter Task Four Results and Discussion

3.4.4.1 Varying Metal Concentration

Figures 3-3, 3-6, and 3-11 in section 3.4.2 show the equilibrium isotherm curves first for the three selected media. There was no obvious correlation between maximum uptake capacity and performance in the low concentration metal uptake tests. An additional batch equilibrium test was performed using zinc at low concentrations for the TP 207 resin, compost and St. Cloud zeolite. These media were selected since they had very different 'ultimate' uptake capacities for zinc.

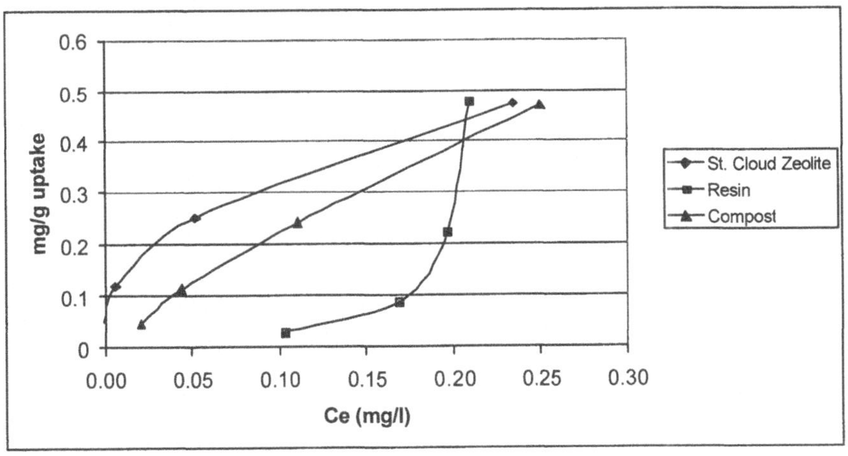

Figure 3-31. Batch Equilibrium for Zinc onto Media at Low Initial Concentrations.

Discussion. The results of the low concentration isotherm test give a very different view of the media than the initial batch equilibrium studies. The results of zinc sorption onto the commercial resin, compost, and the St. Cloud zeolite (mg/g uptake vs. equilibrium zinc concentration) are shown in Figure 3-31. The four points shown in this graph represent initial concentrations of 0.25, 0.6, 1.3, and 2.5 ppm. At the lower concentrations, the zeolite had the best uptake capacity, followed by the compost, and then the resin. At an initial zinc concentration of 0.25 ppm the zeolite reduced the zinc concentration to approximately 0 ppm (below the limits of detection of the ICP), the compost reduced zinc to 0.02 ppm, and the resin reduced zinc to 0.10 ppm. As the concentrations increase, this order begins to reverse. This has important implications for filter design: Determining the range of metal concentrations to be treated is crucial to selecting the best media, since the removal efficiencies of the media relative to each other change as metal concentration increases or decreases.

3.4.4.2 Impact of Major Ions

During equilibrium tests, the ions Ca, Mg, K, and Na were quantified to determine if there was any correlation between metal sorption and ion desorption. Figures 3-32 through 3-34 show zinc sorbed and major ions desorbed from peat-sand, St. Cloud zeolite, and compost. For comparison, a few batch equilibrium isotherm tests were also performed with copper. Figures 3-35 and 3-36 show copper sorbed and ions desorbed for St. Cloud zeolite and compost.

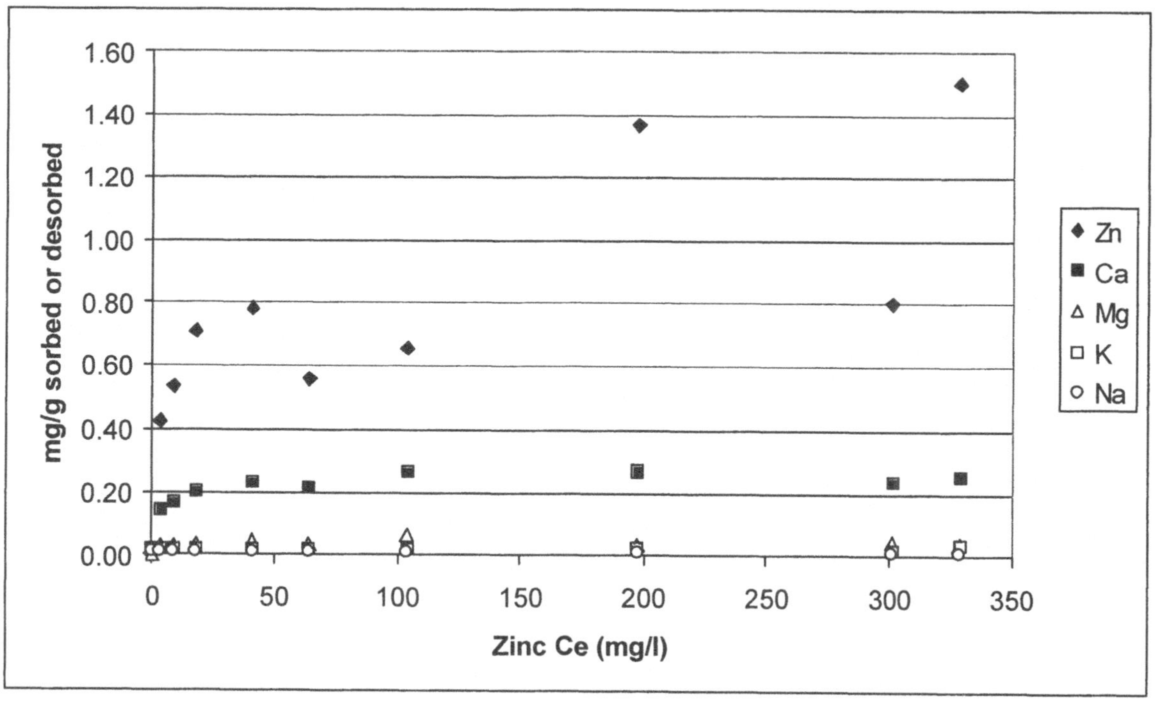

Figure 3-32. Zinc Sorbed and Major Ions Desorbed during Batch Equilibrium Tests for Zinc onto Peat-Sand

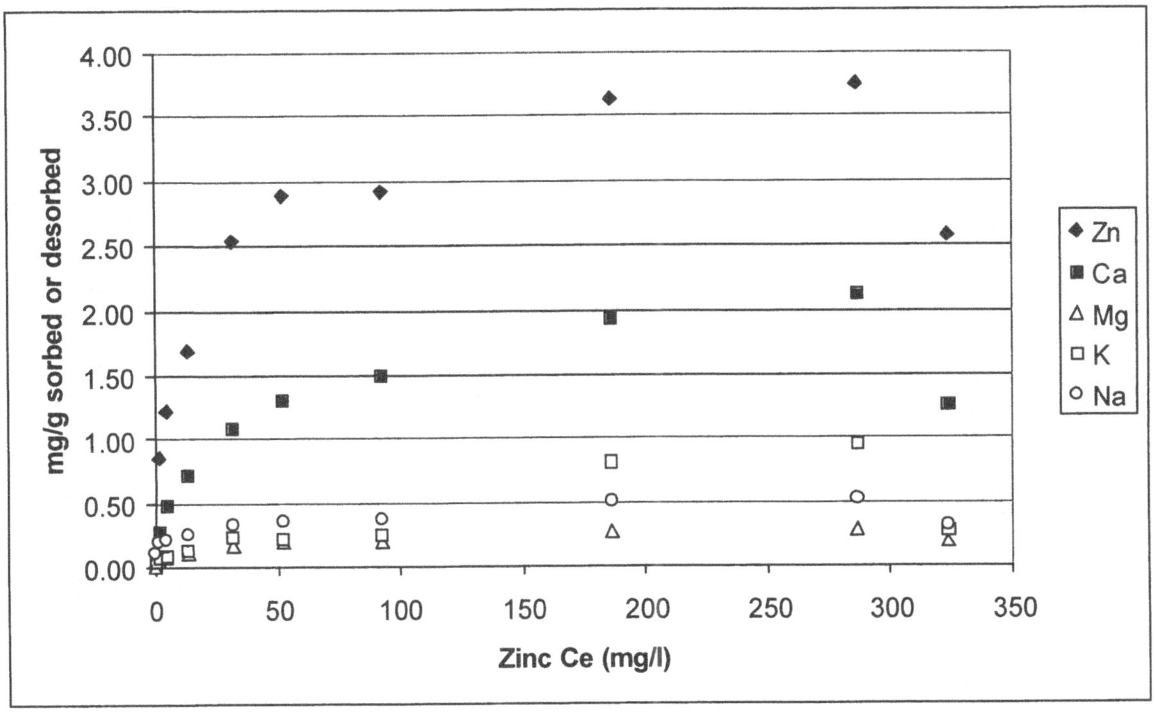

Figure 3-33. Zinc Sorbed and Major Ions Desorbed during Batch Equilibrium Tests for Zinc onto St. Cloud Zeolite

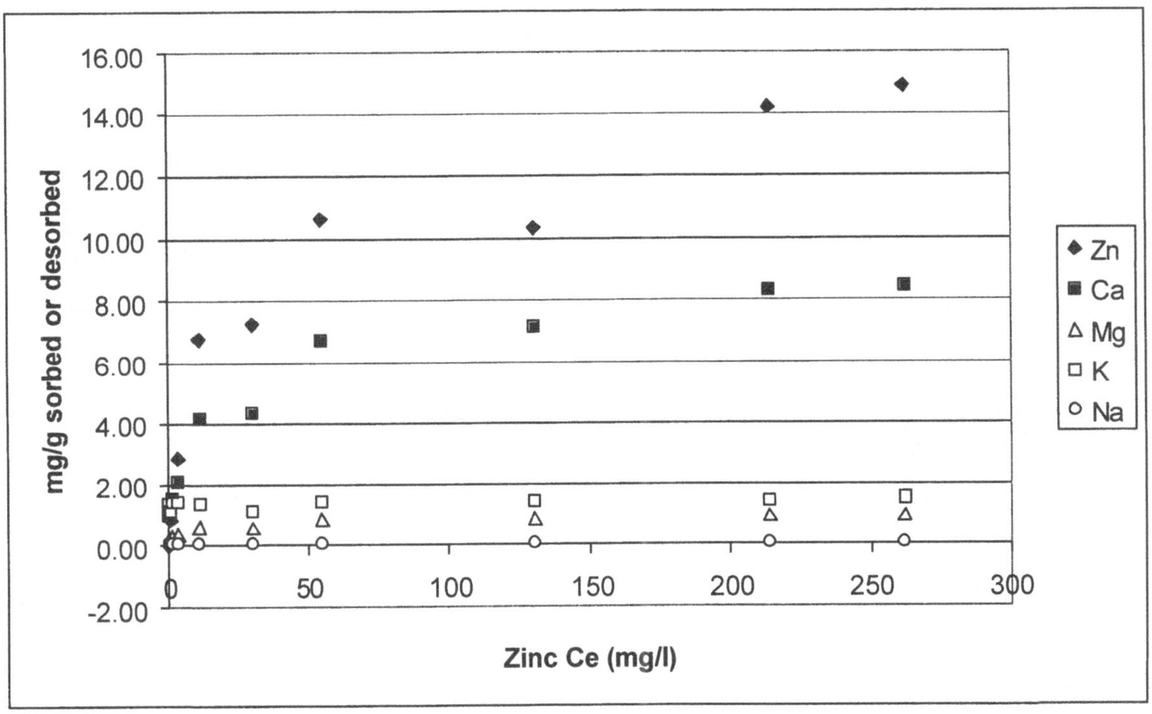

Figure 3-34. Zinc Sorbed and Major Ions Desorbed during Batch Equilibrium Tests for Zinc onto Compost

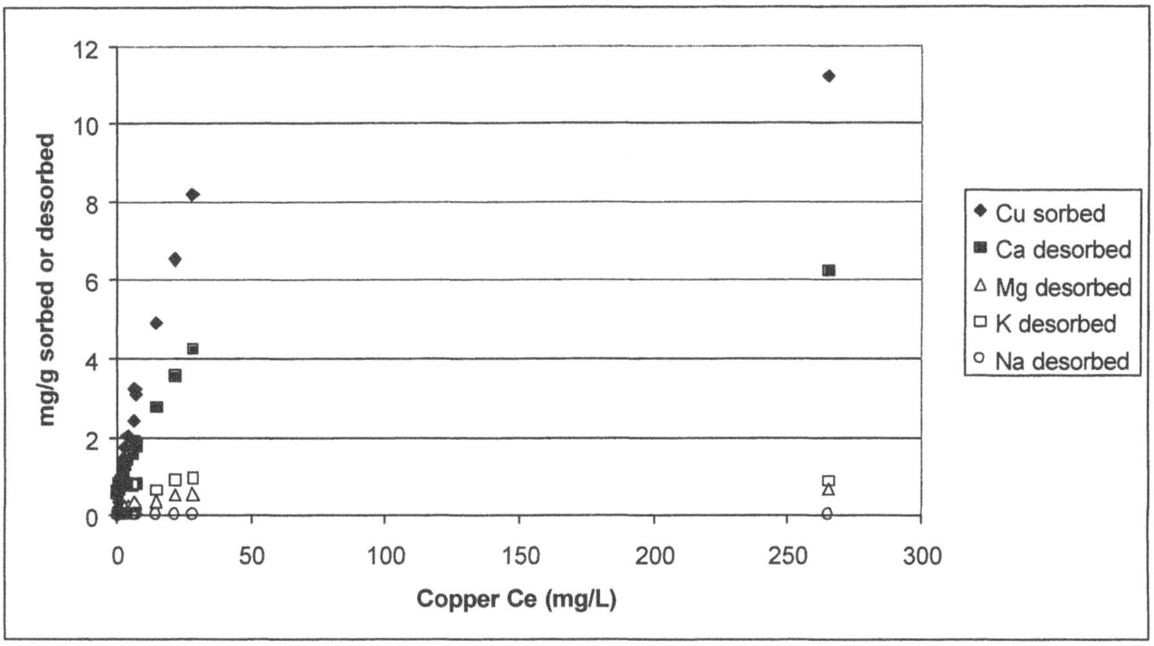

Figure 3-35. Copper Sorbed and Major Ions Desorbed during Batch Equilibrium Tests for Copper onto St. Cloud Zeolite

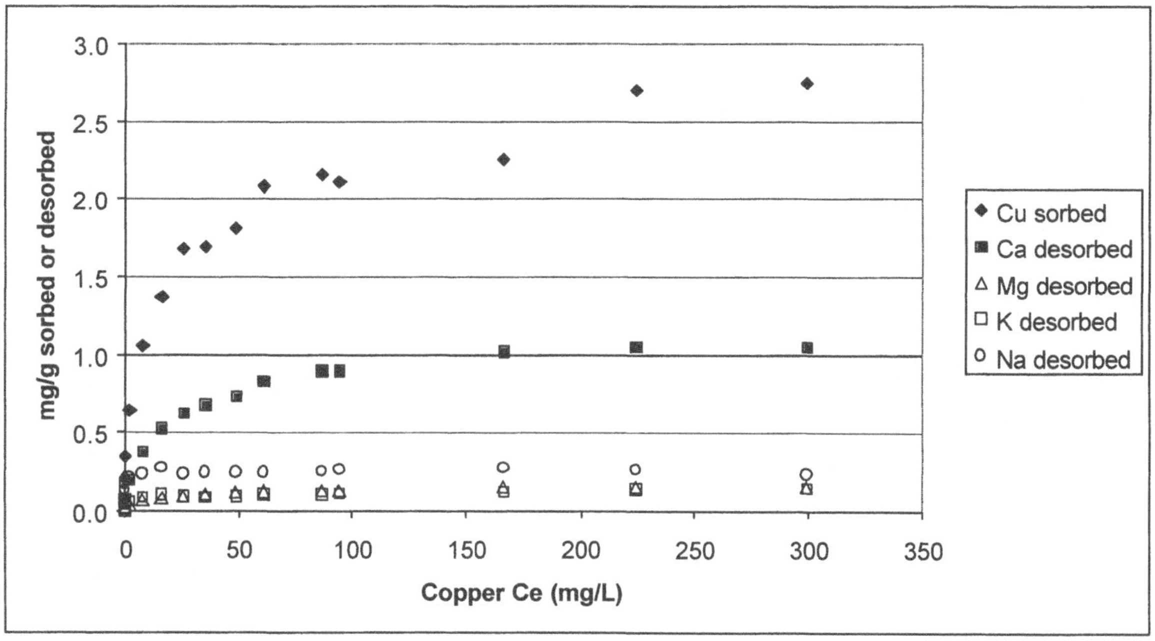

Figure 3-36. Copper Sorbed and Major Ions Desorbed during Batch Equilibrium Tests for Copper onto Compost

Discussion. The amount of Ca that desorbed from peat-sand appeared to increase as the quantity of zinc sorbed increased, indicating the possibility that Ca participates in ion exchange with the metals. The amount of Mg, K, and Na that desorbed were comparatively small and any correlation with zinc sorption was uncertain.

The quantities of all four ions that desorbed from the St. Cloud zeolite appeared to increase as the quantity of the zinc that sorbed increased, indicating the possibility that all these cations participate in ion exchange with zinc. For Cu sorption, the amount of Ca, K, and Mg desorbed increased, as the amount of Cu that sorbed increased. In the lower solution ranges, it looked as if Na desorption may have been increasing with copper sorption, but then Na sorption leveled off while copper sorption continued to increase.

The quantities of Ca, Mg, and K that desorbed (mg/g) from the compost were higher than the amounts desorbed from zeolite. However, less Na (mg/g) was desorbed from the compost than was desorbed from the St. Cloud zeolite. Comparing the major ions that desorbed from the compost to the amount of copper uptake, Ca and Mg desorption (mg/g) increased as zinc and copper sorption (mg/g) increased, an indication that ion exchange was occurring. Na and K were also desorbed from the compost, but the amount of Na and K desorbed appeared to hold roughly constant and did not appear to be related to zinc and copper uptake.

3.4.4.3 Comparison of Competing Metals

The results from the kinetic experiments can be used to examine which metals were removed the fastest and to the greatest degree for each media under the given test conditions. Figures 3-37 through 3-39 show the fraction of the initial metal concentrations (C_t/C_o) remaining in solution verses time for the three final media.

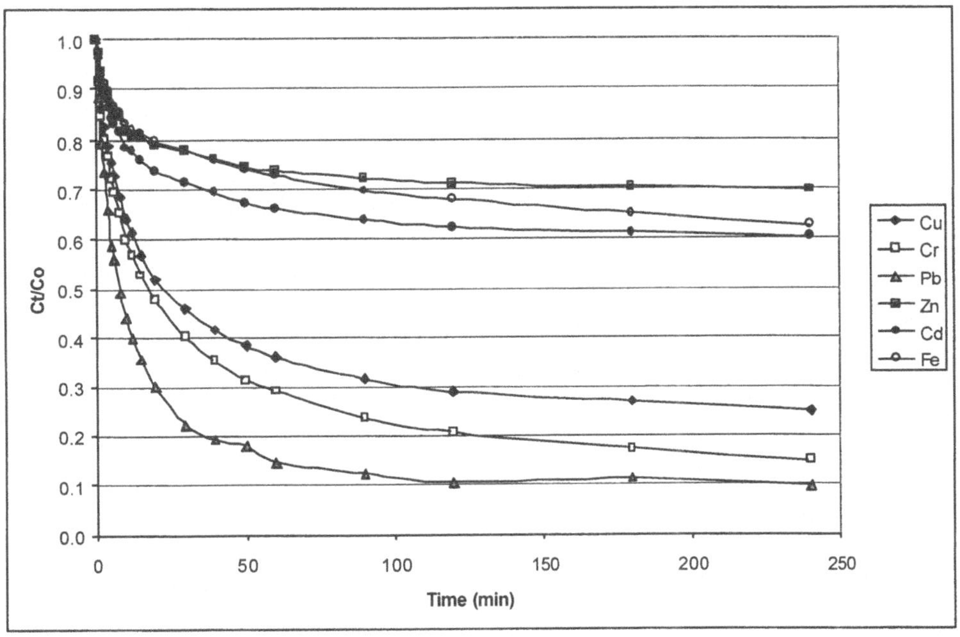

Figure 3-37. Fraction of Initial Metal Concentration Remaining verses Time for Metals onto Peat-Sand for a Mixed Metal Solution.

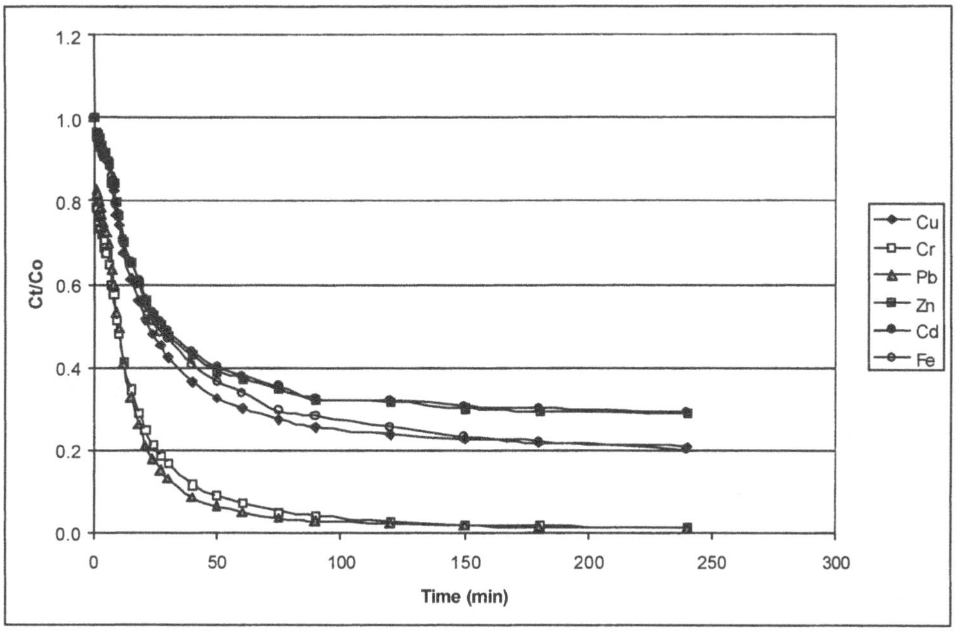

Figure 3-38. Fraction of Initial Metal Concentration Remaining verses Time for Metals onto St. Cloud Zeolite for a Mixed Metal Solution.

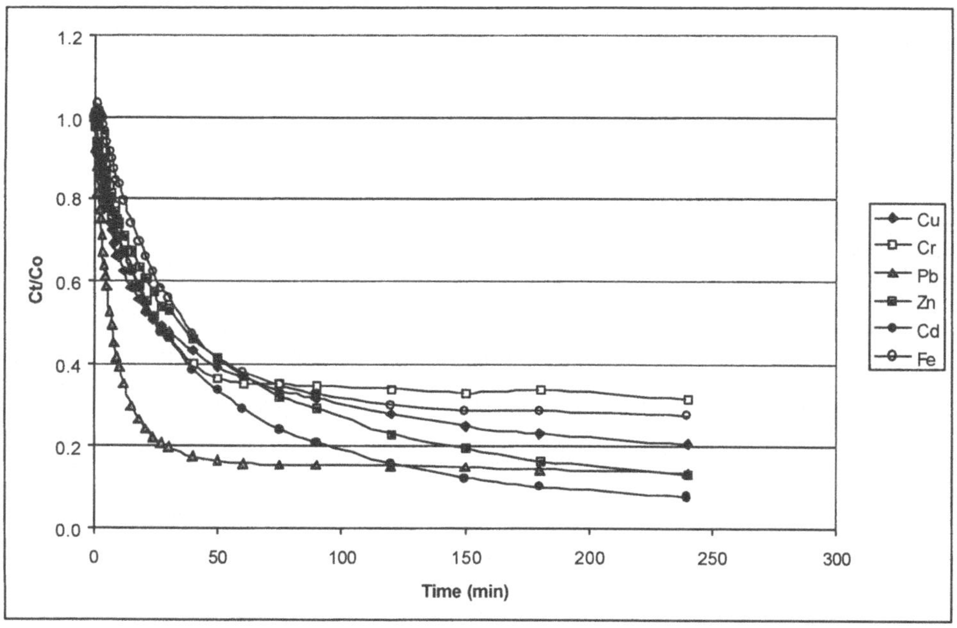

Figure 3-39. Fraction of Initial Metal Concentration Remaining verses Time for Metals onto Compost for a Mixed Metal Solution.

Discussion: Based on the results from the kinetic tests the order of preference for removal was Pb>Cr>Cu>Cd,>Zn, Fe for the peat-sand, Pb,Cr>Cu,Fe>Zn,Cd for the St. Cloud zeolite, and

Cd>Zn>Pb>Cu>Fe>Cr for the compost. The order of preference for the compost varied with time, the final order (at 4 hours) is given. The results from the low concentration metal (LCM) uptake tests (Section 3.4.2.2) can also be analyzed to examine the order of preference for uptake when competing metals are present. The order of preference for removal, based on the LCM uptake tests, was Cd, Pb>Zn,Cu>Cr>Fe for peat-sand, Zn,Cd>Pb>Cu>Cr>Fe for St. Cloud zeolite, and Cd>Zn>Pb>Cu>Cr>Fe for compost. Only compost had a similar order of preference for both tests. The differences between the results from the kinetic and the LCM uptake tests were probably due to the different solutions used for the tests. The kinetic tests used distilled, de-ionized water spiked with metals that had a pH of 4.5, the LCM uptake tests used actual stormwater runoff spiked with metals that had pH values between 6.6 and 7.0. These results emphasize the importance of characterizing media capability under the conditions that are expected in actual applications. As expected, subsequent column tests (Section 3.4.5), which used metal spiked stormwater runoff from the same location as that used in the LCM uptake tests and that had similar pH values, found that the results from the LCM uptake tests were more accurate than the results from the kinetic tests in predicting the order of preference for dissolved metal removal in the columns.

3.4.5 Filter Task Five Results and Discussion

The impact of media and flow rate on dissolved metal removal, total metal (dissolved plus particulate bound) removal, total dissolved and suspended solids removal, pH and color was examined for three media (peat-sand mix, St. Cloud zeolite, and compost) and a control column (empty) for five different residence times: ten minutes, five minutes, three minutes, one minute, and 30 seconds.

The wet bed volumes determined for each media column were 61.3 mL for peat-sand, 96.7 mL for St. Cloud Zeolite, 86 mL for compost, and 100ml for the control. 30 bed volumes were pumped through each column for each run and an additional 5 to 10 bed volumes were pumped through the columns before each run while calibrating the pumps to the correct speed. Therefore, over the course of the three series of runs, approximately 500 to 600 bed volumes were pumped through each column.

3.4.5.1 Dissolved Metal Removal by Columns

Table 3-22 list the dissolved metal concentrations for the influent for all runs by series and approximate hydaulic residence time –Actual residence times were very close to the desired value for all but the slowest flow rate - actual residence time for these runs was sometimes greater than 10 minutes. Actual residence times are shown when plotting data and are given in section 3.4.5.4.

Table 3-22. Influent dissolved metal concentrations (milligrams/liter) for column runs.

Series	Approximate Residence Time (min)	Cu mg/L	Cr mg/L	Pb mg/L	Zn mg/L	Cd mg/L	Fe mg/L
1	10	0.127	0.100	0.087	0.296	0.198	0.248
	5	0.146	0.120	0.122	0.359	0.237	0.323
	3	0.135	0.110	0.101	0.348	0.226	0.268
	1	0.134	0.110	0.099	0.344	0.223	0.266
	0.5	0.137	0.112	0.104	0.361	0.233	0.271
2	10	0.168	0.116	0.086	0.319	0.249	0.173
	5	0.151	0.095	0.062	0.313	0.236	0.143
	3	0.163	0.104	0.070	0.302	0.236	0.155
	1	0.159	0.101	0.064	0.300	0.233	0.150
	0.5	0.159	0.101	0.066	0.297	0.232	0.151
3	10	0.135	0.097	0.056	0.280	0.235	0.137
	5	0.168	0.128	0.086	0.329	0.263	0.174
	3	0.133	0.097	0.054	0.254	0.217	0.135
	1	0.165	0.129	0.088	0.334	0.265	0.174
	0.5	0.133	0.092	0.050	0.274	0.224	0.132

Figures 3-40 through 3-58 show percent dissolved metal removal verses residence time over all three series.

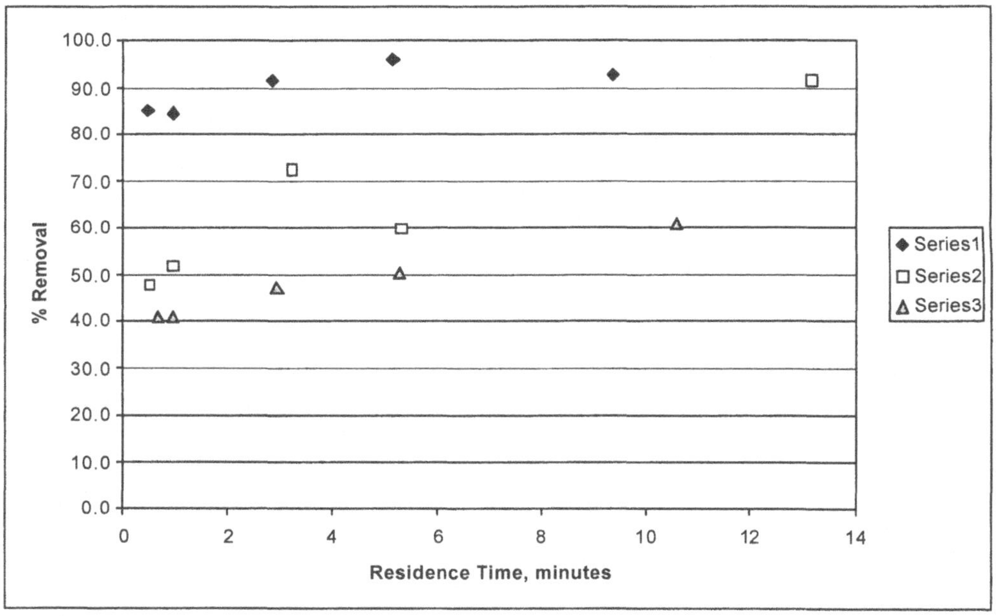

Figure 3-40. Percent Removal of Dissolved Copper by Peat-Sand Filter Column verses Residence Time for 3 Consecutive Series of Column Runs.

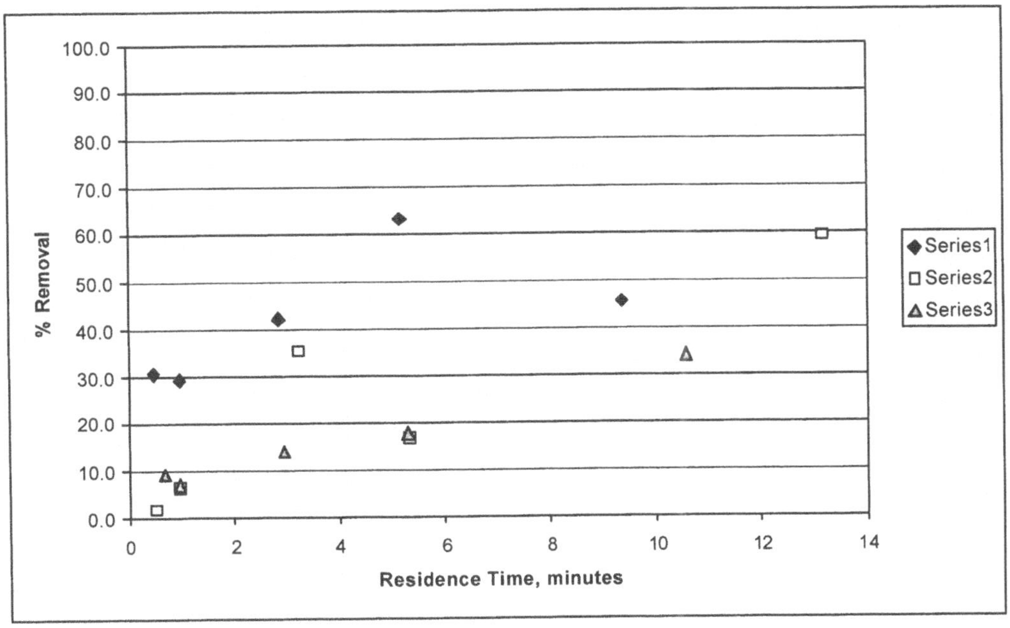

Figure 3-41. Percent Removal of Dissolved Chromium by Peat-Sand Filter Column verses Residence Time for 3 Consecutive Series of Column Runs.

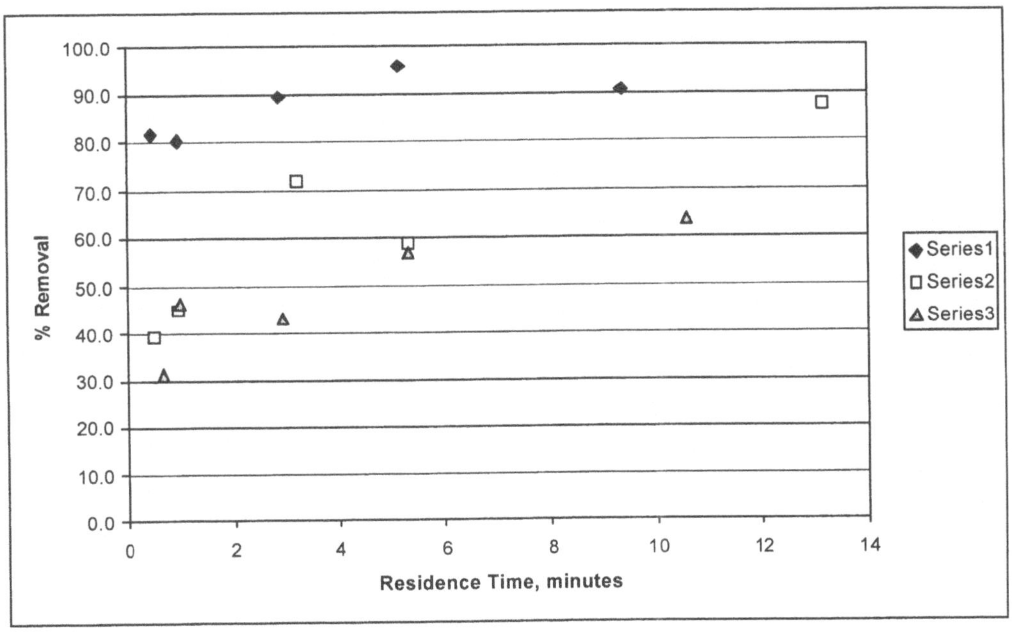

Figure 3-42. Percent Removal of Dissolved Lead by Peat-Sand Filter Column verses Residence Time for 3 Consecutive Series of Column Runs.

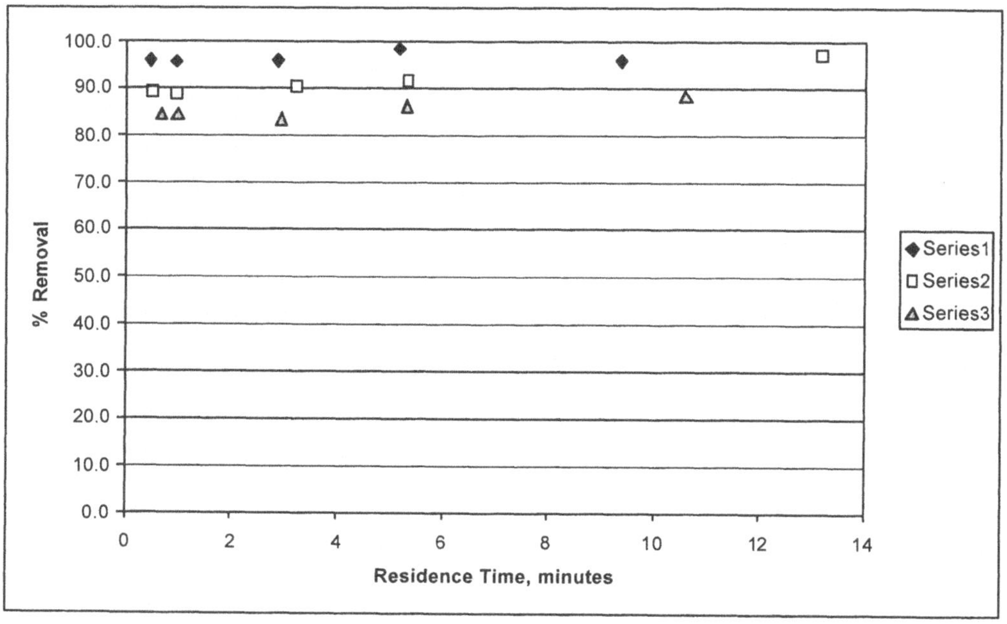

Figure 3-43. Percent Removal of Dissolved Zinc by Peat-Sand Filter Column verses Residence Time for 3 Consecutive Series of Column Runs.

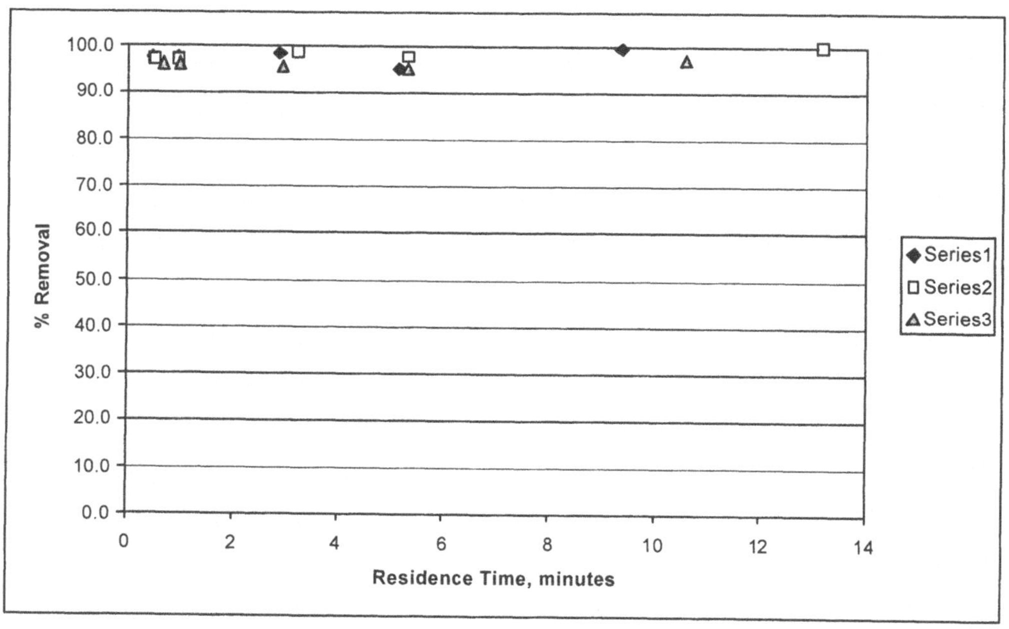

Figure 3-44. Percent Removal of Dissolved Cadmium by Peat-Sand Filter Column verses Residence Time for 3 Consecutive Series of Column Runs.

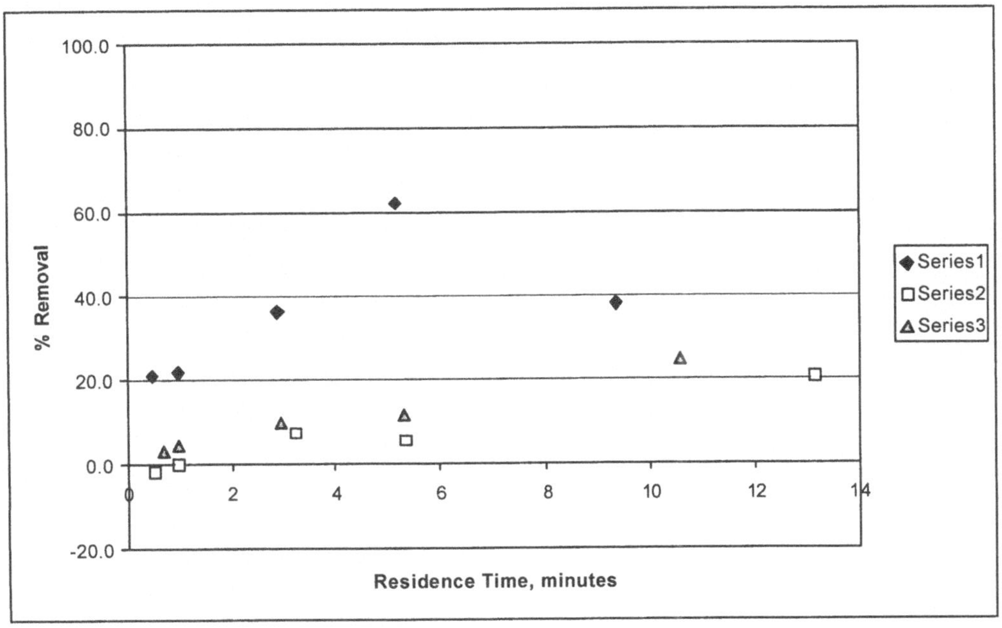

Figure 3-45. Percent Removal of Dissolved Iron by Peat-Sand Filter Column verses Residence Time for 3 Consecutive Series of Column Runs.

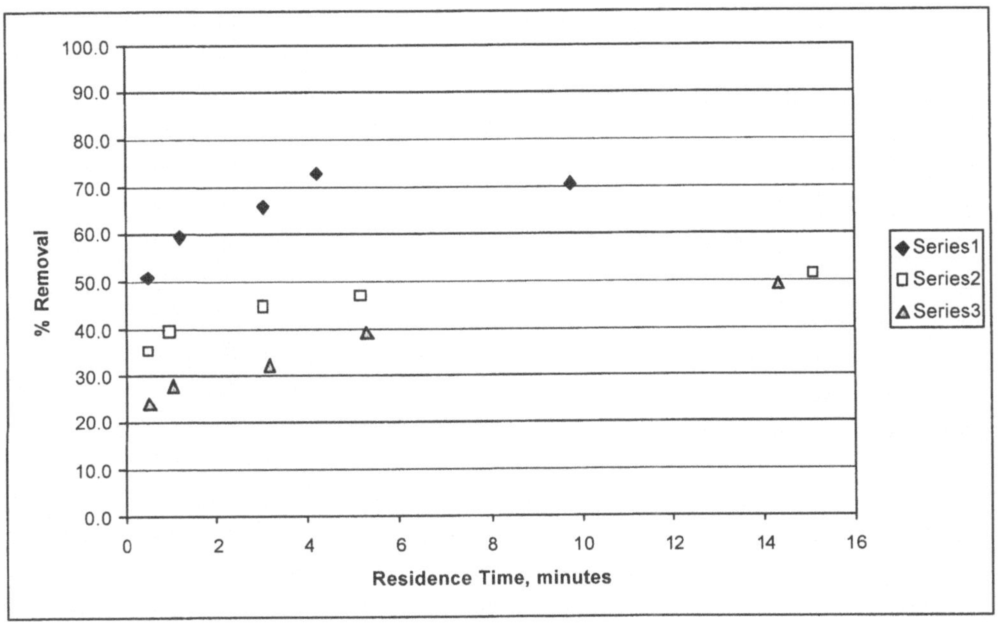

Figure 3-46. Percent Removal of Dissolved Copper by Compost Filter Column verses Residence Time for 3 Consecutive Series of Column Runs.

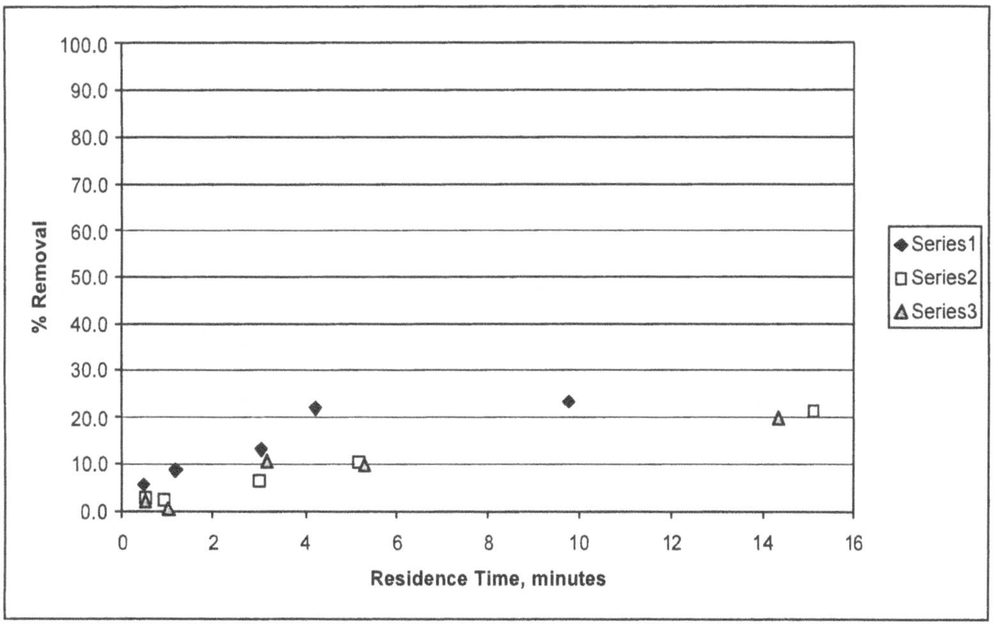

Figure 3-47. Percent Removal of Dissolved Chromium by Compost Filter Column verses Residence Time for 3 Consecutive Series of Column Runs.

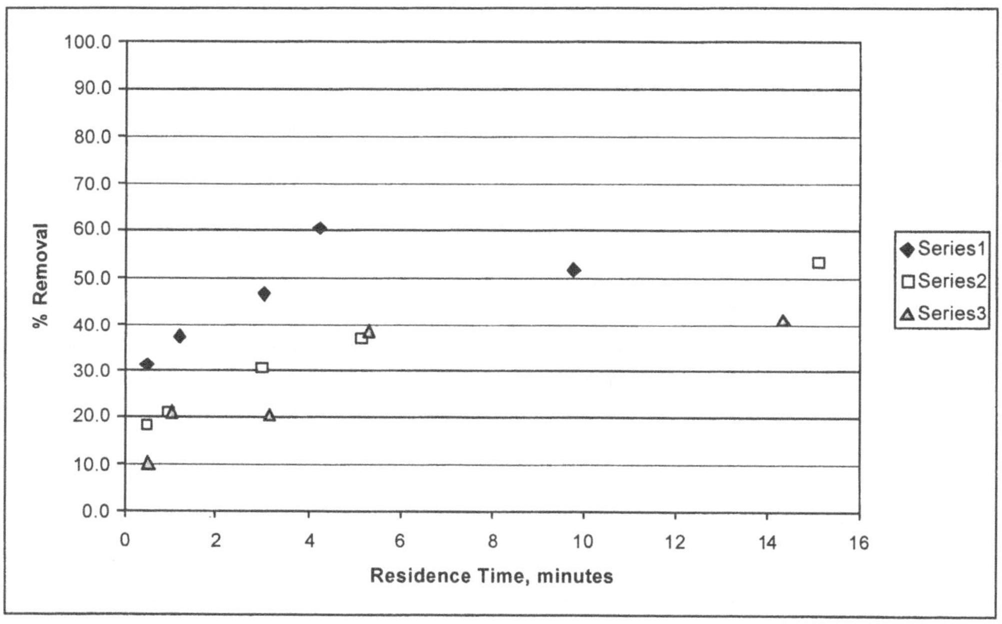

Figure 3-48. Percent Removal of Dissolved Lead by Compost Filter Column verses Residence Time for 3 Consecutive Series of Column Runs.

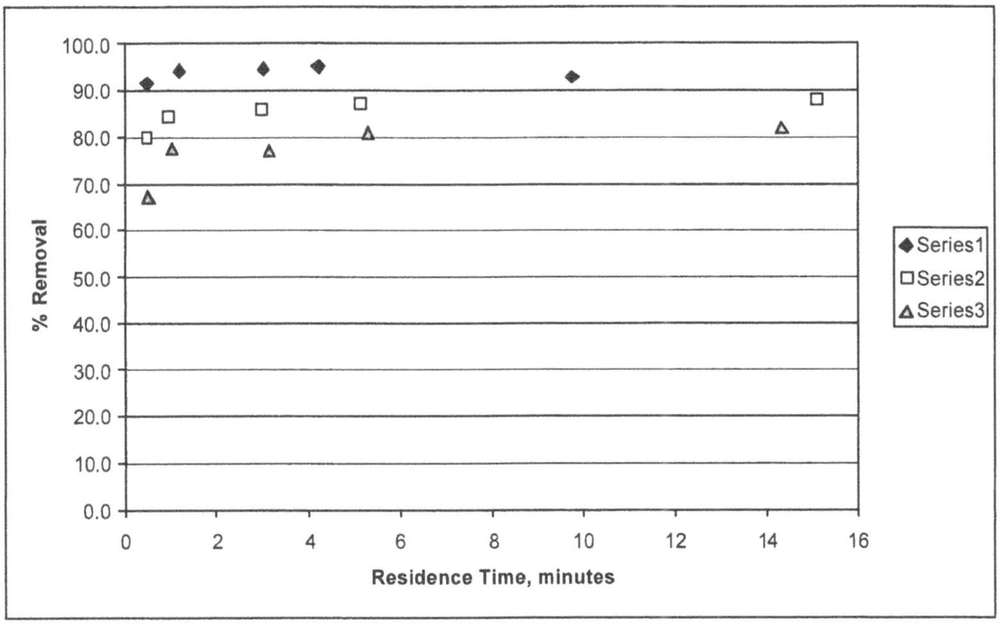

Figure 3-49. Percent Removal of Dissolved Zinc by Compost Filter Column verses Residence Time for 3 Consecutive Series of Column Runs.

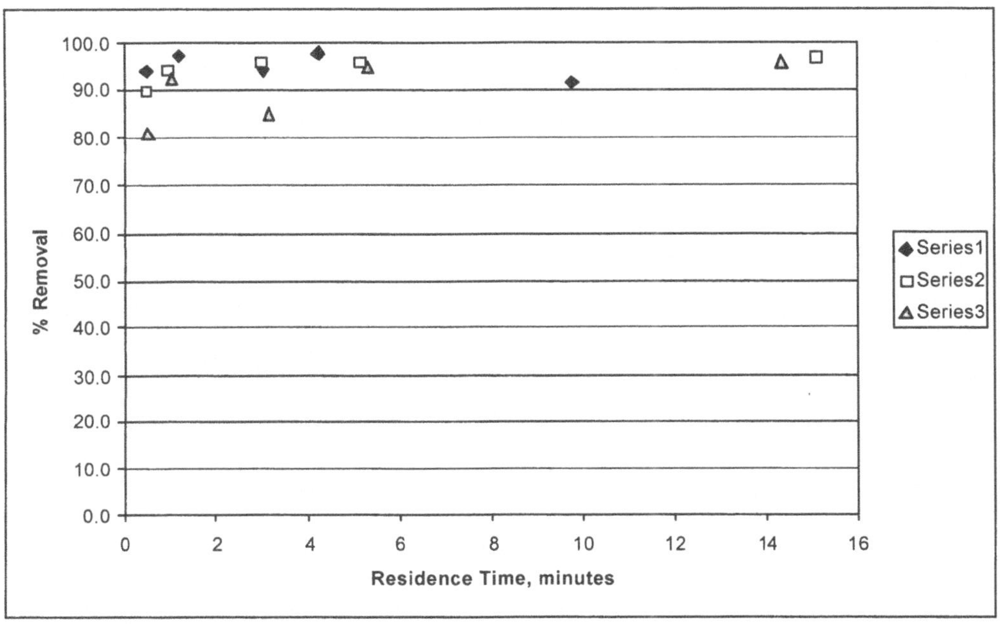

Figure 3-50. Percent Removal of Dissolved Cadmium by Compost Filter Column verses Residence Time for 3 Consecutive Series of Column Runs.

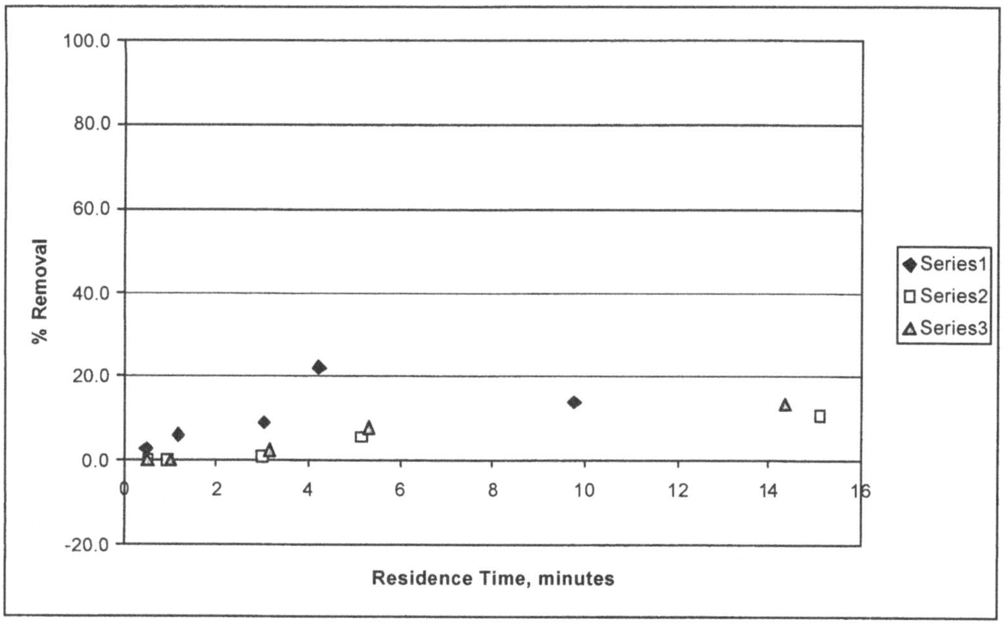

Figure 3-51. Percent Removal of Dissolved Iron by Compost Filter Column verses Residence Time for 3 Consecutive Series of Column Runs.

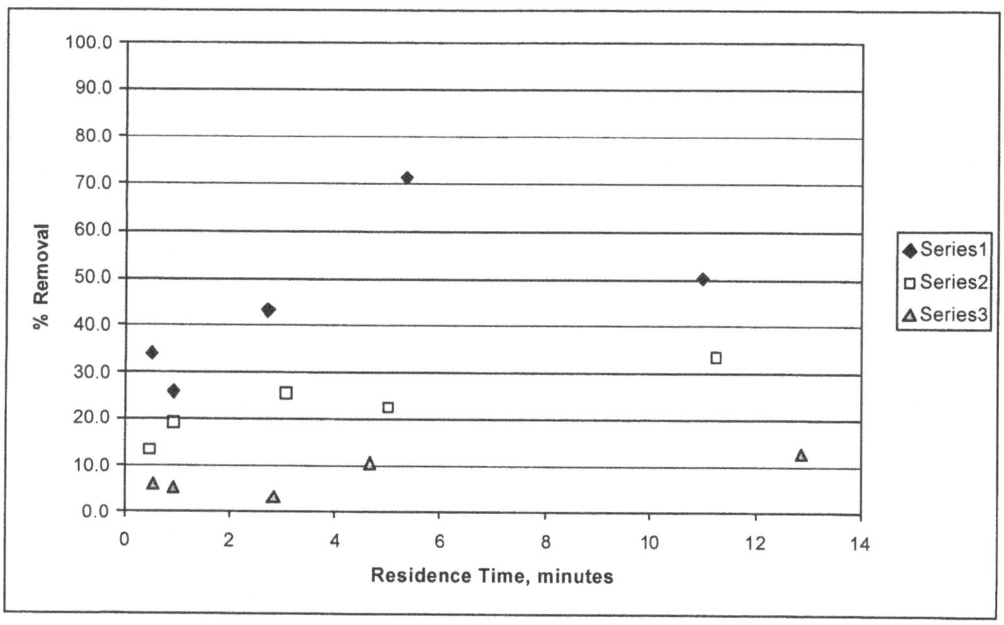

Figure 3-52. Percent Removal of Dissolved Copper by St. Cloud Zeolite Filter Column verses Residence Time for 3 Consecutive Series of Column Runs.

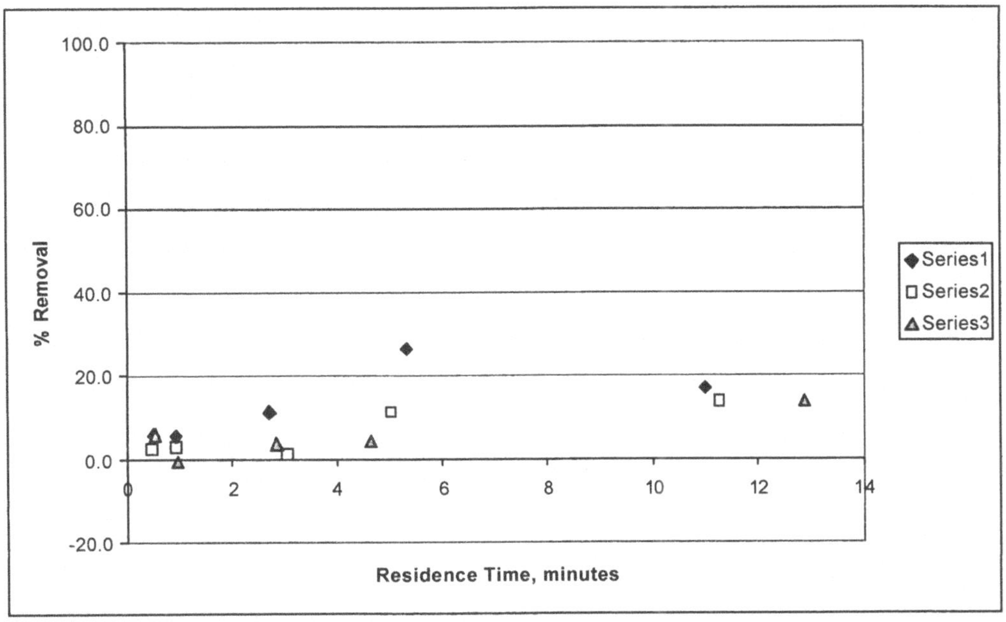

Figure 3-53. Percent Removal of Dissolved Chromium by St. Cloud Zeolite Filter Column verses Residence Time for 3 Consecutive Series of Column Runs.

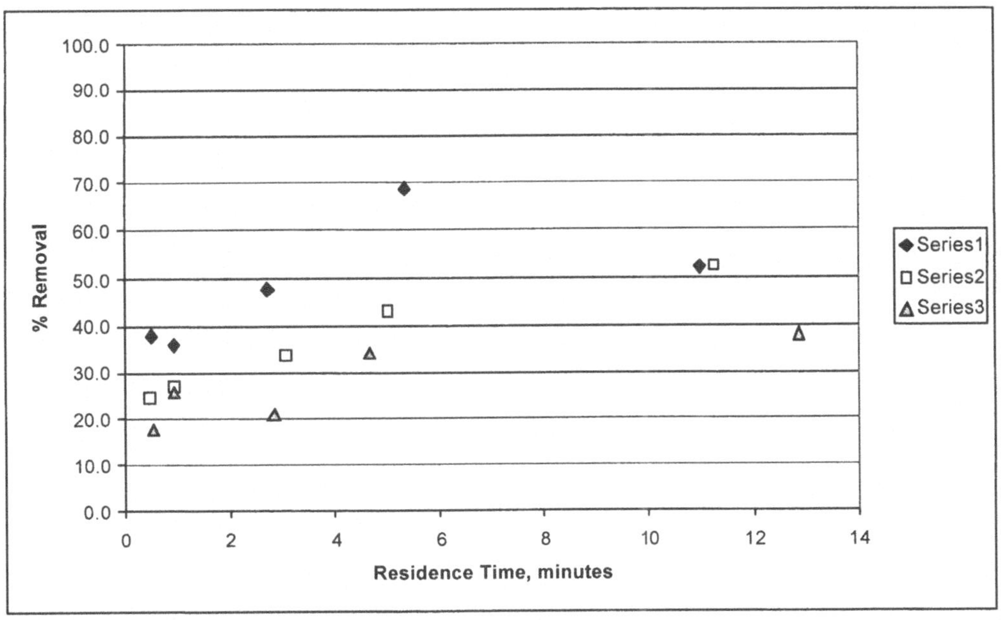

Figure 3-54. Percent Removal of Dissolved Lead by St. Cloud Zeolite Filter Column verses Residence Time for 3 Consecutive Series of Column Runs.

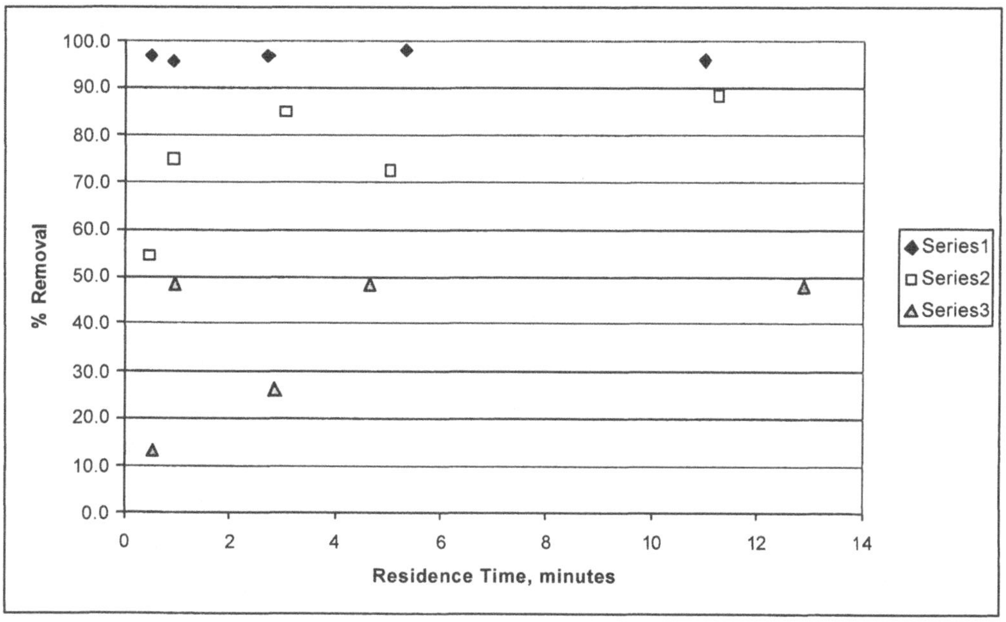

Figure 3-55. Percent Removal of Dissolved Zinc by St. Cloud Zeolite Filter Column verses Residence Time for 3 Consecutive Series of Column Runs.

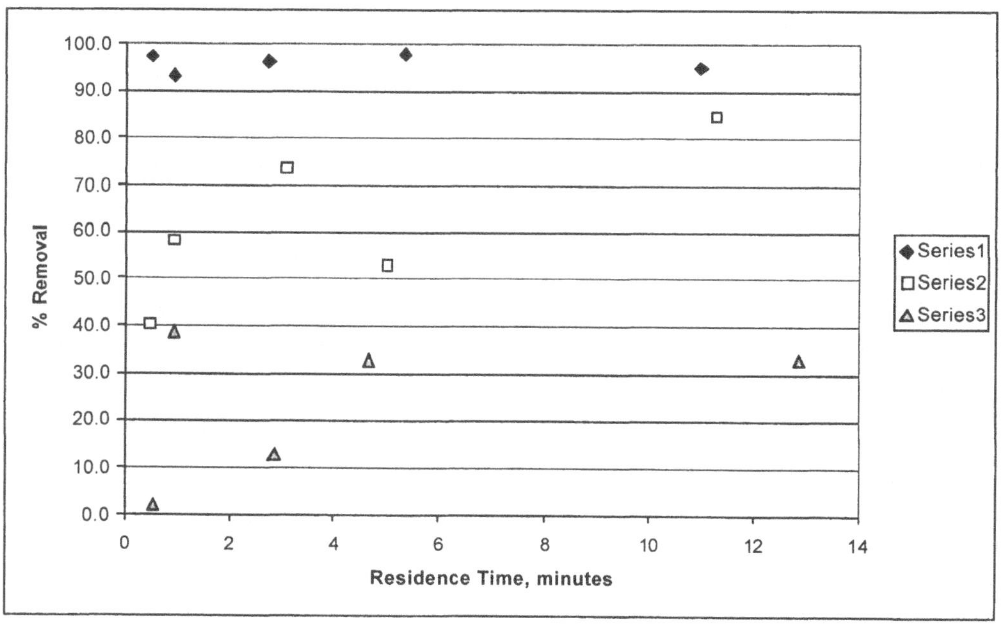

Figure 3-56. Percent Removal of Dissloved Cadmium by St. Cloud Zeolite Filter Column verses Residence Time for 3 Consecutive Series of Column Runs.

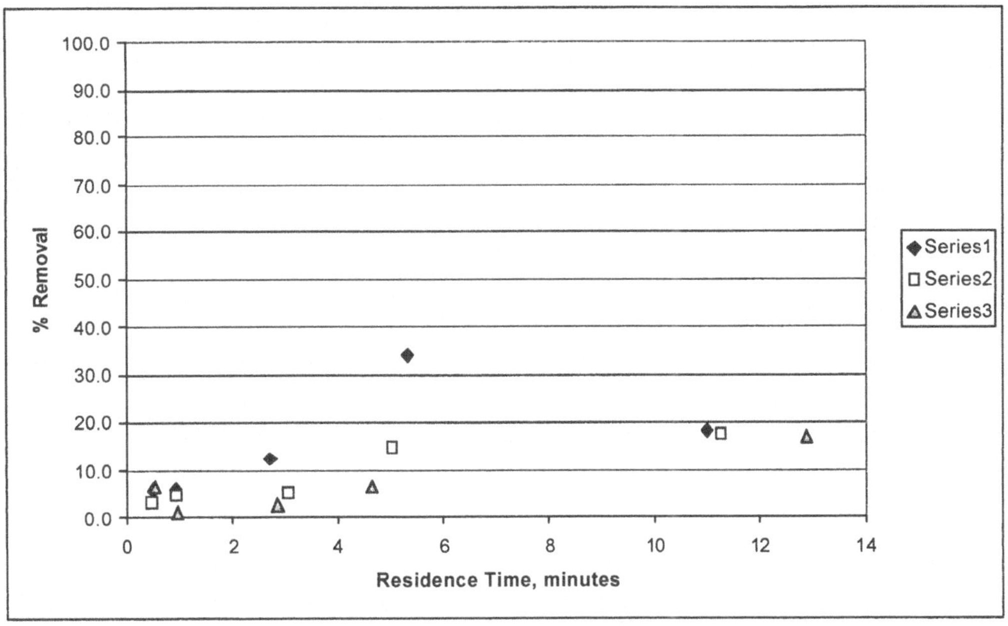

Figure 3-57. Percent Removal of Dissolved Iron by St. Cloud Zeolite Filter Column verses Residence Time for 3 Consecutive Series of Column Runs.

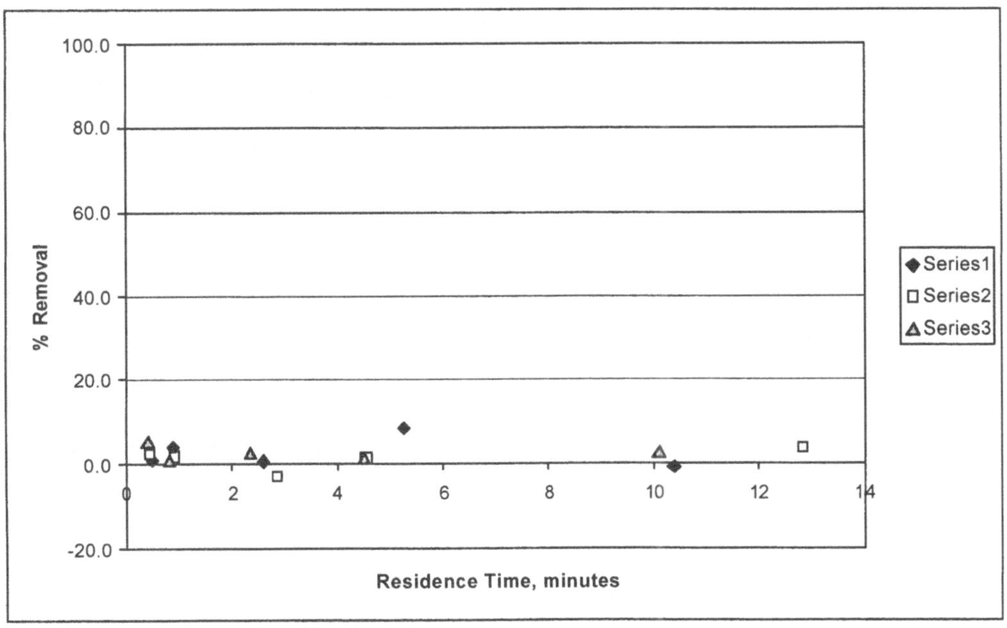

Figure 3-58. Percent Removal of Dissolved Copper by Control Filter Column verses Residence Time for 3 Consecutive Series of Column Runs.

Only the chart for percent dissolved copper removal is shown for the control column since the charts for the other metals were essentially the same as for copper.

Discussion: Dissolved Metal Removal by Filter Columns. There was an observable relationship between dissolved metal removal and residence time. An increase in dissolved metal removal was observed as residence time increased. However, the impact of residence time on removal varied depending on the metal and the media. Lead, chromium, and copper had the most notable increase in dissolved metal removal as residence time increased. Zinc and cadmium, which had the greatest percent removal for all the media (above 90% initially), showed no improvement in removal efficiency as residence time increased during the first series of column tests. During the second two series of tests an impact by residence time on zinc and cadmium percent removal was noted for the zeolite, and to a slight extent for the compost and on zinc removal for the peat-sand. This change may have been due to the metal uptake capacity of the media beginning to be saturated during later series.

Comparing the three media, the peat-sand mix performed best at removing dissolved metals. Peat-sand had better removal of metals for almost all runs, only on a few runs did compost or zeolite have equivalent (but not better) removal. Compost and zeolite showed similar removal efficiency for all metals but copper during the first series of runs. Compost removed copper better. However, during the second and third series of runs, compost out-performed zeolite on zinc and cadmium removal, with the gap between the removal efficiencies of the two media increasing between the second and third series of runs.

3.4.5.2 Particulate and Particulate Bound Metal Removal by Columns

Table 3-23 lists the approximate particulate bound metal concentration for the influent for all runs by series and the approximate hydraulic residence time. Particulate bound metal concentration was calculated by subtracting the dissolved metal concentration from the total metal concentration. Table 3-24 lists total suspended and dissolved solids in the influent. Figures 3-59 through 3-66 show percent total suspended and dissolved solids removal by the filter columns.

Table 3-23. Influent particulate bound metal concentrations (milligrams/liter) for column runs.

Series	Approximate Residence Time (min)	Cu mg/L	Cr mg/L	Pb mg/L	Zn mg/L	Cd mg/L	Fe mg/L
1	10	0.040	0.036	0.040	0.033	0.008	0.623
	5	0.040	0.032	0.046	0.035	0.008	1.017
	3	0.062	0.047	0.078	0.059	0.020	1.340
	1	0.068	0.052	0.088	0.071	0.024	1.533
	0.5	0.056	0.049	0.072	0.050	0.013	1.371
2	10	0.090	0.063	0.120	0.119	0.041	1.714
	5	0.096	0.073	0.126	0.115	0.039	2.016
	3	0.093	0.067	0.101	0.102	0.038	1.469
	1	0.096	0.059	0.099	0.103	0.044	1.251
	0.5	0.086	0.061	0.107	0.108	0.039	1.566
3	10	0.315	0.209	0.411	0.452	0.152	7.411
	5	0.201	0.144	0.278	0.275	0.093	4.860
	3	0.365	0.241	0.476	0.553	0.194	9.178
	1	0.207	0.146	0.280	0.284	0.100	4.892
	0.5	0.320	0.216	0.418	0.468	0.165	7.826

Table 3-24. Influent total suspended and dissolved solids concentrations (milligrams/liter) for column runs.

Series	Approximate Residence Time (min)	Total Suspended Solids mg/L	Total Dissolved Solids mg/L
1	10	29.3	116.0
	5	48.3	114.9
	3	71.0	117.2
	1	81.0	119.3
	0.5	47.0	117.7
2	10	85.3	150.1
	5	60.0	171.6
	3	36.7	170.5
	1	40.0	140.8
	0.5	47.3	172.7
3	10	439.0	168.9
	5	221.3	171.6
	3	460.0	160.6
	1	210.0	169.9
	0.5	410.0	172.2

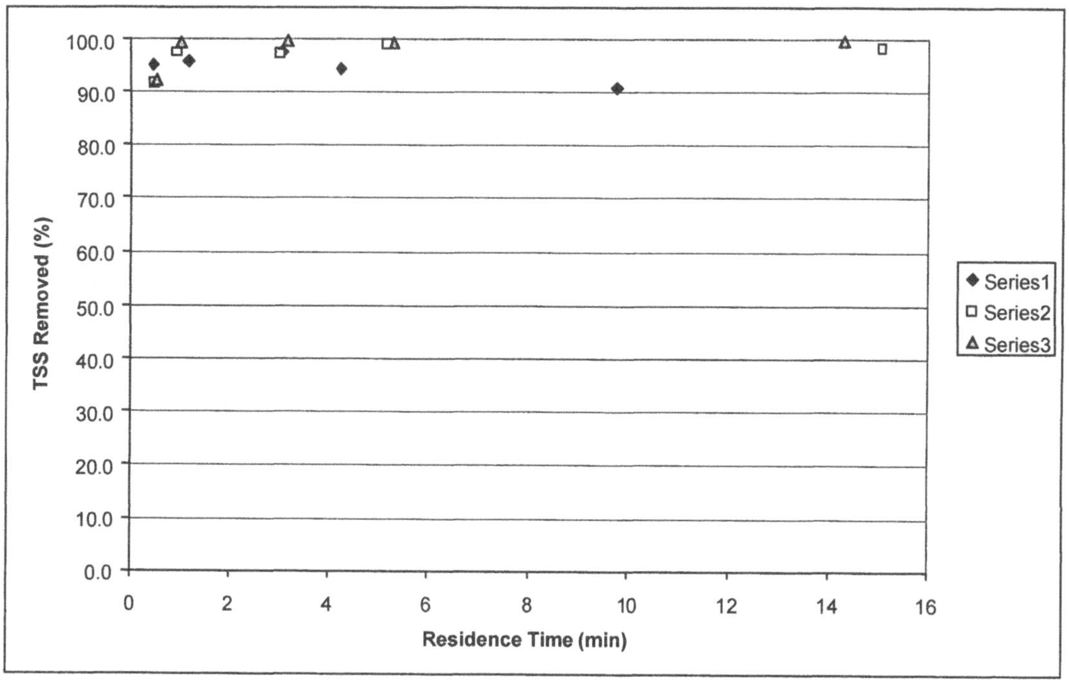

Figure 3-59. Percent Removal of Total Suspended Solids by Compost Filter Column verses Residence Time for 3 Consecutive Series of Column Runs.

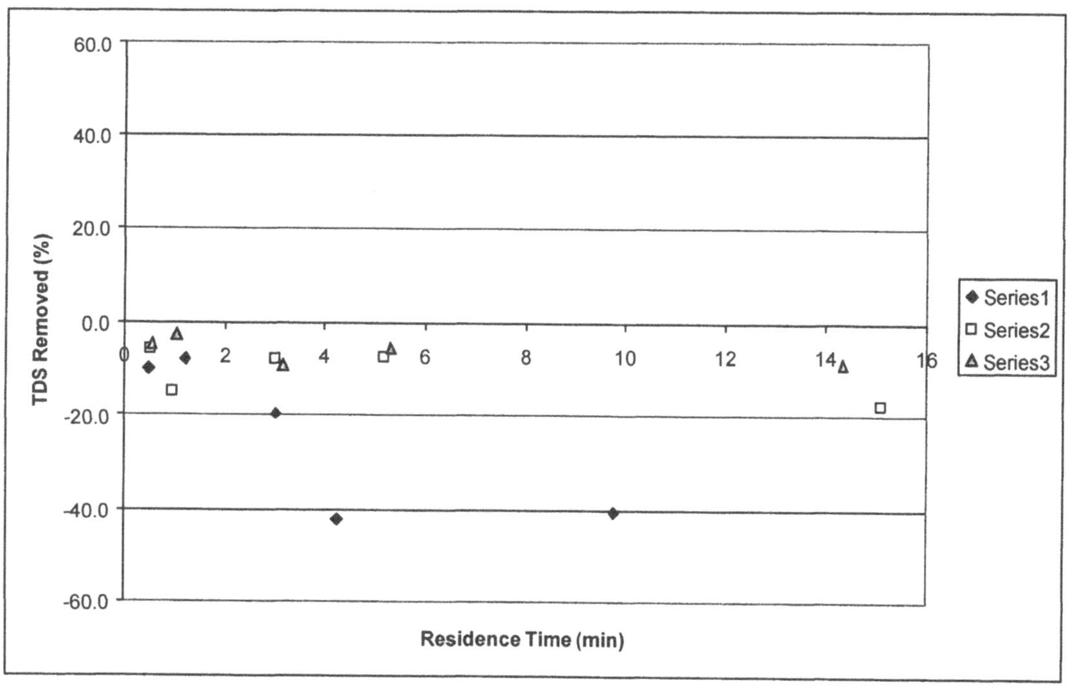

Figure 3-60. Percent Removal of Total Dissolved Solids by Compost Filter Column verses Residence Time for 3 Consecutive Series of Column Runs.

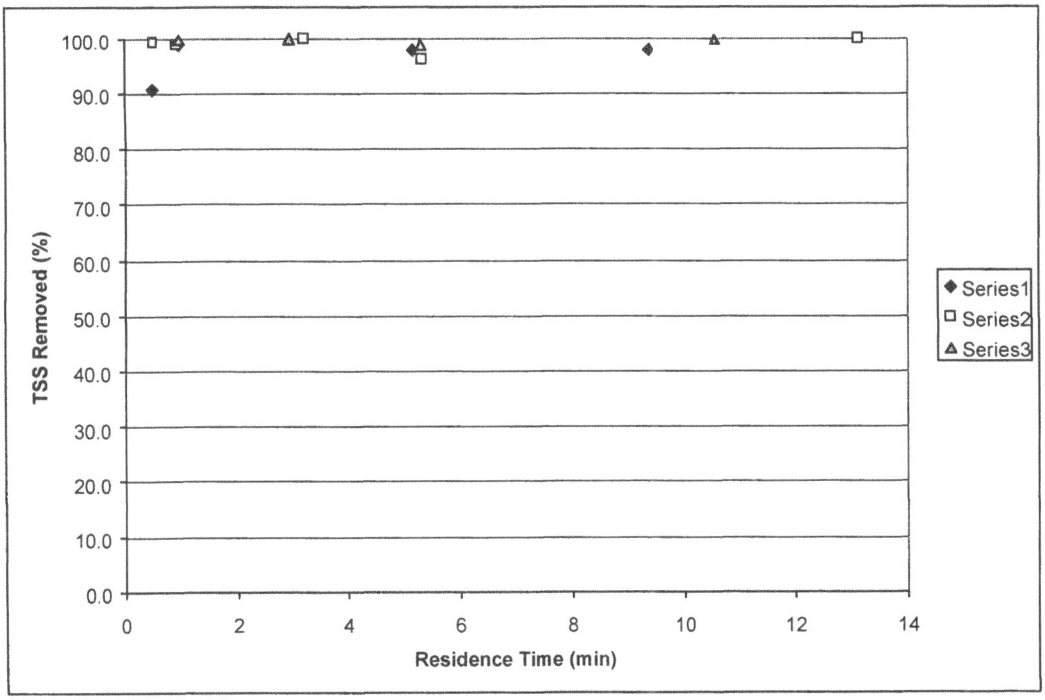

Figure 3-61. Percent Removal of Total Suspended Solids by Peat-Sand Filter Column verses Residence Time for 3 Consecutive Series of Column Runs.

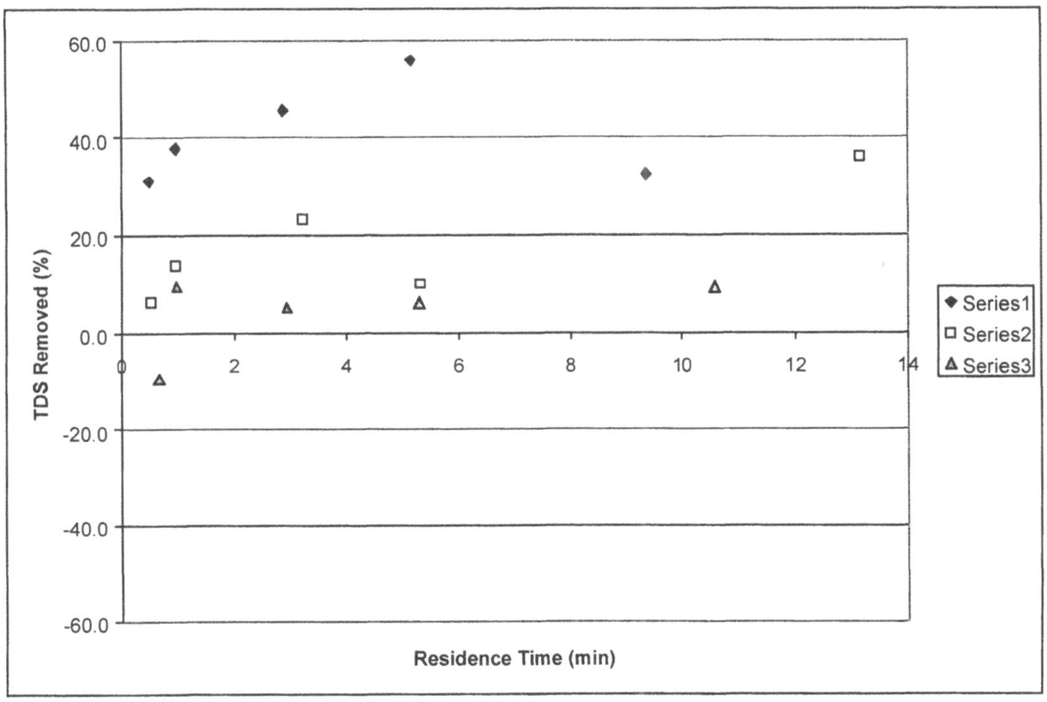

Figure 3-62. Percent Removal of Total Dissolved Solids by Peat-Sand Filter Column verses Residence Time for 3 Consecutive Series of Column Runs.

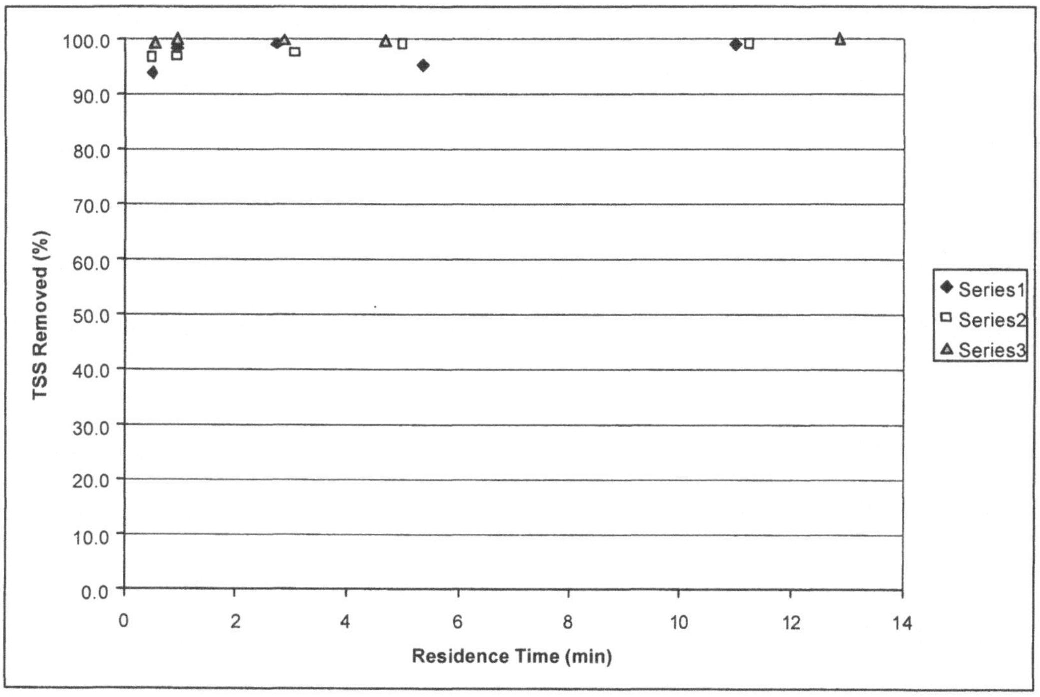

Figure 3-63. Percent Removal of Total Suspended Solids by St. Cloud Zeolite Filter Column verses Residence Time for 3 Consecutive Series of Column Runs.

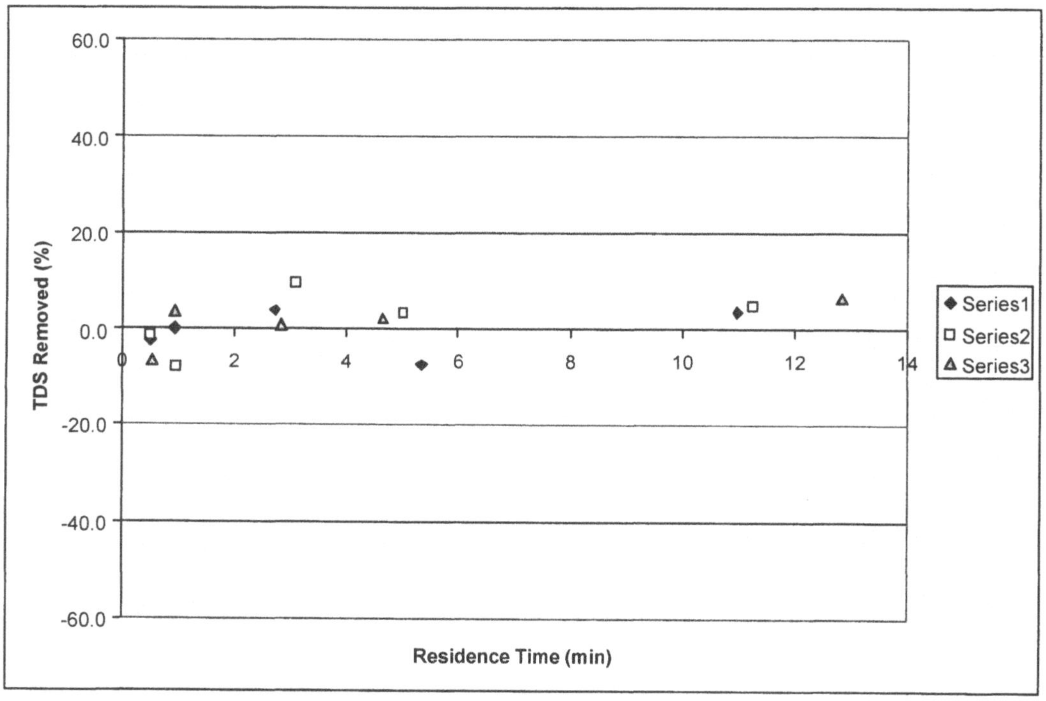

Figure 3-64. Percent Removal of Total Dissolved Solids by St. Cloud Zeolite Filter Column verses Residence Time for 3 Consecutive Series of Column Runs.

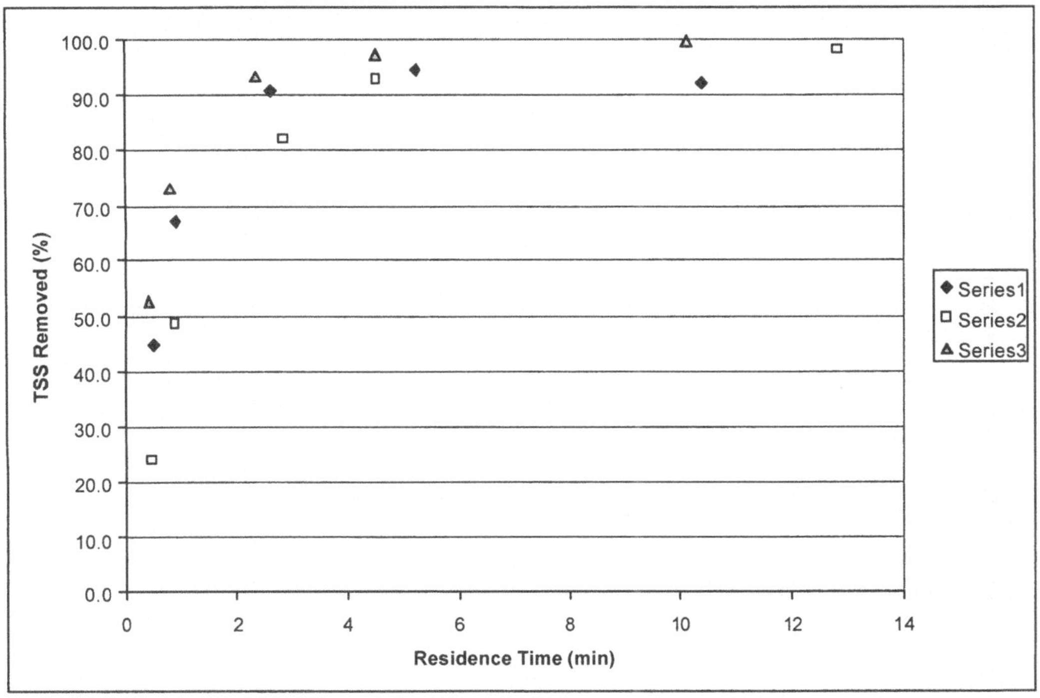

Figure 3-65. Percent Removal of Total Suspended Solids by Control Filter Column verses Residence Time for 3 Consecutive Series of Column Runs.

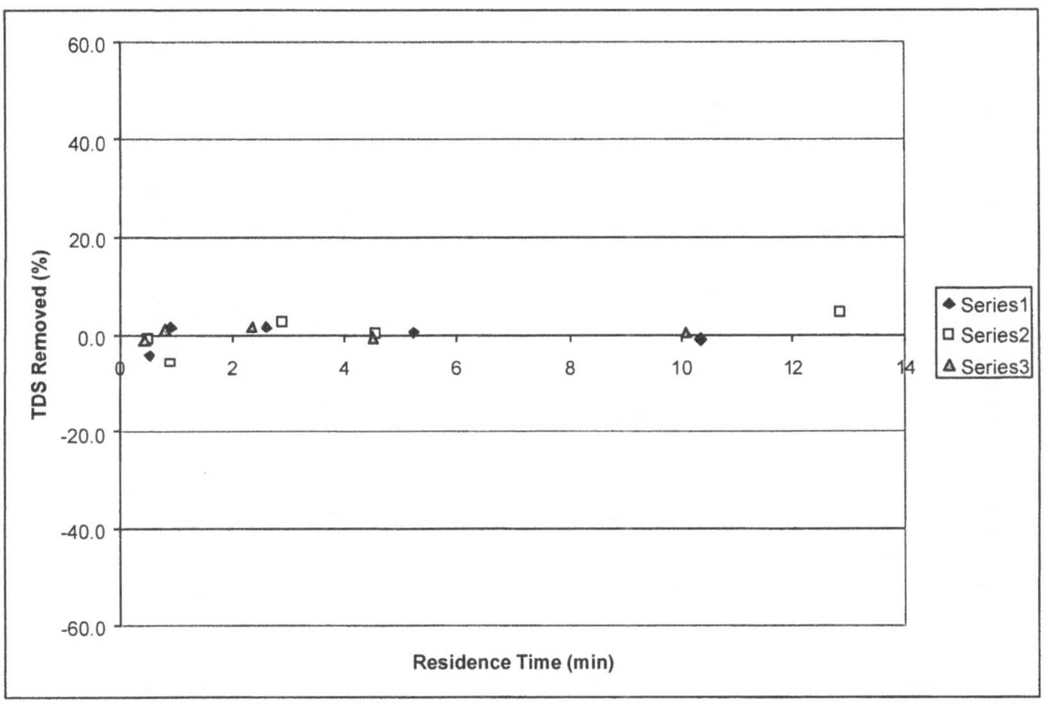

Figure 3-66. Percent Removal of Total Dissolved Solids by Control Filter Column verses Residence Time for 3 Consecutive Series of Column Runs.

Figures 3-67 through 3-90 present the percent removal of particulate bound metals by the columns.

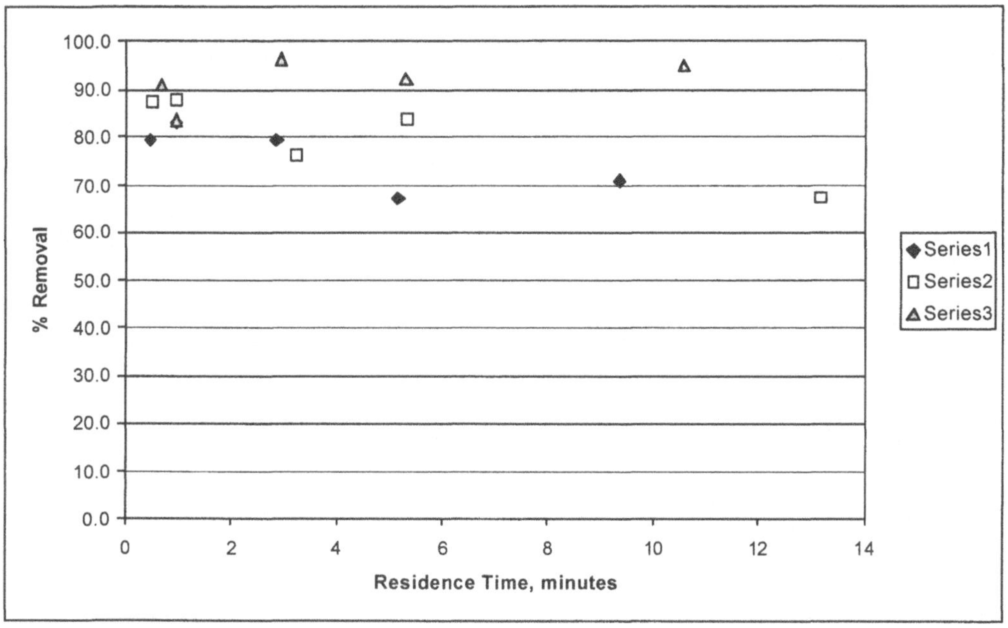

Figure 3-67. Percent Removal of Particulate Bound Copper by Peat-Sand Filter Column verses Residence Time for 3 Consecutive Series of Column Runs.

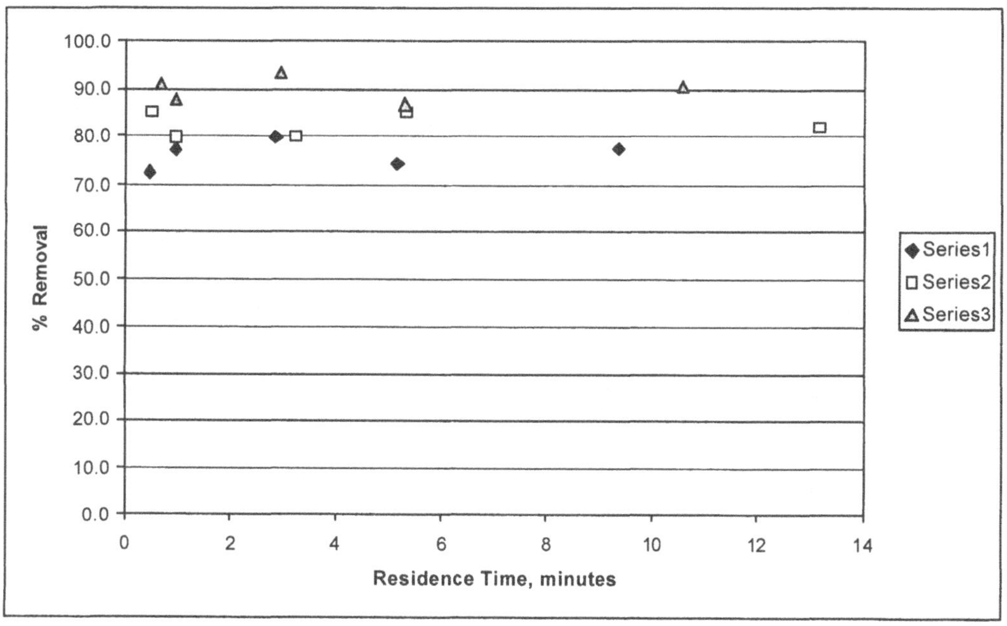

Figure 3-68. Percent Removal of Particulate Bound Chromium by Peat-Sand Filter Column verses Residence Time for 3 Consecutive Series of Column Runs.

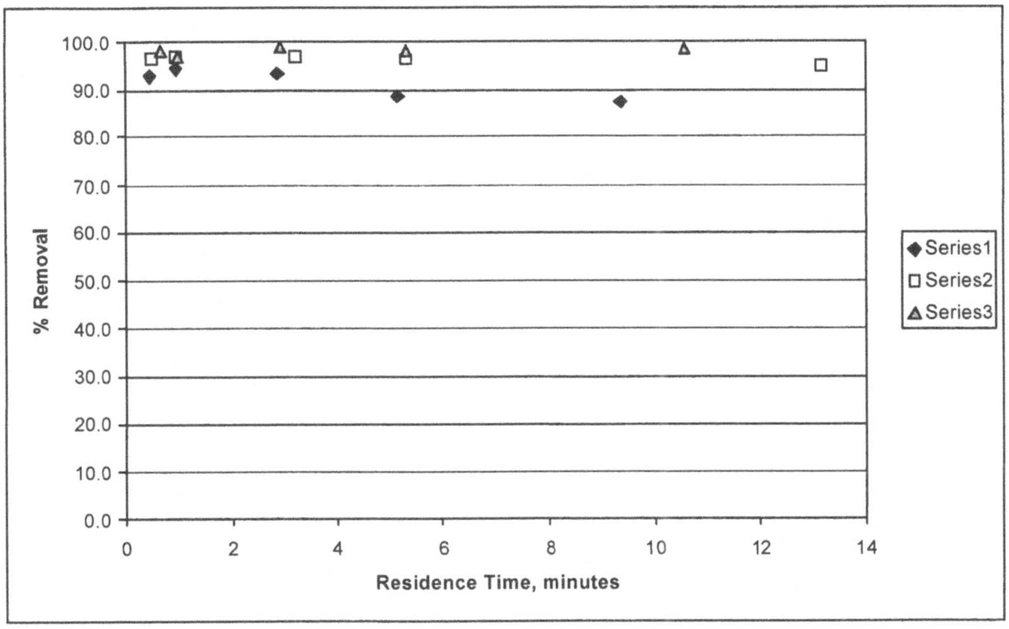

Figure 3-69. Percent Removal of Particulate Bound Lead by Peat-Sand Filter Column verses Residence Time for 3 Consecutive Series of Column Runs.

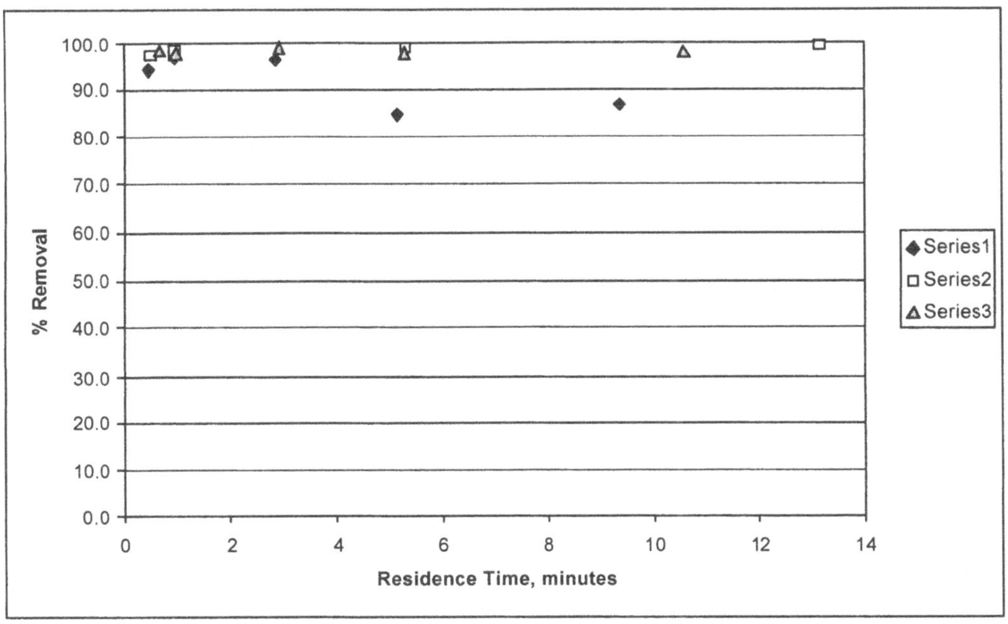

Figure 3-70. Percent Removal of Particulate Bound Zinc by Peat-Sand Filter Column verses Residence Time for 3 Consecutive Series of Column Runs.

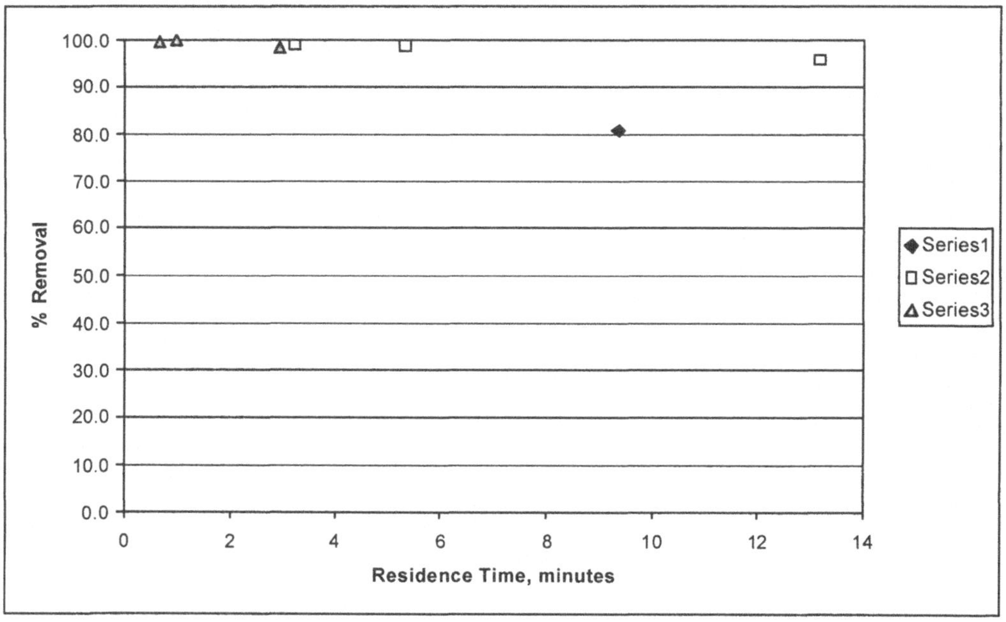

Figure 3-71. Percent Removal of Particulate Bound Cadmium by Peat-Sand Filter Column verses Residence Time for 3 Consecutive Series of Column Runs.

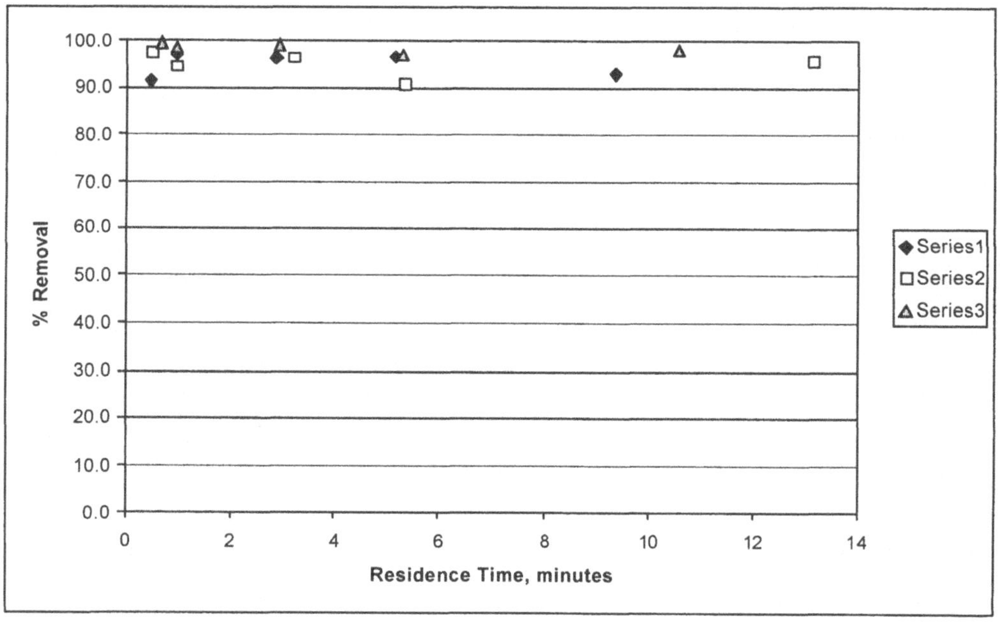

Figure 3-72. Percent Removal of Particulate Bound Iron by Peat-Sand Filter Column verses Residence Time for 3 Consecutive Series of Column Runs.

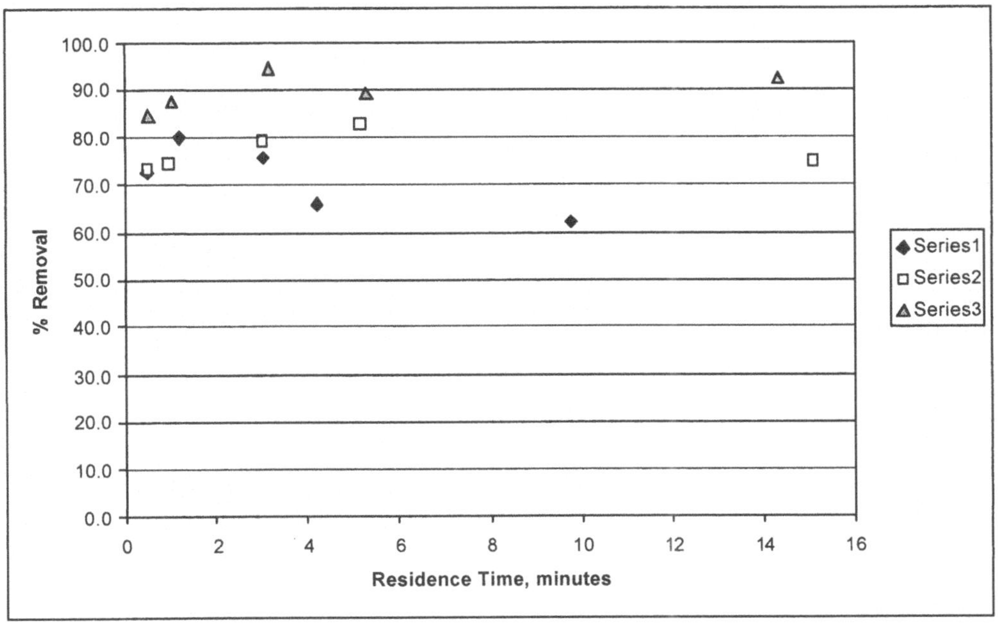

Figure 3-73. Percent Removal of Particulate Bound Copper by Compost Filter Column verses Residence Time for 3 Consecutive Series of Column Runs.

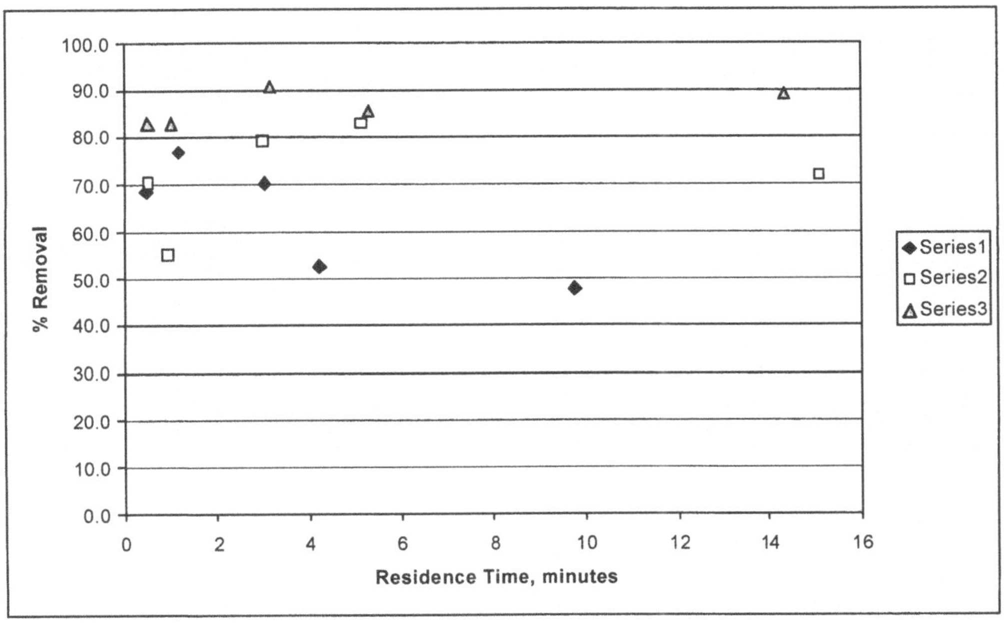

Figure 3-74. Percent Removal of Particulate Bound Chromium by Compost Filter Column verses Residence Time for 3 Consecutive Series of Column Runs.

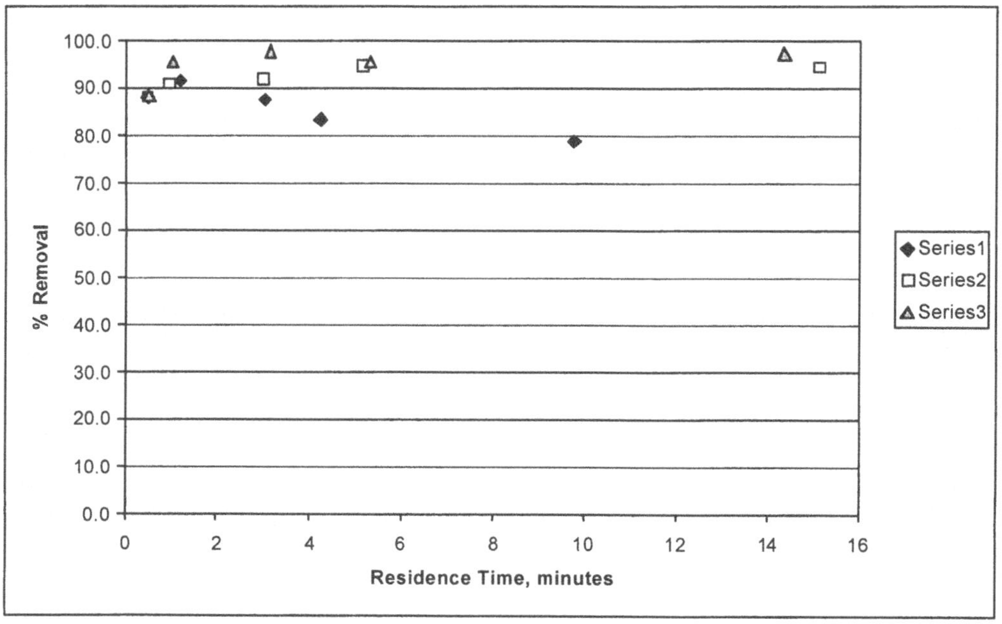

Figure 3-75. Percent Removal of Particulate Bound Lead by Compost Filter Column verses Residence Time for 3 Consecutive Series of Column Runs.

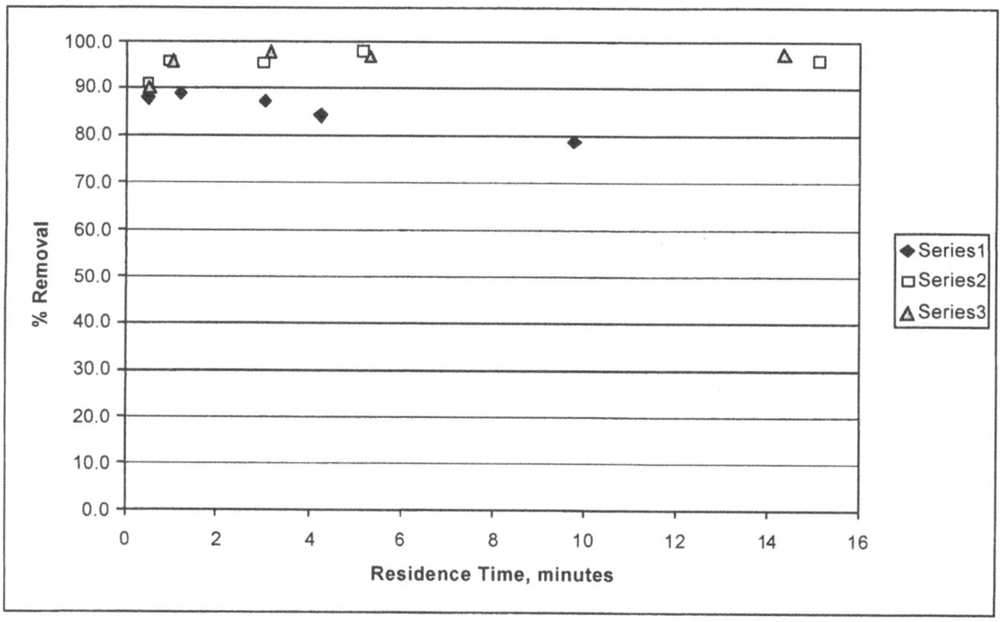

Figure 3-76. Percent Removal of Particulate Bound Zinc by Compost Filter Column verses Residence Time for 3 Consecutive Series of Column Runs.

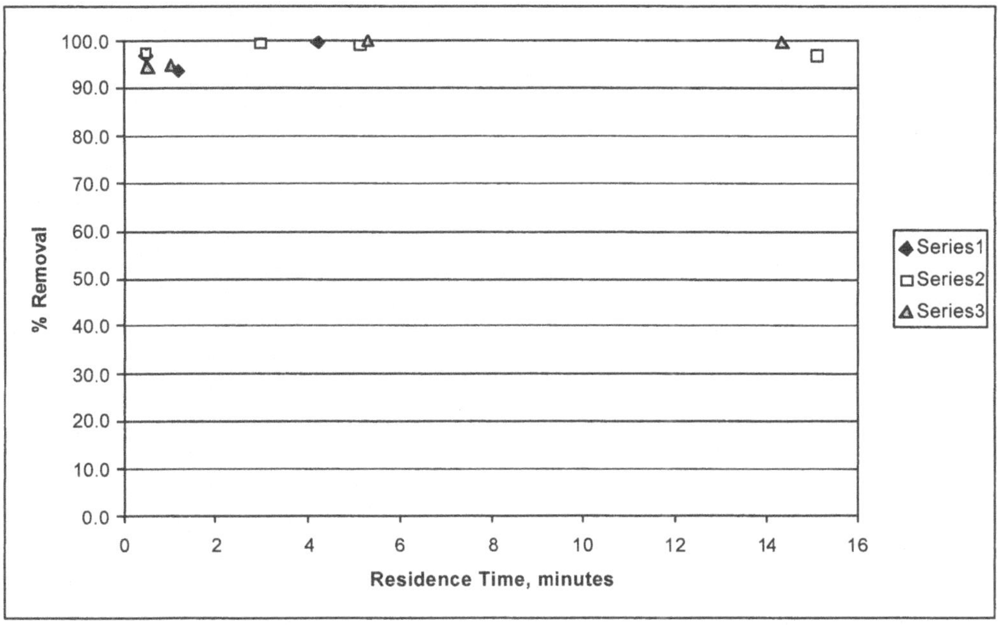

Figure 3-77. Percent Removal of Particulate Bound Cadmium by Compost Filter Column verses Residence Time for 3 Consecutive Series of Column Runs.

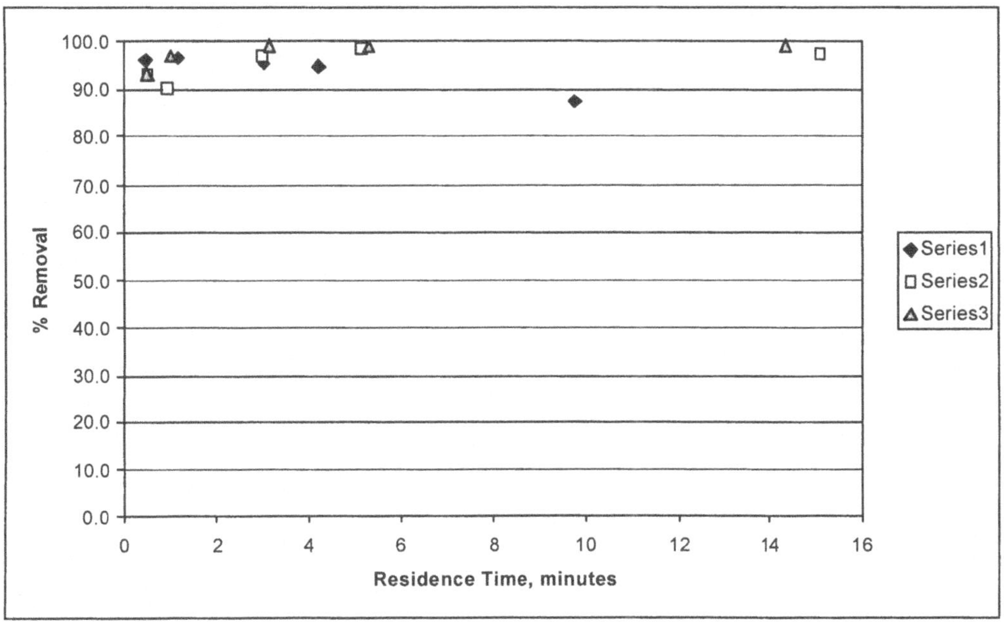

Figure 3-78. Percent Removal of Particulate Bound Iron by Compost Filter Column verses Residence Time for 3 Consecutive Series of Column Runs.

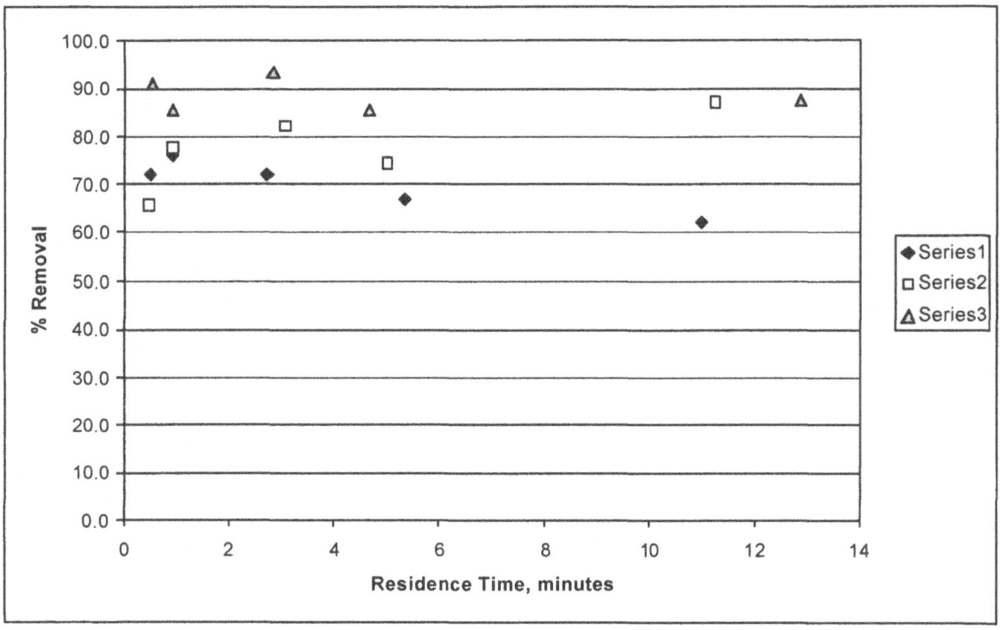

Figure 3-79. Percent Removal of Particulate Bound Copper by St. Cloud Zeolite Filter Column verses Residence Time for 3 Consecutive Series of Column Runs.

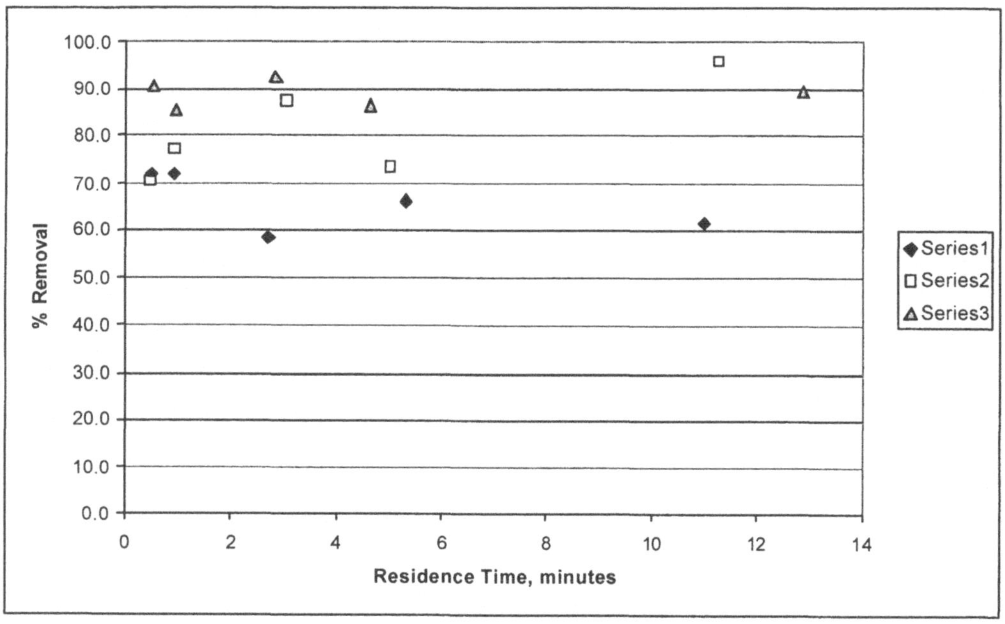

Figure 3-80. Percent Removal of Particulate Bound Chromium by St. Cloud Zeolite Filter Column verses Residence Time for 3 Consecutive Series of Column Runs.

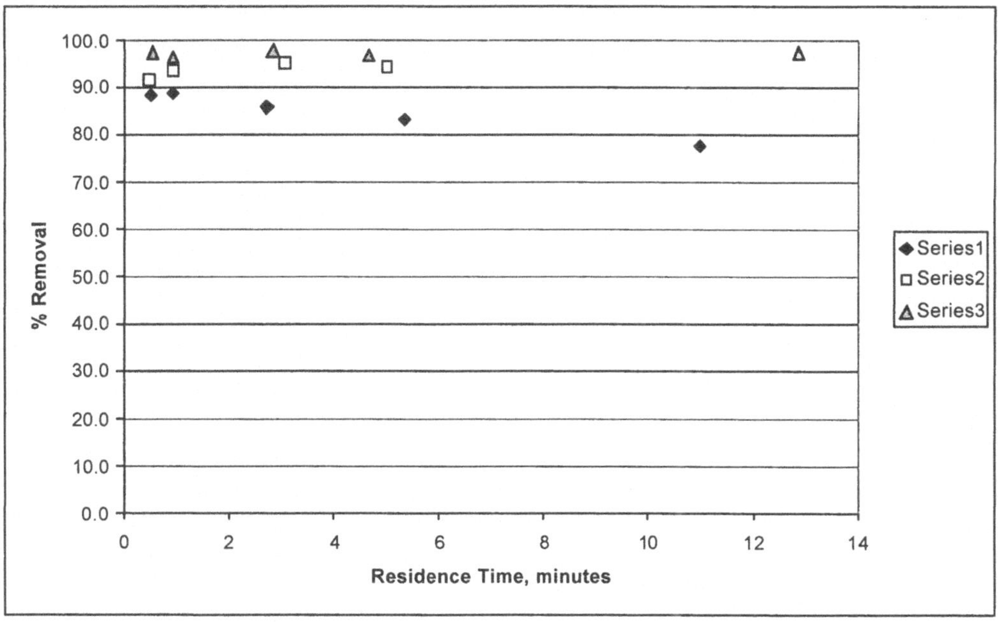

Figure 3-81. Percent Removal of Particulate Bound Lead by St. Cloud Zeolite Filter Column verses Residence Time for 3 Consecutive Series of Column Runs.

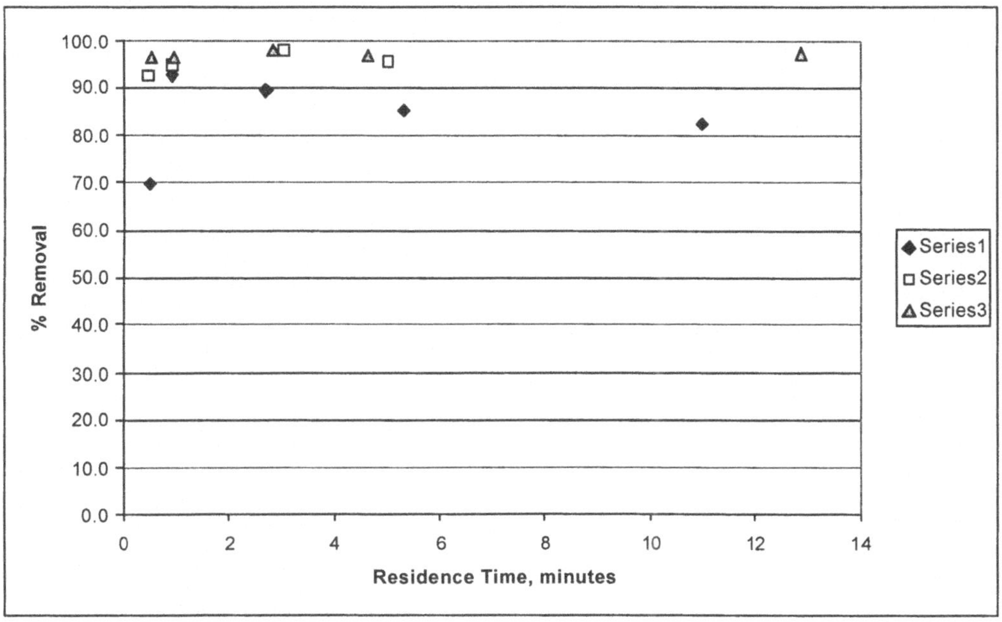

Figure 3-82. Percent Removal of Particulate Bound Zinc by St. Cloud Zeolite Filter Column verses Residence Time for 3 Consecutive Series of Column Runs.

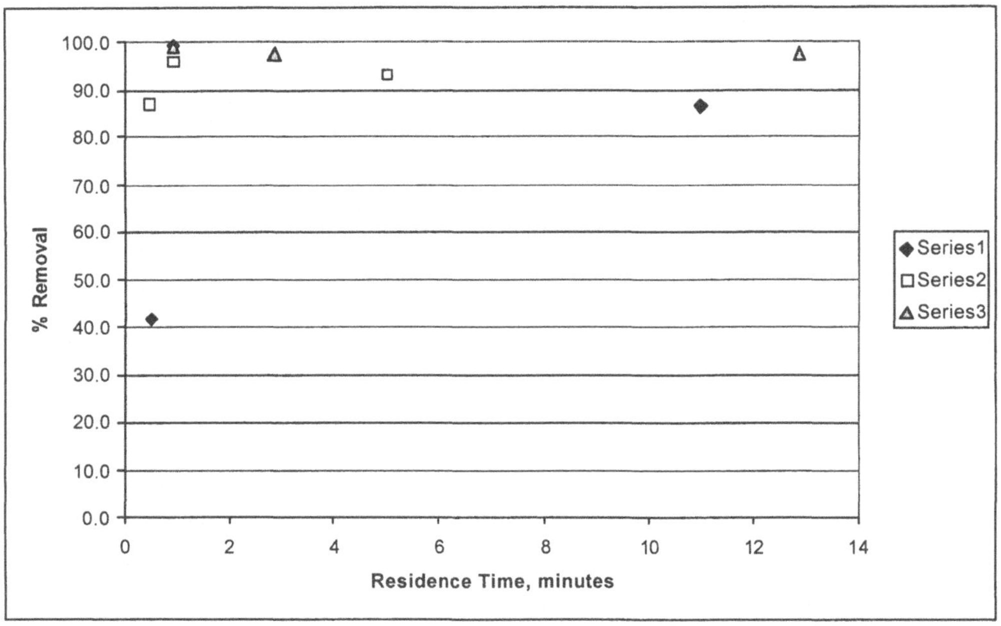

Figure 3-83. Percent Removal of Particulate Bound Cadmium by St. Cloud Zeolite Filter Column verses Residence Time for 3 Consecutive Series of Column Runs.

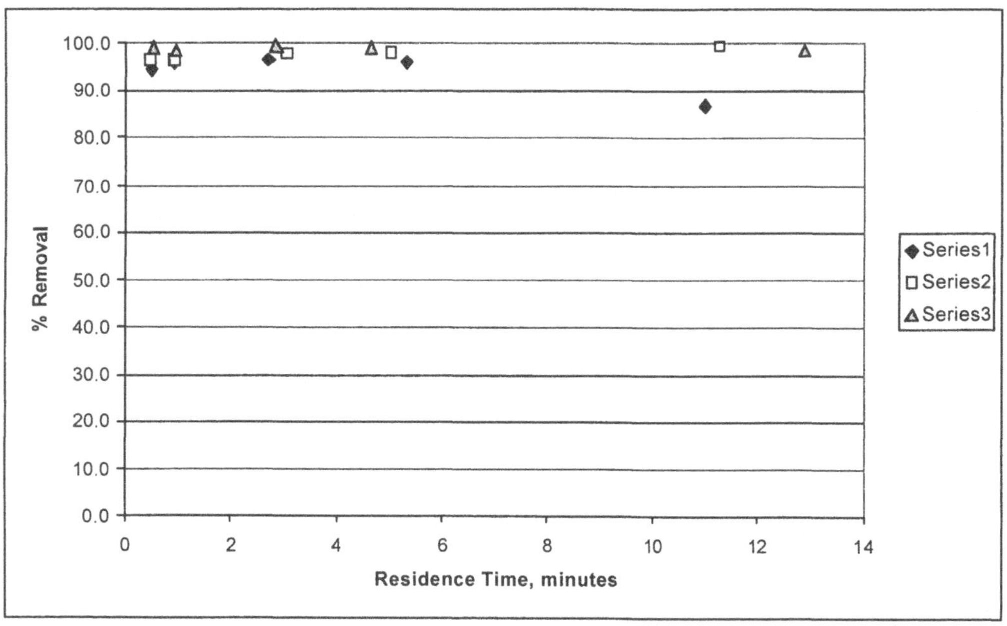

Figure 3-84. Percent Removal of Particulate Bound Iron by St. Cloud Zeolite Filter Column verses Residence Time for 3 Consecutive Series of Column Runs.

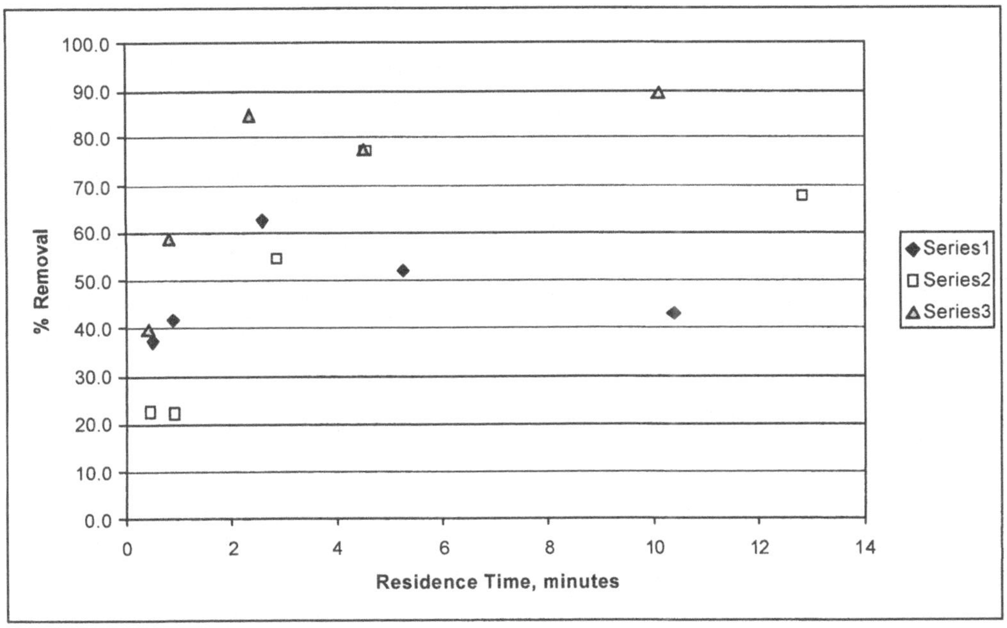

Figure 3-85. Percent Removal of Particulate Bound Copper by Control Filter Column verses Residence Time for 3 Consecutive Series of Column Runs.

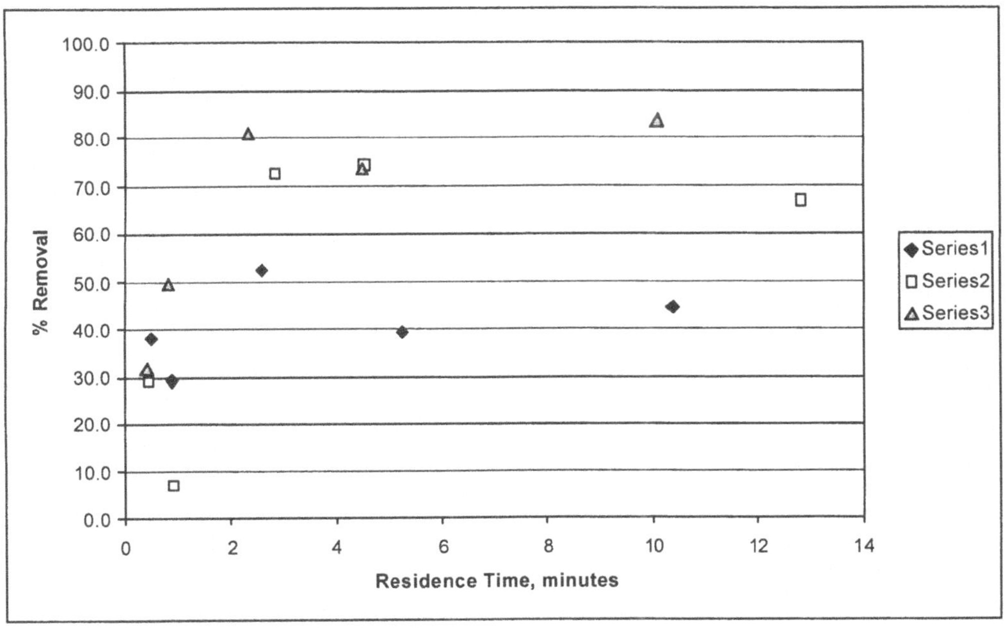

Figure 3-86. Percent Removal of Particulate Bound Chromium by Control Filter Column verses Residence Time for 3 Consecutive Series of Column Runs.

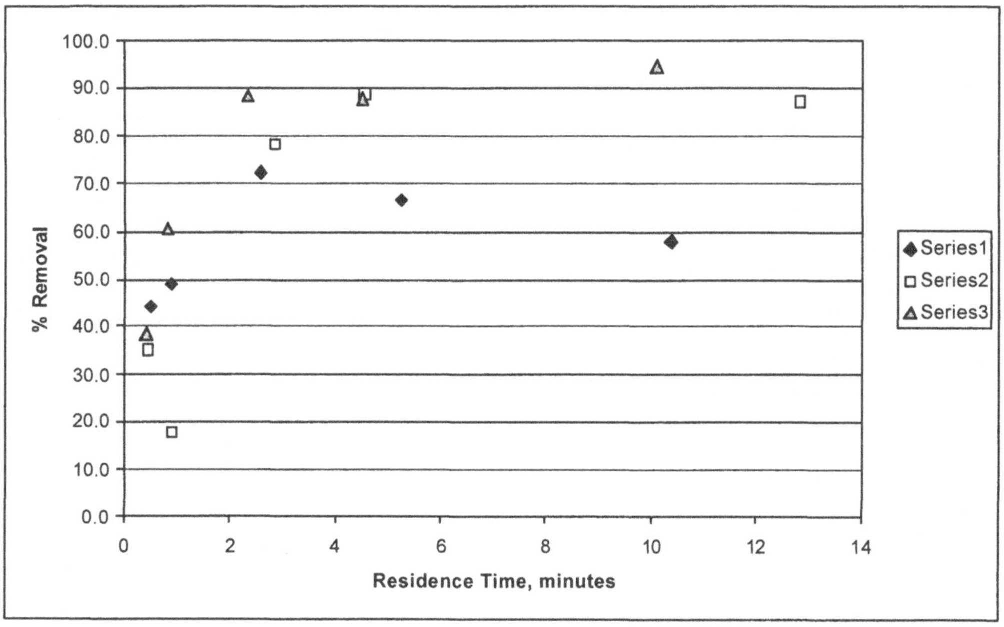

Figure 3-87. Percent Removal of Particulate Bound Lead by Control Filter Column verses Residence Time for 3 Consecutive Series of Column Runs.

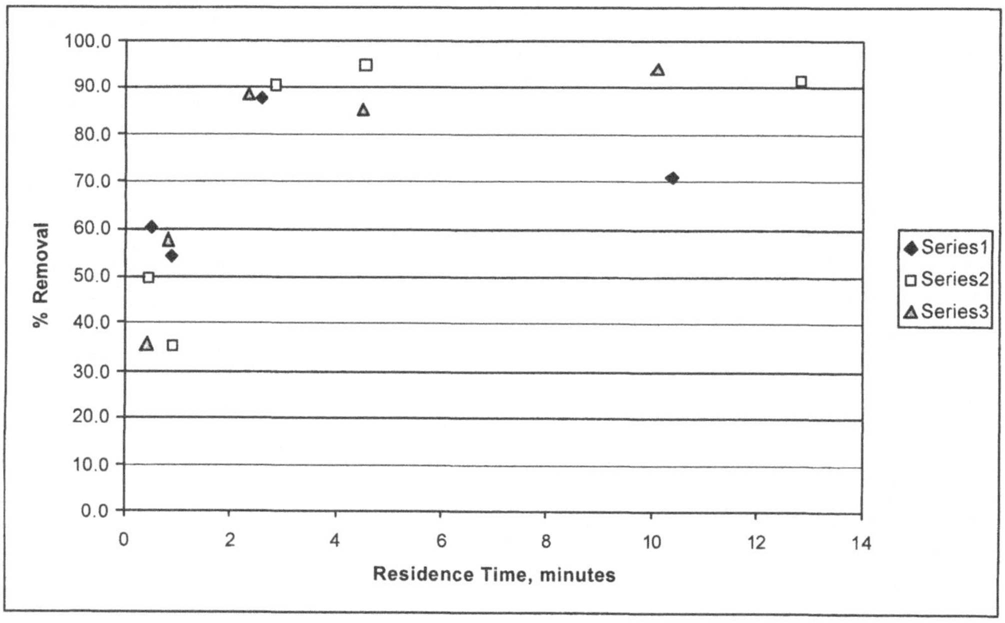

Figure 3-88. Percent Removal of Particulate Bound Zinc by Control Filter Column verses Residence Time for 3 Consecutive Series of Column Runs.

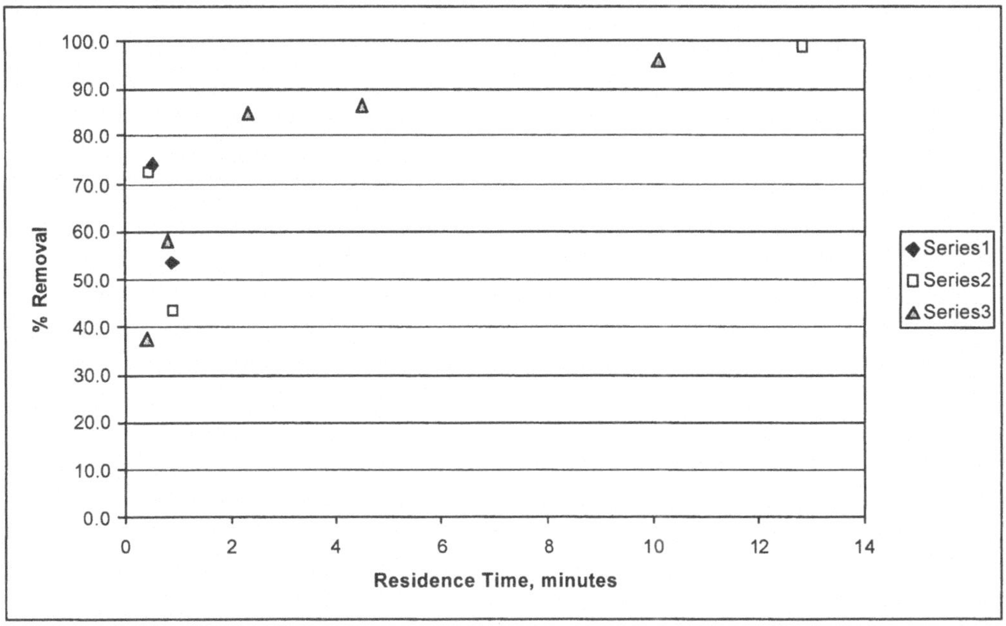

Figure 3-89. Percent Removal of Particulate Bound Cadmium by Control Filter Column verses Residence Time for 3 Consecutive Series of Column Runs.

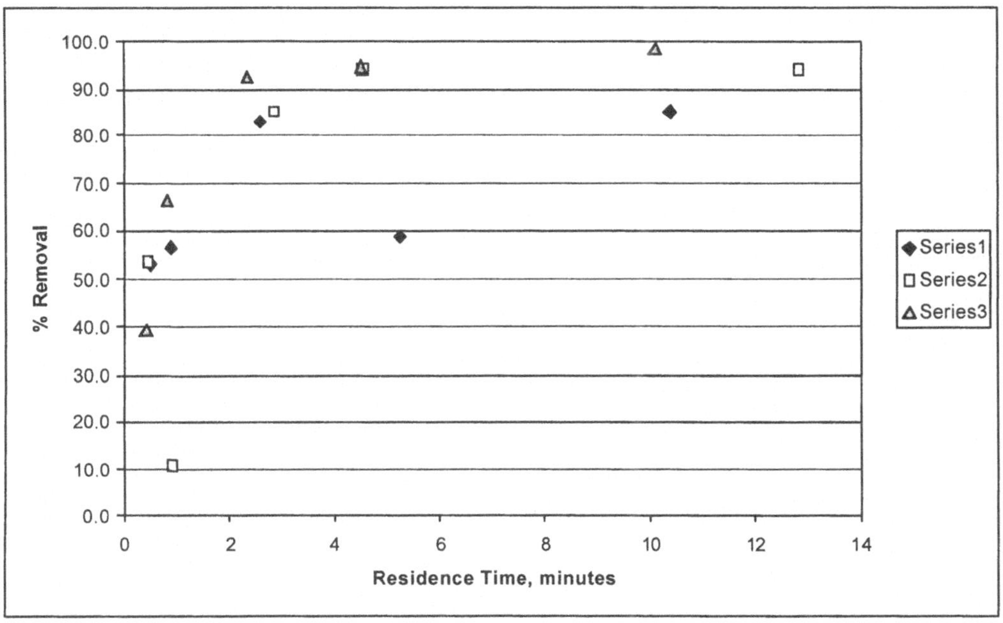

Figure 3-90. Percent Removal of Particulate Bound Iron by Control Filter Column verses Residence Time for 3 Consecutive Series of Column Runs.

Discussion: Total suspended solids, total dissolved solids, and particulate bound metal removal. All the media removed total suspended solids well over all runs, but only peat showed any removal of total dissolved solids (the removal of TDSs dropped as the runs progressed). Compost appeared to increase the total dissolved solids concentration. As expected, the behavior of the total suspended solids and particulate bound metals paralled each other. Total suspended solids removal and particulate bound metal removal by peat-sand and zeolite had no observable relationship to residence time. Some improvement in total suspended solids and particulate bound metal removal efficiency was noted for the compost as residence time increased for the second and third series of runs. The 0.5 and 1 minute residence times had slightly worse removal efficiencies than the 3, 5, and 10 minute residence times for compost – there was no observable improvement in particulate bound removal efficiency as residence time increased passed 3 minutes. The behavior of the particulate metal removal in the first series of runs was erratic, perhaps due to particulate bound metal concentrations being lower in the influent during the first series than the influent concentrations in the second and third series. As would be expected, the control column (no media) had a very noticeable increase in total suspended solids and particulate bound metal removal efficiency as residence time increased. Copper and chromium typically had lower removal efficiencies than the other metals; one possible explanation is that these metals had a larger fraction of their particulate bound metal concentration associated with smaller particulates which passed more easily through the columns. Peat had the best removal efficiencies for particulate bound metals. Removal efficiencies of compost and zeolite were approximately the same.

3.4.5.3 Impact of Filter Columns on pH and Color

Tables 3-25 and 3-26 list the values of the influent and effluent pH and color. On the longer column runs (residence time 3 minutes and longer) influent pH was taken at several intervals over the course of the run. For many of these runs, the pH varied slightly over time. The highest and lowest pH values for these runs are listed, as well as the average pH over the runs.

Table 3-25. Influent and effluent pH

Series	Residence Time minutes	Influent			Effluent			
		Low	High	Average	Peat	Zeolite	Compost	Control
1	10	7.65	7.74	7.70	3.99	7.57		7.48
	5	6.93		6.93	4.18	7.34	7.54	7.21
	3	7.00	7.10	7.05	3.94	7.30	7.51	7.23
	1	7.10		7.10	3.95	7.29	7.53	7.22
	0.5	6.90		6.90	4.10	7.18	7.35	7.15
2	10	6.94	7.01	6.98	5.11	7.40	7.58	7.24
	5	6.95	7.29	7.12	6.69	7.21	7.38	7.19
	3	6.90	6.95	6.93	4.39	7.05	7.31	6.99
	1	6.97	7.03	7.00	6.08	7.10	7.26	6.97
	0.5	7.10		7.10	6.60	7.13	7.28	7.13
3	10	7.09	7.18	7.14	6.95	7.30	7.52	7.34
	5	7.08	7.18	7.13	6.89	7.31	7.35	7.30
	3	7.32		7.32	6.91	7.31	7.35	7.32
	1	7.12		7.12	6.76	7.17	7.30	7.15
	0.5	7.22		7.22	6.93	7.19	7.39	7.28

Table 3-26. Influent and effluent color (Apparent color in APHA Platimum Cobalt Units)

Series	Residence Time minutes	Influent	Effluent			
			Peat	Zeolite	Compost	Control
1	10	45	45	32	225	35
	5	45	38	30	175	45
	3	45	35	35	75	45
	1	45	38	32	55	45
	0.5	45	38	32	55	45
2	10	45	50	40	150	45
	5	45	50	32	60	45
	3	45	45	40	65	45
	1	45	45	40	55	45
	0.5	47	47	40	53	45
3	10	35	48	32	70	35
	5	35	40	32	52	35
	3	35	39	32	50	35
	1	35	39	34	42	35
	0.5	40	37	34	47	35

Discussion: Color and pH. Some increase in pH was noted for the effluent from the control column. There was a tendency for the influent pH to increase during the runs, this change most likely continued as the runoff moved through the control column. The peat-sand mix had the most dramatic impact on pH, dropping the pH from around 7 down to approximately 4 during the first series of runs. Peat-sand continued to cause a drop during the second and third series of runs, but the degree of change in pH decreased as the runs progressed, until only a slight drop was occurring during the third series of runs. No relationship between pH change and residence time was noted for peat. The zeolite and compost caused a slight increase in pH. However, the pH of the zeolite effluent was only marginally higher than that of the control column effluent in the initial series and this margin decreased as the later series of runs continued. Compost increased the pH to a slightly greater degree, but again its impact on pH decreased as the runs progressed. The data was not conclusive enough to determine if residence time had an impact on the pH change.

Compost caused an increase in runoff color, with the degree of change decreasing as residence time decreased. Zeolite slightly decreased color, with no observable relation to residence time. The impact of the peat-sand varied. In several runs with a longer residence time, peat-sand slightly increased color, while in some of the runs with short residence times, peat-sand slightly decreased color.

3.4.5.4 Headloss and Clogging of Filter Columns

Tables 3-27 through 3-30 list headloss (ft) across each of the columns at the start of each run and at the end (30 bed volumes pumped for each run). The increase in headloss over the course of each run is also listed.

Table 3-27. Initial and final headloss (ft) across peat-sand column.

Series	Actual Residence Time minutes	Flow Rate, m/day	Headloss in feet		
			Start	End	Increase during run
1	9.36	8.28	0.58	1.39	0.81
	5.15	15.04	1.62	2.08	0.46
	2.87	26.96	3.70	3.93	0.23
	0.96	81.01	9.12	9.70	0.58
	0.49	159.24	13.40	13.51	0.12
2	11.25	6.89	0.69	0.81	0.12
	5.00	15.49	1.62	1.62	0.00
	3.07	25.28	2.19	3.70	1.50
	0.94	82.53			
	0.49	158.73	16.05	16.05	0.00
3	10.57	7.33		1.04	
	5.31	14.61	2.08	2.43	0.35
	2.94	26.33	3.41	3.81	0.40
	0.98	79.37	7.16	9.59	2.43
	0.68	113.74	12.94	28.18	15.25

Table 3-28. Initial and final headloss (ft) across zeolite column.

Series	Actual Residence Time minutes	Flow Rate, m/day	Headloss in feet		
			Start	End	Increase during run
1	10.99	11.12	0.23	0.23	0.00
	5.34	22.87	0.35	0.35	0.00
	2.73	44.73	0.46	0.46	0.00
	0.94	129.28	1.04	1.16	0.12
	0.50	242.63	1.96	1.96	0.00
2	13.15	9.29	0.12	0.12	0.00
	5.33	22.92	0.35	0.35	0.00
	3.22	37.91	0.46	0.46	0.00
	0.96	127.89	0.92	0.92	0.00
	0.52	236.82	1.73	1.96	0.23
3	12.86	9.50	0.12	0.12	0.00
	4.66	26.21	0.46	0.52	0.06
	2.85	42.84	0.46	0.46	0.00
	0.95	128.52	0.81	1.62	0.81
	0.55	221.65	1.96	4.62	2.66

Table 3-29. Initial and final headloss (ft) across compost column.

Series	Actual Residence Time minutes	Flow Rate, m/day	Headloss in feet		
			Start	End	Increase during run
1	9.77	11.12	0.12	0.12	0.00
	4.23	25.68	0.12	0.12	0.00
	3.04	35.76	0.46	0.46	0.00
	1.19	90.99	0.81	0.92	0.12
	0.48	224.95	1.62	1.62	0.00
2	15.09	7.20	0.12	0.12	0.00
	5.16	21.05	0.23	0.23	0.00
	3.01	36.10	0.12	0.12	0.00
	0.94	115.25	0.69	0.69	0.00
	0.50	215.59	1.50	1.85	0.35
3	14.33	7.58	0.12	0.12	0.00
	5.30	20.52	0.12	0.12	0.00
	3.17	34.33	0.12	0.12	0.00
	1.04	104.64	0.81	0.92	0.12
	0.53	205.99	1.27	2.54	1.27

Table 3-30. Initial and final headloss (ft) across control column.

Series	Actual Residence Time minutes	Flow Rate, m/day	Headloss in feet		
			Start	End	Increase during run
1	10.36	12.20	0.00	0.00	0.00
	5.24	24.14	0.00	0.00	0.00
	2.61	48.44	0.00	0.00	0.00
	0.90	140.78	0.00	0.00	0.00
	0.52	241.37	0.00	0.00	0.00
2	12.82	9.86	0.00	0.00	0.00
	4.52	27.93	0.00	0.00	0.00
	2.86	44.23	0.00	0.00	0.00
	0.90	140.27	0.00	0.00	0.00
	0.46	274.23	0.00	0.00	0.00
3	10.10	12.51	0.00	0.00	0.00
	4.50	28.05	0.00	0.00	0.00
	2.35	53.71	0.00	0.00	0.00
	0.81	156.07	0.00	0.00	0.00
	0.41	304.56	0.00	0.00	0.00

Discussion: Headloss. Peat-sand had the greatest headloss and the greatest change in headloss over the course of each run. Zeolite and compost had considerably less headloss, and typically only showed an increase in headloss for the two shortest residence times (1 and 0.5 minutes). The degree of change in headloss increased as the total suspended solids concentration of the influent increased. TSS concentration was much higher in the third series of runs than in the previous two.

Composite collection on the peat sand column had to be stopped after 6 bed volumes were collected (26 bed volumes pumped over entire run instead of 30) because the pressure on the column was starting to cause leaking at the plastic couplings on the columns.

3.4.6 Filter Tasks Six Results and Discussion

Microbial growth was observed on the peanut hulls media where copper (Cu II) was present at concentrations of >100 mg/l, despite the fact that, copper sulfate ($CuSO_4$), the salt chosen for these investigations, is a commonly used biocide. Attempts were made to characterize these bacteria and evaluate their potential role in bioremediation in the filter. As an organism capable of surviving in a metals-saturated environment, it was hypothesized that it could affect metals uptake through physical or biological processes.

In initial testing, columns were run in parallel with the same concentrations of $CuSO_4$ (10 mg/l and 100 mg/l respectively) but in sterile and non-sterile conditions. Figure 3-91 shows the experimental set-up for these columns.

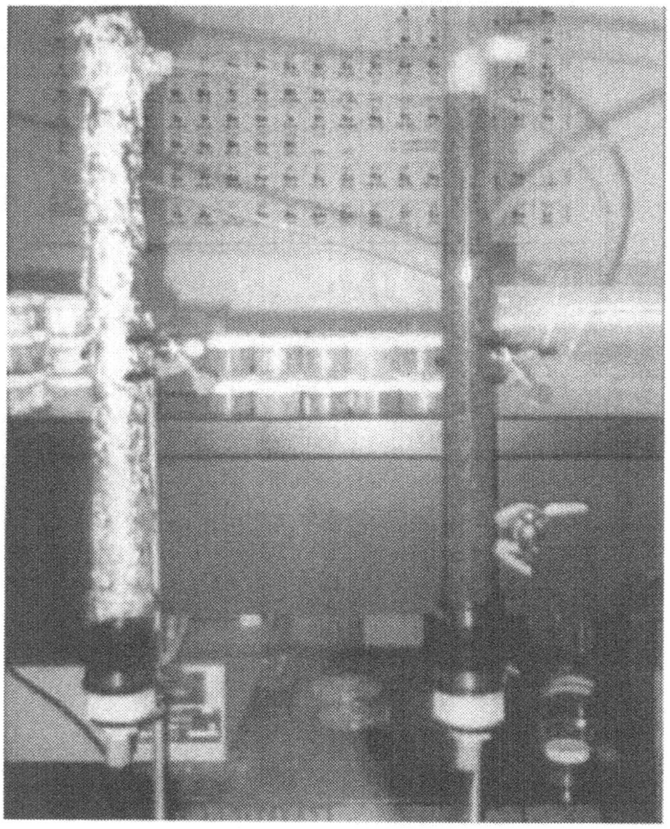

Figure 3-91. Experimental configuration for sterile versus non-sterile columns.

Results of these tests indicated the microbial growth had no effect on Cu adsorption onto the media, with typical results shown in Figure 3-92. Figure 3-93 shows the absorbance due to microbial growth in the columns.

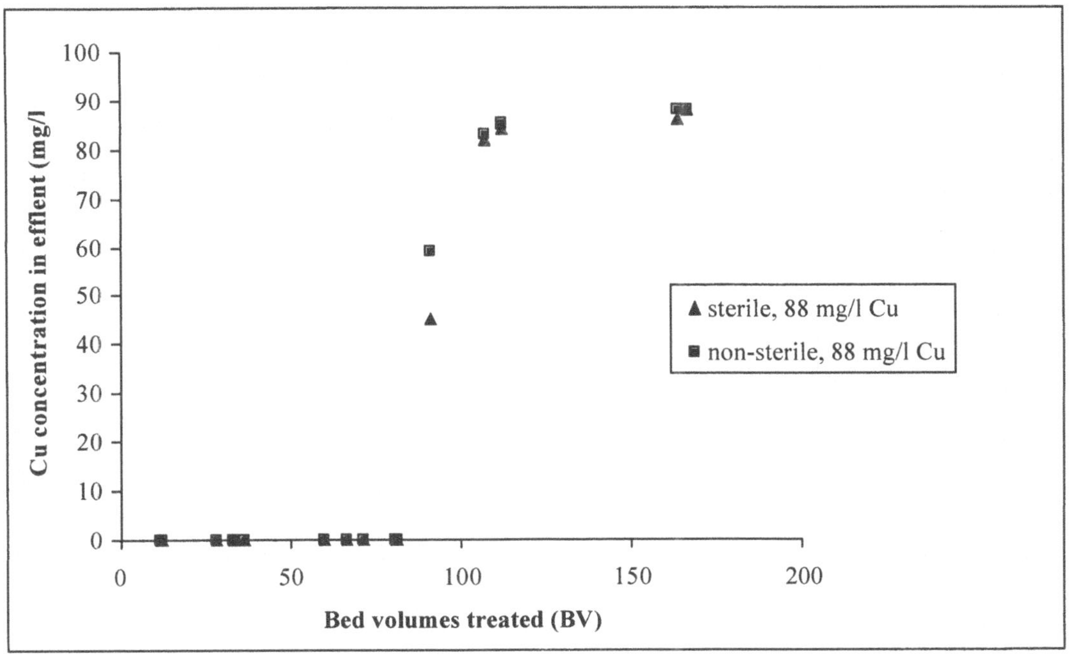

Figure 3-92. Behavior of sterile versus non-sterile columns.

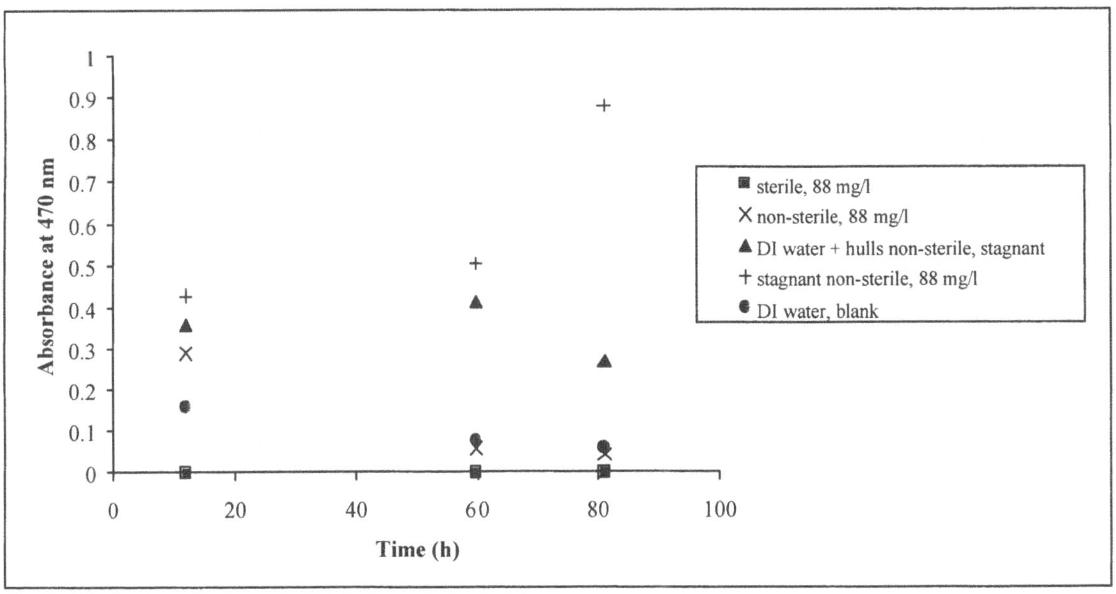

Figure 3-93. Microbial growth on the hulls during the above column test.

Several zero-flow stagnant columns were also tested to compare growth without flow conditions. Microbial growth on the non-sterile columns was approximately equal to the blank.

3.4.7 Filter Tasks Eight Results and Discussion

Under the Land Disposal Restrictions program, a restricted waste may be land disposed only if a TCLP extract of the waste or a TCLP extract of the treatment residue of the waste does not exceed the regulatory levels for hazardous constituents listed for that waste. The regulatory levels are 1 mg/L for cadmium and 5 mg/L for chromium and lead. Preliminary batch equilibrium tests (see Section 3.3.8) to examine the adsorption of Cr, Pb, and Cd by the three column media from multi- metal solutions were used to select the initial mg/l concentrations of metal solutions (spiked runoff) for generating a set of samples for TCLP analysis. The purpose was to create media samples with a range of metal concentrations for TCLP analysis.

As discussed in section 3.3.8 the lowest mg/g metal concentration was to be 0.1 mg/g for Cr and Pb and 0.02 mg/g for Cd, since these concentrations are the minimum that might result in a TCLP extract with concentrations that will cause the media to be classified as a RCRA hazardous waste. With the minimum mg/g concentrations as a start point, the initial solution concentrations of the metal spiked runoff were chosen to generate a series of metal concentrations on the media such that a reasonable range was represented. Several factors were important in selecting the initial metal concentrations of the spiked runoff.

One issue was the concentration of the metal in solution at equilibrium. An equilibrium concentration within the range of what might be found in actual runoff was one goal when selecting the initial concentrations for the metal solutions. For this reason, an equilibrium concentration of less than 1mg/l was desired with 2 mg/l being the maximum allowed.

The most important factor was the capacity of the media. Although column tests showed that peat-sand works better than the other two media at low concentrations, its overall capacity is less. Of the three media the compost had the highest overall capacity. Preliminary test showed that all three metals could be adsorbed onto the compost at the same time to levels several times higher that the minimum. However, when a multi-metal solution was used to saturate the peat-sand, only a limited range of variation was apparent for Cr and Pb compared to compost and Cd, which had a lower concentration than Cr and Pb in the initial metal solutions, never reached the minimum of 0.02 mg/g. Zeolite did better than peat-sand but not as well as compost. Because of this, having identical metal adsorption ranges for all three media was not possible, but these variations between the media are also representative of what must be considered in field conditions. However, since actual runoff treated by the media may easily have only one of the three metals, resulting in less competition for adsorption sites, it was necessary to have a good range of adsorption concentrations for all metals for the peat-sand and zeolite. Therefore, some samples of these media were exposed to only one or two metals instead of all three. To maintain some degree of comparability between the adsorption on the three media compost and most of the zeolite samples were exposed to all three metals simultaneously so that the individual metal adsorption concentrations for these samples were not drastically higher than those of peat-sand.

Table 3-31 lists the TCLP samples generated with the metal concentration initially in the solution, the equilibrium metal concentration, and the mg/g of metal adsorbed on each media. The blanks for each media were run using un-spiked runoff. The seventeen samples generated were then taken to TTL Inc. in Tuscaloosa, Alabama to have the TCLP analysis performed.

Table 3-31. Initial and final metal solution concentrations and metal adsorption on media samples generated for TCLP analysis

Sample #	Media	Initial mg/l metal concentration			Final mg/l metal concentration			mg/g of metal adsorbed onto media		
		Cr	Pb	Cd	Cr	Pb	Cd	Cr	Pb	Cd
1	Peat-Sand	0.00	0.00	0.00	0.00	0.00	0.00	0.000	0.000	0.000
2		6.21	6.22	0.00	0.15	0.09	0.00	0.120	0.121	0.000
3		7.22	11.43	0.00	0.21	0.21	0.00	0.139	0.222	0.000
4		8.30	13.65	0.00	0.22	0.23	0.00	0.160	0.266	0.000
5		15.38	0.00	0.00	0.54	0.00	0.00	0.294	0.000	0.000
6		0.00	0.00	1.25	0.00	0.00	0.07	0.000	0.000	0.023
7		0.00	0.00	4.11	0.00	0.00	0.16	0.000	0.000	0.078
8		0.00	0.00	8.24	0.00	0.00	0.52	0.000	0.000	0.153
9	St. Cloud	0.00	0.00	0.00	0.00	0.00	0.00	0.000	0.000	0.000
10	Zeolite	6.10	6.10	1.28	0.05	0.01	0.04	0.120	0.121	0.024
11		14.46	14.86	2.56	0.14	0.05	0.28	0.284	0.293	0.045
12		24.77	24.66	5.33	0.25	0.10	1.32	0.486	0.486	0.079
13		0.00	0.00	6.41	0.00	0.00	0.10	0.000	0.000	0.125
14	Compost	0.00	0.00	0.00	0.00	0.00	0.00	0.000	0.000	0.000
15		6.21	6.20	1.30	0.29	0.04	0.02	0.117	0.122	0.025
16		15.53	41.79	10.47	0.42	0.21	0.11	0.299	0.823	0.205
17		31.14	84.01	15.51	0.77	0.31	0.19	0.601	1.657	0.303

Note: Samples are currently being analyzed at TTL Inc.

3.5 Conclusions

In the column tests, the upflow column design functioned well. Pumping the water allowed contact time to be controlled. At residence times of three to ten minutes, the compost and zeolite columns showed little sign of increased headloss during the runs, even when influent suspended solids concentrations were around 400 mg/l. At these lower flow rates, most of the suspended solids appeared to settle out in the sump area of the columns.

Overall, peat had the best removal capabilities of the media in the column test, but also had the most detrimental impact on pH, the greatest headloss, and showed the most potential for clogging. Compost had the second best metal removal capabilities, while zeolite had the lowest. Compost also added color to the effluent. However, compost and zeolite had less impact on the pH of the effluent, less headloss, and exhibited less potential for clogging when compared to the peat-sand mix.

The main application of the sorption results to field conditions is in providing a comparison between the characteristics of different media, which can serve as guidelines for media selection. While the LCM and column test give insight into how the media function for actual runoff with metal concentrations that might be encountered in field situations, it is important to remember that the type and quantity of metals, pH, and other runoff characteristics can vary a great deal between sites. The results of this project emphasize the importance of characterizing the wastewater to be treated before selecting a media. Determining the range of metal concentrations to be treated is crucial to selecting the best media, since the removal efficiencies of the media relative to each other change as metal concentration increases or decreases (see Figure 3-31). The TP 207 commercial ion-exchange resin had far better removal of zinc than any of the other media at high initial zinc concentrations, but had lower removal than many of the media at the low concentrations typically encountered in stormwater and it had a dramatically decreased the pH of the treated runoff.. Conversely, the peat-sand mix had a very low removal of zinc at high initial zinc concentrations, but had better removal than many of the media at the low concentrations typically encountered in stormwater. This factor needs to be considered very carefully. At the metal concentrations used in the column tests in this study peat performed very well, it appears to be able to remove dissolved metals from solution better than the other two column media when there is a low concentration gradient between the media and influent. But if the runoff to be treated had metal concentrations at the higher extremes of what is found in runoff (see Table 3.10) the 'ultimate' sorption capacity of the peat-sand mix could quickly be exhausted. Runoff with high-end metal concentrations might be better treated with another media, such as compost. Another possibility is the use of multiple media. One example would be to use a compost filter followed by a peat-sand filter to treat runoff with high-end metal concentrations. The compost filter could remove the bulk of the metal load, reducing the metal concentrations down to the range where peat-sand works best, then the peat-sand could 'polish' the runoff to lower metal concentrations than the compost can achieve. However, further work would need to be performed to examine the potential benefits and difficulties of using the media in this study in a multi-media filter.

CHAPTER 4.0

PILOT-SCALE FILTRATION TESTS

4.1 Objectives

- See chapter 3 section 1

4.2 Introduction

Filtration can be a good solution for the control of heavy metals in stormwater. This chapter discusses a series of pilot-scale field tests that examined several different filtration media. Also described are some laboratory scale tests used to design or verify the larger tests. In addition, this chapter also describes some tests using an upflow filtration arrangement that shows promise for much more rapid treatment of stormwater heavy metals. Much of the actual data analysis can be found in Appendix B, with representative graphs and tables shown in this chapter.

In order for filtration to be accepted for use, design equations/modeling algorithms that account for all significant processes occurring within the filters must be developed and validated. In a stormwater filter, two major processes are occurring: (1) the physical removal of particulate pollutants from the water by straining the water and trapping the solids on the surface or in the pores of the media, and (2) the chemical binding reaction between the pollutants and the surface of the media, i.e., adsorption/ion-exchange. Investigators have been examining the pollutant straining mechanism for many years in water treatment and wastewater treatment filters and have developed design equations for predicting the pollutant removal ability of a filter over time for these continuous flow applications. These modeling equations typically only discuss the removal of solids and the pollutants associated with the particulate matter.

The adsorption/ion-exchange process also has been investigated and modeled both on a macro- (filter)-scale and on a micro-scale. However, these adsorption investigations have primarily been concerned with the separation of a product from a waste stream and the treatment of industrial wastewater. These traditional applications typically involve very high influent concentrations of the chemical(s) of concern and tend to have a matrix that, once characterized, will be somewhat predictable in terms of constituent concentrations and impacts on the adsorption process. The influent water to a stormwater filter will be highly variable (Pitt, *et al.*

1995). The major difference for the sorption process between the traditional applications and stormwater treatment is the significantly lower concentration of most of the pollutants in stormwater runoff (typical concentrations of metals less than 1 mg/L).

Filtration performance depends on the desired treatment rate, the source water quality (types and concentrations of pollutants), any required chemical treatment, and the physical characteristics of the media (type, size distribution, depth, and hydraulic loading rate) (Clark and Pitt 1999). In general, filter performance is measured by one or more of the following criteria, all of which likely will change over time: (1) effluent water quality (traditionally, turbidity and suspended solids concentration; possibly particle sizing and dissolved organic carbon concentration [DOC]); (2) heavy metal concentration; (3) water production (unit filter run volume); and (4) head-loss development (rate and time to backwash).

The potential for using filtration media to treat stormwater runoff has been investigated by numerous researchers (City of Austin 1988; Claytor and Schueler 1996; Clark and Pitt 1999; Urbonas 1999). The Clark and Pitt (1999) research indicated that filtration/sorption media are potentially useful for treating stormwater runoff containing heavy metals. However, many questions could not be answered under the scope of that initial research. The data needs specifically included the development and evaluation of design equations that accounted for the chemical capacity of the media, the rate of the sorption reaction between the pollutant and the media, and the need for maintenance due to potential physical clogging and/or consumption of the available chemical 'retention' capacity. Also, since this prior work was performed using laboratory-scale setups, it was unknown how the different filters would perform under normal field conditions.

One concern raised during the laboratory work of Clark and Pitt (1999) was short-circuiting/bypassing of the filter media. The literature indicated that, with a width/diameter of the filter surface greater than 20 times the average grain size, wall effects such as short-circuiting would be minimal. During the early studies, the filter diameter was approximately 100 times the average grain size (filter diameter = 45.3 mm, average particle size = 0.45 mm). However, some of the results (particularly for solids) indicated that bypassing of the media may have occurred. The bypassing could be a result of the development of 'wormholes,' i.e., holes that are approximately the size of a filter grain and that generally continue, or are linked to other holes that continue, through the complete depth of the filter.

A second concern was the effect on pollutant capture and retention of intermittent filtration where part of the filter may dry out between runoff events. This drying cycle would be expected to affect filter performance, including possibly increasing flow rate in a slow-flowing filter. The cycle of wetting and drying of the filter also could affect the growth and particularly the composition of any developing biomass. The deeper media also would remain in contact with the pore water for significant periods of time, thus allowing long-term chemical interactions and the possible development of micro-anaerobic environments in the filter. It was noted with the small columns used in the laboratory work that the complete depth of media never completely dried out (not even after being unused for several months). While it would not be expected that filters in the field would dry completely (inter-event period for Birmingham: average 3 days with a typical range of <1 day to 14 days), it would be expected that the top several centimeters would dry. In order to more accurately predict actual large-scale performance and to clarify design questions, pilot-scale setups were used during this research to better examine these effects.

The third concern is the ability of a filter to perform adequately when treating a variable influent. Most of the previous work with filtration/sorption/ion-exchange has involved the use of these adsorbers to treat wastewater with a relatively constant influent concentration. However, research has shown that stormwater runoff from the same site may have extremely variable concentrations of metals between storms. This variable influent, when combined with the intermittent operation and the potential for the filter to dry out near the surface and go anaerobic in the depth of the filter, may cause filter performance to be highly variable, although average removals and uncertainties about this removal should be able to be estimated. For this reason, many designers are looking to include a stormwater filter as part of a combination of treatment processes (Pitt, *et al.* 1999). The filter would be expected to behave similarly to other unit treatment processes that have been proposed for urban stormwater treatment, i.e., variable treatment efficiencies for a single unit process treating runoff, even when the runoff is from the same site.

The overall goal of this research was to further the understanding of the potential use of filtration media to treat stormwater runoff containing heavy metals. In order for filtration to be useful, the design would need to be sufficiently robust to cope with the variety of conditions that could be seen. Specifically, this research investigated several natural and synthetic media to determine both their applicability and their appropriateness for use in stormwater treatment (including potential future maintenance problems). Natural media are of interest because they usually are low cost and readily available. Also of interest are potential waste materials, such as a cotton textile waste, which are currently being landfilled. The purpose of this research was twofold: first, to develop a simple modeling algorithm that could describe the behavior of a stormwater filter both as a particulate removal device and as a dissolved pollutant adsorber, and second, to evaluate the filter media using a pilot-scale device and an influent of pre-settled stormwater runoff. This research investigated (using actual stormwater runoff) the effectiveness of the following media, in addition to sand: (1) activated carbon, (2) peat moss, (3) municipal leaf-based compost, (4) zeolite (ammonia-removing), (5) cotton textile waste, and (6) shredded kenaf fibers.

The research phase included the following major elements: (1) physical and chemical characteristics of the media; (2) suspended solids loading needed to clog the filters (for hydraulic design and maintenance requirements); (3) capacity of the media for various pollutants of interest in stormwater runoff, including heavy metals; (4) contact time effects on pollutant removal and retention, i.e., rate constants for sorption/ion exchange; (5) development of an algorithm to describe the removal of both particulate and dissolved pollutants by the filter media; (6) performance of pilot-scale filter columns for treating intermittent stormwater runoff flows; (7) effect of upflow filtration on the ability of the filters to retain suspended solids; and (8) effect of anaerobic conditions on the filter's ability to retain captured pollutants.

This chapter includes the following topics: the model development and includes a discussion of the methodology used to validate the models; details about these media, including sieve size and cation exchange capacity analyses, and about the filter construction for the laboratory-scale experiments; reviews the work performed in the laboratory, both the batch kinetics and capacity testing and the laboratory-scale breakthrough curve testing; information regarding the pilot-scale test setup and the results obtained from these setups; a review of the research, including the model equations and model verifications; and preliminary testing regarding upflow filtration and the effects of long-term anaerobic versus aerobic exposure.

The model algorithm investigated in this research is a combination of equations that could predict the removal efficiency of a stormwater filter for several of the heavy metals of interest. The first equation is based on work by Urbonas (1999), who tested several sand filters for stormwater treatment. Urbonas' equation states that the flow rate (measured as a loading rate with units of length/time) is a function of the suspended solids loading (measured as mass/area). The second equation of the model is based on the early equation proposed by Bohart and Adams (1920) for the removal of chlorine by charcoal and is termed the Bed Depth Service Time (BDST) analysis of fixed-bed adsorbers. As it is traditionally applied, the BDST model relates the breakthrough time of the filter to the depth of the filter bed. The integrated equation proposed by Bohart and Adams (1920) was modified based upon the known characteristics of the stormwater filter media into an equation that relates the breakthrough curve (shown as the ratio of the effluent concentration to the influent concentration) to the number of bed volumes treated. The applicable equations are the following:

$$u = k_i \bullet L_m^{-c} \qquad (4.2\text{-}1)$$

where u = unit flow/loading rate (m/day), k_i = flow rate reduction rate constant (day/m), L_m = suspended solids loading, and c = power of the flow rate reduction with suspended solids loading.

$$\ln\left(\frac{C_0}{C_e} - 1\right) = -1x10^6\, kx \sum_{j=1}^{n}\left(\frac{C_0(BV)}{u}\right) \qquad (4.2\text{-}2)$$

where C_0 = influent concentration (mg/L), C_e = effluent concentration (mg/L), k = rate constant predicted from laboratory breakthrough curves (day^{-1}), x = bed depth (m), u = unit flow/loading rate (m/day), and BV = bed volumes treated.

The first activity performed under this research project was to evaluate the Urbonas model for relating flow rate to the suspended solids loading using laboratory-scale columns. For the non-fibrous filter media, Urbonas' model could be used to describe the decrease in flow rate. However, for the fibrous media (the cotton textile waste and the agrofiber), the decrease in flow rate was not found to be a power function of the suspended solids loading; it appeared to be a linear one. The fibrous filters differed significantly from the non-fibrous ones for solids removal in that the fibrous filters appeared to use more of the filter depth for solids removal. For the non-fibrous media, surface and near-surface straining were found to be the primary removal mechanisms for the particulates in stormwater. The equations developed from this activity are shown in Table 4-1.

Table 4-1. Model Equations for the Effect of Solids Loading on Flow Rate

Filtration Media	Equation for Effect of Suspended Solids Loading on Flow Rate*
Sand	$u = 44500 \bullet L_{m,sand}^{-1.02}$
Carbon-Sand	$u = 14800 \bullet L_{m,carbon}^{-0.77}$
Peat-Sand	$u = 2000 \bullet L_{m,peat}^{-0.71}$
Compost-Sand	$u = 1.6 \times 10^{13} \bullet L_{m,compost}^{-4.09}$
Zeolite-Sand	$u = 60 \bullet L_{m,zeolite}^{-0.23}$
Agrofiber-Sand	$u = 205 - 0.09 \bullet L_{m,agrofiber}$
Cotton-Sand	$u = 106 - 0.01 \bullet L_{m,cotton}$

*Statistically significant coefficients ($\alpha = 0.05$).

The second activity was to evaluate the effects of contact time on pollutant removal in a batch setup. The test water was pre-settled stormwater runoff. Estimates of media capacity were also obtained in this series of experiments. For most pollutants and most media, the batch tests showed that a contact time of approximately 15 minutes was sufficient to remove the majority of the pollutants that the medium could adsorb. Longer contact times did not seem to significantly enhance pollutant removal. The experimental data were analyzed to find rate constants for the adsorption/ion exchange. Media capacities were measured based on an exposure of 3 days to the polluted runoff water.

The third activity used laboratory-scale filter columns to measure physical clogging and chemical breakthrough. The data from the breakthrough curves were used to estimate an operational capacity for the media, as well as to predict rate constants based upon the modified BDST model equation (Equation 4-2). The results of the breakthrough testing using laboratory-scale columns are shown in Table 4-2 for media and pollutant combinations (statistically-significant results only). One important result from the comparison of the batch capacities and rate constants to the breakthrough curve capacities and rate constants is that the batch data were not equivalent and tended to over-predict the ability of the filter. This information is important for filter/sorber designers who want to extrapolate batch data, especially from tests in which stormwater runoff was not the matrix.

Table 4-2. Design Criteria for Modified BDST Equation* (capacity and rate constants)

Pollutant	Media	Capacity (mg/g)	k (day^{-1})
Phosphate (PO_4)	Carbon-Sand	0.65	5.15×10^{-9}
	Peat-Sand	0.06	3.91×10^{-8}
	Compost-Sand	0.06	9.94×10^{-8}
	Sand	0.06	7.25×10^{-9}
Nitrate (NO_3)	Carbon-Sand		
	Peat-Sand	0.03	5.17×10^{-8}
	Compost-Sand	0.03	1.03×10^{-7}
	Sand	0.03	
Hardness	Carbon-Sand	20	4.96×10^{-10}
	Peat-Sand	20	2.30×10^{-9}
	Compost-Sand	1.2	1.20×10^{-9}
	Sand	1.2	4.13×10^{-10}
Calcium (Ca)	Carbon-Sand	0.6	1.12×10^{-8}
	Peat-Sand	>10 (clogged prior to breakthrough)	4.91×10^{-9}
	Compost-Sand		
	Sand		
Magnesium (Mg)	Carbon-Sand	0.2	2.14×10^{-8}
	Peat-Sand	>1 (clogged prior to breakthrough)	2.95×10^{-8}
	Compost-Sand	0.4	3.06×10^{-8}
	Sand		
Zinc (Zn)	Carbon-Sand		8.17×10^{-9}
	Peat-Sand		
	Compost-Sand		
	Sand		2.91×10^{-8}
Sodium (Na)	Carbon-Sand		
	Peat-Sand	0.001	4.37×10^{-8}
	Compost-Sand		
	Sand		
Potassium (K)	Carbon-Sand		
	Peat-Sand	0.53	1.99×10^{-8}
	Compost-Sand		
	Sand		

*Statistically significant rate constants ($\alpha = 0.05$)

The second result from these breakthrough tests was the growth of a biomass in the top 5 cm of the filter media. It is unknown based upon these experiments what impact the biomass had on pollutant removal and retention by the filter media. The biomass could be expected to block access to sorption sites; however, the bacteria may be incorporating some of the captured pollutants into its biomass and freeing up some sorption sites.

Pilot-scale tests were then run to compare the results obtained in the laboratory-scale columns with filters more closely sized to those that would be used in a full-scale design. For some parameters and some media, the predictions were excellent; for others, the predictions were poor. The lack of ability to predict removal for all pollutants may be due to the inability of a given medium to remove a sufficient fraction of the pollutants in the laboratory-scale breakthrough experiments so that relatively accurate rate constants could be calculated. It also may be due to the extremely low influent concentrations of some pollutants to the pilot-scale filters. The influent concentration of the pollutants may already be at the point where an equilibrium adsorption-desorption activity controls the removal ability of the filter. However, the pilot-scale testing did confirm the laboratory breakthrough curve studies in those batch tests of capacity and kinetics are not applicable to the operation of a filter in the field.

The confirmation of the modeling equations for some pollutants for some media indicates that this modeling approach has the potential to provide an estimate of the life of the filter, i.e., the time until media replacement is needed, especially for applications where the influent concentration is not quite as low as it was in the two detention ponds used in this pilot-scale study. Filtration may not be a useful best-management practice for stormwater treatment from residential areas where the pollutant loadings are relatively low, especially if the filters are used to polish detention pond effluent. One excellent application for these filters, however, would be to treat the effluent from a detention pond that receives runoff from scrap metal recyclers.

Much of the information in this chapter also was supported by the U.S. EPA (in conjunction with WERF) and has been published previously in the Ph.D. dissertation of Shirley Clark (Clark, S. *Urban Stormwater Filtration: Optimization of Design Parameters and a Pilot-Scale Evaluation*. Ph.D. Dissertation. University of Alabama at Birmingham, Birmingham, Alabama. 450 pages. 2000).

4.3 Modeling of the Stormwater Filtration Process

The ability of a filter to remove influent pollutants is a function of the likelihood of a successful interaction occurring between the medium and the pollutant. In general, for fixed-bed adsorbers and for filters whose primary action is straining the water, successful interactions are more likely to occur when the flow rate through the filter is slow. However, drainage design, of which these filters will be a component, is concerned with removing the water from the parking lot, street, etc., quickly. When using an in-line filter, this means that the flow rate through the filter should not be slower than the allowable (as set by the drainage designer) slowest flow rate. If the flow rate drops below a pre-set point, either the filter will be bypassed or the water will pond at the entrance to the filter.

The flow rate through a gravity-fed filter is a function of the suspended solids loading on the filter, i.e., the greater the suspended solids loading, the lower the flow rate of water through the filter. This is due to the capture of particles at or near the surface of the filter and the corresponding reduction in pore spaces accessible to flow. The first part of the model development included the determination of the effect of suspended solids loading on the flow rate through a gravity-fed filter (see Particulate Removal Modeling later in this section). The second part of the model development included the modification of the BDST model (as derived from the Bohart-Adams equations) and the determination of the capacity and rate constants that are needed as part of the modified BDST model (see Dissolved Pollutant Modeling later in this section). The BDST model was chosen because of its simplicity of use in this application. Since the BDST model included the flow rate through the filter, the flow rate through the filter can be modified for each successive storm event by substituting in the flow rate calculated based on the suspended solids loading. This incorporates the effects of the decreasing flow rate on the contact time between the pollutant and the media, which is expected to improve the removal ability of the filter for the dissolved pollutants. An example of the modeling algorithm, which incorporates both the effects of solids loading on the filter and the decrease in capacity of the filter for dissolved pollutants during the life of the media, is given later.

The following are brief descriptions of (1) the method proposed by Urbonas (1999) to predict flow rate in a filter loaded with suspended solids; and (2) the methods used to design a traditional fixed bed adsorber, including the modifications developed for use in this application. The research performed here includes the development of the modeling algorithm that is potentially applicable for predicting treatment efficiency, the life of stormwater filters, and the evaluation of the algorithm to predict chemical breakthrough using pilot-scale filter set-ups. This algorithm was compared with the experimental data to determine the effects of the changing influent conditions and of scale-up on the model coefficients and to evaluate the ability of the different filtration media to effectively treat stormwater.

4.3.1 Particulate Removal Modeling

Suspended solids removal in a filter is generally a straining process, i.e., physical capture of the particles at the surface or in pores whose diameter is smaller than the diameter of the particle(s) in question. An ideal filter would have a cut-off pore size that would be based on the packing density of the material, and this pore size would provide an absolute cut-off to the largest size particle that can be removed from the influent water. However, void formation (wormholes, or pores larger than a sand particle's diameter) has been observed in traditional sand filters. These wormholes start with holes in the surface deposit and remain open despite the continuing solids flow into them. Aggregates that enter the wormholes, even if they are larger than the hole, may be deformed or disrupted by the hole, yet they do not completely stop flow through the hole (Ives 1989). Preferential flow (macropore, fingering, or funneled flow) also has been observed in sand filters, as it has in many other filters and soils. During preferential flow, the fluid follows the local wetting front in wormholes and bypasses the matrix pore space. Filtration efficiency for preferential flow pathways is much smaller than it is for matrix pore flow (Steenhuis, *et al.* 1990).

The potential for preferential flow and/or void development in the filter and the resulting impact on both solids removal and flow rate through the filter forces the designer to take one of two approaches. First, flow through the matrix and voids could be modeled separately and then the results combined into an overall suspended solids removal. This approach has been applied to the modeling of water flow through a landfill; however, it requires a detailed understanding of the flow characteristics in both regimes, as well as knowing when one versus both regimes are in operation (Uguccioni and Zeiss 1997). The second approach would be to use a 'lumped-parameter' model, where the individual removal and flow mechanisms are not described in detail, and instead, the overall pattern of behavior is modeled.

In an ideal filter, drying of the filter between events should not affect its future removal ability. However, cyclic wetting and drying does affect performance. This phenomenon of increased performance with cyclic operation has been observed at sites where treated sanitary wastewater is used for groundwater recharge (National Research Council 1996). In constrast, at a Superfund site, the sand filter breakthrough occurred earlier than expected even for the colloidal-sized particles, likely because of the intermittent use of the filter (Dahab, *et al.* 1991). Flow rate recovery has been noted in laboratory scale filters used to treat stormwater runoff after a down-time of several days to several weeks (Clark and Pitt 1999). The impact of wetting and drying cycles, as well as the variability in the construction of mixed-media filters that would greatly affect preferential versus matrix flow properties, encourage the lumped-parameter-type model.

Urbonas (1999) developed a series of design equations to predict the unit flow rate through a sand filter based on the suspended solids load. They also could be used to estimate the required size (surface area) of the filter based on the required removal capacity per area of filter and on the filter's predicted maintenance interval. For a sand filter, the suggested design equation for predicting the unit flow rate, u, through the filter in feet per day $(ft^3/day)/ft^2$ based on the suspended solids load, L_m, accumulated on/in the filter in pounds per square foot of filter area is Equation 4.3.1-1 This equation indicates that, for a gravity-driven sand filter, the vast majority of the suspended solids removal occurs on the surface of the filter and therefore, removal is not a function of the media depth, but instead a function of the filter's surface area.

$$u = k_i \bullet L_m^{-c} \qquad (4.3.1\text{-}1)$$

The linear form of Equation 4.3.1-1 is as follows:

$$\ln (u) = \ln (k_i) - c \bullet \ln (L_m) \qquad (4.3.1\text{-}2)$$

Although Urbonas developed these equations for the sand filters installed in Lakewood, Colorado, Equation 4.3.1-1 and/or its linear form (Equation 4.3.1-2)) above appear to be conceptually and theoretically reasonable for modeling the suspended solids removal for any site for any filter medium whose primary solids removal mechanism is surface and near-surface straining. The coefficients calculated for each filtration media and column set-up could then be statistically compared (description of statistical work provided later) to see if the choice of media significantly affects the clogging of the filter.

4.3.2 Breakthrough Curve Theory for Dissolved Pollutant Removal Modeling

Many applications of adsorbers to waste streams use a fixed bed where the stream flows through a stationary absorbent mass. While the wastewater flows through the absorbent, many of the pollutants are expected to come into contact with active surface sites and thus be retained on the surface of the adsorbing media. When investigating the desirability of using adsorption to treat a waste stream, the approach has been to use the Langmuir and/or the Freundlich isotherms to predict the adsorption capacity of the adsorption, which then can be used to predict the life of the adsorber. This design approach must assume an instantaneous, reversible equilibrium, since it is assumed that all the sites where adsorption occurs during a lengthy isotherm test are available to a pollutant during its passage through the bed. This likely would be true if the contact time between the contaminated waste stream and the adsorbent were at least the same as the time used in the isotherm experiments. A contact time of many hours-to-days is not practical in fixed-bed adsorber applications, and, for many media, not all of the active sites are available, especially those in micropores below the media surface, when the contact time is seconds or minutes.

For a practical design, an optimum contact time must be determined. Like capacity measurements, it often has been determined through a series of batch kinetics tests in which a known mass of media is exposed to a polluted waste stream of known concentration for specific amounts of time. At the end of each time increment, the concentration of the pollutant remaining in solution is measured. The time at which the curve of time vs. the ratio of pollutant remaining in solution to initial concentration, begins to flatten is the minimum contact time.

The kinetics modeling still has as its basic assumption that adsorption is instantaneous once the pollutant reaches the active site, and, therefore, it is the rate of travel of the pollutant to the active site that controls the effectiveness of pollutant removal in a fixed-bed adsorber. Several models of one or more of the diffusion processes have been proposed and evaluated for the design of fixed bed adsorbers. The problem with many diffusion-limited models is that the diffusion coefficients are difficult to determine without the appropriate laboratory set-ups, and the results are not easily transferable to different applications. The concern for modeling stormwater filters using such a detailed approach is that the different types of diffusion are sorption-limiting at different flow rates through the filter, and it is known that the filter flow rate will vary both as the suspended solids loading on the filter increases and as the length of drying time between filtering events potentially causes a recovery in the filter's ability to remove pollutants from its influent water.

The solution to the use of detailed models is the use of a model that lumps all the rate-limiting steps into one rate constant. Bohart and Adams (1920) proposed such a model as part of their analysis on the removal of chlorine from air by charcoal. The assumption behind the Bohart and Adams equation is that equilibrium is not instantaneous, and, therefore, the rate of the sorption reaction is proportional to the fraction of sorption capacity still remaining on the media. It is based on the solution of the following two partial differential equations (Equation 4.3.2-1 and 4.3.2-2.)

$$\frac{\partial c}{\partial x} = -\frac{k}{v} ac \qquad (4.3.2\text{-}1)$$

$$\frac{\partial a}{\partial t} = -kac \qquad (4.3.2\text{-}2)$$

where a = capacity remaining on the sorbent for the pollutant (dimensionless), c = pollutant concentration at a depth x (dimensionless), x = depth of media (units of length), t = time (units of time), v = depth of media that has been saturated per unit time (units of length/time), and k = lumped-parameter rate constant (units of time^{-1}).

Bohart and Adams integrated the above partial differential equations with the following boundary conditions: (1) at time zero, the remaining capacity in the media is equal to the initial capacity of the media; and (2) at a zero filter depth, no removal occurs and the concentration at the exit is equal to the influent concentration. The result of that integration is Equation 4.3.2-3 with units appropriate for stormwater filtration design:

$$\frac{C}{C_I} = \frac{\exp(kC_I t)}{\exp\left(\frac{kN_0 x}{v}\right) - 1 + \exp(kC_I t)} \qquad (4.3.2\text{-}3)$$

where t = time (day), C_I = initial concentration of pollutant (dimensionless: parts/part), C = concentration of pollutant (dimensionless: parts/part), v = depth saturated at time t (m/day), k = lumped-parameter rate constant from Bohart and Adams theory (day^{-1}), N_0 = capacity of the media for each pollutant in a multi-component solution (dimensionless: parts/part), and x = depth of the filter bed (m).

Inverting and simplifying results in Equations 4.3.2- 4 and 4.3.2-5

$$\frac{C_I}{C} = \frac{\exp\left(\frac{kN_0x}{v}\right) - 1 + \exp(kC_It)}{\exp(kC_It)} \tag{4.3.2-4}$$

$$\frac{C_I}{C} = \exp\left(\frac{kN_0x}{v} - kC_It\right) - \frac{1}{\exp(kC_It)} + 1 \tag{4.3.2-5}$$

If it is assumed that the middle term is significantly less than the other terms (approximately zero) (Equation 4.3.2-6) as Bohart and Adams (1920) do, then the following equations result (Equations 4.3.2-7 through 4.3.2-10).

$$\frac{1}{\exp(kC_It)} \approx 0 \tag{4.3.2-6}$$

$$\frac{C_I}{C} - 1 = \exp\left(\frac{kN_0x}{v} - kC_It\right) \tag{4.3.2-7}$$

$$\ln\left(\frac{C_I}{C} - 1\right) = \frac{kN_0x}{v} - kC_It \tag{4.3.2-8}$$

$$t = -\frac{1}{kC_I}\left[\ln\left(\frac{C_I}{C} - 1\right) - \frac{kN_0x}{v}\right] \tag{4.3.2-9}$$

$$t = \frac{N_0x}{C_Iv} - \frac{1}{kC_I}\ln\left(\frac{C_I}{C} - 1\right) \tag{4.3.2-10}$$

The last equation above (Equation (4.3.2-10) is the BDST model as it is typically written (McKay 1996; Cooney 1999). The parameters k and N_0 are determined through experiments where breakthrough times are found for the filter medium of interest using a constant flow rate and a constant influent concentration and a predetermined breakthrough concentration/percent removal. Since this is a linear equation (the independent variable is the bed depth and the dependent variable is the time until the breakthrough concentration is reached), the results can be regressed and N_0 can be calculated from the slope value and k can be calculated from the intercept value (McKay 1996; Watts 1998; Cooney, 1999). When intraparticle diffusion resistance is the dominant rate-controlling factor, the rate constant k is independent of the fluid velocity in the column. However, if external film diffusion controls the rate of sorption or ion exchange, k may be found to be dependent on the velocity, u (Ko, *et al.* 1996).

The BDST model has been used successfully to model the removal of cadmium, copper and zinc from a wastewater by a cation-exchange resin (bone char) (Ko, *et al.* 1999). It also has been used to model the removal of copper and nickel from industrial wastewater (Gupta 1998).

However, the standard BDST model equation is not useful for the stormwater modeling because the parameters in the equation are not in a format that is easy for designers to use. Also, it is preferred that the depth of the filter be fixed and that the dependent variable be the ratio of the effluent concentration to the influent concentration, i.e., a true model for the breakthrough curve for a fixed depth of the adsorber bed. In order to make the model more easily applicable, the typical BDST model equation is modified as described in the following paragraphs.

Bohart and Adams (1920) defined the term, v, as the rate of the supply of pollutant expressed as the filter depth saturated per unit time, i.e., for any given time (Equation 4.3.2-11).

Saturated Depth = Volume saturated/Filter cross-sectional area (4.3.2-11)

Since the flow rate divided by the cross-sectional area is the unit-depth loading rate, u, of the filter (Equation 4-16), the saturated depth equation for a total filtering time, t, can be simplified using u and resulting in Equation 4-17.

$$Q_{flow} = uA_f \tag{4.3.2-12}$$

$$Depth\,saturated = \frac{(1{,}000\,L/m^3)(C_0)(u)(t)}{Capacity_{mass}(\rho_{media})} \tag{4.3.2-13}$$

The definition of v is the depth saturated for a unit time:

$$v = [1000(C_0)(u)]/[(N_M)(\rho_{media})]$$
[4-18]

where v = saturated depth per time (m/day), u = unit loading rate (m/day), N_M = mass capacity of media (mg/g), and ρ_{media} = density of media (mg/m^3)

Since the conversion factor between the C's in the initial Bohart equation is the same for the term on the left side of the equation, the ratio (C_I/C) with both terms dimensionless can be expressed as (C_0/C) where these concentrations are in typical concentration units of mg/L. The results of substituting for v (Equation 4.3.2-14) and replacing the C_I/C with C_0/C, along with the resulting simplifications, are shown in Equations 4.3.2-15 through 4.3.2-19.

$$\frac{C_I}{C} - 1 = \exp\left(\frac{kN_0x}{v} - kC_It\right) \tag{4.3.2-14}$$

$$\frac{C_0}{C}-1=\exp\left(\frac{kN_0x}{\frac{1{,}000C_0u}{N_M\rho_{media}}}-kC_It\right) \qquad (4.3.2\text{-}15)$$

$$\frac{C_0}{C}-1=\exp\left(\frac{kN_0N_M\rho_{media}x}{1{,}000C_0u}-kC_It\right) \qquad (4.3.2\text{-}16)$$

$$\frac{C_0}{C}-1=\exp\left(\frac{kN_M{}^2\rho_{media}x}{\left(1{,}000\frac{mg}{g}\right)\left(1{,}000\frac{L}{m^3}\right)C_0u}-kC_It\right) \qquad (4.3.2\text{-}17)$$

$$\frac{C_0}{C}-1=\exp\left(\frac{kN_M{}^2\rho_{media}x}{1{,}000{,}000C_0u}-kC_It\right) \qquad (4.3.2\text{-}18)$$

$$\ln\left(\frac{C_0}{C}-1\right)=\frac{kN_M{}^2\rho_{media}x}{1{,}000{,}000C_0u}-kC_It \qquad (4.3.2\text{-}19)$$

Where C_0 = influent concentration (mg/L), C = effluent concentration (mg/L), k = rate constant (day^{-1}), N_M = capacity of media for pollutant of interest (mg pollutant/g media), x = depth of filter (m), and u = unit loading rate (m/day).

Equation 4.3.2-19 above relates the duration of filtration to the ratio of the influent concentration to the effluent concentration. For stormwater, since the concentration of any given pollutant in water is small enough to not affect the density of water, the relationship between C_I and C_0 can be expressed using the following conversion factor, the density of water at 1 mL/g (Equations 4.3.2-20 and 4.3.2-21).

$$\text{mg/L} = (\text{mg/L})(1\ \text{mL/g})(0.001\ \text{L/mL})(0.001\ \text{g/mg}) \qquad (4.3.2\text{-}20)$$

$$C_0\ (\text{mg/L}) = 1 \times 10^{-6}\ C_I\ (\text{dimensionless}) \qquad (4.3.2\text{-}21)$$

One traditional method of comparing adsorbents is using the concept of bed-volumes treated. The use of "bed-volumes treated" as an indirect measure of filtration duration expressed as time removes the problems inherent in comparing tests performed at different flow rates and different bed depths. Substituting for time (as shown in Equation 4.3.2-22) and C_I (from Equation 4.3.2-21) leads to Equation 4.3.2-23, with all variables defined below;

$$t=\frac{(BV)x}{u} \qquad (4.3.2\text{-}22)$$

$$\ln\left(\frac{C_0}{C_e}-1\right)=\frac{kN_M{}^2\rho_{media}x}{1000000\left(\frac{mg-L}{g-m^3}\right)C_0u}-\frac{1000000\,kC_0x}{u}(BV) \tag{4.3.2-23}$$

where C_0 = influent concentration (mg/L), C_e = effluent concentration (mg/L), k = rate constant (day^{-1}), N_M = capacity of media or mix of media for pollutant of interest (mg pollutant/g media), ρ_{media} = density of media or mix of media (g/m^3), x = depth of filter (m), u = unit loading rate (m/day), and BV = bed volumes treated (dimensionless) – calculated as total volume treated divided by the volume of the media

Since many of the proposed stormwater filters are actually combinations of media, it is necessary to modify the capacity and the density to reflect the media combination. Therefore, the N_M in Equation 4.3.2-23 is the capacity of the mix, and it can be calculated as a weighted average (Equation 4.3.2-24).

$$N_{M_{mix}}=\frac{N_{M_i}(M_i)+N_{M_{sand}}(M_{sand})}{M_i+M_{sand}} \tag{4.3.2-24}$$

The density is calculated as a weighted average of the density of the sand layer and the density of the mix layer (Equation 4.3.2-25).

$$\rho_{mix}=\frac{x_{sand}}{x_{total}}\rho_{sand}+\frac{x_{mix}}{x_{total}}\left(\frac{M_{sand_{mix}}+M_i}{V_{mix}}\right) \tag{4.3.2-25}$$

4.3.3 Statistical Analysis of the Model

Based upon the discussion by Berthouex and Brown (1994) and Burton and Pitt (2001), the following list contains the steps that should be followed when building and evaluating a model. These steps were the ones followed in evaluating the various models proposed and/or used in this research.

1. Define the objectives of the curve-fitting exercise.
2. Identify alternative models from the literature that have been successfully applied for similar problems.
3. Evaluate the data to ensure that regression is applicable, including making suitable data transformations.
4. Apply regression procedures to the selected alternative models.
5. Evaluate the regression results as described below for the model(s) and select the most appropriate model(s).
6. If not satisfied, examine alternate models, especially based on data patterns and modification of the theoretical basis of existing models.

The models used in this research were ones that could be fitted by linear regression. According to Berthouex and Brown (1994), regression analyses are best used to fit data to a theoretically-derived equation that has some physical meaning.

4.3.3.1 Equation Parameter Determination

The determination of the model parameters for each of the design equations was performed using linear regression (least-squares estimation using the statistical package in SigmaPlot 5.0 SPSS, Inc.]). The Kolmogorov-Smirnov test is used by SigmaPlot to test whether the source data is normally distributed around the regression. For each attempted fit of the model equation to the data, the hypotheses tested were that the calculated slope of the equation was significantly different from zero, as was the calculated intercept of the equation. Expressed mathematically, these hypotheses were as follows:

For slopes: H_0: slope = 0
H_1: slope ≠ 0

For intercepts: H_0: intercept = 0
H_1: intercept ≠ 0

Comparison of Different Media

Slopes. The slopes of the regression lines of the media can be compared to each other to see if they are significantly different statistically (Kleinbaum, *et al.* 1998). The hypothesis testing for the model equation would be the following (the subscripts 1 and 2 denote the two different media being compared):

H_0: $slope_1 = slope_2$
H_1: $slope_1 \neq slope_2$

The test statistic for a normal distribution regression is then given by Equation (4.3.3.1-1) (Kleimbaum, *et al.* 1998, p. 323):

$$T = \frac{\beta_{11} - \beta_{12}}{\left[\frac{(n_1-2)S^2_{Y|X1} + (n_2-2)S^2_{Y|X2}}{n_1+n_2-4}\right]\left[\frac{1}{(n_1-1)\,S^2_{X1}} + \frac{1}{(n_2-1)\,S^2_{X1}}\right]} \qquad (4.3.3.1\text{-}1)$$

where n = number of observations used to construct the breakthrough curve, β_{1i} = slope of the regression line for media i, $S^2_{Y/Xi}$ = residual mean-square error for media i, and S^2_{Xi} = variance of X's for media i.

The test statistic, T, should be distributed as a Student's t with ($n_1 + n_2 - 4$) degrees of freedom when H_0 is true. The statistical test is then $|T| > t_{n_1+n_2-4,1-(\alpha/2)}$, where α = 5%.

Intercepts. The intercepts of the regression lines of the media can be compared to each other to see if they are significantly different statistically. The hypothesis testing for this using Equation (4.3.3.1-2) would be the following (the subscripts 1 and 2 denote the two different media being compared):

H_0: intercept$_1$ = intercept$_2$
H_1: intercept$_1$ ≠ intercept$_2$

The test statistic is then given by the Equation 4.3.3.1-2 (Kleinbaum, *et al.* 1998, p. 325):

$$T = \frac{\beta_{01} - \beta_{02}}{\left[\frac{(n_1 - 2)S^2_{Y|X1} + (n_2 - 2)S^2_{Y|X2}}{n_1 + n_2 - 4}\right]\left[\frac{1}{n_1} + \frac{1}{n_2} + \frac{\overline{X}_1^2}{(n_1 - 1)S^2_{X1}} + \frac{\overline{X}_2^2}{(n_2 - 1)S^2_{X2}}\right]} \qquad (4.3.3.1\text{-}2)$$

where n = number of observations used to construct the breakthrough curve, β_{0i} = intercept of the regression line for media i, $\overline{X}$ = mean of the X values for each equation, $S^2_{Y/Xi}$ = residual mean-square error for media i, and S^2_{Xi} = variance of X's for media i.

The test statistic, T, should be distributed as a Student's t with ($n_1 + n_2 - 4$) degrees of freedom when H_0 is true. The statistical test is then $|T| > t_{n_1+n_2-4,1-(\alpha/2)}$, where α = 5%.

The use of the t-distribution generally assumes that the samples are normally distributed. If it is assumed that the samples are actually log-normally distributed but with a low COV, the normal approximation is still reasonable.

4.3.4 Removal of Colloidal-Bound Pollutants

The drawback to the use of gravity filters for the treatment of stormwater runoff is their potential lack of ability to remove colloidal-sized and/or colloidal-bound pollutants. In general, colloidal-sized particulates are small enough to pass through most depth filters, at least until a significant fraction of the pore is blocked and the openings in the filter become smaller than the colloid itself. Therefore, their behavior may be hard to predict using a traditional surface-straining model. However, the pollutants associated with colloids also may not behave as a true "dissolved" fraction. Since these pollutants are strongly bound to the colloids, the potential sorption/ion-exchange ability of both the pollutant and the colloid may be used in their interaction and may not be available for an interaction with the filter media. The potential of removal of colloidal-bound pollutants was investigated as part of breakthrough curve testing in the laboratory; however, the separate effect of colloidal-bound pollutants was not determined during this research activity.

4.3.5 Laboratory-Scale Set-Up to Support Model Development

This section summarizes the set-up of the laboratory work in support of the model parameter determinations. The first part of this section describes the actual media used during these tests, including their particle/grain size distribution and expected cation exchange capacities. Second, a description of the laboratory set-up is provided. Finally, the development and confirmation of the model using laboratory-scale filters are included in this section.

4.3.5.1 Filter Media

The media selected for testing during this research were commercially available, with the exception of the two fibrous materials. The cotton is a garment-manufacturing waste material that needed a beneficial application in lieu of landfilling, while the agrofiber is derived from an alternative agricultural crop and is still under development.

Sand. The sand was purchased from Porter Warner Industries in Birmingham, Alabama. It is listed as the very fine grade (sandblast) sand. It was selected because it was the commercially available sand that was closest in size distribution to the sand used in the Jefferson County, Alabama, wastewater treatment plants in their polishing filters. The density of the sand was measured as 1,400,000 g/m^3.

***Activated* carbon.** The activated carbon, a high-purity, lignite-based carbon, was purchased from Aquatic Eco-Systems, Inc., in Apopka, Florida. It is advertised as having a small particle size (approximately 1/16 to 1/8-inch or 1.7 to 4.7 mm) in order to reduce the water travel distances necessary to enter absorption surfaces. It is designed to adsorb/absorb a full range organic contaminants, pesticides, odors, colors, chlorine, dissolved organics, ozone, and many heavy metals. The surface area is listed as approximately 625 m^2 of surface area per gram. The density of the activated carbon was measured as 370,000 g/m^3.

Peat moss. The peat moss was purchased from the K-Mart Garden Center in Homewood, Alabama. It is a *Sphagnum* moss, a fibric peat. Chopped twigs are still visible in the peat moss mixture. The density of the peat moss was measured as 190,000 g/m^3.

Compost. The compost was supplied by Stormwater Management, Inc., of Portland, Oregon. It is the compost that is sold commercially as the CSF media in their StormFilter system. The compost is from the decomposition of deciduous leaves collected in the Portland, Oregon, area. It is advertised as being excellent at removing oil, grease and heavy metals from stormwater runoff. The density of the compost was measured as 530,000 g/m^3.

Fibrous materials. The cotton fibers are a textile waste supplied by Russell Athletic in Alexander City, Alabama. It was derived from scraps of fabric by mechanical picking and combing the material apart. The agrofiber is a kenaf-based wood sliver that has been introduced into a web-like matting. It was supplied by the Forest Products Research Lab in Madison, Wisconsin. The density of the cotton textile waste was measured as 26,400 g/m^3, and the density of the agrofiber as 46,000 g/m^3.

Ammonia-removing zeolite. This clinoptilolite media was purchased from Aquatic Eco-Systems, Inc., in Apopka, Florida, and is described as an ammonia remover. The advertised capacity of this zeolite for ammonia is 1.5 g NH_4/g zeolite. However, high salinities will reduce ammonia-removing capacity, as will high pH and water temperatures (due to the generation of unionized ammonia which will quickly foul the zeolite). The zeolite is regenerated by immersing it in a salt-water solution. The density of the zeolite was measured as 800,000 g/m^3.

Lightweight "sand" filter media (used only in the pilot-scale set-up). The lightweight "sand" was purchased from Aquatic Eco-Systems, Inc., in Apopka, Florida. It is intended to be used in household-scale swimming pool filters with built-in backflush capabilities. According to the manufacturer, bed expansion with the lightweight "sand" during backwashing is significantly enhanced because it expands 30 to 50 percent with only 8 to 10 gpm per square foot backflush. It is of interest in stormwater filtration because of its weight, only 25 lb per ft^3 (400 kg/m^3). It has a weight approximately one-fourth that of regular sand. It is made of non-hydrous silicon dioxide, and its irregular shape provides extra void space (thus reducing the development of head loss). It is designed to remove particles down to 20 to 40 μm diameter sizes.

4.3.5.2 Characterization of the Non-Fibrous Filter Media by Sieve Size Analysis

One traditional method of characterizing filtration media, a sieve-size analysis, was performed on the non-fibrous media used in this project. Approximately 100 g of each media was placed in the top sieve of a stack (the sieves had openings of 4, 750; 3, 260; 1, 180; 425; 250; 150; and 75 μm) on a Ro-Tap machine. The sieve size analysis results are shown in Figure 4-1.

The d_{10} and uniformity coefficients (d_{60}/d_{10}) for each granular medium are given in Table 4-3. Traditional filter design in water and wastewater treatment encourages the use of media with a small uniformity coefficient because the pores in the filter would then tend to be more evenly sized and also the fine particles in the media are much less likely to be washed through the filter during operation. Except for the compost and peat, all these non-fibrous materials had small uniformity coefficients.

Table 4-3. Sieve Size Analysis of Non-Fibrous Media

Media	d_{10} (mm)	Uniformity Coefficient
Sand	0.27	2.52
Activated Carbon	1.3	2.54
Peat	0.08	14.4
Compost		9.09
	0.11	
Zeolite	1.6	2.31

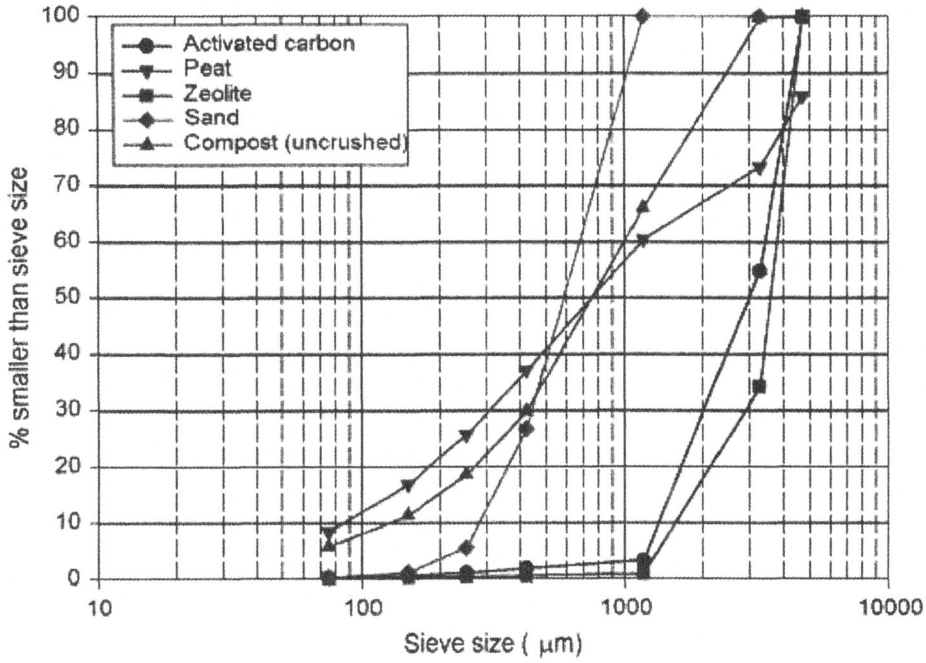

Figure 4-1. Sieve size analysis of granular media.

4.3.5.3 Characterization of the Media by Cation-Exchange Capacity (CEC)

For media that may be able to remove pollutants from a waste stream by adsorption or ion exchange, one method of characterization is through the determination of the CEC (Meier and Kahr 1999). Watts (1998) states that sandy soils generally have a CEC of between 2 and 7 milliequivalents per 100 g, silt loam soils have a CEC of between 9 and 27 milliequivalents per 100 g, and clays have a CEC of between 5 and 60 milliequivalents per 100 g.

The results of the cation exchange capacity analyses are given in Table 4-4. As expected based on the literature review, the media with the highest cation exchange capacity are the compost and the peat. All of the selected media, even the cotton textile waste, have CECs greater than the sand. This indicates that the use of a mixed media filter, rather than a pure sand filter, will provide greater removal of cations, such as metals. The CECs determined for peat and compost are in the range given by Watts (1998) for a silt loam soil, as would be expected with their organic content. The other media are generally in the range of sandy soils, also as expected.

Table 4-4. Cation Exchange Capacity of Filtration Media

Medium	CATION EXCHANGE CAPACITY (MEQ/100 G)
Peat Moss	21.47
Compost	18.83
Agrofiber	9.36
Zeolite	6.87
Activated Carbon	5.40
Cotton Waste	3.77
Lightweight Sand (used in Jacuzzi Filter)	3.20
Sand	3.49

4.3.5.4 Potential Leaching of Pollutants by Media

One concern with the use of filtration/sorption/ion exchange media is that the media itself will contribute undesirable pollutants to its effluent water, similar to the concerns regarding a solid waste. For a solid waste, a Toxicity Characteristic Leaching Procedure (TCLP) test typically is used to determine if it leaches hazardous components. The difficulty with using a traditional TCLP approach to evaluate the potential for media degradation is the selection of the test solution. Conventional TCLP testing recommends the use of a weak acid. However, a weak acid is not indicative of the effects on the media from stormwater runoff, nor are 18-MΩ water (completely deionized) or traditional deionized or distilled water. Stormwater runoff often is highly buffered, and none of these typical test waters are buffered to the level of runoff. Tap water, because of its chlorine content, would be expected to activate sites on several of the media. Therefore, the 'best' water to test the stability of the media for pollutant leaching would be stormwater runoff itself.

As part of Clark's (1996) and Clark and Pitt's (1999) tests, most of these media were tested for their removal ability for the pollutants in pre-settled stormwater runoff. The results from those experiments were examined again as part of this research to determine if the media were likely to add pollutants to the effluent water. Two sets of experiments were performed. The first was similar to the breakthrough curve testing in this research, except that the stormwater runoff used as a test water was not spiked with additional pollutants. The columns were constructed as described in the following section, with a similar depth of media and the same ratio of filtration/sorption/ion exchange medium to sand. The results from that series of experiments (turbidity [unfiltered fraction], turbidity [filtered fraction], ammonium, sodium, potassium, magnesium, calcium, total hardness, carbonate, bicarbonate, nitrate, sulfate, copper [unfiltered fraction], copper [filtered fraction], zinc [unfiltered fraction], and zinc [filtered fraction]) should indicate if the media are likely to contribute pollutants to the water being filtered. The point of interest for this re-analysis of the data is the first sample collected in the run because it is this sample that is likely to reflect any degradation of the media that would be contributing pollutants to the effluent water. Samples collected later in the run and whose C_e/C_0 is greater than 1 may reflect a wash-off of the previously captured pollutants, not necessarily a degradation of the media. The resulting C_e/C_0 values from these experiments for the first sample on the run are shown in Table 4-5.

Most of the media had effluent turbidities greater than the influent. The influent turbidity for the unfiltered fraction ranged between 0.74 and 4 NTU (all very low), while, for the filtered fraction, the influent turbidity was between 0.27 and 3 NTU (even lower). Compared to typical filtration operations, these influent turbidities are low, and the demonstrated lack of removal was not unexpected.

The compost-sand column appeared to leach calcium and magnesium and, therefore, total hardness. For all the other media, no noticeable effect was seen for calcium, magnesium, and total hardness (C_e/C_0 between 0.9 and 1.5). Potassium also appeared to be an exchangeable ion ($C_e/C_0 > 6$) for the compost-sand media. Sodium was the likely exchangeable cation for the carbon ($C_e/C_0 > 4$). For the anions, compost-sand added significant amounts of carbonate and bicarbonate to the effluent water. It also added between 1 and 2 mg/L of phosphate to the effluent of the compost-sand. The compost-sand appeared to leach nitrate. The exchangeable anion for the carbon was the sulfate.

The media contributed a significant amount (C_e/C_0 between 2 and 8) of dissolved copper to the effluent water. None of the media appeared to contribute zinc to their effluent waters. This is potentially due to the high influent zinc concentrations seen in these samples.

These results certainly indicated a potential for some of the media to degrade and release some pollutants when they are challenged with extremely low influent pollutant concentrations.

A second set of experiments conducted on these media (Clark 1996; Clark and Pitt 1999) examined the effects on influent pH and ionic strength on the removal ability of the media. Below is a re-examination of those results (in particular, the first sample of each run) in light of the potential for the media to degrade under non-normal/neutral influent pH and ionic strength. The effects of the influent pH and ionic strength on turbidity, chemical oxygen demand (COD), color, conductivity, hardness, copper, and zinc are summarized below.

Table 4-5. Potential Leaching from Media

Pollutant	Media	$C_e/C_0 > 1$ for first sample
Turbidity (unfiltered)	Sand	1.3
	Carbon-Sand	No
	Peat-Sand	4.9
	Compost-Sand	4.9
	Agrofiber-Sand	4.5
	Cotton-Sand	2.8
Turbidity (filtered)	Sand	1.4
	Carbon-Sand	No
	Peat-Sand	2.7
	Compost-Sand	4.0
	Agrofiber-Sand	3.4
	Cotton-Sand	1.5
Carbonate	Sand	1.3
	Carbon-Sand	No
	Peat-Sand	No
	Compost-Sand	3.5
	Agrofiber-Sand	2.0
	Cotton-Sand	1.6
Bicarbonate	Sand	1.1
	Carbon-Sand	No
	Peat-Sand	No
	Compost-Sand	1.5
	Agrofiber-Sand	1.0
	Cotton-Sand	1.0
Nitrate	Sand	1.0
	Carbon-Sand	No
	Peat-Sand	No
	Compost-Sand	2.6
	Agrofiber-Sand	No
	Cotton-Sand	No
Sulfate	Sand	No
	Carbon-Sand	2.9
	Peat-Sand	No
	Compost-Sand	1.0
	Agrofiber-Sand	No
	Cotton-Sand	No
Ammonium	Sand	1.8
	Carbon-Sand	No
	Peat-Sand	2.1
	Compost-Sand	15
	Agrofiber-Sand	1.6
	Cotton-Sand	1.1
Potassium	Sand	No
	Carbon-Sand	No
	Peat-Sand	1.8
	Compost-Sand	9.1
	Agrofiber-Sand	1.2
	Cotton-Sand	1.1
Magnesium	Sand	No
	Carbon-Sand	No
	Peat-Sand	No
	Compost-Sand	1.8
	Agrofiber-Sand	No
	Cotton-Sand	No
Calcium	Sand	1.0
	Carbon-Sand	No
	Peat-Sand	No
	Compost-Sand	1.1
	Agrofiber-Sand	No
	Cotton-Sand	No
Hardness	Sand	No
	Carbon-Sand	1.0
	Peat-Sand	No
	Compost-Sand	1.1
	Agrofiber-Sand	No
	Cotton-Sand	No

Table 4-5. (Continued)

Pollutant	Media	$C_e/C_0 > 1$ for first sample
Color (unfiltered)	Sand	1.0
	Carbon-Sand	No
	Peat-Sand	1.6
	Compost-Sand	2.0
	Agrofiber-Sand	No
	Cotton-Sand	No
Color (filtered)	Sand	No
	Carbon-Sand	No
	Peat-Sand	No
	Compost-Sand	2.2
	Agrofiber-Sand	No
	Cotton-Sand	No
Copper (unfiltered)	Sand	No
	Carbon-Sand	No
	Peat-Sand	No
	Compost-Sand	1.9
	Agrofiber-Sand	1.7
	Cotton-Sand	1.2
Copper (filtered)	Sand	No
	Carbon-Sand	2.6
	Peat-Sand	8.1
	Compost-Sand	1.9
	Agrofiber-Sand	8.3
	Cotton-Sand	5.4

Influent pH and ionic strength affected turbidity for peat-sand, compost-sand, cotton-sand, and zeolite-sand. For peat-sand, a high influent pH had the greatest effect on turbidity ($C_e/C_0 > 3$). High pH also affected the effluent turbidity for the compost-sand, although not to the extent of the effect on peat-sand. The combination of low pH and a normal ionic strength (approximately 200 – 300 μS/cm) affected the effluent turbidities for the peat-sand and compost-sand. For zeolite-sand, the normal ionic strength affected the effluent turbidity, C_e/C_0 was greater than 1. For the cotton-sand, the low pH, normal ionic strength influent caused the ratio of C_e/C_0 to exceed 1. In general, no effect of influent pH and ionic strength on effluent turbidities was seen for the carbon-sand and the sand media, i.e., if there was a wash-off effect, it was masked by the excellent removal ability of the filter media for particles.

No effect was seen for COD for carbon-sand, cotton-sand, and zeolite-sand. For compost-sand, only the neutral pH (with added ionic strength) influent had no visible effect on effluent chemical oxygen demand. The high pH influent caused the ratio of the effluent COD concentration to the influent COD concentration to exceed 1 for the peat-sand media. For sand, only the high pH, low ionic strength influent caused the ratio of C_e/C_0 to exceed 1.

A neutral pH, high ionic strength influent caused C_e/C_0 to exceed 1 for sand, peat-sand, compost-sand, and cotton-sand for color. The addition of color likely results from the leaching of organic acids from the media. These organic acids are released to a greater degree when the test water pH is less than neutral. For the compost-sand, the low pH of the influent appeared to affect the effluent color to the greatest degree when compared to the high influent pH. The leaching of color-causing compounds was greatest for peat-sand for the low pH, normal ionic strength influent. The ratio of C_e/C_0 exceeded 1 for the peat-sand also for both the neutral pH and the high pH (high ionic strength) influents.

The neutral-pH, high-ionic-strength influent condition resulted in media degradation for the peat-sand, compost-sand, and zeolite-sand. The ionic strength modifications were made with an evaporated seawater salt, which, in addition to sodium chloride, had a significant amount of calcium and magnesium (as calcium and magnesium chloride). Therefore, it is likely that at a neutral pH, where these chloride compounds are potentially most soluble, the removal capacity of the media was overwhelmed. For the compost-sand and zeolite-sand, at the elevated pH, precipitation of calcium and magnesium compounds may have occurred, rather than adding the compounds to the passing water, as is seen for the other influent conditions.

Because metal binding to these media has been shown to be very strong (and often irreversible), the effects of influent pH and ionic strength would be expected to be smaller than has been seen for some of the other constituents. For total copper, the neutral pH, high ionic strength influent degraded the media to the extent that C_e/C_0 was greater than 1 for the peat-sand and the zeolite-sand. A potential slight degradation was observed for the zeolite-sand, normal ionic strength condition for total copper. The low pH influent conditions had no visible degradation effect on the media.

A summary of the potentially leachable pollutants for each media is given in Table 4-6.

Table 4-6. Summary of Potentially Leachable Compounds from Media

Media	Compounds Potentially Leached
Sand	Turbidity, ammonium
Carbon-Sand	Sulfate, dissolved copper
Peat-Sand	Turbidity, ammonium, potassium, color, dissolved copper
Compost-Sand	Turbidity, carbonate, bicarbonate, phosphate, ammonium, potassium, magnesium, calcium, hardness, color, total copper, dissolved copper
Agrofiber-Sand	Turbidity, carbonate, ammonium, potassium, total copper, dissolved copper
Cotton-Sand	Turbidity, carbonate, ammonium, potassium, total copper, dissolved copper

4.3.5.5 Construction of Test Filters for Laboratory-Scale Experiments

Filter columns containing the various media were constructed in glass, Kimax-brand, one-liter, graduated burets (inner diameter [ID] = 48 mm), giving a cross-sectional area for filtration of 18 cm^2. Filter columns were constructed first by placing a square piece of fiberglass window screen in the bottom of the buret to prevent the gravel from clogging the drain hole and then filling the buret to the "1000 mL" line with epoxy-coated fish tank gravel (3-4 cm depth). A 15-cm layer of very fine sand (sandblast grade from Porter Warner Industries, Birmingham, Alabama) was added on top of the gravel (filled to "700 mL" line) before a 30-cm layer of the filtration media was placed in the column (filled to "150 mL" line). The columns were constructed using the recommended depths for the Austin sand filter and preparatory times and steps of compost filter system as supplied by CSF Treatment Systems, Inc. (1994). The preparation guidance for the compost filter system recommended that the media should be added in increments with a thorough washing with tap water between increment additions. After the filter system was completely constructed, a final rinse of at least 2 bed volumes of tap water was recommended prior to a drying time of approximately 12 to 24 hours before the filter's first use. Compaction/compression of the media occurred during this incremental filling and rinsing.

All sorption media were mixed with an equal volume of sand in order to maintain a relatively consistent hydraulic conductivity for the different media. The added sand slowed the water passage (increasing the contact time) for the larger granular media (carbon, zeolite), while providing structural support to minimize compaction and premature clogging for the non-granular material (peat, compost, cotton, agrofiber). The masses of sorption media used in the laboratory-scale breakthrough tests are given below in Table 4-7. The masses listed in Table 4-7 do not include the 400-g sand underlayer, nor the 400 g of sand mixed with the media. A diagram of the construction is shown in Figure 4-2.

The filter columns were placed on a specially constructed carousel (Figure 4-3). The water was delivered to a funnel-type flow splitter machined from Delrin plastic, using a Masterflex Peristaltic Pump with Masterflex Tygon tubing.

Table 4-7. Medium Sorption Mass Used in Laboratory-Scale Columns

Column Description	15-cm Sorption Medium Mixed with Sand (g)
Carbon-Sand	43
Peat-Sand	14
Compost-Sand	79
Cotton-Sand	3
Agrofiber-Sand	4
Zeolite-Sand	75
Sand	400

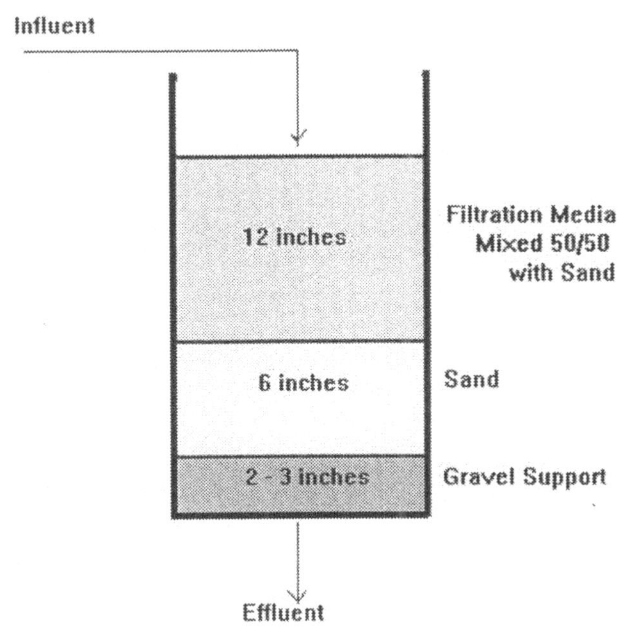

Figure 4-2. Filter column construction.

Figure 4-3. Carousel with columns constructed for laboratory-scale experiments.

4.3.5.6 Leaching of Construction Materials used in Laboratory and Pilot-Scale Treatment Test Apparatus

An important consideration when constructing the treatability apparatus was potential contamination of the test solutions by materials used in the construction of the device (Robertson 1995; Clark 1996; Clark and Pitt 1999; Pitt, *et al.* 1999). Table 4-8 lists potential contaminants from some materials that were considered for use in the bench-scale and pilot-scale test equipment (Cowgill 1988). Cowgill found that extensive steam cleaning (at least five washings using steam produced from distilled water) practically eliminated all contamination problems for sampling equipment. Cemented materials should probably be avoided, as is evident from this table. Threaded or bolted together components are preferable.

This research tested the leaching potentials for many materials that may be used in bench-scale and pilot-scale stormwater treatment units. Samples of the various materials were immersed for a period of 72 hours in approximately 500 mL of laboratory grade 18-MΩ water. Analyses conducted on each of these samples and the sample blank were the same that are used for many stormwater investigations, with the exception of solids analysis. Table 4-9 presents the contaminants that were found in the leaching water at the end of the test in concentrations sufficiently large enough to affect the test results. The most serious problems occur with plywood, including both treated and untreated plywood. Attempting to seal the wood with Formica and caulking was partially successful, but toxicants were still leached. Covering of the Formica clad plywood with polyethylene plastic sheeting seemed to eliminate any potential serious problem. Fiberglass screening material, especially before cleaning, also causes a potential problem with plasticizers and other organics. Polyvinyl chloride (PVC) plastic and aluminum may be acceptable materials, if phthalate esters and aluminum contamination can be tolerated.

The plywood used for the carousel was sealed with an enamel paint, although leaching of pollutants into the filter effluents was not anticipated because the water was not in direct contact with the plywood. The use of Teflon reduces most of the problems, but it is quite expensive. Delrin is almost as effective, is much less expensive, and is much easier to machine when manufacturing custom equipment. Glass is not typically used.

Table 4-10 is a summary of the basic materials considered for construction of the test apparatus, indicating the relative problems associated with each material and the constituents of greatest concern. Results indicated that the plastic screen used to support the filter media to be the only material to be of potential concern. Prior to installation, the screen was rinsed with tap water, which was shown by further testing to reduce leaching of toxicants.

Table 4-8. Potential Sample Contamination from Sampler Material (Data derived from Cowgill 1988)

Material	Contaminant(s)
PVC – threaded joints	Chloroform
PVC - cemented joints	Methylethyl ketone, toluene, acetone, methylene chloride, benzene, ethyl acetate, tetrahydrofuran, cyclohexanone, organic tin compounds, and vinyl chloride
Teflon	Nothing
Polypropylene and polyethy lene	Plasticizers and phthalates
Fiberglass reinforced epoxy material (FRE)	Nothing
Stainless steel	Chromium, iron, nickel, and molybdenum
Glass	Boron and silica

Table 4-9. Potential Sample Contamination from Treatability Test Apparatus Construction Materials (Data derived from Pitt, *et al.* 1999; Clark and Pitt 1999)

Material	Contaminant Observed
Untreated plywood	Toxicity, Cl, SO_4, Na, K, Ca, 2,4-dimethylphenol, benzylbutyl phthalate, bis(2-ethyl-hexyl) phthalate, phenol, N-nitro-so-di-n-propylamine, 4-chloro-3-methylphenol, 2,4-dinitrotoluene, 4-nitrophenol, α-BHC, γ-BHC, 4,4'-DDE, endosulfan II, methoxychlor, endrin ketone
Treated plywood (CCA)	Toxicity, Cl, SO_4, Na, K, hexachloroethane, 2,4-dimethylphenol, bis(2-chloroethoxyl) methane, 2,4-dichlorophenol, benzylbutyl phthalate, bis(2-ethylhexyl) phthalate, phenol, 4-chloro-3-methylphenol, acenaphthene, 2,4-dinitrotoluene, 4-nitrophenol, α-BHC, γ-BHC, β-BHC, 4,4'-DDE, 4,4'-DDD, endosulfan II, endosulfan sulfate, methoxychlor, endrin ketone, Cu, Cr, As (likely)
Treated plywood (CCA) and Formica	Toxicity, Cl, SO_4, Na, K, bis(2-chloroethyl) ether*, diethylphthalate, phenanthrene, anthracene, benzylbutyl phthalate, bis(2-ethylhexyl) phthalate, phenol*, N-nitro-so-di-n-propylamine, 4-chloro-3-methylphenol*, 4-nitrophenol, pentachlorophenol, α-BHC, 4,4'-DDE, endosulfan II, methoxychlor, endrin ketone, Cu, Cr, As (likely)
Treated plywood (CCA), Formica and silica caulk	Lower pH, toxicity, bis(2-chloroethyl) ether*, hexachlorocyclopentadiene, diethyl phthalate, bis(2-ethylhexyl) phthalate, phenol*, N-nitro-so-di-n-propylamine, 4-chloro-3-methylphenol*, α-BHC, heptachlor epoxide, 4,4'-DDE, endosulfan II, Cu, Cr, As
Formica and Silica caulk	Lower pH, toxicity, 4-chloro-3-methylphenol, aldrin, endosulfan I
Silica caulk	Lower pH, toxicity, heptachlor epoxide
PVC pipe	N-nitrosodiphenylamine, and 2,4-dinitrotoluene
PVC pipe – cemented joint	Bis(2-ethylhexyl) phthalate*, acenaphthene, endosulfan sulfate
Plexiglass & cement	Naphthalene, benzylbutyl phthalate, bis(2-ethylhexyl) phthalate, endosulfan II
Aluminum	Toxicity, Al (likely)
Plastic aeration balls	2,6-dinitrotoluene
Filter fabric material	Acenaphthylene, diethylphthalate, benzylbutyl phthalate, bis(2-ethylhexyl) phthalate, pentachlorophenol
Sorbent pillows	Diethylphthalate, bis(2-ethylhexyl) phthalate
Black plastic fittings	Pentachlorophenol
Reinforced PVC tubing	Diethylphthalate, benzylbutyl phthalate
Fiberglass window screening	Toxicity, dimethylphthalate, diethylphthalate*, bis(2-ethylhexyl) phthalate, di-n-octyl phthalate, phenol, 4-nitrophenol, pentachlorophenol, 4,4'-DDD
Delrin™	Benzylbutyl phthalate
Teflon™	Nothing (likely)
Glass	Zn (likely from borosilicate glass)

Table 4-10. Summary of Leach Potential Commonly Used in Treatability Test Apparatus (Data derived from Pitt, *et al.* 1999)

MATERIAL	LEACH POTENTIAL
Teflon	LOW
Delrin	LOW
	LOW (but potential for zinc problem)
GLASS	
PVC pipe and cement	LOW
Polyethylene sheeting	LOW (n-nitroso-di-n-propylamine)
Plexiglas and cement	LOW (conductivity, chloride, sodium)
Formica and caulk	LOW (toxicity, conductivity, pH, nitrobenzene, 4-chloro-3-methylphenol)
Aluminum	LOW (toxicity, conductivity, chloride, calcium, pentachlorophenol)
Wood	MODERATE (OK for non-contact uses, but needs to be sealed)
Plastic screen (un-washed)	HIGH (toxicity)
	LOW (phenol, 4-nitrophenol, pentachlorophenol, di-n-octylphthalate)
Treated plywood	HIGH (toxicity, hexachloroethane, 2,4-dimethylphenol, 4-chloro-3-methylphenol, 4-nitrophenol; likely heavy metals)

4.3.6 Model Parameter Determination through Laboratory-Scale Experimentation

4.3.6.1 Physical Capacity of the Media (Clogging Tests) Using Downflow Filtration

The purposes of this research task were to determine (1) the amount of suspended solids loading on the media that will reduce the flow rate to certain end points; and (2) the effect of maintenance, such as disturbing or removing the top 3 to 5 cm of the mixed media, on flow rate. This task was performed using laboratory-scale columns for compost, activated carbon, zeolite, cotton textile waste, agrofiber, peat moss, and sand. The data generated under this task will be compared with the results from the pilot-scale field studies and shall be used to calculate estimated scale-up factors that will enable designers to use bench-scale data to design full-scale filters for actual installation and use. The results of these experiments also can be used to estimate the removal of particulate-bound pollutants since their behavior should mimic the behavior of the solids to which they are attached.

The columns required were constructed as described earlier in this chapter. The test water used was a mixture of red clay and tap water. The red clay was selected because it was easily visible in the column (and, therefore, the depth of filter being used for physical removal can be determined). To mimic particle sizes found in stormwater, the clay used in this series of tests was first passed through a 250-μm sieve. The concentration of clay in the test water was approximately 4 g/L in the first series of clogging tests and 1.5 g/L in the second series of tests. Because of the tendency of the clay to settle, the test water was continuously stirred throughout the experiments. For comparison, the bench-scale breakthrough tests and the pilot-scale tests used actual stormwater during these prolonged tests to examine the applicability of the equations calibrated during these experiments.

Once the initial flow rate through the column was determined using tap water, the clay-water mixture was passed through the column. The flow rate through the column was determined in the same manner that the initial flow rate was determined (measure effluent volume and the amount of time required to collect the volume). Flow rate measurements and effluent turbidity measurements were collected periodically throughout the filter run. The endpoints of interest in this experiment (where maintenance may be required) were when the flow rate reached 10 m/day, 5 m/day, and 1 m/day. When the flow rate through the column reached 5 m/day, maintenance (performed only during the first series of tests), such as disturbing the top 3 to 5 cm of media, was performed. Immediately after any maintenance activity was performed, a flow rate measurement was taken. The filter surface was disturbed whenever a medium's flow rate reached 5 m/day. Eventually, the maintenance of disturbing the media was no longer effective, and the flow rate continued to decrease. When the flow rate reached 1 m/day, the visible depth of penetration of the red clay was also measured.

The results of the downflow experiment for all the media except zeolite are shown in Figure 4-4. The graphs show flow rate versus suspended solids loading (g solids/m^2 filter area) for the two downflow series of tests. Zeolite was not included in this graph because the size distribution of the media was significantly changed by the manufacturer between the batches used in the first and the second series of downflow tests. The zeolite in the first series of tests was approximately twice as large as the zeolite pellets that were used in the second series of tests.

The scatter in the data confirms the effect that column preparation has on the flow of water through the column (re: initial flow rate) and on the particulate removal ability of the media. This scatter was apparent even in the sand column in which the scatter should have been minimal. However, the general trend of carbon-sand having the greatest loading capacity prior to clogging and peat-sand having the least capacity is the same between the two series of tests.

Along with the flow rate information, the visible depth of penetration at clogging (flow rate ≤ 1 m/day) was noted during the first series of downflow clogging tests. Table 4-11 gives the penetration depth at clogging both in centimeters and in percent of the filter depth that contained visible red clay.

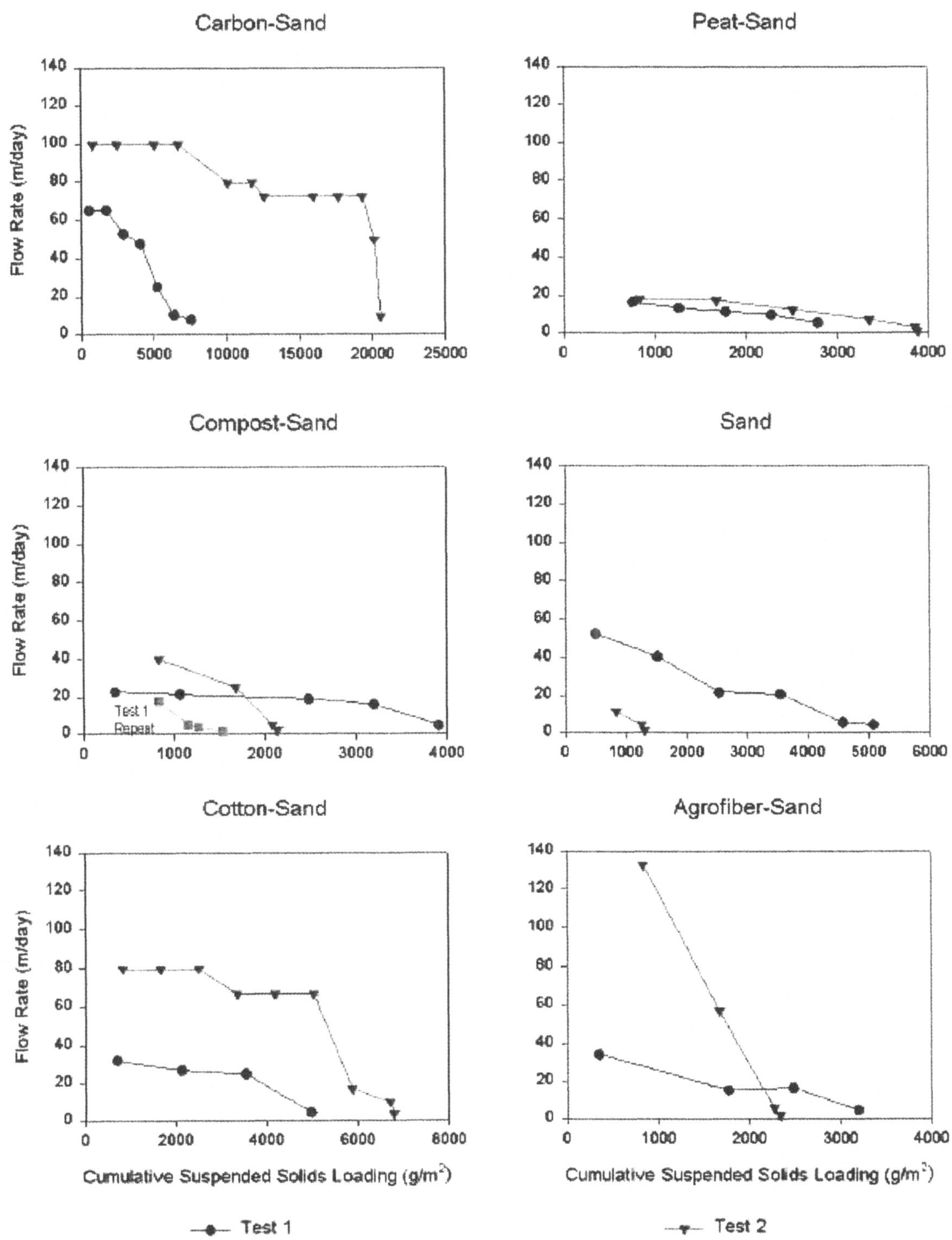

Figure 4-4. Effect of cumulative suspended solids loading from a clay-water mixture on the filter flow rate.

Table 4-11. Visible Depth of Red Clay Penetration When Filters Clogged ($u \le 1$ m/day).

MEDIA	Penetration Depth at Clogging (cm)	Penetration Depth as Percent of Filter Depth
Sand	8.89	19
Carbon-Sand	4.76	10
Peat-Sand	1.9	4
Compost-Sand	3.81	8
Zeolite-Sand	3.81	8
Cotton-Sand	1.27	3
Agrofiber-Sand	8.26	18

As Table 4-11 indicates, no more than 20 percent of the filter depth contained visible quantities of the test clay at the time of clogging. The removal of approximately the top 1 cm was found to be ineffective at significantly recovering flow rate, even in the cotton-sand filter where the visible depth of penetration was not significantly greater than the depth of media removed. This clogging and lack of recovery, even when the layer with visible solids buildup is removed, indicated that a stormwater filter may no longer be usable well before the chemical capacity of the media is exhausted. Penetration of small solids was apparently greater than that seen with the visible red clay penetration.

Regression equation parameters for each medium. The linear form of the modeling algorithm equation (Equation 4.3.6.1-1, shown below) was used to determine the parameters defining solids removal in a downflow filter for each medium.

$$\ln q = \ln k_i - c \bullet \ln (L_m) \qquad (4.3.6.1\text{-}1)$$

Equation 4.3.6.1-1 indicated that a linear relationship between the flow rate and the suspended solids loading existed once both sets of data have been log-transformed and, therefore, the transformed data could be analyzed using linear regression techniques. The hypotheses that are tested for the linear regression of the data for each medium are as follows.

For the intercept of the linear equation:

H_0: $\ln k_i = 0$
H_1: $\ln k_i \neq 0$

For the slope:

H_0: $c = 0$
H_1: $c \neq 0$

Coefficients that were statistically significant from zero (at $\alpha \le 0.05$) in the regression analysis are shown in Table 4-12. If the intercept was found to not be significant, the data were regressed using Equation 4.3.6.1-2

$$\ln u = -c \bullet \ln (L_{mi}) \qquad (4.3.6.1\text{-}2)$$

For the data regressed using Equation 4.3.6.1-2, no intercept data is shown in Table 4-12 (designated as "not applicable"). If neither equation was found to have statistically significant coefficients, the entry in the table is "not significant." Residual plots for the model equations whose coefficients are given in Table 4-12, except for the Test 1 results for compost-sand, cotton-sand, and agrofiber-sand, show a normal random distribution and thus indicate that assumptions of the linear regression are upheld for the data. Also, for all data presented in Table 4-12, the assumption of normality around the regression was upheld.

For the Test 1 results for compost-sand, cotton-sand and agrofiber-sand, neither the slope nor the intercept was found to not be statistically significant. The data were reanalyzed using Equation 4.3.6.1-2. While the slopes of the new regression were found to be significant (new slopes shown in italicized type in Table 4-12), no statistical correlation between the suspended solids loading and the flow rate was found. As can be seen by examination of Equation 4.3.6.1-2, the intercept for the revised regression was zero.

Table 4-12. Model Equation Coefficients for the Effect of Suspended Solids Loading on Flow Rate (95% confidence intervals for coefficients)

Media	$\ln k_i$		Corresponding k_i		c	
	Test 1	Test 2	Test 1	Test 2	Test 1	Test 2
Sand	10.7 (4.1, 17.3)	Not significant	44500	Not significant	1.02 (0.17, 1.9)	Not significant
Carbon-Sand	9.6 (3.9, 15.4)	Not significant	14800	Not significant	0.77 (0.06, 1.5)	Not significant
Peat-Sand	7.6 (3.1, 12.2)	Not significant	2000	Not significant	0.71 (0.1, 1.3)	Not significant
Compost-Sand	Not applicable	30.4 (26.3, 34.5)	Not applicable	1.6×10^{13}	*0.35*	4.09 (3.5, 4.7)
Zeolite-Sand	4.07 (2.9, 5.3)	71.1 (48, 94)	60	7.6×10^{30}	0.23 (0.08, 0.39)	0.04 (0.02, 0.05)
Cotton-Sand	Not applicable	12.8 (3.8, 22)	Not applicable	3.6×10^{5}	*0.36*	1.14 (0.04, 2.2)
Agrofiber-Sand	Not applicable	Not significant	Not applicable	Not significant	*0.39*	Not significant

However, the general trend of the data was directly opposite that provided by the new regressions. Therefore, in the further evaluations of the model equations and the future calculations with these modeling equations, only the parameters listed in Table 4-12 (not italicized) were used. The result was that no modeling equation of the form proposed by Urbonas (1999) for the particulate removal by agrofiber could be developed.

However, as can be seen by the agrofiber-sand data, a relationship existed between the suspended solids loading on the media and the flow rate measured as a unit-loading rate. Therefore the agrofiber-sand loading versus flow rate data was regressed assuming a linear relationship between the two parameters, i.e., the design equation for the agrofiber-sand was Equation (4.3.6.1-3)

$$u = c \bullet L_m + b \qquad (4.3.6.1\text{-}3)$$

where u = loading rate (m/day), c = rate of loading rate decrease (m^3/g day), L_m = suspended solids loading on the media (g/m^2), and b = intercept of the regression line (m/day).

The results of the regression analysis for the first series of tests of clogging in agrofiber showed a significant intercept and a non-significant slope, i.e., the suspended solids loading and the flow/loading rate were not related in a statistically significant manner. However, the results of the regression analysis for the second series of agrofiber clogging tests showed a significant result (with normally distributed residuals) of the form of Equation 4.3.6.1-4.

$$u_{ag} = 205 - 0.09 \bullet L_{m,agrofiber} \qquad (4.3.6.1\text{-}4)$$

The 95% confidence intervals were calculated for the intercept (193, 217) and for the slope (0.08, 0.093). Equation (4.3.6.1-5) is the equation that should be used in the future model evaluation for agrofiber. This result indicates that more of the filter depth is being used for particulate removal, as opposed to a surface straining mechanism that is seen in the other media. Due to the void space decreasing deeper in the filter due to compaction, the media clogs quickly as the suspended solids front moves deeper in the filter and occupies more fully the smaller pore space at the bottom of the filter.

A second check of the validity of the relationship between the suspended solids loading on the media and the unit flow/loading rate was to evaluate the coefficients of determination (R^2) for each regression. Table 4-13 gives the correlation coefficients for each regression. Since the hypothesis behind the equation proposed by Urbonas (1999) is that the flow rate through a filter is controlled by the flow rate entering the filter and that the flow rate entering the filter is controlled by the quantity of suspended solids on the top of the filter clogging/blocking the pores, high R^2 values might indicate that his hypothesis likely is valid. High R^2 values were seen for the peat-sand, the compost-sand, and the zeolite-sand (second series where the zeolite particles were significantly smaller). The R^2 value for cotton-sand indicated that less than half the variance of the flow rate data in the second series of tests could be explained by the relationship with the suspended solids loading. Therefore, these data were regressed using a linear relationship rather than a power relationship between the unit flow/loading rate and the suspended solids loading. The results show that the R^2 value for the new regression is 0.80, indicating that the relationship between the two variables for cotton-sand was more of a linear relationship than a power relationship. Thus, the equation that should be used in future modeling of the cotton-sand media is the following (slope and intercept statistically significant; residuals normally distributed) (Equation 4.3.6.1-5).

$$u_{cotton} = 106 - 0.01 \bullet L_{m,cotton} \qquad (4.3.6.1\text{-}5)$$

The 95% confidence intervals were calculated for the intercept (79, 134) and for the slope, (0.007, 0.02).

Table 4-13. Coefficients of Determination (R^2) for the Effect of Suspended Solids Loading on Flow Rate

Media	R^2	
	Test 1	Test 2
Sand	0.73	Not significant
Carbon-Sand	0.61	Not significant
Peat-Sand	0.82	Not significant
Compost-Sand	Not applicable	0.99
Zeolite-Sand	0.69	0.99*
Cotton-Sand	Not applicable	0.46 (using power relationship) 0.80 (using linear relationship)
Agrofiber-Sand	Not applicable	0.99 (using linear relationship)

*Intercept and associated confidence intervals were not reasonable and the model was therefore rejected, based on the graphical analysis of the residuals.

Summary of the evaluation of the particulate removal models. The proposed models for the modeling of particulate removal from a stormwater runoff are shown in Table 4-14.

Table 4-14. Significant Models of Effects of Particulate Loading on Flow Rate

Filtration Media	Equation for Effect of Suspended Solids Loading on Flow Rate
	$u = 44500 \bullet L_{m,sand}^{-1.02}$
SAND	
Carbon-Sand	$u = 14800 \bullet L_{m,carbon}^{-0.77}$
Peat-Sand	$u = 2000 \bullet L_{m,peat}^{-0.71}$
Compost-Sand	$u = 1.6 \times 10^{13} \bullet L_{m,compost}^{-4.09}$
Zeolite-Sand	$u = 60 \bullet L_{m,zeolite}^{-0.23}$
Agrofiber-Sand	$u = 205 - 0.09 \bullet L_{m,agrofiber}$
Cotton-Sand	$u = 106 - 0.01 \bullet L_{m,cotton}$

The first test for model validity is to answer the questions: (1) Does the model make sense? (2) Does it have a theoretical basis? For these proposed model equations, the answers to the two above questions are "yes." The model makes sense when it states that the decrease in flow rate through the media is a function of the suspended solids loading on the surface of the media (for the power-decay equations) and throughout the depth of the media (for the fibrous media where the relationship is a linear one with suspended solids loading).

The second test of the models is included in the residual analysis. The results of this evaluation of the models are shown in Table 4-15. The coefficients of the regression equation are shown along with the probability of them being significant. Also shown in the table is the probability of the Durbin-Watson statistic showing an autocorrelation between the data (p values greater than the statistically significant level indicate that either autocorrelation does not exist or that the existence of autocorrelation cannot be shown in the data set). The Kolmogov-Smirnov test for normality attempts to determine if the residuals are normally distributed (normality is assumed if the p value exceeds the criteria for statistical significance). The p value for constant variance is similar to the test for normality and the Durbin-Watson statistic test in that the null hypothesis is that the data have a constant variance, and therefore, the assumption of constant variance is made when the p value exceeds the statistical significance criteria.

Table 4-15. Evaluation of Particulate Loading Effects on Flow Rate Models

Media	ln k_i or k_i	α of k_i	c	α of c	p of Durbin-Watson statistic	Normality test p	Constant variance test p	R^2	Model p
Sand	10.7068	0.0109	1.0171	0.0293	> 0.1	0.1418	0.0600	0.734	0.0293
Carbon-Sand	9.6369	0.0076	0.7741	0.0378	> 0.1	0.1931	0.6602	0.611	0.0378
Peat-Sand	7.6225	0.0130	0.7125	0.0348	>0.05	0.5266	0.0500	0.818	0.0348
Compost-Sand	30.4229	0.0010	4.0880	0.0011	>0.05	0.6549	<0.0001	0.998	0.0011
Zeolite-Sand	4.0728	0.0002	0.2340	0.0104	>0.05	0.1523	0.2604	0.692	0.0104
Agrofiber-Sand	204.8698	0.0002	0.0870	0.0003	>0.05	0.5595	<0.0001	0.999	0.0003
Cotton-Sand	106.5686	0.0001	0.0133	0.0011	>0.1	0.4637	0.6434	0.802	0.0011

The results in Table 4-15 indicate that the requirements for the model equations are met for all of the media except agrofiber-sand and compost-sand. For these two media, the constant variance test of the regression failed. These two media also demonstrate that, as Berthouex and Brown (1994) and Burton and Pitt (2001) discuss, a very high R^2 value is not necessarily an indication of an acceptable model. A review of the data for those two regressions shows that the regression was based on fewer data points than the other regressions (less than five points each for the agrofiber-sand and compost-sand). The small number of sample points affects the likelihood that the model will pass the test of constant variance (generally, a minimum of 10 to 15 data points is preferred for a regression and residuals analysis).

Comparison of media for particulate removal models. The statistical analysis to compare between the media (to see if the choice of the filtration medium affects the time until the media is 'clogged' or the time to breakthrough for chemical capacity) was two-fold: (1) testing of slopes and (2) testing of intercepts. Should groups of equations (such as between media) not be statistically significant from each other, the media could be categorized according to groupings of the regression parameters. For the design engineer, these groupings would be very useful for estimating the design equations for a media combination that was not tested but that can be assumed to behave like a specific category of media (for example, possible groupings might be organic media, granular inorganic media, fibrous media).

The hypothesis for comparing the slopes of the physical clogging model equation (based on the linear Equation 4-3) was the following (the subscripts 1 and 2 denote the two different media being compared):

$H_0: c_1 = c_2$

$H_1: c_1 \neq c_2$

For the physical modeling equation, the regression results showed that the regressions passed the normality test. Therefore it was appropriate to use the test statistics given above. The results of the analysis for the 'c' value in the physical model equation showed that the slope of the linear equation for the impact of physical impact of clogging on the flow rate through the media was statistically significantly different for the media comparisons shown in Table 4-16. The table data indicated that establishing groupings of the media would not be valid because the models are statistically different from each other based on the slope of the regression line (the power term in the equation).

Table 4-16. Statistically Significant Slopes of the Regression (p value of t-test shown)

	Sand	Carbon-Sand	Peat-Sand	Compost-Sand (Old)	Compost-Sand (New)	Zeolite-Sand (1)	Zeolite-Sand (2)	Cotton-Sand	Agrofiber-Sand
Sand	--		<0.01	< 0.01		<0.01		NA	NA
Carbon-Sand		--	<0.01			<0.01		NA	NA
Peat-Sand	<0.01	<0.01	--		<0.01	0.01	<0.01	NA	NA
Compost-Sand (Old)	<0.01			--	0.02		0.03	NA	NA
Compost-Sand (New)			<0.01	0.02	--	<0.01		NA	NA
Zeolite-Sand (1)	<0.01	<0.01	0.01		<0.01	--	<0.01	NA	NA
Zeolite-Sand (2)			<0.01	0.03		<0.01	--	NA	NA
Cotton-Sand	NA	NA	NA	NA	NA	NA	NA	--	<0.01
Agrofiber-Sand	NA	NA	NA	NA	NA	NA	NA	<0.01	--

NA = not applicable. The agrofiber-sand and cotton-sand slopes and intercepts were only compared against each other since they were based on linear relationships between the variables, rather than a power relationship.

The intercepts of the regression lines of the media (regression equation form: $Y = \beta_0 + \beta_1 X$) were compared to each other to see if they were statistically significantly different. The hypothesis testing for Equation 4-3 would be the following (the subscripts 1 and 2 denote the two different media being compared):

H_0: $\ln k_{i1} = \ln k_{i2}$
H_1: $\ln k_{i1} \neq \ln k_{i2}$

This comparison was performed on all the media except for agrofiber and cotton, since these two media had a different form of the relationship between suspended solids loading and flow rate. Cotton and agrofiber were compared to each other in a similar test and with the following hypothesis:

H_0: $b_{cotton} = b_{agrofiber}$
H_1: $b_{cotton} \neq b_{agrofiber}$

The results of the comparison of the intercepts showed no statistically significant difference for all the media for which the Urbonas equation was applicable. For the cotton and the agrofiber, however, their intercepts were found to be statistically significantly different from each other.

The lack of repeatability between the tests for each medium indicated that the filter construction and homogeneity of the mixture was crucial in determining the ability of the filter to remove suspended solids from the influent water. In general, the performance of the peat-sand filter was significantly different from that of the other media. This was likely due to the small size of many particles that will fill the pore spaces in the sand. Peat was defined as a well-graded material, i.e., with a wide distribution of pore sizes, unlike in the other media. Peat also exhibited a significant shrink-swell behavior when wet. This greatly affected the hydraulic conductivity of a peat-sand filter. Part of the purpose of this activity was to categorize the media for design purposes. However, it appeared from this data that no category could be established for any of these filters operated in a downflow mode.

4.3.6.2 Dissolved Pollutant Removal Modeling

In general, the removal of the dissolved fraction of pollutants in stormwater will occur through ion exchange or sorption. In order to predict the effluent quality after filtration, two parameters of the sorption process need to be quantified, the capacity of the media for the pollutant in question and the rate at which the pollutant 'binds' to the medium.

Kinetics and capacity investigation using batch experiments. Van Benschoten, *et al.* (1994) demonstrated that metal sorption kinetics has both a fast reaction, where metals bind to surface sites, and a slow reaction, where metals bind to interior sites. When designing a stormwater filter, the design flow rate should, at least, be based on the time required for most of the pollutants participating in the 'fast' reaction to be removed from the influent water. Generally, a contact time that is longer than that needed for the 'quasi-equilibrium' to occur is not economically feasible for stormwater filters whose secondary function is to remove the runoff from the parking lot, street, etc. to the drainage system quickly.

Capacity analysis assumes that the system has reached equilibrium. When a mass of adsorbent and a waste stream are in contact for a sufficiently long time, an equilibrium between the amount of pollutant adsorbed and the amount remaining in solution will develop. For any system under equilibrium conditions, the amount of material adsorbed onto the media, its capacity, can be calculated using the mass balance of Equation 4.3.6.2-1, which will calculated the amount sorbed at any given time.

$$\frac{X}{M} = (C_0 - C_e)\frac{V}{M} \qquad (4.3.6.2\text{-}1)$$

where X/M = mass of pollutant per mass of media (mg pollutant/g media), C_0 = initial pollutant concentration in solution (mg/L), C_e = pollutant concentration in solution after equilibrium has been reached (mg/L), V = volume of the solution to which the media mass is exposed (L), and M = mass of the media (g).

The purpose of this phase, the batch kinetics and capacity investigation, was twofold: (1) the determination of the optimum residence/contact time between the water and media, and (2) the capacity of the media for the metals of interest. One result was the development of a curve of removal efficiency versus contact time (similar to a traditional rate-of-sorption experiment), allowing an efficient contact time to be selected, but which may not be equal to the time required for the media and runoff water to truly achieve equilibrium. Longer contact times require larger filters, and, therefore the goal is to provide sufficient contact time to achieve the required removal but not to increase the contact time if the gain in removal efficiency is small. For an exponential relationship between removal efficiency and time, the result is that the desired contact time is immediately before the curve begins to level off.

The second result was the determination of the sorption capacity of each medium and heavy metals after a 3-day exposure period. The results of the batch kinetics and capacity experiments would then be compared with the results of the breakthrough curve testing to determine if rate constants and capacities determined in the batch experiments were statistically different from those determined using the breakthrough curves. Batch testing would be desirable because the testing generally is cheaper and easier and may require less sample analysis.

This series of experiments was performed by immersing known quantities of media in spiked pre-settled stormwater for different periods of time and then analyzing the concentrations of the various pollutants remaining in the water. In order to confirm repeatability of these results, triplicates were run of each test increment (each time for the kinetics and each weight for the capacity). The spiking concentrations for the initial test water are shown in Table 4-17.

For the contact time testing, known quantities of each medium (10 g sand, 5 g compost, 5 g carbon, 2 g peat, 1 g agrofiber, and 1 g cotton) were placed in contact with 500 mL of the spiked stormwater runoff for a specified period of time: 1 minute, 5 minutes, 15 minutes, 30 minutes, and 60 minutes. At the end of each designated time, the water was filtered using a 0.45-μm MSI MicronSep nitrocellulose filter and analyzed. For the capacity testing, three weights of each media were placed in contact with the spiked stormwater runoff for a period of 3 days. The weight ratios used were one-hundredth of, one-tenth of, and the same mass of media used in the kinetics experiments. At the end of 3 days, the water was filtered through a 0.45-μm nitrocellulose filter and analyzed. For each medium, the kinetics and capacity tests were run at the same time, thus allowing the 3-day sample to be compared with the results from the shorter exposure periods. The filtrate waters were analyzed for the pollutants listed in Table 4-18. Triplicates were run for each medium for each test condition. Blank samples (samples with no medium) were also run for each series of tests to provide the influent concentration values for the later calculations of removal efficiency. The purpose of testing blank samples rather than relying upon analysis of the initial batch of spiked stormwater runoff was to account for the effects of any potential sorption to the test jars.

Table 4-17. Spike Concentrations and Standards Used for Kinetics and Capacity Experiments

Pollutant	Approximate Concentration (μg/L)	Standard Used to Provide Spike
Lead	500	Lead Nitrate in Dilute (2%) Nitric Acid, 1000 ppm
Zinc	2000	Zinc Sulfate ($ZnSO_4 \bullet 7H_2O$)
Copper	1000	Cupric Sulfate ($CuSO_4 \bullet 5H_2O$)
Iron	1500	Ferrous Sulfate ($FeSO_4 \bullet 7H_2O$)
Phosphate	2000	Potassium Dihydrogen Phosphate, 1000 ppm

Table 4-18. Analysis Methods for Kinetics and Capacity Testing

POLLUTANT	Analysis Method (EPA Method number shown if available; equivalent Standard Methods method in parentheses, if available)
PH	EPA Method 150 (Standard Methods 4500-H^+.B.)
Conductivity	EPA Method 120.6 (Standard Methods 2510.B.)
Phosphate	EPA Method 365.2 (Standard Methods 4500-P.E.)
Nitrate	EPA Method 353.3 (Standard Methods 4500-NO_3^-.E.)
Ammonia	EPA Method 350.2
Hardness	EPA Method 130.2 (Standard Methods 2340.C.)
Alkalinity (bicarbonate and carbonate Ions)	Based on Standard Methods 4500-CO2. D.
Lead, cadmium, zinc, copper, chromium, iron, calcium and magnesium	ICP (Standard Methods 3500)

The kinetics results are presented graphically for sand, activated carbon, compost, peat moss, cotton textile waste, and agrofiber for the parameters for which significant removal (> 10 percent) is seen. The graphs illustrate percent removal (concentration in water at time t divided by the initial concentration as measured in the bottles with no media) versus contact time for the groups of major cations and heavy metals in contact with the designated medium. Since each test was performed in triplicate, error bars for each contact time increment are shown in these figures. For several of the media, one or more of the analytes are not shown because the ratio of effluent-to-influent concentrations does not show greater than 10 percent removal for the pollutants, or the ratio is higher than the range on the graphs, thus indicating that the media are leaching these pollutants into the water (ammonia: compost, peat, and agrofiber; iron: sand, cotton, and agrofiber; lead: agrofiber; sulfate: carbon). Lead also is not shown on the cotton figure because no lead was detected in the blank water. The leaching likely resulted either from a degradation of the media during the experiment or to the exchange of these ions during the removal of other pollutants. These results could be used by a designer to provide an estimate of the desirable contact time for each filter medium depending on the pollutant(s) of interest.

Figure 4-5 shows the removal ability of activated carbon for the heavy metals. The figures for the peat moss and agrofiber are in Appendix B. The only metal for which activated carbon showed significant removal (10 to 20 percent) was copper. The optimum contact time for copper removal was 10 minutes. Longer contact times appeared to result in the release of approximately half of the trapped copper. Removal of all four heavy metals (copper, lead, zinc and iron) was excellent for the peat moss and in general between 50 percent (copper and lead) and 90 percent (zinc and iron) was removed immediately. Another 15 to 20 percent of the lead and copper in solution can be removed by the peat moss if the contact time is approximately 15 minutes. After 15 minutes, the removal curve for those two metals levels off. For agrofiber and zinc, it appears that the longer the contact time, the higher the removal efficiency, with a contact time of 60 minutes giving approximately 10 to 15 percent removal.

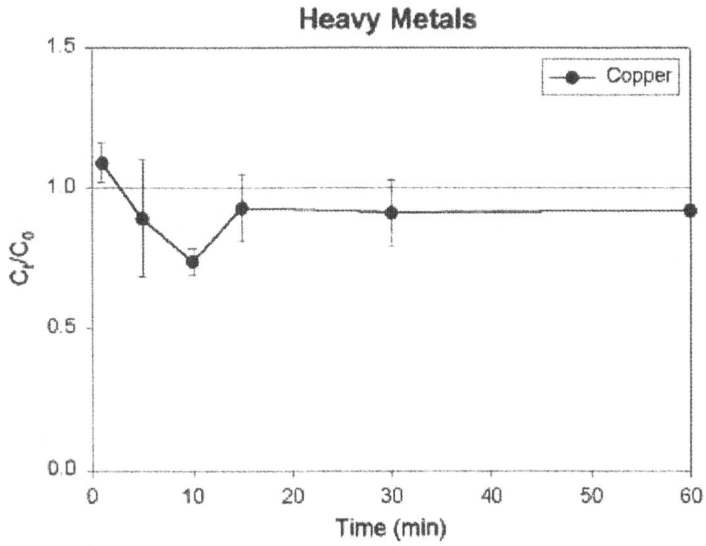

Figure 4-5. Contact time experiments for heavy metals with activated carbon.

Sand appeared to have minimal, or no, impact on the major ion concentration in solution and it possibly may have increased slightly the heavy metals concentration in solution. These results were anticipated given the literature review of un-amended sand and the low initial pollutant concentrations. For activated carbon and the heavy metals, a contact time of approximately 10 minutes demonstrated small (10 to 20 percent) removals of the metals. Longer contact times appeared to cause a release of approximately half of the metals sorbed during the 10 minutes of contact. Activated carbon appeared to have a minimal impact on the removal of the major cations and anions, with the exception of sulfate. The sulfate concentration increased by two to three times during the experiment. To minimize sulfate addition, short contact times would be desirable.

Because of its organic nature and comparatively high cation exchange capacity, it was anticipated that the removal of the heavy metals would be better in the compost than was observed. The increase in the heavy metal concentration in the water possibly was due to the degradation of the media. Since the source of the material in the compost is unknown (except that it was stated that the compost was made from deciduous leaves), it is feasible to assume that heavy metals were present in the original material and were retained in the compost. Heavy metal contamination would be likely if the compost truly was based on a municipal leaf waste. The leaves could be coated with metals through atmospheric deposition and then by automobile debris from the street once they had fallen off the trees. Research by Kao (1993) has shown that, when metals are present in the original material, they likely will be incorporated into the compost.

No significant removal occurred for the cotton textile waste. It was found to leach copper, zinc, and iron into the stormwater runoff. Although the agrofiber was advertised as having the ability to remove copper from wastewater, at these low concentrations, the medium had a tendency to leach copper into solution, possibly due to the use of copper compounds to activate the sorption sites on the media. The agrofiber also added significant quantities of lead, iron, and ammonia to the runoff. Removal of the nutrients, cations, and anions was inconsistent and/or nonexistent. The data were extremely scattered, and, therefore, no conclusions about the potential of agrofiber to remove the nitrate, phosphate, the major cations, or the major anions would be valid.

Based upon the results (Figure 4-5 and in the appendix) from the batch kinetics and capacity tests, it appeared that designing the filters for a contact time of approximately 10 to 15 minutes was best for most of the pollutants. A contact time of 15 minutes is calculated to correspond to a daily flow/loading rate of 44 m/day for a 0.46-m (18-inch total media depth). However, this decision should be based on the media selected for the filter and on the pollutants requiring removal.

Given the desire to evaluate the filters to remove a wide variety of pollutants (a general investigation rather than a design for a specific application), the contact time of 15 minutes was used to determine the maximum flow rate (44 m/day) for the columns both in the laboratory breakthrough curve study and the pilot-scale study. It would be expected that the suspended solids loading on the filter would increase over time to the point where it was sufficiently great to restrict the flow to less than the 44 m/day calculated based on a 15-minute contact time. Flow rates less than 44 m/day would result in longer contact time between the influent water and the media, potentially enhancing the removal of some pollutants and increasing the leaching of others.

Bohart and Adams (1920) developed their series of equations for fixed-bed adsorbers (the concentration equation is Equation 4.3.2-9). These kinetics and capacity tests were performed in order to see if laboratory kinetics and capacity experiments could be used to predict the rate constants and capacities seen in the operation of a fixed-bed adsorber. Rate constants, "k"s, were calculated from the contact time/kinetics studies. The data from the plots of the ratio of effluent concentration to influent concentration versus exposure time were regressed using a linearized version (Equation 4.3.6.2-2):

$$\ln\left(\frac{C_e}{C_0}\right) = \ln b - kt \qquad (4.3.6.2\text{-}2)$$

The regression results are given in Tables 4-19 through 4-21. The tables contain the statistically significant coefficients (data not shown were not found to be statistically significant at $\alpha = 0.05$) and the coefficients of determination, R^2, for each regression. As the data in the tables show, very few statistically significant relationships were found where the relationship of time on the ratio of effluent-to-influent concentrations was found to explain most of the variance in the C_e/C_0 ratio.

Statistical evaluation of the dissolved pollutant removal model based on batch testing. The model makes sense when it states that the removal/sorption/ion-exchange is first order, i.e., concentration dependent. The model equation includes the same logic that is used by Bohart and Adams (1920) in their discussion of the removal of chlorine on charcoal, i.e., that the ability of the filter media to remove additional pollutants is controlled by the number of available sites on the media.

Table 4-19. Rate Constants and Intercepts for Sorption/Ion-Exchange: Batch Study for Hardness, Calcium, and Magnesium (95% confidence intervals for coefficients)

Filter Medium	Hardness			Calcium			Magnesium		
	ln b	K (min^{-1})	R^2	ln b	k (min^{-1})	R^2	ln b	k (min^{-1})	R^2
Sand	NSS	0.0013 (0.0002, 0.002)	0.063	NSS	NSS	NSS	NSS	NSS	NSS
Activated Carbon	NSS	NSS	NSS	NSS	NSS	NSS	NSS	NSS	NSS
Peat Moss	NSS	0.0074 (0.002, 0.01)	0.118	-0.3006 (-0.4, -0.2)	0.0082 (0.004, 0.013)	0.870	-0.1697 (-0.3, -0.1)	0.0050 (0.002, 0.008)	0.864
Compost	NSS	NSS	NSS	NSS	-0.0058 (-0.01, -0.001)	0.000	NSS	-0.0026 (-0.005, - 0.0004)	0.000
Cotton Textile Waste	NSS	NSS	NSS	NSS	NSS	NSS	NSS	NSS	NSS
Agrofiber	NSS	NSS	NSS	NSS	0.0008 (0.0002, 0.001)	0.000	NSS	0.0013 (0.0001, 0.003)	0.000

Table 4-20. Rate Constants and Intercepts for Sorption/Ion-Exchange: Batch Study for Copper, Lead, and Zinc (95% confidence intervals for coefficients)

Filter Medium	Copper			Lead			Zinc		
	ln b	k (min^{-1})	R^2	ln b	k (min^{-1})	R^2	ln b	k (min^{-1})	R^2
Sand	NSS	NSS	NSS	0.3437 (0.003, 0.07)	0.0151 (0.003, 0.03)	0.753	NSS	NSS	NSS
Activated Carbon	NSS	NSS	NSS	NSS	NSS	NSS	NSS	NSS	NSS
Peat Moss	-1.0193 (-1.2, -0.8)	0.0079 (0.001, 0.01)	0.725	-3.1694 (-3.4, -3)	-0.0085 (0.0007, 0.02)	0.700	-0.5112 (-0.7, -0.3	0.0105 (0.002, 0.02)	0.765
Compost	NSS	-0.0093 (-0.02, -0.001)	0.000	NSS	NSS	NSS	NSS	NSS	NSS
Cotton Textile Waste	NSS	NSS	NSS	NA	NA	NA	NSS	NSS	NSS
Agrofiber	0.5661 (0.51, 0.62)	-0.0057 (-0.008, -0.004)	0.944	1.2853 (1, 1.6)	-0.0142 (-0.03, -0.004)	0.775	NSS	0.0023 (0.002, 0.003)	0.964

Note: NA – No regression available for lead since there was no initial concentration of lead in the water.

Table 4-21. Rate Constants and Intercepts for Sorption/Ion-Exchange: Batch Study for Iron, Potassium, and Sodium (95% confidence intervals for coefficients)

Filter Medium	Iron			Potassium			Sodium		
	ln b	k (min^{-1})	R^2	ln b	k (min^{-1})	R^2	ln b	k (min^{-1})	R^2
Sand	NSS	NSS	NSS	NSS	NSS	NSS	NSS	-0.0014 (-0.003, 0)	0.000
Activated Carbon	NSS	NSS	NSS	NSS	NSS	NSS	NSS	NSS	NSS
Peat Moss	-3.1988 (-3.4, -3)	-0.0155 (-0.01, -0.02)	0.935		0.0012 (0, 0.02)	0.000	NSS	NSS	NSS
Compost	NSS	-0.0180 (-0.02, 0.04)	0.000	0.5055 (0.3, 0.7)	-0.0076 (-0.02, -0.0007)	0.693	NSS	-0.0011 (-0.002, -0.0003)	0.000
Cotton Textile Waste	NSS	NSS	NSS	NSS	NSS	NSS	NSS	NSS	NSS
Agrofiber	1.4825 (1.1, 1.8)	-0.0134 (-0.03, -0.002)	0.716	NSS	NSS	NSS	NSS	-0.0065 (-0.0002, 0.01)	0.000

The additional statistical analyses of the residuals are contained in Appendix B.

When examining the tabular data (Tables 4-22 through 4-24), it can be seen that, for several media and several pollutants, the regression does not pass one or more of the tests for applicability, although all the coefficients reported in these tables are statistically significant. Based upon a graphical analysis of the residuals, the regressions for the sand appeared to meet the requirements of a normal distribution of the errors. For activated carbon, the residual analysis also appeared to meet the requirements of a normal distribution of errors. However, like the sand, the R^2 values for activated carbon indicate that the regression does not explain much of the variability of the data. An analysis of the residuals for compost showed that the regression requirements were met for all pollutants where statistically significant regression coefficients were calculated, except for calcium, magnesium, iron, and copper. This was confirmed by the R^2 value of zero for these parameters and compost. For peat moss, the residuals analysis showed that the regression requirements likely were not met for potassium. The R^2 value of zero for these three regressions confirmed the graphical analysis of the residuals for peat moss and these three pollutants. For the cotton, none of the regressions gave a statistically significant slope. However, the requirements for the regression likely were not met as seen in the graphical analysis of the residuals and as confirmed by the R^2 value of zero. An analysis of the residuals for the agrofiber indicated that the regression requirements were met, except for sodium, calcium, and magnesium. Plots of the residuals versus both time and the predicted value showed a trend and indicated that the regressions were not valid. For these three pollutants and agrofiber, the R^2 values were zero, indicating that, although the regression coefficients were found to be significant, the regression could explain none of the variability in the data (the fitted line was very flat). Although some of the regressions were not valid (see tables in Appendix B), the coefficients will be compared with the rate constants calculated from the breakthrough curves.

Table 4-22. Evaluation of Dissolved Pollutant Removal Models for Zinc in Batch Testing

Media	Intercept	α of Intercept	Slope	α of Slope	p of Durbin-Watson statistic	Normality test p	Constant variance test p	R^2	Model p
Sand									
Carbon-Sand									
Peat-Sand	-0.5112	0.0035	0.0105	0.0225	>0.05	0.1284	0.0600	0.765	0.0225
Compost-Sand									
Cotton-Sand									
Agrofiber-Sand			0.0023	0.0001	>0.05	0.6262	0.0600	0.964	0.0001

Table 4-23. Evaluation of Dissolved Pollutant Removal Models for Lead in Batch Testing

Media	Intercept	α of Intercept	Slope	α of Slope	p of Durbin-Watson statistic	Normality test p	Constant variance test p	R^2	Model p
Sand	0.3437	0.0488	0.0151	0.0252	>0.05	0.0827	0.0600	0.753	0.0252
Carbon-Sand									
Peat-Sand	-3.1694	0.0001	-0.0085	0.0377	>0.05	0.2601	0.0600	0.700	0.0377
Compost-Sand									
Cotton-Sand									
Agrofiber-Sand	1.2853	0.0003	-0.0142	0.0207	>0.05	0.4115	0.0600	0.775	0.0207

Table 4-24. Evaluation of Dissolved Pollutant Removal Models for Iron in Batch Testing

Media	Intercept	α of Intercept	Slope	α of Slope	p of Durbin-Watson statistic	Normality test p	Constant variance test p	R^2	Model p
Sand									
Carbon-Sand									
Peat-Sand	-3.1988	<0.0001	0.0155	0.0016	>0.1	0.3195	0.0600	0.935	0.0016
Compost-Sand			0.0180	0.0502	>0.05	0.3370	0.0600	0.000	0.0502
Cotton-Sand									
Agrofiber-Sand	1.4825	0.0002	0.0134	0.0336	>0.05	0.1650	0.0600	0.716	0.0336

The media's capacity for any given pollutant was estimated as part of the kinetics study. The same set-up as described above for the kinetics experiments was used for the equilibrium/capacity batch experiments in which three different masses of each media were exposed (in triplicate) to the spiked stormwater runoff for 3 days. Sample preparation and analysis techniques were identical to the kinetics tests. The amount of pollutant adsorbed on any media sample was calculated as mg of pollutant per g of media. The capacities for each media for each pollutant are in Table 4-25.

Table 4-25. Capacity of Media (mg pollutant/g media) by Pollutant: Batch Multicomponent Isotherm Study (COVs for non-zero capacities)

Pollutant	Sand	Carbon	Peat	Compost	Cotton Textile Waste	Agrofiber
Hardness	0.2 (21)	115 (0.11)	105 (1.4)	0	110 (0.29)	0
Copper	0.002 (6.5)	0.03 (0.37)	0.4 (0.78)	0.03 (1.9)	0.3 (0.45)	0
Lead	0.002 (3.1)	0.003 (0.02)	0.008 (1.7)	0	NA	0
Zinc	0.05 (2.3)	0.2 (0.23)	5.7 (0.36)	1.1 (0.77)	1.1 (0.02)	0.02 (1.8)
Iron	0.006 (1.5)	0.001 (2.1)	0	0	0	0.04 (0.16)
Sodium	0	0	0	0	0	0
Potassium	0.003 (5.3)	0.6 (0.83)	0	0	0	0
Calcium	0.08 (0.38)	8 (0.50)	38 (0.46)	0	2.1 (0.52)	0.6 (1.1)
Magnesium	0.04 (0.30)	2 (1.1)	10 (0.53)	1.2 (1.4)	0.2 (0.78)	0.2 (1.2)

As can be seen in Table 4-25, with the exception of zinc (and slightly iron), sand was nearly inert to most of the heavy metals and major cations. The compost, peat, cotton, and carbon all had some capacity for removal of metals, and in general they had the greatest capacity for zinc. These results were not in agreement with most lists of binding strength and sorbent preference for metals; however, this likely resulted from the fact that these tests were performed with comparatively extremely low concentrations of lead and low concentrations of copper. Lead and copper would be expected to sorb preferentially over zinc given sufficient time for 'equilibrium' to be reached, unless there was insufficient concentrations of these metals to fill the sorption sites.

Another possible explanation was the stormwater matrix with its colloids competing with the media for pollutants interfered with the normal preferential sorption order. If colloidal matter in the runoff sorbed the spiked lead and copper, this would leave more sites open on the media for zinc. The pre-settled stormwater was spiked to provide approximate concentrations of lead and copper of 500 µg/L and 1500 µg/L, respectively. However, even in the samples in which no media was exposed, the concentration in the filtered water (through a 0.45-µm nitrocellulose filter) was less than 10 µg/L and 500 µg/L for lead and copper, respectively. These low concentrations in solution at the start of the kinetics and capacity experiments were likely due to the rapid sorption of these metals onto the particulate/colloidal matter found in the pre-settled runoff, with realistic dissolved metal concentrations compared to typical urban runoff. Part of the concern with the use of adsorbents (fixed-bed adsorbers) for treating urban runoff was that the concentrations of many of the pollutants were extremely low compared to the typical concentrations in industrial wastewater treated using adsorption, and as was seen in most of the research studies reviewed as part of this literature search. These batch capacity tests allowed for capacities to be determined based on the typical pollutants and concentrations found in urban runoff.

The capacity calculations confirmed the literature that indicated that peat moss, since it is often formed in calcium-poor conditions, had a high exchange/sorption capacity for calcium and for hardness. These data were transformed into units identical to the cation exchange capacities calculated earlier (with the results shown in Table 4-26). For the carbon, peat, and cotton, the quantity of cations exchanged is much greater than the CEC calculated earlier. This likely is a result of the size of the test molecule for the CEC experiments, a copper trielthylenetetramine complex, which may not have been able to penetrate some of the micropores that the ionic forms of the metals and major ions could penetrate.

Table 4-26. Capacity of Media (meq pollutant/100 g media) by Pollutant: Batch Multicomponent Isotherm Study

Pollutant	Sand	Carbon	Peat	Compost	Cotton Textile Waste	Agrofiber
Hardness	0.40	230	210	0	220	0
Copper	0.01	0.09	1.2	0.09	0.08	0
Lead	0	0	0.01	0	0	0
Zinc	0.15	0.46	17	3.4	3.4	0.06
Iron	0.03	0.01	0	0	0	0.22
Sodium	0	0	0	0	0	0
Potassium	0.01	1.5	0	0	0	0
Calcium	0.40	40	190	0	11	3
Magnesium	0.33	17	83	10	1.7	1.7
CATION EXCHANGE CAPACITY (calculated from batch tests)	1.41	59.9	292	13.5	23.9	5.71
CATION EXCHANGE CAPACITY (CEC study)	3.49	5.40	21.47	18.83	3.77	9.36

4.3.6.3 Testing of the Breakthrough Curve Model Using Laboratory-Scale Filters

The purpose of this research task was to confirm the breakthrough model using laboratory-scale filters of sand, peat-sand, compost-sand and carbon-sand, and multi-component mixtures of pollutants. The influent water for these experiments was pre-settled stormwater runoff that had been spiked in the same manner as the spikes for the kinetics and capacity investigations. Water was filtered through the columns continuously, and the effluent was collected as a grab sample after a cumulative volume of 2 L, 4 L, 12 L, 20 L, 28 L, 36 L, etc. of water had passed through the column (each 8 L increment would be equal to a loading on the filter of approximately 4.4 m). The effluent samples were filtered through 0.45-µm nitrocellulose filters and analyzed for the parameters listed for the kinetics and capacity investigations. A column that contained no media was used as the blank ('influent concentration' or C_0) for this series of experiments in order to account for any potential effects of the pollutants in the water interacting with the glass sides of the columns.

During these tests, the flow rate for each medium was noted when the samples were collected (as time to collect a known volume of effluent). While the runoff was assumed to be low in particulate matter after several days of settling, it has been well documented that some particulates, especially those less than 1 µm in diameter, exist in settled stormwater. These particulates, along with the potential shrinking or swelling of the media, could potentially affect the flow rate in the media.

The data for this aspect of the research were analyzed in the following two ways: (1) capacities of each medium were calculated using the point where C_e/C_0 became equal to or greater than 1, and (2) linear regression was used to fit curves to the log-transformed C_e/C_0 data versus the untransformed bed volumes treated. The capacity reported from the breakthrough curves is the 'ultimate' capacity, i.e., the mass loading per gram of media when the ratio of C_e/C_0 is equal to or greater than 1. Since all of the media were mixed with sand to improve their hydraulic performance, the sand capacity was calculated first, and the capacity data were then adjusted for the pollutant removal accounted for by the sand in the column (both the 0.15-m layer at the bottom of the column [approximately 400 g] and the approximately 400 g used in the mixed-media layer). The result of the capacity calculations is given in Table 4-27.

Table 4-28 compares the capacities calculated from the batch and the breakthrough curve (continuous flow) studies. In general, the batch-study capacities are at least equal to, if not greater than, the breakthrough-curve capacities. For determining the field capacities of the filters, it is the operational capacity that is the field capacity. The result of this comparison is that, for determining capacities of the media, batch studies are not applicable.

Table 4-27. Capacity of Media (mg pollutant/g media) by Pollutant: Multicomponent Breakthrough Curve Study (continuous flow)

Pollutant	Sand	Carbon	Peat	Compost
Hardness	1.2	20	0	0
Copper				
Lead				
Zinc				
Iron				
Sodium	0	0	0.02	0
Potassium	0.02	0.53	0	0
Calcium	0.08	0.51	>10	0
Magnesium	0.04	0.12	> 1	0.31

Table 4-28. Comparison of Capacities (mg pollutant/g media) Calculated by Batch and Continuous-Flow Studies (BTC)

Pollutant	SAND BTC	Sand Batch	Carbon BTC	Carbon Batch	Peat BTC	Peat Batch	Compost BTC	Compost Batch
Hardness	1.2	0.2	20	115	20	121	0	0
Copper		0.002		0.03		0.2		0.03
Lead		0.002		0.003		0.4		0
Zinc		0.05		0.2		5.7		1.1
Iron		0.006		0.001		0		0
Sodium	0	0	0	0	0.02	0	0	0
Potassium	0.02	0.003	0.53	0.6	0	0	0	0
Calcium	0.08	0.08	0.51	8	>10	38	0	0
Magnesium	0.04	0.04	0.12	2	> 1	10	0.31	1.2

The equation used to model the dissolved pollutant removals (Equation 4.3.6.3-1) is the following:

$$\ln\left(\frac{C_0}{C_e}-1\right)=\frac{kN_M{}^2\rho_{media}x}{1{,}000{,}000\left(\frac{mg-L}{g-m^3}\right)C_0u}-\frac{1{,}000{,}000kC_0x}{u}(BV) \qquad (4.3.6.3\text{-}1)$$

where C_0 = influent concentration (mg/L), C_e = effluent concentration (mg/L), k = rate constant (day^{-1}), N_M = capacity of media or mix of media for pollutant of interest (mg pollutant/g media), ρ_{media} = density of media or mix of media (g/m^3), x = depth of filter (m), u = unit loading rate (m/day), and BV = bed volumes treated (dimensionless) – calculated as total volume treated divided by the volume of the media.

Given the slope from the regressions, the rate constants for the dissolved pollutant removal model can be calculated using Equation 4.3.6.3-2

$$Slope=\frac{1{,}000{,}000kC_0x}{u} \qquad (4.3.6.3\text{-}2)$$

The statistically significant slopes for the regressions are in Tables 4-29 through 4-31. The calculation of the rate constant assumed that the flow rate through the filter is 44 m/day, which is a contact time of 15 minutes as measured in the batch kinetics study. The constants shown in Tables 4-32 through 4-34 are only for media and pollutants where removal occurred and for which statistically significant coefficients could be obtained. For many pollutants, the intercept was not significant since initial removal efficiencies were approximately 50 percent. Therefore, for the pollutants in the tables below, the slopes shown are the regression without an intercept term. Although the data with leaching could be regressed, a negative removal rate constant is meaningless in this application.

Table 4-29. Rate Constants for Modified BDST Model Using Laboratory Breakthrough Curve Data for Hardness, Calcium, and Magnesium

Filter	HARDNESS		Calcium		Magnesium	
	Slope	k (day^{-1})	Slope	k (day^{-1})	Slope	k (day^{-1})
Sand	0.0496 (0.02, 0.08)	4.13×10^{-10}				
Carbon-Sand	0.0595 (0.02, 0.1)	4.96×10^{-10}	0.1122 (0.09, 0.14)	1.12×10^{-8}	0.1282 (0.09, 0.17)	2.14×10^{-8}
Peat-Sand	0.2764 (0.06, 0.5)	2.30×10^{-9}	0.0491 (0, 0.1)	2.14×10^{-8}	0.1770 (0.05, 0.31)	2.95×10^{-8}
Compost-Sand	0.1439 (0.06, 0.23)	1.20×10^{-9}			0.1833 (0.07, 0.30)	3.06×10^{-8}

Table 4-30. Rate Constants for Modified BDST Model Using Laboratory Breakthrough Curve Data for Copper, Lead, and Zinc

Filter	Copper		Lead		Zinc	
	Slope	k (day^{-1})	Slope	k (day^{-1})	Slope	k (day^{-1})
Sand					0.0523 (0.04, 0.07)	2.91×10^{-8}
Carbon-Sand					0.0147 (0.012, .018)	8.17×10^{-9}
Peat-Sand						
Compost-Sand						

Table 4-31. Rate Constants for Modified BDST Model Using Laboratory Breakthrough Curve Data for Iron, Sodium, and Potassium

Filter	IRON		Sodium		Potassium	
	Slope	k (day^{-1})	Slope	k (day^{-1})	Slope	k (day^{-1})
Sand						
Carbon-Sand					0.0498 (0.04, 0.06)	1.99×10^{-8}
Peat-Sand			0.1224 (0.05, 0.19)	4.37×10^{-8}		
Compost-Sand						

Table 4-32. Comparison of Rate Constants from Batch Kinetics and Continuous-Flow Breakthrough Curve Studies for Hardness, Calcium, and Magnesium

Filter	Hardness		Calcium		Magnesium	
	Batch k (day^{-1})	BTC k (day^{-1})	Batch k (day^{-1})	BTC k (day^{-1})	Batch k (day^{-1})	BTC k (day^{-1})
Sand	1.8	4.7×10^{-8}				
Carbon-Sand		5.7×10^{-8}		1.1×10^{-6}		3.7×10^{-5}
Peat-Sand	11	2.1×10^{-7}	12	2.5×10^{-6}	7.2	5.1×10^{-5}
Compost-Sand		1.4×10^{-7}				5.4×10^{-5}

Table 4-33. Comparison of Rate Constants from Batch Kinetics and Continuous-Flow Breakthrough Curve Studies for Copper, Lead, and Zinc

Filter	Copper		Lead		Zinc	
	Batch k (day^{-1})	BTC k (day^{-1})	Batch k (day^{-1})	BTC k (day^{-1})	Batch k (day^{-1})	BTC k (day^{-1})
Sand			22			2.6×10^{-6}
Carbon-Sand						7.3×10^{-7}
Peat-Sand	11				15	
Compost-Sand						

Table 4-34. Comparison of Rate Constants from Batch Kinetics and Continuous-Flow Breakthrough Curve Studies for Iron, Sodium, and Potassium

Filter	Iron		Sodium		Potassium	
	Batch k (day^{-1})	BTC k (day^{-1})	Batch k (day^{-1})	BTC k (day^{-1})	Batch k (day^{-1})	BTC k (day^{-1})
Sand						
Carbon-Sand						1.4×10^{-5}
Peat-Sand			1.7	6.0×10^{-6}		
Compost-Sand						

The rate constants calculated for the batch kinetics studies are significantly (several orders of magnitude) greater than those calculated using a breakthrough curve analysis of pollutant removal. This, along with the decreased operational capacity seen in the breakthrough curves, indicates that laboratory-scale breakthrough curve testing will be necessary to predict the behavior of a media in the field.

The growth of a biofilm on the media during the breakthrough testing is one possible explanation for the very large differences between the rate constants observed in the batch testing and those observed in the breakthrough curve studies. Biofilm growth likely would tend to decrease the rate at which pollutants were removed from the solution because of the increase in the diffusional resistance and the blocking of sorption sites by the bacteria. However, unless nearly 100 percent removal is required by the application, the growth of a biofilm may be beneficial because many of the pollutants in urban stormwater runoff are either macro- or micronutrients for the bacteria. Therefore, the bacteria may absorb into its biomass the pollutants and thus reopen previously occupied sorption sites. Also, some bacteria act as if they carry a charge, thus making available more charged sites for adsorption. In addition, some bacteria secrete an exopolymer which may aid in removing pollutants from water. The increase in diffusional resistance (which means that a longer contact time is necessary for better performance) is counter-balanced by the potential reopening/availability of sorption sites, i.e., the increase in capacity seen in the breakthrough curve tests as compared to the batch equilibrium studies.

Once the breakthrough testing was complete (for all filters in the laboratory, clogging caused failure before all pollutants had demonstrated chemical breakthrough), samples of the upper 4 cm of the filter media were sent to the University of Alabama for analysis of adenosine triphosphate (ATP) and protein content. The results of this bacteriological testing are given in Table 4-35. According to Urrutia (personal communication), ATP content in natural mineral soils generally ranges between 0.37 and 7.5 μg ATP/g dry soil. The sand filter ATP content is at the lower end of that range, as expected, since it does not have an organic substrate available for the biomass. The bacteria on the sand filter must obtain their nutrients from the pollutants in the stormwater runoff. This likely is a limiting factor on the growth of the biomass in the sand filter. The carbon-sand and peat-sand filters also appear to have a similar bacterial activity to the sand filter. The compost filter, however, has an excellent organic substrate for biomass growth. Compost has been found to be an excellent substrate for the growth of heterotrophic bacteria.

Table 4-35. Bacteriological Analysis of Filter Media after Breakthrough

Filter	ATP (μg ATP/g dry soil)	Protein content (mg protein/ g dry media)
Sand	0.80 +/- 0.08	0.20 +/- 0.02
Carbon-Sand	≈ 0.80	
Peat-Sand	≈ 0.80	
Compost-Sand	1.61 +/- 0.07	13.7 +/- 0.2

The biological activity in the filters when combined with the cycling of the filters likely explains much of the scatter in the data in the breakthrough curves.

Statistical evaluation of the dissolved pollutant removal model based on breakthrough curves. The model makes sense when it states that the number of bed volumes treated before chemical breakthrough is a function of both the flow rate (and hence contact time), the rate constant for sorption/ion-exchange/removal, and the initial concentration of the pollutant.

The primary test of the models is in the residual analysis. The tabular results of the model evaluation are in Tables 4-36 through 4-41 for nitrate, magnesium, calcium, potassium, hardness, sodium, and zinc. The coefficient of the regression equation is shown along with the probability of it being significant. Also shown in the table is the probability of the Durbin-Watson statistic showing an autocorrelation between the data (p values greater than the statistically significant level indicate that either autocorrelation does not exist or that the existence of autocorrelation cannot be shown in the data set). The Kolmogov-Smirnov test for normality attempts to determine if the residuals are normally distributed (normality is assumed if the p value exceeds the criteria for statistical significance). The p value for constant variance is similar to the test for normality and the Durbin-Watson statistic test in that the null hypothesis is that the data has a constant variance, and, therefore, the assumption of constant variance is made when the p value exceeds the significance criteria. The traditional graphical residual analysis reveals the problem found with evaluating a model where only three or four data points were collected before breakthrough was achieved, i.e., evaluating trends in the residuals with only a few data points. When examining the tabular data, the lack of number of data points before breakthrough likely is the reason that several combinations of media and pollutants failed the constant variance test.

Table 4-36. Evaluation of Dissolved Pollutant Removal Models for Magnesium

Media	Slope	α of Slope	p of Durbin-Watson statistic	Normality test p	Constant variance test p	R^2	Model p
Sand	0.1801	0.1155	0.05	0.2149	<0.0001	0.295	0.1155
Carbon-Sand	0.1282	0.0006	>0.1	0.5979	0.0600	0.387	0.0006
Peat-Sand	0.1770	0.0485	>0.1	0.1082	<0.0001	0.827	0.0485
Compost-Sand	0.1833	0.0372	0.05	0.5595	<0.0001	0.743	0.0372

Table 4-37. Evaluation of Dissolved Pollutant Removal Models for Calcium

Media	Slope	α of Slope	p of Durbin-Watson statistic	Normality test p	Constant variance test p	R^2	Model p
Sand							
Carbon-Sand	0.1122	<0.0001	>0.05	0.1738	0.2965	0.692	0.0001
Peat-Sand	0.0491	0.0703	>0.1	0.2323	<0.0001	0.635	0.0703
Compost-Sand							

Table 4-38. Evaluation of Dissolved Pollutant Removal Models for Potassium

Media	Slope	α of Slope	p of Durbin-Watson statistic	Normality test p	Constant variance test p	R^2	Model p
Sand							
Carbon-Sand	0.0498	<0.0001	>0.1	0.2609	0.3873	0.821	0.0001
Peat-Sand							
Compost-Sand							

Table 4-39. Evaluation of Dissolved Pollutant Removal Models for Hardness

Media	Slope	α of Slope	p of Durbin-Watson statistic	Normality test p	Constant variance test p	R^2	Model p
Sand	0.0496	0.0120	>0.05	0.4385	0.0061	0.392	0.0120
Carbon-Sand	0.0595	0.0114	>0.05	0.2535	0.0061	0.522	0.0114
Peat-Sand	0.2764	0.0555	0.05	0.5811	<0.0001	0.791	0.0555
Compost-Sand	0.1439	0.0346	0.05	0.6319	<0.0001	0.834	0.0346

Table 4-40. Evaluation of Dissolved Pollutant Removal Models for Sodium

Media	Slope	α of Slope	p of Durbin-Watson statistic	Normality test p	Constant variance test p	R^2	Model p
Sand							
Carbon-Sand							
Peat-Sand	0.1224	0.0105	>0.05	0.7456	0.0500	0.276	0.0105
Compost-Sand							

Table 4-41. Evaluation of Dissolved Pollutant Removal Models for Zinc

Media	Slope	α of Slope	p of Durbin-Watson statistic	Normality test p	Constant variance test p	R^2	Model p
Sand	0.0523	<0.0001	>0.05	0.4235	0.9477	0.752	0.0001
Carbon-Sand	0.0147	<0.0001	>0.05	0.6863	0.7924	0.801	0.0001
Peat-Sand							
Compost-Sand							

4.3.6.4 Combining Physical and Chemical Removal in a Single Series of Design Equations

The final algorithm for a stormwater filter design must incorporate both the physical straining and the chemical/biological 'adsorption' occurring in the filter. Failure for a filter will need to be defined by the designer. In general, it will be defined as the time at which either the effluent concentration of one of the pollutants has exceeded the design criteria, often a regulatory concentration limit, or when the flow through the filter is reduced to the point below which it is not acceptable, i.e., when either most of the flow bypasses the filter or, if bypassing is not possible, the water accumulates to an undesirable depth in an undesirable location.

The two general design equations needed for the algorithm are Equation 4.3.6.4-1 and Equation 4.3.6.4-9.

$$u = k_i \bullet L_m^{-c} \qquad (4.3.6.4\text{-}1)$$

Where u = unit flow/loading rate (m/day), k_i = flow rate reduction rate constant (day/m), L_m = suspended solids loading, and c = power of the flow rate reduction with suspended solids loading.

For the media:

Sand: $u = 44{,}500\ L_m^{-1.02}$ (4.3.6.4-2)

Carbon-Sand: $u = 14{,}800\ L_m^{-0.77}$ (4.3.6.4-3)

Peat-Sand: $u = 2{,}000\ L_m^{-0.71}$ (4.3.6.4-4)

Compost-Sand: $u = 1.6 \times 10^{13}\ L_m^{-4.09}$ (4.3.6.4-5)

Cotton-Sand: $u = 160 - 0.01\ L_m$ (4.3.6.4-6)

Agrofiber-Sand: $u = 205 - 0.09\ L_m$ (4.3.6.4-7)

Zeolite-Sand: $u = 60\ L_m^{-0.23}$ (4.3.6.4-8)

$$\ln\left(\frac{C_0}{C_e} - 1\right) = \frac{kN_M{}^2\rho_{media}x}{1{,}000{,}000\left(\frac{mg-L}{g-m^3}\right)C_0u} - \frac{1{,}000{,}000kC_0x}{u}(BV) \qquad (4.3.6.4\text{-}9)$$

Where C_0 = influent concentration (mg/L), C_e = effluent concentration (mg/L), k = rate constant (day^{-1}), N_M = capacity of media or mix of media for pollutant of interest (mg pollutant/g media), ρ_{media} = density of media or mix of media (g/m^3), x = depth of filter (m), u = unit loading rate (m/day), and BV = bed volumes treated (dimensionless) – calculated as total volume treated divided by the volume of the media.

However, Equation 4.3.6.4-9 can be simplified as follows through inspection of the slope and the intercept term to see if they are of the same order of magnitude and canceling out shared variables:

$$\frac{kN_M{}^2\rho_{media}x}{1{,}000{,}000\left(\frac{mg-L}{g-m^3}\right)C_0u} \overset{?}{=} \frac{1{,}000{,}000kC_0x}{u}(BV) \qquad (4.3.6.4\text{-}10)$$

$$\frac{N_M{}^2\rho_{media}}{1{,}000{,}000\left(\frac{mg-L}{g-m^3}\right)C_0} \overset{?}{=} 1{,}000{,}000C_0(BV) \qquad (4.3.6.4\text{-}11)$$

The densities of the media examined are shown in Table 4-42.

Table 4-42. Densities of Filter Media

Filter Media	Density (g/m^3)
Sand	1,400,000
Carbon	370,000
Compost	530,000
Peat	190,000
Cotton	26,400
Agrofiber	46,000
	800,000
ZEOLITE	

These densities are of the same order of magnitude as the conversion factor in the denominator, and for a mixed media with sand occupying approximately two-thirds of the volume, the density would be approximately 1,000,000 g/m^3 and would cancel out the conversion factor.

$$\frac{N_M{}^{2}}{C_0} \overset{?}{=} 1{,}000{,}000 C_0 (BV) \qquad (4.3.6.4\text{-}12)$$

None of the media have a capacity approaching 1,000 mg pollutant/g media for any pollutant, except potentially for some of the metals and major cations, as can be seen from the capacity data in Table 4-27 and 4-28. Therefore, for pollutants other than the metals, the first term in the dissolved pollutant model equation can be ignored as approximately equal to zero, and the dissolved pollutant model equation reduces to the following:

$$\ln\left(\frac{C_0}{C_e}-1\right) = -\frac{1{,}000{,}000\, k C_0 x}{u}(BV) \qquad (4.3.6.4\text{-}13)$$

The rate constants for this equation are given in Tables 4-29 through 4-30.

The modeling algorithm using this series of design equations will be applied to the pilot-scale data to see if the model can predict the removal ability of the filters at very low challenge concentrations.

4.4 Pilot-Scale Testing

4.4.1 Pilot-Scale Testing Site

One of the two locations chosen for testing of the pilot-scale device was the stormwater detention pond known as Star Lake in Hoover, Alabama. This pond is fed by stormwater runoff, primarily from a medium-density residential area. This area (off Deo Dara and Patton Chapel Rds.) was under development by the time of the aerial update to the topographic maps in 1978 (aerial photographs taken 1970 and 1975). In addition to the residential area, runoff may be entering the lake from one or more of the gasoline stations along the west side of U.S. Highway 31 contained in the drainage area. The approximate surface drainage area for Star Lake was calculated to be 71 ha, with a lake surface area of 1.9 ha. Figure 4-6 depicts the approximate Star Lake drainage area.

The second detention pond used to provide influent water to the filters was Georgetown Lake, also in Hoover, Alabama. Like Star Lake, this pond is fed by stormwater runoff, primarily from a medium-density residential area. This area (off Lorna Rd.) was under development by the time of the aerial update to the topographic maps in 1978 (aerial photographs taken 1970 and 1975). In addition to the residential area, runoff may be entering the lake from the gasoline station along the east side of Lorna Rd. contained in the drainage area. The approximate surface drainage area for Georgetown Lake was calculated to be 51 ha, with a lake surface area of 1.2 ha. Figure 4-7 depicts the approximate Georgetown Lake drainage area.

4.4.2 Filter Construction

The filtration columns used during the pilot-scale tests were constructed in large (55 gallon [0.21 m3]) polyethylene tanks purchased from Aquatic Eco-Systems, Inc. in Apopka, Florida. The catalog description of these tanks is as follows, "Semi-translucent natural polyethylene...The 55-gallon tank comes with a three-quarter inch spigot valve already fitted. All tanks have 5/32-inch thick walls" (Aquatic Eco-Systems, Inc. 1999). The outer diameter of the tanks are 0.53 m and the depths are 0.91 m. The inner diameter of the tanks are 0.53 m and the surface area of the tank at any cross-section is 0.217 m2.

The media selected for study in the pilot-scale filters included those previously used in the laboratory experiments (sand, activated carbon, peat moss, compost, cotton textile waste, and agrofiber), as well as a loamy topsoil and a lightweight "sand." The two additional filter containers included a Jacuzzi filter set (sold by Aquatic Eco-Systems, Inc., Apopka, Florida) and the StormFilter cartridge test tank (for use with the compost) supplied by Stormwater Management, Inc., of Portland, Oregon. The filtration columns were constructed using the same guidelines as was used in the bench-scale tests, i.e., the City of Austin (1988) and CSF Treatment Systems, Inc. (1994), requirements. The inside of the valve was covered with a washed fiberglass window screen (purchased at a local hardware store), and gravel was placed in the bottom of the tank to a depth above the top of the spigot (approximately 5 cm). Above the gravel was a 0.15 m layer of sand (the same as used in the sand filter). A 0.3-m layer of the medium of choice (usually a mixed media, 50/50 v/v with sand) was placed above the sand. The filters were then rinsed thoroughly with tap water, and the initial flow rates through the media were measured. The ratio of column diameter to median filter grain particle size for the sand filter (the media used to determine filter height and column diameter) was significantly greater than 100, which, according to other researchers (Clark, *et al.* 1992), should be sufficient to avoid significant wall effects. The filters then were mounted on a trailer for easy transport to the sampling locations.

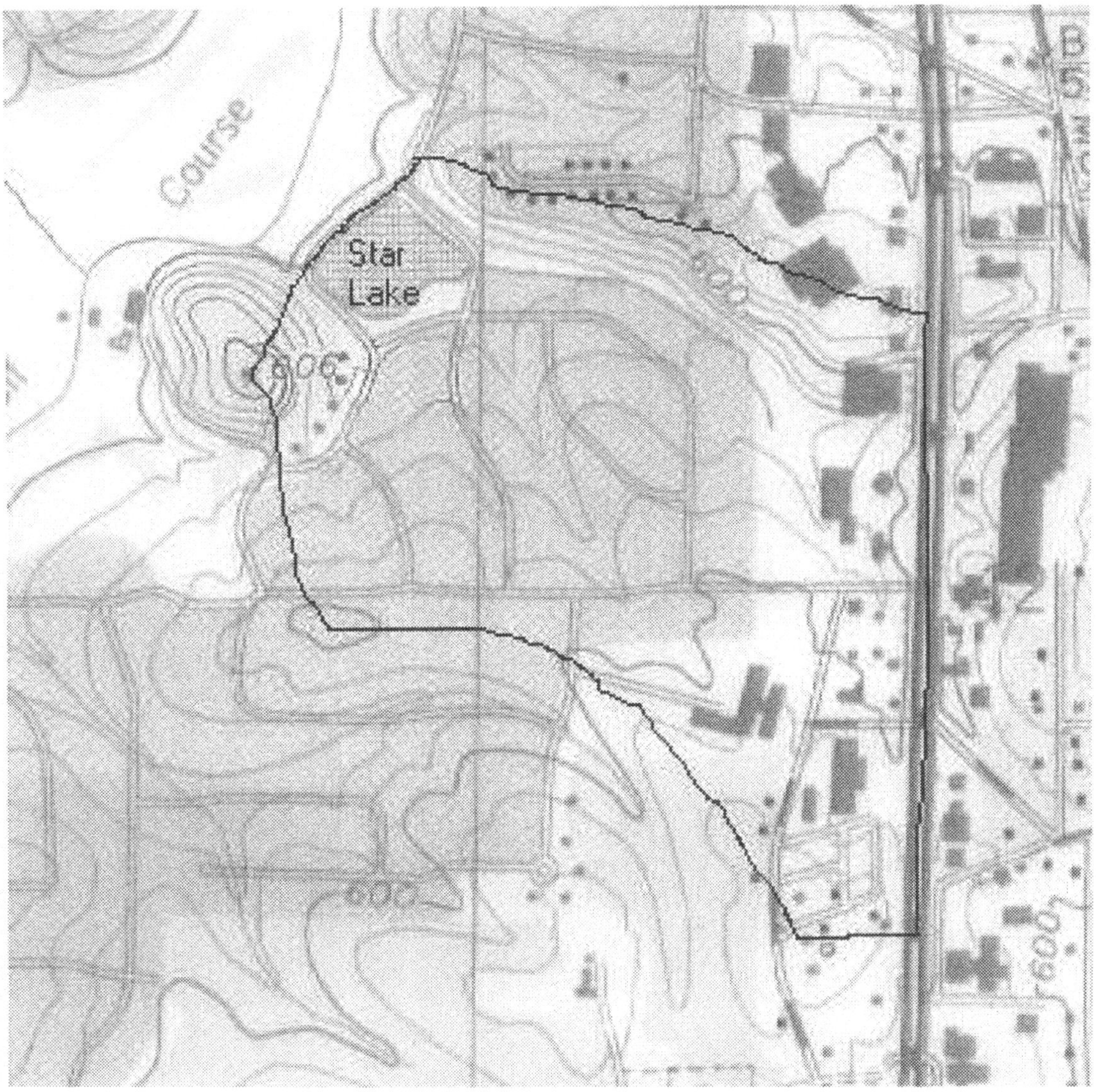

Figure 4-6. Star Lake and its surface drainage area, Hoover, Alabama.

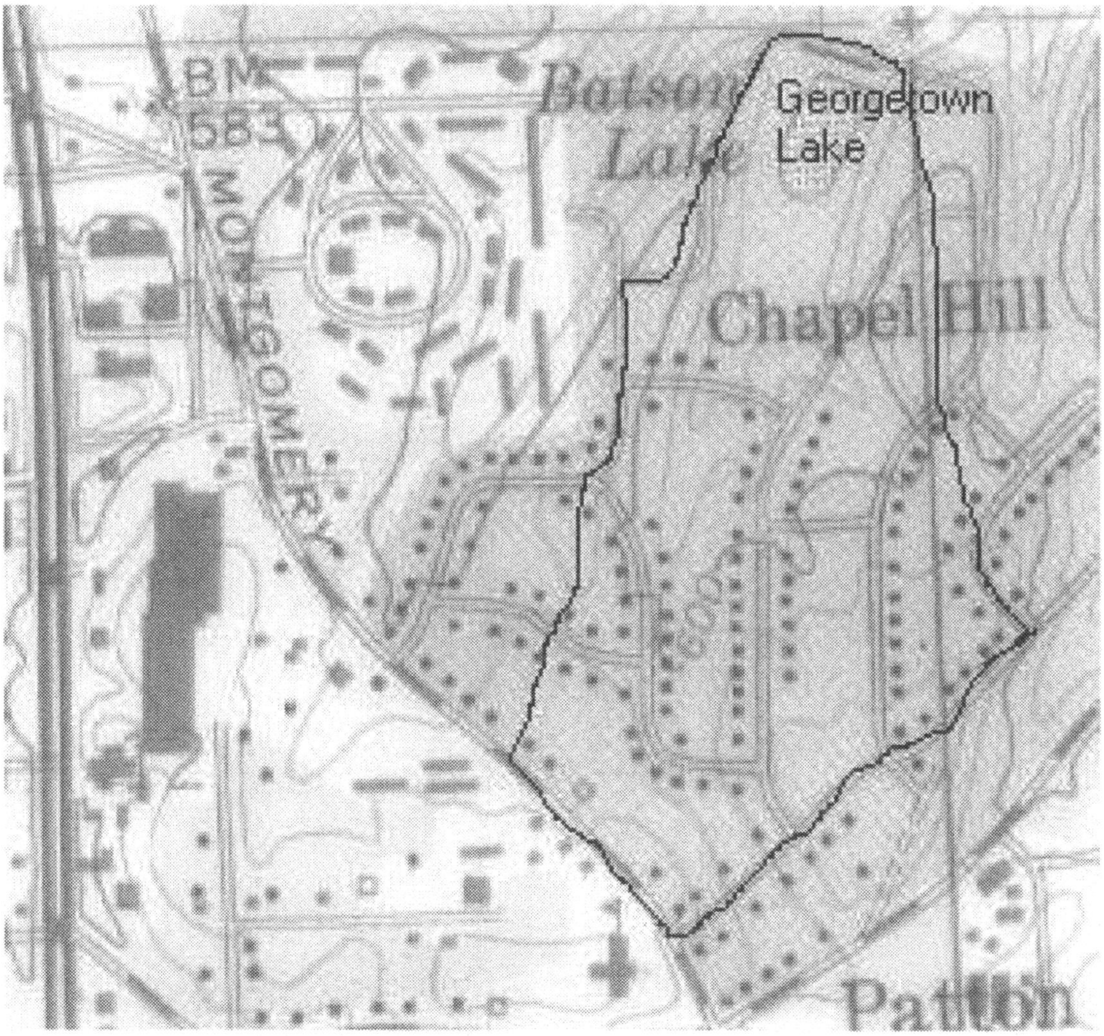

Figure 4-7. Georgetown Lake and its surface drainage area, Hoover, Alabama.

Nonwoven, synthetic fabrics were also placed on the surface of the filters to improve run times and make cleaning of the filters easier. The fabric used has a greater porosity and has been shown to be a more efficient filter for particles larger than what is typical in pre-settled stormwater runoff from a residential area. Collins, *et al.* (1992) showed that a filter cover may not improve a sand filter's ability to treat raw waters of varying solids quality and that no suitable cleaning method exists for the fabrics in large-scale installations. The primary reason for using the filter fabric was that, during the design of the filter chamber of the MCTT (Pitt, *et al.* 1999), bypassing of large amounts of filter area was noted. Prior testing of several geotextiles had found that the Amoco 4557 filter did not allow flow through it until a head of approximately 5 cm had built up on it, but that, once flow began, the fabric did not appear to impede flow. Because of the bypassing problem and the potential for the fabric to catch larger particles, a layer of the Amoco 4557 fabric was added to the top of the filter. It was cut slightly larger than the top surface of the filter to ensure that the complete surface of the filter is covered.

Test water to the pilot-scale filters was supplied through two submersible pumps. The pump on the Jacuzzi filter had a 70 gpm maximum capacity; however, it was not able to overcome the approximately 3.3 m head requirement. Therefore, it was attached directly to one of the two submersible pumps. The other submersible pump was attached to a distribution manifold (Figure 4-8). The purpose of the manifold was two fold: (1) to ensure that the water entering any specific filter plus the sample collected as the influent was a random subsample of the water received from the submersible pump, and (2) the flow to any filter could be regulated through the use of gate valves to maintain a flow rate no greater than that dictated as the optimum contact time (approximately 10 to 15 minutes for most pollutants for the media tested). The water flow path for the pilot-scale testing from the manifold to the sample collection point is shown in Figures 4-8 through 4-10.

Figure 4-8. Manifold (with flowmeters) used to distribute influent water to individual filters.

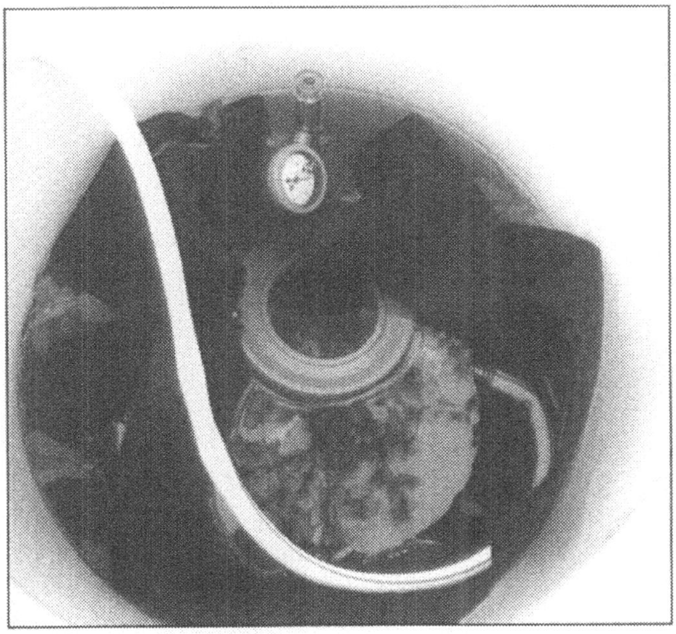

Figure 4-9. Ring distributor for influent water (note: water is evenly distributed across filter surface, thus reducing bypassing).

Figure 4-10. Sample effluent collection hose.

4.4.3 Experimental Design: Determination of the Required Number of Composite Samples

For this phase of the research, the experimental objective was to test the ability of various filtration media to remove a variety of pollutants from pre-settled stormwater runoff. However, it has been shown in past research (Clark, 1996) that statistically significant increases in specific pollutant concentrations could occur in certain media, potentially to a level that could cause problems in the receiving waters. Therefore, the statistical hypothesis for the pilot-scale testing would be the following for each filter column (a two-tailed test of significance):

H_0: influent concentration$_i$ = effluent concentration$_i$
H_1: influent concentration$_i$ ≠ effluent concentration$_i$

where i = pollutant of interest.

Based upon past work and a literature review, the removal efficiency of the filter should remain constant for a specified period of time (usually until a certain portion of sorption sites fill), and then it will begin to decrease. The reason for this decrease is that, once a certain fraction of sites fill, the availability of sites for new pollutants to sorb are limited, and a contact between the pollutant and the media may result in the pollutant bumping into another previously-sorbed pollutant, rather than into an empty site. Until this fraction of the total number of sorption sites is filled, the removal efficiency is expected to depend only on the likelihood of a favorable contact between the pollutants in the influent water and the sorption site on the filter medium. The number of samples required for the first part of the breakthrough curve can be estimated using a paired sample calculation, as described in the following paragraphs.

For the pilot-scale work, the filter capacity may not be exhausted for each medium for each pollutant. For this set-up, it is the initial removal efficiency that will be tested as part of the hypothesis testing. It was assumed that the samples in the two populations (influent and effluent) were independent of each other during the first part of the filter run. A systematic random sampling scheme was then used to calculate the number of samples required during the first part of the filter run. The calculation of the number of samples needed for a two-tailed test for systematic random sampling of either the influent or effluent was performed using the following equation (Equation 4.4.3-1) (Gilbert 1987):

$$n = (Z_{1-\alpha/2} + Z_{1-\beta/2})^2 \eta^2 / d_r^2 \qquad (4.4.3\text{-}1)$$

where n = number of samples needed; $Z_{1-\alpha/2}$ and $Z_{1-\beta/2}$ = x axis coordinates for the two-tailed test that give areas under the normal curve equal to 1-α and 1-β, respectively; η = coefficient of variation (standard deviation/average); and d_r = relative error ([sample mean - population mean]/population mean).

Because this hypothesis testing is based on the idea that the probability of the null hypothesis being true is less than a pre-specified value, designated α, the possibility exists that rejection of a true null hypothesis or acceptance of a false null hypothesis may occur. (Walpole and Myers 1993). Assuming a degree of confidence of 95% ($\alpha = 0.05$) and a power of 90% ($1 - \beta = 0.9$), the appropriate Z values are for the two-tailed test, 1.96 and 1.645, respectively.

Substituting into Equation 4.4.3-2, the result is as follows:

$$n = (1.96 + 1.645)^2\eta^2/d_r^2 \qquad (4.4.3\text{-}2)$$

and assuming an acceptable coefficient of variation of 0.5 (refer to Table 4-43) and an acceptable relative error of 1, the number of samples required of influent or effluent from each filter media (prior to the start of breakthrough) equals 3.245 samples or 4 samples each.

The underlying assumptions of the previous equations were that the populations were normally distributed and that both populations (influent and effluent) had the same variance. The assumption of normality was likely not met, since most environmental water quality populations are log-normally distributed and since there is an upper limit on 100-percent removal of pollutants. However, when the coefficient of variation is small (less than 0.4-0.5), the results for a log-normally distributed population can be approximated by using the equations for a normally distributed population. The assumption that both the influent and effluent populations had the same variance also may not be strictly true, but, based on prior sampling (Clark 1996), the variances are likely very close. The third assumption of the above equation was that the size of the population was very large relative to the variance.

Table 4-43. Average Percent Pollutant Removal (COV): Laboratory-Scale Filters (Source: Clark 1996)

Parameter	Carbon-Sand	Peat-Sand	Zeolite-Sand	Compost-Sand	Agrofiber-Sand	Sand
Hardness		52 (0.26)				
Potassium	15 (1.11)		39 (0.33)			
Calcium		88 (0.02)	17 (0.57)			
Total Solids		17 (1.28)				
Dissolved Solids		45 (0.29)				
Volatile Solids						23.4 (1.06)
Zinc (unfiltered)	76 (0.24)	71 (0.24)	73 (0.19)	76 (0.14)	77 (0.06)	71 (0.17)
Zinc (filtered)	48 (0.78)	58 (0.57)	62 (0.46)	82 (0.24)	84 (0.23)	88 (0.26)
COD (unfiltered)	96 (0.06)					
COD (filtered)	85 (0.4)					
2,4-Dinitrophenol	43 (1.19)	36 (1.54)				
Di-n-butylphthalate		66 (0.60)				
Bis(2-ethy hexyl) phthalate	90 (0.05)	39 (0.56)	82 (0.16)			94 (0.09)
Pentachlorophenol			79 (0.53)		92 (0.18)	100 (0)
Dieldrin		68 (0.58)				

*Only percent removals greater than zero are shown in this table.

While neither sampling method was wholly satisfactory for calculating the needed numbers of samples, both methods indicated that a minimum four samples each of influent and effluent from each medium were required. For the pilot-scale testing, a minimum of eight sampling events was conducted, twice the minimum number of samples indicated by these techniques.

4.4.4 Testing of the Pilot-Scale Device

Testing on the pilot-scale filters was performed in a manner expected to simulate the intermittent use that they would receive in a full-scale application. Eight filtering events were performed, four at Star Lake and four at Georgetown Lake. Each filtering event lasted approximately 8 hours. At the start of each filtering event, the effluent from the filters immediately upon starting was collected. Although not quantifiable in terms of sorbed mass lost from the filters, this initial effluent was expected to indicate any potential loss of previously sorbed material from the filter media. Grab samples of approximately 1 L were collected every hour from the effluent of each filter and combined into a cleaned, 8-L Nalgene jug (one per media) to form a composite sample of the effluent from that day's sampling. A similar composite sample was of the influent water to the filters. In addition, grab samples were collected three to four times per day and analyzed on-site for turbidity, pH, and conductivity. The purpose of these grab samples was to indicate potential changes in the effluent composition throughout the sampling event and to obtain an early indicator of filter failure, should it occur. Finally, periodically during a sampling event, the effluent flow rate and depth of water on top of the filter were noted for each media. Sample collection and handling was in accordance with *Standard Methods for the Examination of Water and Wastewater* (American Public Health Association [APHA], *et al.* 1995). At the end of the day's sampling, the composite and the initial samples were transported back to the laboratory for analysis.

This task lasted for approximately 4 weeks, from September 19, 1999 through October 22, 1999. Prior to performing the pilot-scale testing, several potential endpoints for the filtering were noted: (1) physical clogging, (2) chemical breakthrough for several pollutants, or (3) end of the project. By testing these filters intermittently for several weeks, the effects of intermittent drying (by comparing the results with the laboratory experiments) on filter performance could be determined.A summary of the influent concentrations for these filters (the water pumped from the two detention ponds) is found in table 4-44

Table 4-44. Influent Concentrations (Unfiltered Fraction) for the Pilot-Scale Filters.

Event	Copper (µg/L)	Lead (µg/L)	Zinc (µg/L)	Iron (µg/L)	Calcium (mg/L)	Magnesium (mg/L)	Total Solids (mg/L)	TSS (mg/L)
1	5.33	8.20	93.40	434.8	21.62	8.32	132	3
2	11.42	0.00	17.80	50.25	17.83	3.99	105	2
3	8.85	0.00	14.29	146.8	19.85	6.94	118	11
4	11.60	0.96	16.47	245.3	18.09	7.01	108	4
5	7.64	0.00	1.15	583.5	13.76	4.01	85	12
6	13.40	0.00	16.34	462.7	13.91	3.30	88	15
7	8.65	0.00	0	386.0	12.88	3.07	86	18
8	6.94	3.35	1.07	268.5	12.18	3.18	80	6
Average	9.23	1.56	20.07	322.22	16.26	4.98	100	9

The protocols followed for analysis and holding time limits are listed in at least one of following documents:

EPA Methods and Guidance for the Analysis of Water (EPA 821/C-97-001; 1997)

Handbook of Sampling and Sample Preservation of Water and Wastewater (EPA 600/4-82-029, 1982, and including later additions)

Methods for Chemical Analysis of Water and Wastes (EPA 600/4-79-020; revised 1983, and including later additions)

Quality Assurance Project Plan: Effects, Sources, and Treatability of Urban Stormwater Toxicants (Parmer and Pitt 1995)

Quality Assurance Project Plan: Natural Media Filtration (Clark 1999)

Standard Methods for the Examination of Water and Wastewater, 18th Edition (APHA, *et al.* 1995)

4.4.4.1 Hydraulic Characterization of Water Flow Through Pilot-Scale Filters

Characterization of the flow hydraulics through the pilot-scale filters using a salt tracer and conductivity as a measure was performed in early May 2000. These tests were performed by passing a slug of salt water through the filters once flow had been established. Because both sodium and chloride will slightly react with several of the filter media and are the regenerant for the zeolite, the dissipation of the slug through all of the filters but the sand was noted. Reported in the table below is the time until the peak conductivity occurred at the filter discharge. The data from these tests are presented in Figure 4-11. The contact time for all filters, based on field observations and flow measurements, was about 15 minutes, close to optimal for pollutant removal and retention.

Media	Time to Peak (sec)	Time to Peak (min)
Zeolite-sand	170	2.83
Compost-sand	195	3.25
Loam	263	4.38
Sand	310	5.17
Agrofiber-sand	370	6.17
Cotton-sand & Carbon-sand	620	10.3
Peat-sand	1260	21

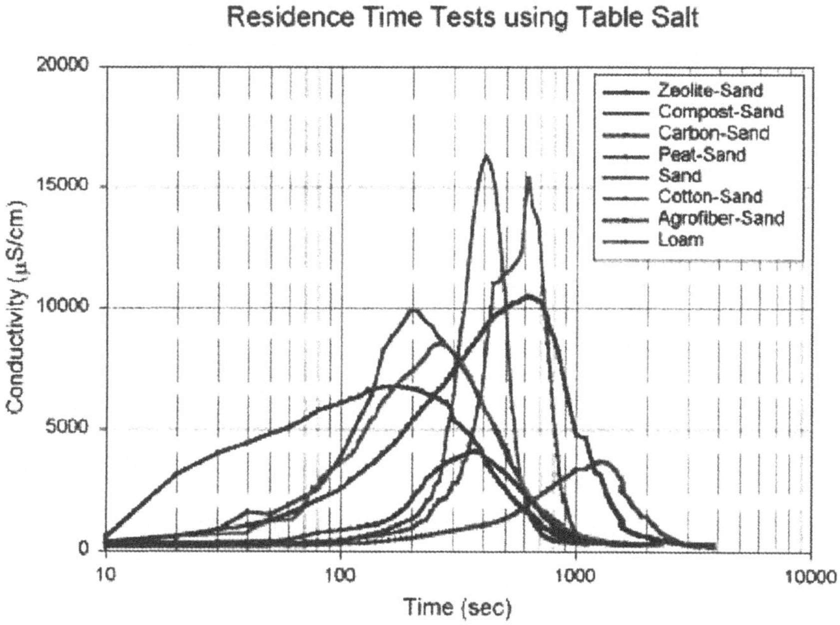

Figure 4-11. Salt tracer study in field filtration units.

4.4.5 Sample Analysis

Samples collected in the different phases of this research were analyzed according to the following analytical protocols (Table 4-45). Additional information regarding these procedures are in the UAB laboratory Standard Operating Procedures (SOPs) (Pitt and Clark 1998).

Table 4-45. Analytical Techniques for Pilot-Scale Samples

Analytical Parameter	Analysis Method (EPA Method number shown if available; equivalent Standard Methods method in parentheses, if available)
PH	EPA Method 150 (Standard Methods 4500-H$^+$.B.)
Conductivity	EPA Method 120.6 (Standard Methods 2510.B.)
Turbidity	EPA Method 180.1 (Standard Methods 2130.B.)
Color	EPA Method 110.3 (Standard Methods 2120.C.)
Hardness	EPA Method 130.2 (Standard Methods 2340.C.)
Toxicity	Microtox Rapid Toxicity Screening (UAB Laboratory SOP)
Pb, Cd, Zn, Cu, Cr, Fe, Ca, Mg Magnesium	ICP (Standard Methods 3500)
Solids (Total, Dissolved, Suspended, Volatile)	EPA Method 160.1 and 160.2 (Standard Methods 2540.B.C.D.E.)
Particle Size Distribution	Standard Methods 2560.B. (Coulter Counter, UAB Laboratory SOP)

4.4.6 Results and Discussion of the Pilot-Scale Testing

The following paragraphs present the results of the pilot-scale study assuming that the influent and effluent samples from each filter are independent of other influent and effluent samples, respectively. This assumption was made because, as can be seen in the section in this chapter on model evaluation, for many of the parameters, this series of tests was performed during the early part of the filter's life where the pollutant loading on the filter is not great enough to interfere significantly with the availability of sorption sites for the pollutants in the influent.

The statistical analysis on the paired influent and effluent data was performed using the nonparametric Wilcoxon signed-rank test, since it could not be assumed that the underlying populations were normally distributed. Another advantage of the Wilcoxon signed-rank test was that it could handle a limited number of 'non-detects' in the laboratory results. This was crucial since it was expected that, when using pre-settled runoff, many of the pollutants present in the water are in concentrations that approach the method detection limit. The procedure for performing the Wilcoxon signed-rank test is explained by Gilbert (1987), Lehmann and D'Abrera (1975), and Hollander and Wolfe (1999). The hypothesis tested for this set of statistical analyses was the following:

H0: influent = effluent
H1: influent ≠ effluent

The two-sided test was selected so that any undesirable effects would be detected, such as the leaching of pollutants from the media into the passing water. The Wilcoxon signed-rank probabilities are shown on the graphs for each media for each pollutant.

Flow rates through filters. Initially, the flow rates were set at 0.004 m^3/min for all filters (a loading rate of or 0.016 m^3/m^2-min [23 m/day]). This flow rate would give a contact/residence time in the media of slightly longer than 15 minutes and was based on the results of the kinetics studies. Since the purpose of the pilot-scale studies was to investigate the removal ability of the filters for dissolved pollutants, a long run life was desired, hence the selection of pre-settled stormwater runoff for the test water and the selection of a flow rate in the optimum contact time range.

For the cotton-sand, agrofiber-sand, carbon-sand, and zeolite-sand, this flow rate could be maintained throughout the entire project, i.e., the suspended solids loading did not impact the flow rate to the extent that the flow rate was noticeably decreased, as measured by the increase in head loss across the filter. Head loss development was noticeable primarily for the peat-sand filter, with some head loss development noticed with the sand filter. It was also observed that drying the filters overnight brought about an initial recovery in flow rates in the sand filter and slightly in the peat filter, but this recovery quickly disappeared during the next sample event unless the filter was dried for several days.

A comparison of the change in flow rate versus suspended solids loading for each of the filters is shown in Figure 4-12, both for the pilot-scale and laboratory-scale experiments, for the peat-sand, compost-sand, carbon-sand, sand, cotton-sand, and agrofiber-sand filters. For the compost-sand and the agrofiber-sand filters, the flow rate in the field is greater than the flow rate in the bench-scale setup at the higher suspended solids loadings. This indicates that a scale-up factor may be needed to predict field flow rates from laboratory tests (k_i in the design equation likely is greater in a field installation than it would be calculated based on laboratory results). The exact opposite was seen in the peat-sand filter. A potential reason for the decrease in field flow rate at a significantly smaller suspended solids loadings may be shrinking and swelling of the media to an extent not seen in the laboratory, especially since the flow rate recovered slightly every time after drying and since the recovery was greater with longer drying periods.

No predictions of scale-up factors can be made for carbon-sand, cotton-sand, and sand based on the present data because the suspended solids loadings to those pilot-scale filters did not approach the suspended solids loading needed for measurable changes to occur in the flow rate in these filters. No data are shown for zeolite-sand or loam because no valid laboratory data are available for comparison. For zeolite, the pilot-scale filter contained a mixture of the large- and small-particle batches. No laboratory work was performed on the loamy topsoil for physical clogging. No loss of performance was noted for either the zeolite-sand or the loam filter or for the compost cartridge or the Jacuzzi's lightweight sand.

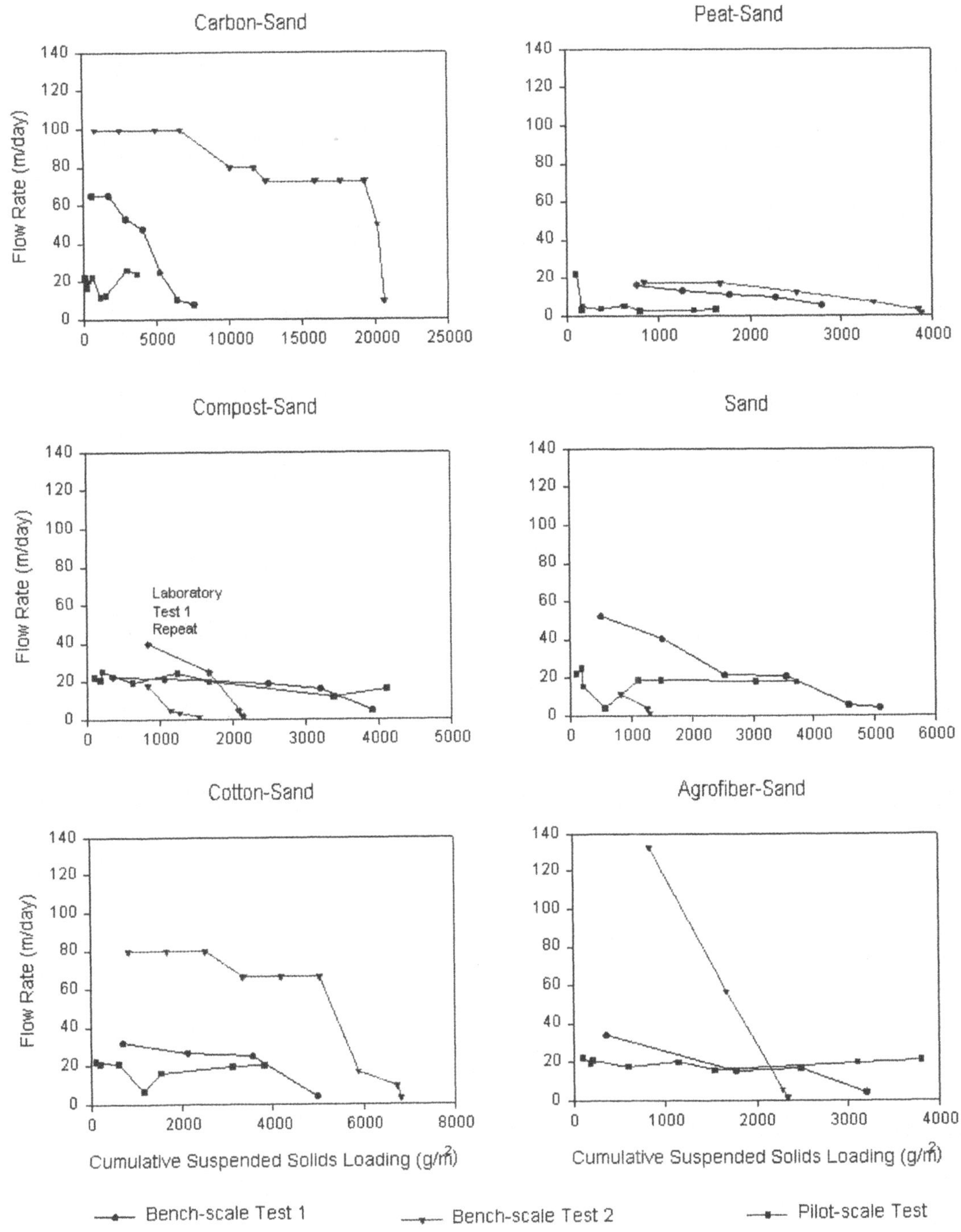

Figure 4-12. Flow rate versus suspended solids loading: comparison of laboratory-scale to pilot-scale.

Heavy metals removal. Figure 4-13 depicts the ability of the filters to remove copper. The graphs for lead, zinc, and iron are located in Appendix B. Copper, lead, and zinc removals were not found to be statistically significant for any of the filter media, even though the media loading was small. It has been assumed in the modeling equations that adsorption is irreversible, assuming that the influent water characteristics, especially pH and conductivity, do not change significantly. However, at a certain low concentration, it would be expected that an equilibrium would be established where removal below that equilibrium concentration would not occur. Concentrations below the equilibrium concentration, whether due to high removal efficiencies or to low influent concentrations, would encourage the reversal of the driving force for the sorption reaction, causing desorption. The locations of this equilibrium for many of the media are indicated on the figures for these three metals. For copper and lead, the minimum water concentration attainable appears to be approximately 5 to 10 μg/L, and approximately 10 to 20 μg/L for zinc, for all of the media.

The influent iron concentrations are significantly greater than the influent copper, lead and zinc concentrations. Therefore, adsorption is the primary removal mechanism occurring. It is overwhelming the desorption and makes it appear that the adsorption is irreversible. Statistically significant removals were seen for the sand, cotton-sand, agrofiber-sand and carbon-sand filters. The peat-sand and compost-sand filters had a probability of 0.054 that the effluent was less than the influent concentrations.

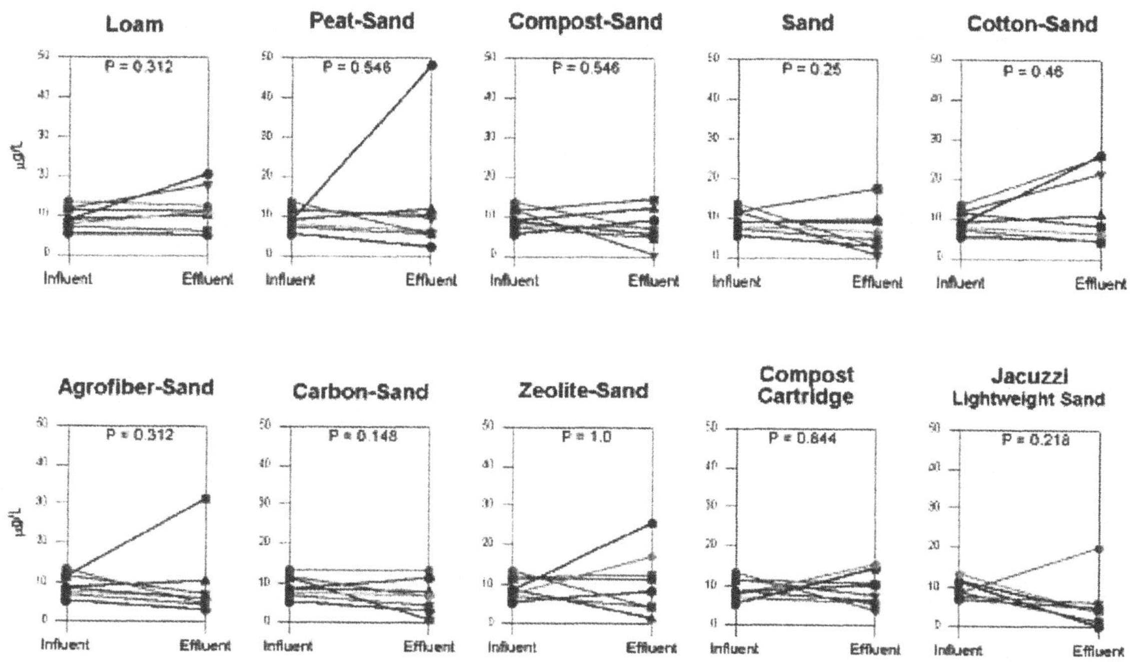

Figure 4-13. Copper removal during pilot-scale filtration experiments.

Major ion removal (calcium, magnesium, hardness). The removal ability is shown in Figure 4-14. The graphs for magnesium and hardness are located in Appendix B. The loam, peat-sand, and cotton-sand filters were able to consistently remove calcium from the influent water. The behavior of the peat-sand filter was not unexpected based upon the literature review and the behavior of the laboratory filter column. The potential of the cotton to remove a small amount of calcium was also seen in the laboratory columns. As can be seen from Figure 4-14, only the peat had the ability to almost completely remove the calcium from the influent water (with one exception). The compost-sand filter consistently added calcium to the runoff water. This also is in agreement with the laboratory column results and with the ability of compost-sand filters to neutralize its influent when the influent pH is significantly different from 7.0 to 7.5.

No filter had the ability to remove magnesium consistently. Hardness removal was only statistically significant with the peat-sand and loam filters, with only the peat-sand filter able to remove large percentages of hardness. This is not unexpected given the ability of the peat filter to remove calcium. Mimicking its behavior with calcium, the compost-sand filter contributed hardness to the runoff water.

pH. The peat-sand filter is the only filter that had a measurable effect on pH. Hydrogen as the hydronium ion, H_3O^+, is one of peat's easily exchangeable ions, and in general, as peat adsorbs other cations, it releases the hydronium ion. pH has been proposed as a method of monitoring remaining life in a peat-filter. The other filter media, when treating an influent at a neutral or near-neutral pH, tend not to affect the pH measurably. Although, previous research (Clark 1996) has shown that when the influent water to a compost-sand column is not near a neutral pH, compost-sand will attempt to neutralize it.

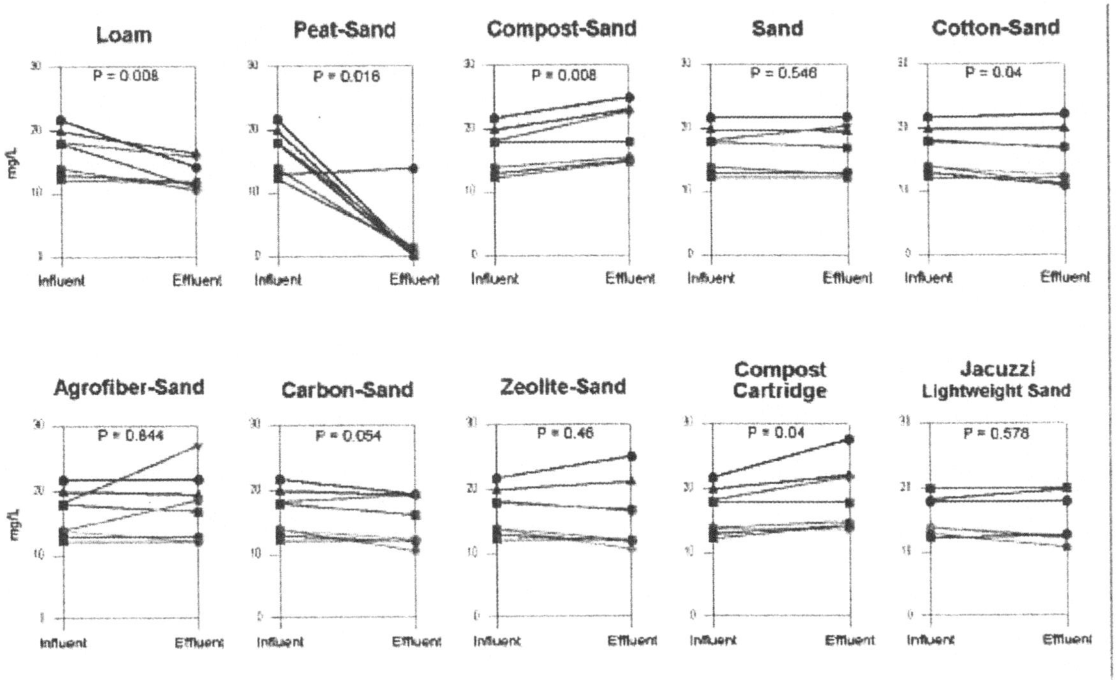

Figure 4-14. Calcium removal during pilot-scale filtration experiments.

Turbidity and solids removal. Figure 4-15 demonstrates the removal ability of the pilot-scale filters for unfiltered turbidity. The removal ability of the pilot-scale filters for turbidity in the filtered fraction, total solids, dissolved solids and suspended solids are shown in Appendix B. All media except for the loam, the compost cartridge and the lightweight-sand Jacuzzi were capable of removing the particles from the influent water that caused the turbidity. The lack of significant removal in the loam was not surprising since it was constructed without a sand underlayer. The loam could best be characterized as having a wide range of particle sizes. Many of the smaller particles near the bottom of the filter or near poreholes that continue through the filter are likely to be washed out during filtration. Although many of the granules of the compost in the cartridge were not as small as those in the loam, small particles were found in the media. These particles were likely to be washed out if they were near a continuous pore formed by the significantly larger granules. The lightweight-sand Jacuzzi filter was a high-rate filter, and the literature review has shown that the higher-rate filters tend to have poorer turbidity removals. As seen in the filtered turbidity figure, no media were capable of removing colloidal-sized particles effectively, i.e., those that would pass through a filter with a nominal pore size between 0.45 μm and 1.0 μm. The loam and lightweight sand both added turbidity in the small particle size range.

The compost medium, whether in the compost-sand filter or in the compost cartridge, increased the total and dissolved solids concentration of the runoff water. The increase in dissolved solids was not unexpected based on the increase in the conductivity. However, since compost-sand decreased turbidity in the unfiltered fraction, it was unexpected that the total solids concentration in the compost-sand effluent was greater than in the influent. This might be caused by the formation and/or release of larger particles in the effluent compared to the influent, which would affect the turbidity. Carbon-sand and peat-sand reduced both the total solids and dissolved solids concentrations in the runoff water. The percent removals in the carbon-sand were smaller than in the peat-sand. The results for the dissolved solids are not surprising since the peat-sand filter is able to soften the influent water by removing calcium. Only the carbon-sand was capable of removing the volatile total solids from the influent water. This is a reflection of the carbon's ability to remove organic molecules.

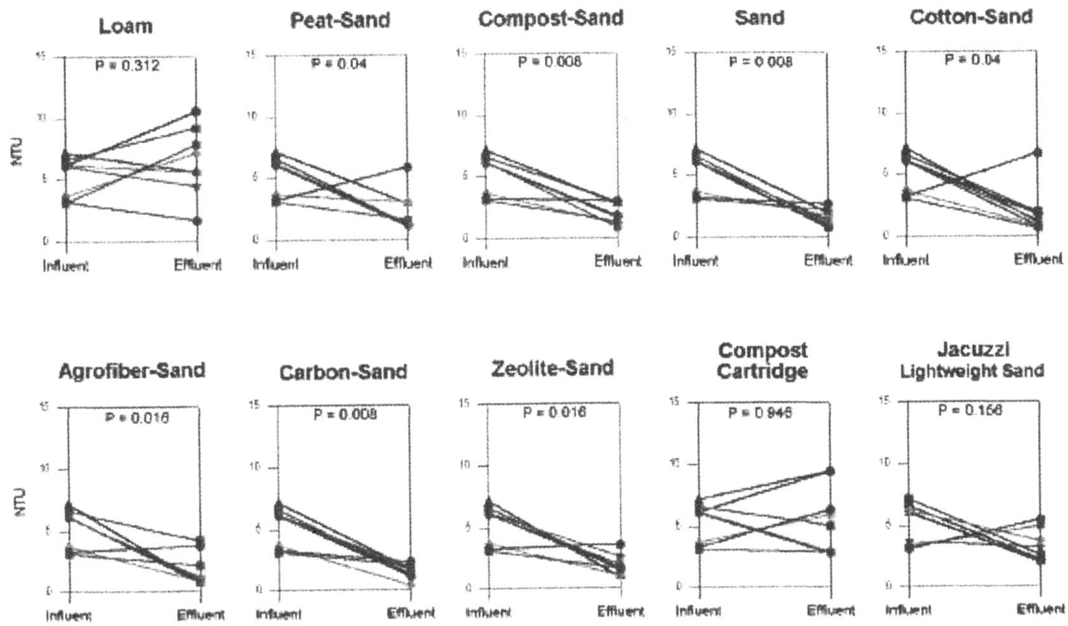

Figure 4-15. Turbidity removal from unfiltered fraction during pilot-scale filtration experiments.

The sand medium lowered the total solids concentration in the runoff. No media could remove total suspended solids to a statistically significant degree, likely due to the small influent concentration of total suspended solids (typically less than 10 mg/L). Total suspended solids removal from pre-settled stormwater runoff, such as in the two detention ponds, could be difficult because the particles that remain in suspension generally are at the limit or below the limit of removal for this filter type.

In other applications, Pitt, *et al.* (1999) found that a sand-peat filter was a very effective control after stormwater was pre-treated by sedimentation. Table 4-46 shows the results from the MCTT (Multi-Chambered Treatment Train) during pilot-scale testing at a university parking area in Birmingham, AL. The combination of this filtration/sorption chamber with the main settling chamber resulted in very high reductions of toxicants: most of the lead was removed by sedimentation, while most of the zinc was removed in the sand/peat chamber. High quality effluent using this combination treatment train was substantiated during full-scale tests in Wisconsin (Table 4-47).

Table 4-46. Median Percent Reductions by MCTT Chamber (Pitt, *et al.* 1999)

Constituent	Main Settling Chamber (percent)	Sand-Peat Chamber (percent)	Overall Device (percent)
Microtox™ toxicity (unfiltered)	***18***	***70***	***96***
Microtox™ toxicity (filtered)	***69***	67	***87***
lead	***88***	***18***	***93***
zinc	***39***	62	***91***
n-Nitro-di-n-propylamine	81	64	92
hexachlorobutadiene	29	97	100
pyrene	100	25	100
bis (2-ethylhexyl) phthalate	***99***	N/A	***99***

[a] Note: Bold italics indicate Wilcoxon 1-sided p value ≤0.05

Table 4-47. Performance Information for Full-Scale MCTT Tests, Compared to Birmingham Pilot-Scale MCTT Results (median reductions and median effluent quality)

	Milwaukee MCTT (15 events)	Minocqua MCTT (7 events)	Birmingham MCTT (13 events)
suspended solids	98 (<5 mg/L)	85 (10 mg/L)	83 (5.5 mg/L)
Microtox® toxicity (total)	na	na	100 (0%)
Microtox® toxicity (filtered)	na	na	87 (3%)
Cadmium (total)	91 (0.1 μg/L)	na	18 (0.6 μg/L)
Cadmium (filtered)	66 (0.05 μg/L)	na	16 (0.5 μg/L)
Copper (total)	90 (3 μg/L)	65 (15 μg/L)	15 (15 μg/L)
Copper (filtered)	73 (1.4 μg/L)	na	17 (21 μg/L)
Lead (total)	96 (1.8 μg/L)	nd (<3 μg/L)	93 (<2 μg/L)
Lead (filtered)	78 (<0.4 μg/L)	na	42 (<2 μg/L)
Zinc (total)	91 (<20 μg/L)	90 (15 μg/L)	91 (18 μg/L)
Zinc (filtered)	68 (<8 μg/L)	na	54 (6 μg/L)
benzo(a)anthracene	>45 (<0.05 μg/L)	>65 (<0.2 μg/L)	nd
benzo(b)fluoranthene	>95 (<0.1 μg/L)	>75 (<0.1 μg/L)	nd
dibenzo(a,h)anthracene	89 (<0.02 μg/L)	>90 (<0.1 μg/L)	nd
Fluoranthene	98 (<0.1 μg/L)	>90 (<0.1 μg/L)	100 (<0.6 μg/L)
indeno(1,2,3-cd)pyrene	>90 (<0.1 μg/L)	>95 (<0.1 μg/L)	nd
Phenanthrene	99 (<0.05 μg/L)	>65 (<0.2 μg/L)	nd
Pentachlorophenol	na	na	100 (<1 μg/L)
Phenol	na	na	99 (<0.4 μg/L)
Pyrene	98 (<0.05 μg/L)	>75 (<0.2 μg/L)	100 (<0.5 μg/L)

na[a] : not analyzed
nd[b]: not detected in most of the samples

4.4.6.1 High Resolution Grab Sample Results During Pilot-Scale Tests

During the field tests, grab samples were obtained every other hour and analyzed for pH, turbidity, and conductivity. Table A-7 in Appendix B contains the results for pH for the grab samples. The influent water had a neutral to slightly above neutral pH. None of the media made significant changes in the pH during filtration, except for peat. Throughout the testing, the effluent pH of the peat kept increasing, potentially indicating that breakthrough for any ions that exchange with the hydrogen ion on the peat surface was beginning to occur. Ions that would be expected to exchange with the hydrogen ion include the cations, such as copper, lead, zinc, cadmium chromium, iron, calcium and magnesium. For the other media, the lack of significant changes may not indicate that adsorption/ion-exchange was not occurring, just that the exchangeable ion was not hydrogen.

Table A-8 in Appendix B shows the percent removals for turbidity during the grab sample analyses. The influent turbidity was approximately 10 NTU (range 7 – 13 NTU). Out of the five media tested for this project and based on the grab sample results, the compost cartridge had the poorest turbidity removal. This is not unexpected since the compost medium is not homogeneous in size. Many of the clumps of compost are several millimeters in width, which will increase the likelihood of channeling and preferential flow through the filter. Also, many of the media had a minimum turbidity (usually between 2 and 3 NTU) below which turbidity-causing particles could not be removed.

Table A-9 in Appendix B contains the percent removals for conductivity for the grab sample analyses. The influent conductivity ranged from 100 – 200 μS/cm. Only the peat-sand media successfully removed the ions that contributed to conductivity. This, in conjunction with the pH data, indicates that when the peat-sand removes the ions from water that contribute to conductivity, it adds the hydrogen ion as the exchangeable ion (this is indicated by the decrease in pH). The other media had an unpredictable effect on conductivity, indicating that most of the exchangeable ions on the other media also impact conductivity, but are not necessarily the hydrogen ion.

4.4.7 Model Evaluation Using the Pilot-Scale Results

The field data also was compared to the model/design equations that were developed during the laboratory work. The results of these comparisons provided an indication of any effects that scale-up from a laboratory (or bench-scale) apparatus to a pilot-scale set-up may have had on the model predictions. It also would show if the model is capable of predicting breakthrough curves for typical, unspiked, pre-settled stormwater runoff.

The pair of equations used in this modeling exercise was Equations 4.4.7-1 and 4.4.7-2 shown below. The second equation, the modified BDST model equation, now includes a summation of the contribution of individual storm events where the influent concentration and the number of bed volumes treated may be different between storms. The summation term also allows for modification of the unit flow/loading rate, u, between storm events, which, as seen in the flow rate calculation, will depend on the suspended solids loading rate once the flow rate decreases below the designer's maximum allowable flow rate.

$$u = ki \bullet Lm^{-c} \qquad (4.4.7\text{-}1)$$

Where u = unit flow/loading rate (m/day), ki = flow rate reduction rate constant (day/m), Lm = suspended solids loading, and c = power of the flow rate reduction with suspended solids loading.

$$\ln\left(\frac{C_0}{C_e}-1\right) = -1x10^6\, kx\sum_{j=1}^{n}\left(\frac{C_0(BV)}{u}\right) \qquad (4.4.7\text{-}2)$$

Where C0 = influent concentration (mg/L), Ce = effluent concentration (mg/L), k = rate constant predicted from laboratory breakthrough curves (day-1), x = bed depth (m), u = unit flow/loading rate (m/day), and BV = bed volumes treated.

The first conclusion that was noticed using the models with the laboratory rate constants and the field data was that the variability in the field data, with a few exceptions, made it impossible to compare. Therefore the model predictions are not shown. However, the analysis of the field data by bed volumes treated, i.e., set up in the same manner as required by the model, highlights the variability in the performance of the filters more than the earlier figures.

Cation removal (hardness, calcium, magnesium). The breakthrough curves for hardness (Figure 4-16) showed the early breakthrough that would be predicted by the model. The shape of the calcium and magnesium graphs for peat indicates that the model should be applicable, provided that the capacity can be accurately estimated. Since peat has a very high capacity for calcium and magnesium (see Appendix B), the intercept term on the model is not insignificant and must be calculated in order to better predict the behavior of the filter for these pollutants. However, during the laboratory breakthrough curve tests, whether due to the small size of the columns and potential short-circuiting or to an undetermined occurrence, the removal of magnesium and hardness never exceeded 50 percent, whereas, in past experiments (Clark 1996), initial removals have exceeded 50 percent for both.

Heavy metals removal. The figures in Appendix B demonstrate the removal ability of the filters for copper, zinc and iron, respectively. Because of the low concentrations of the influent water for copper and zinc, no modeling of the removal could be done. Inspection of the data for the metals removal visually indicates that, for most of the pollutants, no trends could be seen. The most visible exception to this is the removal ability of the lightweight sand in the Jacuzzi for total copper.

A summary of the median removal efficiencies for the pilot-scale testing is shown in Table 4-48.

Table 4-48. Median Removal Efficiencies Based on Pilot-Scale Testing

	Loam	Peat-Sand	Compost-Sand	Compost Cartridge	Cotton-Sand	Agrofiber-Sand	Carbon-Sand	Zeolite-Sand	Lightweight Sand	Sand
Turbidity (unfiltered)		68 (0.04)	65 (0.01)		78 (0.04)	78 (0.02)	74 (0.01)	69 (0.02)		75 (0.01)
Total Solids		35 (0.05)					10 (0.01)			4 (0.01)
Dissolved Solids		40 (0.02)					10 (0.04)			
Volatile Total Solids							32 (0.01)			
Hardness	13 (0.04)	68 (0.01)								
Calcium (total)	20 (0.01)	96 (0.02)			7 (0.04)		8 (0.05)			
Iron (total)		42 (0.05)	44 (0.05)		55 (0.02)	69 (0.02)	67 (0.04)			

*Note: the p values for the statistically significant removals, based on the Wilcoxon sign-rank test, are shown in parentheses.

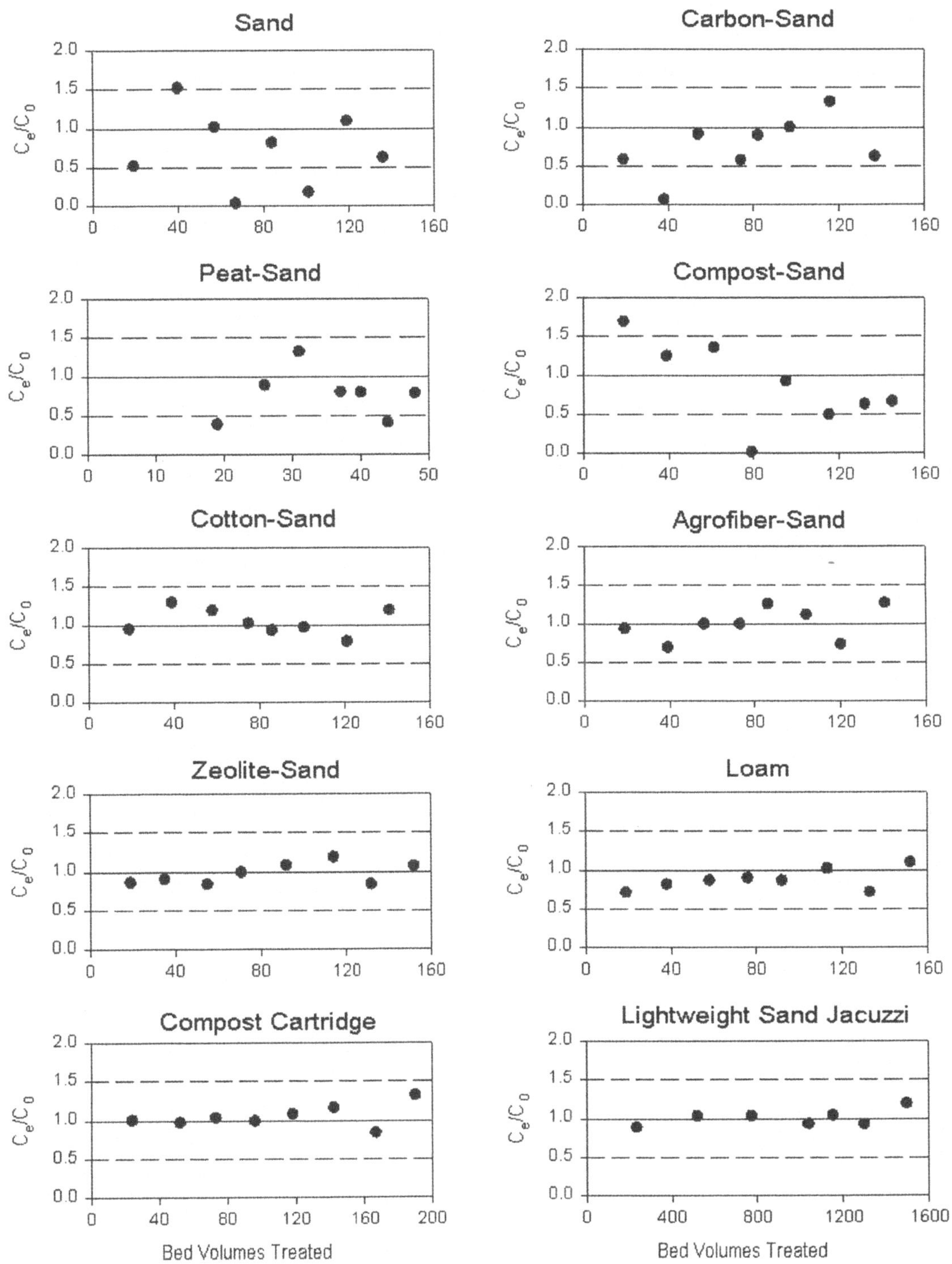

Figure 4-16. Pilot-scale data for hardness.

Analysis of initial samples. From the second sampling event on, samples were collected of the effluent initially leaving the filter. The purpose of this was to monitor any potential release/leaching of pollutants from the media. For many of the pollutant/media combinations, the initial effluent contains very high concentrations of that contaminant. Whether this is a release of previously trapped pollutants or a degradation of the media causing leaching of the pollutants is unknown since it was impossible to determine the length of the 'first flush.' It is possible, based on laboratory work where the effluent water appeared anaerobic even after several bed volumes of water had passed through the media, that the release of these pollutants occurs longer than the replacement of the water trapped in the pores during the time between the filtering events.

The leachings/releases that occurred are consistent with the potential leaching seen in the earlier data. These releases appear to add to the body of data that supports potential degradation of the media. They confirm that many of the previously trapped pollutants are likely released into the effluent water during start-up of the filter. Because the length of the start-up flush could not be determined based on this experimental design, the potential recovery of capacity of the sorption media due to the release of the previously trapped pollutants is unknown. However, the visually degraded start-up flush water only lasted for about 15 minutes during these tests, and would constitute a small fraction of the total amount of treated water.

4.5 Upflow Filtration Tests

Preliminary laboratory tests were performed on two potential areas of future research on stormwater filters. One of the primary problems with downflow filtration is the clogging of the filter prior to the exhaustion of the chemical capacity. Upflow filtration using a siphon control may improve the life of the filters. The potential of upflow filtration was evaluated for suspended solids removal. However, upflow filtration, as well as a long interevent time, may cause a second problem: the water trapped in the pore space going anaerobic. The behavior of bound pollutants during long-term aerobic and anaerobic exposure conditions was also examined for four of the filter media.

4.5.1 Upflow Filtration

Figure 4-17 shows the results from the upflow clogging experiments. It was found that because the complete height of the media was being used to trap pollutants, the flow rate through the media was not affected prior to breakthrough of the solids (measured as turbidity). Since high effluent turbidities were not desirable, it was decided to use turbidity breakthrough as the endpoint for the upflow experiments. Table 4-49 shows the cumulative suspended solids loading (g/m^2) and loading of water (m) until breakthrough was observed (turbidity > 100 NTU).

Table 4-49. Cumulative Suspended Solids Loading and Flow at Breakthrough (Failure) in Upflow Filtration

Filter Media	Cumulative Suspended Solids at Breakthrough/Failure (g/m²)	Cumulative Water Loading at Breakthrough/Failure (m)
Sand	4000	2.7
Activated Carbon-Sand	7000	4.7
Peat-Sand	13000	8.3
Compost-Sand	7500	4.7
Cotton-Sand	4500	2.9
Agrofiber-Sand	15000	9.6

Based upon the behavior during these experiments, the media can be classified into the following three categories.

1. Sudden breakthrough with no decrease in flow rate (agrofiber-sand, cotton-sand, and carbon-sand): The clay front moved up through the media until it reached the top of the column and breakthrough occurred, i.e., true depth filtration. Taller filters should have a longer life.
2. Gradual breakthrough with decreasing flow rate (compost-sand and sand): Much of the clay was trapped at the bottom of the column and did not penetrate the column, i.e., surface filtration. Flow rate was reduced due to the blocking of the pores at the bottom by the trapped clay. No large clay front was observed. A larger sump to trap the non-penetrating clay particles off-line may be necessary. The behaviors of these media are similar to their behaviors during downflow filtration in that the predominant mechanism occurring is surface filtration. These media may be amenable to modeling using the physical clogging equation used for downflow filtration.
3. Gapping media (peat-sand): The media failed when a gap developed between the sand underlayer and the mixed media layer when the fluidized bed velocity was exceeded or when the media floated in water. The media failed when the pressure at the bottom of the filter pushed the peat-sand layer up to an unacceptable height and added peat to the effluent water. Gapping media may not be effective in upflow filtration unless contained by a plate installed across the top of the filter at the maximum allowable bed height.

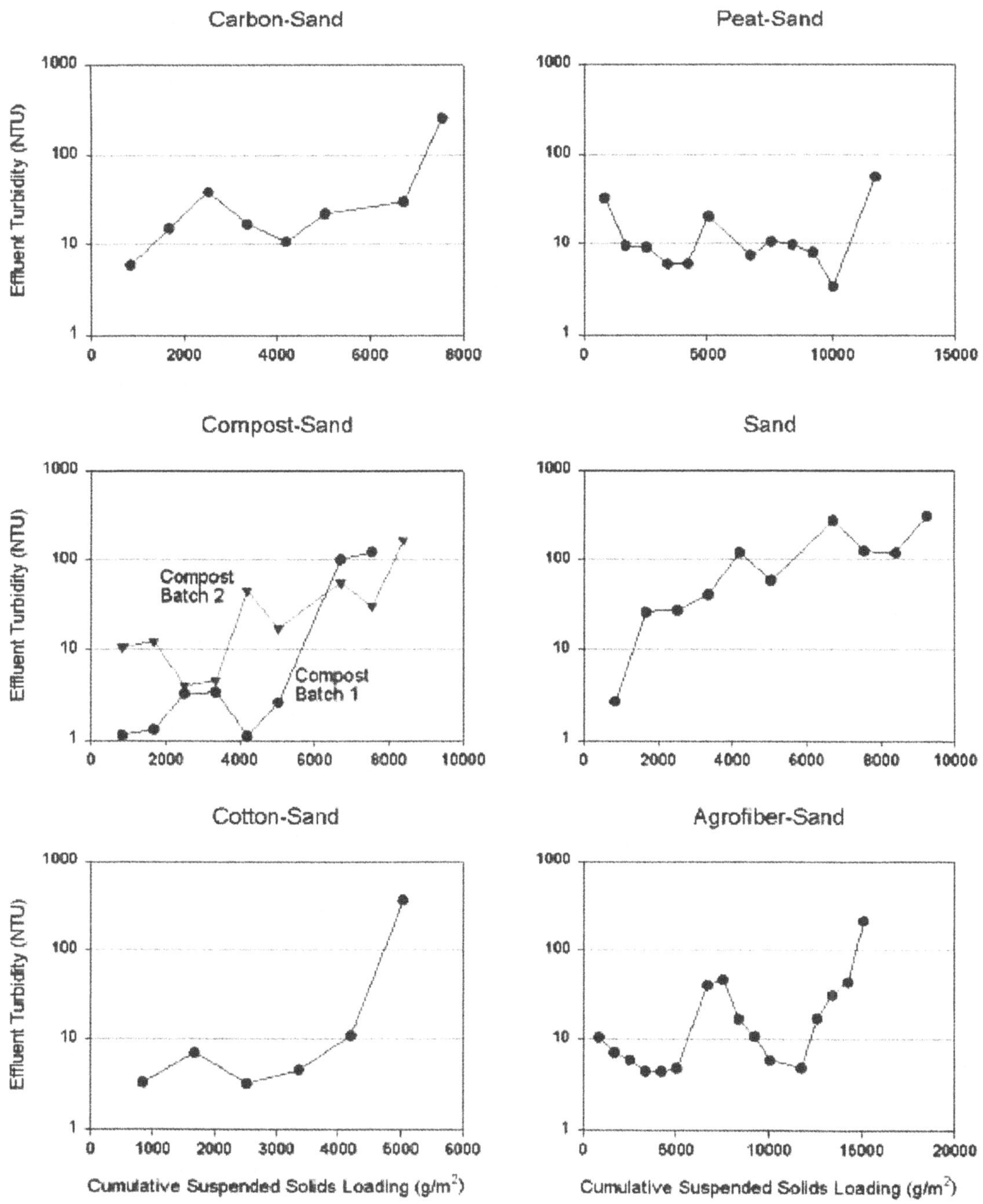

Figure 4-17. Effect of suspended solids loading on effluent turbidity in upflow columns.

4.5.2 Aerobic Versus Anaerobic Exposure Conditions

The behavior of filters under anaerobic conditions is of interest when considering upflow filtration and other low head-loss designs. Standing water in the filter, i.e., inner pore water, could go anaerobic with subsequent release of previously captured pollutants. For upflow filtration with siphon control for entering water, it would be expected that only the top of the filter would be exposed to air (unlike a gravity filter where the top and bottom are potentially exposed to air between storms), and an anaerobic environment could develop. The purpose of this series of tests was to determine if the media were able to retain trapped pollutants even if the filter developed an anaerobic environment, either throughout the filter or only in sections of the filter, between storm events. This test was performed under two different pH conditions for loading and rinsing. The sorption pH conditions (first run, pH 2; second run, pH 3.5) were varied between the two runs in order to load the media under conditions close to optimal for the different pollutants in the water. However, since the pre-settled stormwater runoff used during the exposure phase was collected from the same detention pond, it was confirmed that the stormwater runoff pH during the lengthy exposure periods was very similar.

Adsorption under anaerobic conditions has been shown to lead to weaker bonding between phenolic compounds and activated carbon. Under aerobic conditions, the molecular oxygen aided in the formation of acidic surface oxides on the carbon, which enhanced dimer and trimer formation on the carbon surface. This increase in adsorption capacity in the presence of dissolved oxygen, however, did not hold for aliphatic organic compounds (Abuzald and Nakhla 1994).

A rough classification of metals in natural lake waters is shown in Table 4-50. Under anaerobic conditions, iron is often released from iron oxides. It has been shown that iron concentrations as Fe^{2+} increase with depth in a lake due to the release of iron from anaerobic sediments (Baker, *et al.* 1994). In general, as the pH decreases up to a certain point, the metals move from complexes or un-ionized forms to more ionized forms, where they may be more tightly bound to the media. The optimum pH for adsorption of metals to iron oxyhydroxides has been given as the following: Pb (pH 3.0), Cu (pH 4.3), Zn (pH 5.3), Ni (pH 5.7), and Cd and Co (pH 5.9). The degree of complexation has also been shown to increase with pH in the range of 5.5 to 6.5. The ability of the metals to complex may dictate their ability to remain sorbed to an organic media during exposure to aerobic and anaerobic conditions (Salbu and Steinnes 1995).

Anaerobic conditions also will potential affect the retention ability of nutrients on a surface. Assuming that a sufficient population of denitrifiers exists, denitrification can significantly reduce the concentration of nitrate in the water, as has been shown in the muds of slow-moving streams where nitrate concentrations have been effectively reduced by up to 200 μeq/L over a 2-km length (Baker, *et al.* 1994).

Table 4-50. Classification of Metals Performance in Natural Lake Waters (Data derived from Baker, *et al.* 1994)

Category	Elements
Mainly present in dissolved form	Zn, Cd, Ni, Mn, As
Mainly present in particulate-bound form	Al, Fe, Pb, Cr
Mainly present as dissolved or colloidal organic complexes	Cu, Pb, Hg, Al, Fe
Very susceptible to pH decrease re: mobility,	Al, Mn, Zn, Cd, Ni, Be
SEDIMENTATION, SPECIATION	
Occurring in organometal complexes by biomethylation	Hg, As, Sn, Se (possibly Pb, Tl, Cd)

For this set of tests, only a representative subset of the project media was used: activated carbon, peat moss, compost, and sand. These media were selected because they provided the best overall results during the preliminary work – with sand being used as a comparison to traditional filter material (Clark 1996). The media were exposed to a concentrated solution using spiked tap water (10 mg/L of lead, copper, zinc, iron, nitrate, phosphate, and ammonia) for several hours. The water was then filtered through a 0.45-μm filter. The difference between the initial sample concentration and the filtrate concentration for each media is equal to the amount of each pollutant that is sorbed onto the media (a blank sample was evaluated in a similar manner, except with no media, so that the effects of sorption onto the exposure containers would be accounted for in the data analysis). After rinsing with a buffered distilled water to remove any loosely bound material and to replace any concentrated pore water, the media were exposed to Star Lake water for a period of several weeks. One sample of each medium was maintained in an aerobic environment with constant aeration to keep the lake water saturated in oxygen. The other sample of each medium was exposed to the Star Lake water while in sealed BOD bottles, where the naturally-occurring matter in the water would consume the oxygen and create an anaerobic environment.

The first activity performed under this task was to load the media with pollutants typically found in stormwater runoff. This was performed by exposing a known weight of media to a spiked tap water which contained known concentrations of the following pollutants (nutrients and metals): ammonia, nitrate, phosphate, potassium, lead, copper, iron, and zinc. Reagent grade standards were used to create the spiked test water. Approximate concentrations of the pollutants were as follows: 20 mg/L for the nutrients and approximately 10 mg/L for each of the metals. For the second run, dihydrogen potassium phthalate was added to buffer the test water to a pH greater than 3.

The weights of the various media are shown in Table 4-51. The approximate volume of media exposed in each sorption jar was approximately 150 mL, and the approximate volume of spiked water in each jar was approximately 175 mL.

Table 4-51. Weights of Media Used in Aerobic-Anaerobic Exposure Tests

Media	Weight (g)
Compost	81.73
Sand	143.00
Peat Moss	19.11
Carbon	46.37

The sorption phase of the experiment was performed in a cleaned 250-mL borosilicate jar. Three jars of each media were exposed. Also, three "blank" jars were tested in the same manner as described for the media so as to account for any sorption or leaching from the jars. One-hundred seventy-five milliliters of spiked tap water was added to the pre-weighed media and allowed to shake for approximately 3 hours. The 3-hour time for sorption was selected based on prior kinetics tests which showed that a significant quantity of pollutants had adsorbed in 2 hours or less. After the exposure time was complete, the samples were filtered through a 0.45-μm membrane filter, and the filtrates for each media and for the blank were combined into single sample for each set-up. Each aliquot of media was rinsed with approximately 450 mL of buffered rinse water (first run: rinse water pH = 10.5; second run: pH rinse water = 7.5). The rinse water was also collected and composited into a "buffer rinse sample" for each medium. The buffer rinses were used to remove loosely bound pollutants and to rinse the residual test/sorption water from the media.

Once each aliquot of media was rinsed, one aliquot was selected randomly and placed into a 1-L cleaned Nalgene high-density polyethylene (HDPE) bottle and labeled "Aerobic." A second aliquot was separated approximately into thirds, and each third was placed into one of three 300-mL BOD bottles and labeled "Anaerobic." The third aliquot was disposed. Approximately 900 mL of water collected that day from Star Lake (pre-settled, unspiked stormwater runoff) was added to each of the aerobic bottles. Approximately 300 mL was added to each anaerobic BOD bottle for each medium. The anaerobic bottles were then sealed according to the procedures used to perform BOD testing. An aeration stone was added to each of the aerobic filter bottles. A "blank" was also performed during this test. The blank consisted of Star Lake water with no media and an aerobic and an anaerobic blank were created. All bottles were placed on a shaking table and allowed to shake for approximately 4 weeks.

At the end of the exposure time, the dissolved oxygen (DO) concentration and the oxidation-reduction potential (ORP) of each aerobic and anaerobic sample were taken. Then the samples were filtered through a 0.45-μm gel membrane filter, and the filtrate were analyzed for the parameters, and using the methods, listed in Table 4-52.

Table 4-52. Analysis Techniques for Aerobic-Anaerobic Exposure Tests

Analytical Parameter	Analysis Method (EPA Method number shown if available; equivalent Standard Methods method in parentheses, if available)
pH	EPA Method 150 (Standard Methods 4500-H^+.B.)
Hardness	EPA Method 130.2 (Standard Methods 2340.C.)
Copper	Standard Methods Method 3500-Cu.E.
Lead, Cadmium, Zinc, Copper, Chromium, Iron, Calcium and Magnesium	ICP (Standard Methods 3500)

The DO and ORP results are shown in Tables 4-53 and 4-54. During the first run, it was uncertain as to whether the sand and the blank (Star Lake water with no media) developed anaerobic conditions in the jars. The potential for a small headspace existed in the anaerobic bottles during the first run if the seal was not tight. However, because the bottles were not strongly agitated, i.e., not well-mixed regularly, it was expected that anaerobic conditions did exist in areas of the bottle that were not near the air-water interface. To ensure that anaerobic conditions were created, the second run was made using stoppered BOD bottles. The DO results clearly show that near-anaerobic conditions were achieved for all samples during the second run. The differences in the ORP between the aerobic and anaerobic samples were not nearly as noticeable, with the exception of compost where strong reducing conditions developed when the water became anaerobic.

Table 4-53. DO Results for Aerobic-Anaerobic Exposure Experiments

Media	Run 1 DO (mg/L)		Run 2 DO (mg/L)	
	Anaerobic	Aerobic	Anaerobic	Aerobic
Compost	0.4	4.2	0.2	6.5
Peat	1.5	4.4	0.1	6.7
Carbon	0.8	4.6	0.5	5.7
Sand	3.6	4.6	0.4	8.7
Star Lake Water	3.8	4.8	0.6	7.7

Table 4-54. ORP Results for Aerobic-Anaerobic Exposure Experiments

Media	Run 1 ORP (mV)		Run 2 ORP (mV)	
	Anaerobic	Aerobic	Anaerobic	Aerobic
Compost	-5	230	< -400	215
Peat	285	360	320	380
Carbon	260	310	325	355
Sand	275	300	205	230
Star Lake Water	280	285	280	325

The results for the copper retention are shown in Figures 4-18. The data for all other metals are in Appendix B. For copper for all the media, sorption capacity increased as the pH neared its ideal sorption pH of 5.0. For lead, the greater sorption capacity was observed at the lower pH (2.0). Zinc showed variable behavior – the media's capacity for it was greater for activated carbon, peat, and sand as the pH increased. However, the capacity of compost for zinc decreased as the pH increased. As expected, once the metals were adsorbed onto the media, only negligible removal occurred during rinsing and exposure. The sorption onto the peat and compost appeared to be permanent under all conditions, likely due to the formation of complexes with the organic compounds on the surface of these materials. These complexes are likely to be 'permanent,' as shown by the lack of desorption under the significantly reducing conditions occurring in the compost.

Calcium and magnesium were leached from the compost and peat media (loss greater than total amount sorbed). It is anticipated that the leaching results from the competition between these ions and the other ions in solution (especially the heavy metals) for sorption sites on these media. The leaching is significantly greater under anaerobic conditions, where a reducing environment has been developed. In the reducing environment, hydrogen/protons are being added to the oxidized species, i.e. the charged sorption sites on the media. The data indicate that hydrogen outcompetes calcium and magnesium for sorption sites under reducing conditions. Minimal sorption of calcium and magnesium was seen on the carbon and the sand.

Iron was adsorbed to all four media. Like the other three heavy metals on carbon and peat, the mechanism of sorption is likely ionic bonding (chemisorption), and these chemical bonds are not broken during extended exposure to either oxidizing or reducing conditions. When the initial sorption occurs at a low pH, iron behaves in a similar manner for compost. However, when the initial sorption pH is closer to neutral, the bonding between the compost and the iron is not as strong. Under anaerobic/reducing conditions, some of the iron that has sorbed to the compost is removed, and additional iron is leached from the media.

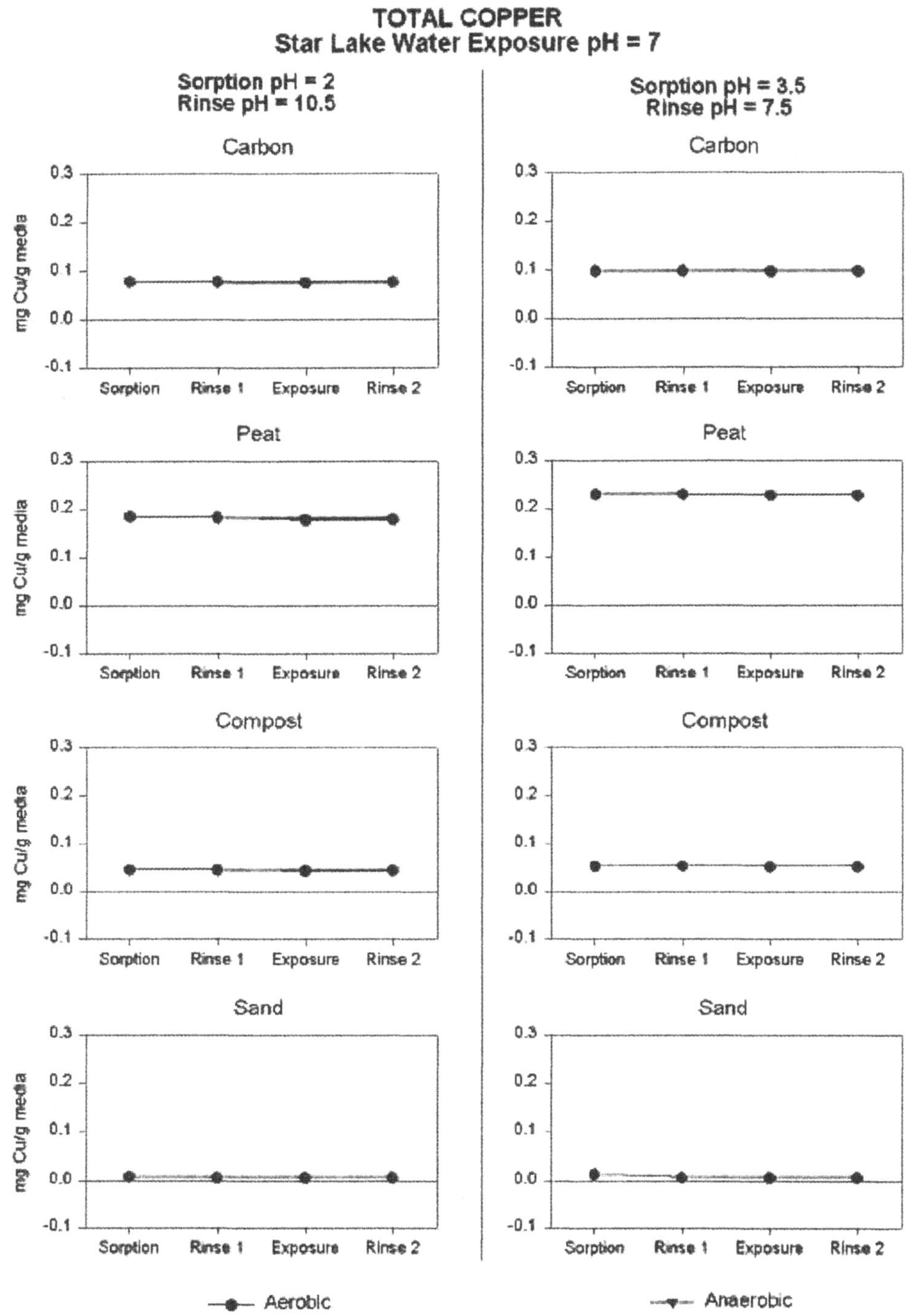

Figure 4-18. Copper behavior during sorption and extended aerobic and anaerobic exposure.

4.6 Design of Filters for the Control of Stormwater Heavy Metals

[Moved to Chapter Six – Filter Design Section]

4.7 Summary of Pilot-Scale Tests

The purpose of this research was to improve the ability of engineers and stormwater managers to design and predict the removal efficiency and life of stormwater filters. Contrary to traditional filter/fixed-bed adsorber design, stormwater filters must be able to treat intermittent flows and low and potentially variable influent concentrations. Depending on the pollutant of interest, concentrations can range from several micrograms per liter (heavy metals) to around 100 mg/L (hardness). Typically, the pollutants that need to be removed from the runoff water are those that are in concentrations of hundreds of micrograms per liter or less, rather than the milligram per liter concentrations found in industrial wastewater. The difficulty with treatment of water whose concentrations are in the range of micrograms per liter is that the treatment efficiency may be highly variable. This was seen when settling, aeration, and photodegradation were used to treat stormwater runoff by Pitt, *et al.* (1999). Removal efficiencies were highly variable for these unit processes not only between sites but between storms at the same site. For this reason, the MCTT was designed to combine a variety of processes (Pitt, *et al.* 1999), including filtration. It was seen during that research that the combination of processes could provide very effective and consistent of stormwater runoff toxicants, although any single treatment process may not have been efficient at a particular site for a particular storm event. The following summarizes the findings associated with each task associated with this research phase:

1) Literature Review

The literature search on filtration/sorption/ion-exchange media showed that traditional filters have been designed to treat water with known and predictable concentrations of the pollutants, and most have been designed to treat pollutant concentrations significantly greater than stormwater. They are also designed for constant flow rates and generally without intermittent flows. Stormwater filters have been designed using a 'black-box' approach to pollutant removal, i.e., the filter has a known range of performance for the various pollutants until they fail. The designs simply have been based on the development of head loss due to the buildup of suspended solids. Since many stormwater filters will fail due to clogging prior to exhausting the chemical capacity of the media especially for heavy metals, this design approach has some merit. However, it is still desirable to develop a modeling equation for the chemical pollutant removal in these filters, especially if there is a need to predict the life of a filter for a compound for which it has a low capacity (i.e., one for which chemical capacity may be exceeded before clogging occurs).

2) Characterization of the Media

The first activity performed as part of this research effort was the characterization of the media using a sieve size analysis and a determination of the cation exchange capacity. The media tested for these two activities were sand, peat moss, municipal leaf compost, activated carbon, zeolite, cotton textile waste (not sieved), and kenaf-based agrofiber (not sieved). It was seen that the carbon, the zeolite, and the sand could be characterized as media with a small uniformity coefficient, while the compost and the peat moss were seen to be 'well-graded,' with a higher uniformity coefficient. Also included in this activity was the re-evaluation of the data collected as part of Clark's (1996) study because of the concern that the media, either through degradation or ion-exchange, could release pollutants to the passing water that were not desirable. This involved the re-evaluation of both the field data (where the influent water was a pre-settled stormwater runoff that was not spiked) and the data collected under the factorial experiment examining the effects of influent pH and ionic strength on the ability of the media to remove and retain pollutants.

When examining the data collected as part of the field exercise in Clark's (1996) study, most of the media had effluent turbidities greater than the influent. The influent turbidity for the unfiltered fraction ranged between 0.74 and 4 NTU, while for the filtered fraction, the influent turbidity was between 0.27 and 3 NTU. Compared to typical filtration operations, these influent turbidities are low, and it was not unexpected to see the demonstrated lack of removal for the media.

The term "weight-of-evidence" has been used frequently during recent environmental assessments. It is now widespread acceptance that multiple approaches (lines of evidence) are essential in order to reach reliable conclusions of whether a problem exists (Burton and Pitt 2001). In general, the weight of evidence produced from the review of the results obtained by Clark (1996) also indicated that a degradation of the media could occur, resulting in ratios of Ce/C0 greater than 1 (Table 4-55). This degradation may occur early in the filter run where it might be expected that degradation would not have had sufficient time to occur.

Table 4-55. Summary of Potentially Leachable Compounds from Media

Media	Compounds Potentially Leached
Sand	Turbidity, ammonium
Carbon-Sand	Sulfate, dissolved copper
Peat-Sand	Turbidity, ammonium, potassium, color, dissolved copper
Compost-Sand	Turbidity, carbonate, bicarbonate, phosphate, ammonium, potassium, magnesium, calcium, hardness, color, total copper, dissolved copper
Agrofiber-Sand	Turbidity, carbonate, ammonium, potassium, total copper, dissolved copper
Cotton-Sand	Turbidity, carbonate, ammonium, potassium, total copper, dissolved copper

3) Impact of Suspended Solids Loadings on Flow Rate

Since it had been shown in the literature review that physical clogging of the media was a significant concern, the first laboratory activity performed under this research project was to evaluate the Urbonas model for relating flow rate to the suspended solids loading using laboratory-scale columns. The modeling equations for the media are given in Table 4-56. The models were evaluated using a complete residual analysis and the model criteria.

Table 4-56. Model Equations for the Effect of Solids Loading on Flow Rate

Filtration Media	Equation for Effect of Suspended Solids Loading on Flow Rate
Sand	$u = 44500 \bullet L_{m,sand}^{-1.02}$
Carbon-Sand	$u = 14800 \bullet L_{m,carbon}^{-0.77}$
Peat-Sand	$u = 2000 \bullet L_{m,peat}^{-0.71}$
Compost-Sand	$u = 1.6 \times 10^{13} \bullet L_{m,compost}^{-4.09}$
Zeolite-Sand	$u = 60 \bullet L_{m,zeolite}^{-0.23}$
Agrofiber-Sand	$u = 205 - 0.09 \bullet L_{m,agrofiber}$
Cotton-Sand	$u = 106 - 0.01 \bullet L_{m,cotton}$

For the non-fibrous filter media, Urbonas' model could be used to describe the decrease in flow rate. However, for the fibrous media, the cotton textile waste and the agrofiber, the decrease in flow rate was not found to be a power function of the suspended solids loading. The relationship between the suspended solids loading and the flow rate for the fibrous media appears to be a linear one. The fibrous filters differ significantly from the non-fibrous ones in relation to solids removal in that the fibrous filters appear to use more of the filter depth for solids removal. For the non-fibrous media, surface and near-surface straining are the primary removal mechanisms. The results of the statistical analysis are shown in Table 4-57.

Table 4-57. Evaluation of Particulate Loading Effects on Flow Rate Models

Media	$\ln k_i$ or k_i	α of k_i	c	α of c	p of Durbin-Watson statistic	Normality test p	Constant variance test p	R^2	Model p
Sand	10.7068	0.0109	1.0171	0.0293	> 0.1	0.1418	0.0600	0.734	0.0293
Carbon-Sand	9.6369	0.0076	0.7741	0.0378	> 0.1	0.1931	0.6602	0.611	0.0378
Peat-Sand	7.6225	0.0130	0.7125	0.0348	>0.05	0.5266	0.0500	0.818	0.0348
Compost-Sand	30.4229	0.0010	4.0880	0.0011	>0.05	0.6549	<0.0001	0.998	0.0011
Zeolite-Sand	4.0728	0.0002	0.2340	0.0104	>0.05	0.1523	0.2604	0.692	0.0104
Agrofiber-Sand	204.8698	0.0002	0.0870	0.0003	>0.05	0.5595	<0.0001	0.999	0.0003
Cotton-Sand	106.5686	0.0001	0.0133	0.0011	>0.1	0.4637	0.6434	0.802	0.0011
Desired Levels		≤ 0.05		≤ 0.05	> 0.05	> 0.05	> 0.05	Close to 1	≤ 0.05

The results in Table 4-57 indicate that the requirements for the model equations are met for all of the media except agrofiber-sand and compost-sand. For these two media, the constant variance test of the regression failed. These two media also demonstrate that, as Berthouex and Brown (1994) and Burton and Pitt (2001) discuss, a very high R^2 value is not necessarily an indication of a 'perfect' model. A review of the data for those two regressions shows that the regression was based on fewer data points than the other regressions (less than five points each for the agrofiber-sand and compost-sand). It is likely that additional sampling points would reveal a similar relationship to that seen in the table but where the regression passed the test of constant variance.

4) Batch Testing for Contact Time Optimization

The second laboratory activity was to evaluate the effects of contact time on pollutant removal in a batch setup. Estimates of the media capacity were also obtained as part of this series of experiments. For most pollutant-media combinations where removal was observed, the batch tests showed that a contact time of approximately 15 minutes was sufficient to remove the majority of the pollutant that the medium was capable of adsorbing and that longer contact times would not enhance pollutant removal significantly. For some of the media/pollutant combinations, it was seen that longer contact times either resulted in the release of previously trapped pollutants or the degradation of the media. The data from these experiments also gave an additional indication of which ions were participating in ion exchange with the media.

The contact time experimental data were analyzed using linear regression (assuming a traditional first-order pollutant removal model) to find rate constants for the adsorption/ion exchange. Negative rate constants were seen as the pollutant being added to the water, i.e., the pollutant being examined was the exchangeable ion for another ion that was being removed. For example, for the peat moss, hydrogen is one of the exchangeable ions. This was demonstrated by the decrease in pH with the increase in contact time. For the activated carbon, ion exchange resulted in the release of sulfate. Calcium, hardness and bicarbonate were released by the compost. The residuals of the regression were analyzed. The data from these analyses are given in Tables 4-58 through 4-65.

Table 4-58. Evaluation of Dissolved Pollutant Removal Models for Calcium in Batch Testing

Media	Intercept	α of Intercept	Slope	α of Slope	p of Durbin-Watson statistic	Normality test p	Constant variance test p	R^2	Model p
Sand									
Carbon-Sand									
Peat-Sand	-0.3006	0.0026	0.0082	0.0066	>0.10	0.3335	0.0600	0.870	0.0066
Compost-Sand			-0.0058	0.0193	>0.05	0.3388	0.0600	0.000	0.0193
Cotton-Sand									
Agrofiber-Sand			0.0008	0.0150	>0.10	0.2292	0.0600	0.000	0.0150

Table 4-59. Evaluation of Dissolved Pollutant Removal Models for Hardness in Batch Testing

Media	Slope	α of Slope	p of Durbin-Watson statistic	Normality test p	Constant variance test p	R^2	Model p
Sand	0.0013	0.0312	>0.05	0.5596	0.0600	0.063	0.0312
Carbon-Sand							
Peat-Sand	0.0074	0.0164	>0.05	0.7465	0.0600	0.118	0.0164
Compost-Sand							
Cotton-Sand							
Agrofiber-Sand							

Table 4-60. Evaluation of Dissolved Pollutant Removal Models for Potassium in Batch Testing

Media	Intercept	α of Intercept	Slope	α of Slope	p of Durbin-Watson statistic	Normality test p	Constant variance test p	R^2	Model p
Sand									
Carbon-Sand									
Peat-Sand			0.0012	0.0344	0.05	0.3688	0.0600	0.000	0.0344
Compost-Sand	0.5055	0.0021	-0.0076	0.0398	>0.05	0.5676	0.0600	0.693	0.0398
Cotton-Sand									
Agrofiber-Sand									

Table 4-61. Evaluation of Dissolved Pollutant Removal Models for Sodium in Batch Testing

Media	Slope	α of Slope	p of Durbin-Watson statistic	Normality test p	Constant variance test p	R^2	Model p
Sand	-0.0014	0.0134	>0.10	0.3498	0.0600	0.000	0.0134
Carbon-Sand							
Peat-Sand							
Compost-Sand	-0.0011	0.0111	>0.05	0.7689	0.0600	0.000	0.0111
Cotton-Sand							
Agrofiber-sand	-0.0065	0.0416	>0.05	0.0203	0.0580	0.000	0.0416

Table 4-62. Evaluation of Dissolved Pollutant Removal Models for Copper in Batch Testing

Media	Intercept	α of Intercept	Slope	α of Slope	p of Durbin-Watson statistic	Normality test p	Constant variance test p	R^2	Model p
Sand									
Carbon-Sand									
Peat-Sand	-1.0193	0.0001	0.0079	0.0315	>0.1	0.0613	0.0600	0.725	0.0315
Compost-Sand			-0.0093	0.0269	0.05	0.6276	0.0400	0.000	0.0269
Cotton-Sand									
Agrofiber-Sand	0.5661	<0.0001	-0.0057	0.0012	>0.05	0.0342	0.0400	0.944	0.0012

Table 4-63. Evaluation of Dissolved Pollutant Removal Models for Zinc in Batch Testing

Media	Intercept	α of Intercept	Slope	α of Slope	p of Durbin-Watson statistic	Normality test p	Constant variance test p	R^2	Model p
Sand									
Carbon-Sand									
Peat-Sand	-0.5112	0.0035	0.0105	0.0225	>0.05	0.1284	0.0600	0.765	0.0225
Compost-Sand									
Cotton-Sand									
Agrofiber-Sand			0.0023	0.0001	>0.05	0.6262	0.0600	0.964	0.0001

Table 4-64. Evaluation of Dissolved Pollutant Removal Models for Lead in Batch Testing

Media	Intercept	α of Intercept	Slope	α of Slope	p of Durbin-Watson statistic	Normality test p	Constant variance test p	R^2	Model p
Sand	0.3437	0.0488	0.0151	0.0252	>0.05	0.0827	0.0600	0.753	0.0252
Carbon-Sand									
Peat-Sand	-3.1694	0.0001	-0.0085	0.0377	>0.05	0.2601	0.0600	0.700	0.0377
Compost-Sand									
Cotton-Sand									
Agrofiber-Sand	1.2853	0.0003	-0.0142	0.0207	>0.05	0.4115	0.0600	0.775	0.0207

Table 4-65. Evaluation of Dissolved Pollutant Removal Models for Iron in Batch Testing

Media	Intercept	α of Intercept	Slope	α of Slope	p of Durbin-Watson statistic	Normality test p	Constant variance test p	R^2	Model p
Sand									
Carbon-Sand									
Peat-Sand	-3.1988	<0.0001	0.0155	0.0016	>0.10	0.3195	0.0600	0.935	0.0016
Compost-Sand			0.0180	0.0502	>0.05	0.3370	0.0600	0.000	0.0502
Cotton-Sand									
Agrofiber-Sand	1.4825	0.0002	0.0134	0.0336	>0.05	0.1650	0.0600	0.716	0.0336

5) Breakthrough Curve Testing for Operational Capacity and Removal Rate Constants

The third laboratory activity was to operate bench-scale filter columns to physical clogging or chemical breakthrough. The data from the breakthrough curves were used to estimate an operational capacity for the media, as well as to predict rate constants based upon the modified BDST model equation. A comparison of the capacities from the batch and breakthrough curve testing is given in Table 4-66. A comparison of the rate constants from the batch and breakthrough curve testing is shown in Tables 4-67 through 4-69.

Table 4-66. Comparison of Capacities (mg pollutant/g media) Calculated by Batch and Continuous-Flow Studies (BTC)

Pollutant	Sand BTC	Sand Batch	Carbon BTC	Carbon Batch	Peat BTC	Peat Batch	Compost BTC	Compost Batch
Hardness	1.2	0.2	20	115	20	121	0	0
Copper		0.002		0.03		0.2		0.03
Lead		0.002		0.003		0.4		0
Zinc		0.05		0.2		5.7		1.1
Iron		0.006		0.001		0		0
Sodium	0	0	0	0	0.02	0	0	0
Potassium	0.02	0.003	0.53	0.6	0	0	0	0
Calcium	0.08	0.08	0.51	8	>10	38	0	0
Magnesium	0.04	0.04	0.12	2	> 1	10	0.31	1.2

Table 4-67. Comparison of Rate Constants from Batch Kinetics and Continuous-Flow Breakthrough Curve Studies for Hardness, Calcium, and Magnesium

Filter	Hardness		Calcium		Magnesium	
	Batch k (day^{-1})	BTC k (day^{-1})	Batch k (day^{-1})	BTC k (day^{-1})	Batch k (day^{-1})	BTC k (day^{-1})
Sand	1.8	4.7×10^{-8}				
Carbon-Sand		5.7×10^{-8}		1.1×10^{-6}		3.7×10^{-5}
Peat-Sand	11	2.1×10^{-7}	12	2.5×10^{-6}	7.2	5.1×10^{-5}
Compost-Sand		1.4×10^{-7}				5.4×10^{-5}

Table 4-68. Comparison of Rate Constants from Batch Kinetics and Continuous-Flow Breakthrough Curve Studies for Copper, Lead and Zinc

Filter	Copper		Lead		Zinc	
	Batch k (day^{-1})	BTC k (day^{-1})	Batch k (day^{-1})	BTC k (day^{-1})	Batch k (day^{-1})	BTC k (day^{-1})
Sand			22			2.6×10^{-6}
Carbon-Sand						7.3×10^{-7}
Peat-Sand	11				15	
Compost-Sand						

Table 4-69. Comparison of Rate Constants from Batch Kinetics and Continuous-Flow Breakthrough Curve studies for Iron, Sodium and Potassium

Filter	Iron		Sodium		Potassium	
	Batch k (day^{-1})	BTC k (day^{-1})	Batch k (day^{-1})	BTC k (day^{-1})	Batch k (day^{-1})	BTC k (day^{-1})
Sand						
Carbon-Sand						1.4×10^{-5}
Peat-Sand			1.7	6.0×10^{-6}		
Compost-Sand						

Rate constants calculated using the breakthrough curves were several orders of magnitude smaller than those predicted in the batch tests. One result from the comparison of the batch capacities and rate constants to the breakthrough curve capacities and rate constants is that the batch data are not truly comparable and may tend to over-predict the ability of the filter. The second result from these breakthrough curve tests in the laboratory was the growth of a biomass in the top 5 cm of the filter media. It is unknown based upon this experimental design and upon the results what impact the biomass had on pollutant removal and retention by the filter media. The biomass could be expected to block access to sorption sites; however, the bacteria may be incorporating some of the captured pollutants into its biomass and freeing up some sorption sites. The scatter in much of the breakthrough curve data may be a result of the activity of the biomass. It also may explain the difference in the capacity calculated based upon batch testing and that calculated from the breakthrough curve testing. For sand, the addition of the biofilm may have enhanced the removal of some pollutants due to the potential for the bacteria to adsorb them as nutrients or due to the exopolymers that may be located on the outside of the bacterial cell and that may be an excellent adsorbent. However, for the other media, the effect is not as clear because the bacterial biofilm may be blocking sorption for a media with a higher capacity for a specific pollutant than the bacteria itself has for that pollutant.

The models generated as part of this third laboratory activity were evaluated using the same criteria as for other models. The residual analyses showed that, when the number of data points was small (generally for a pollutant for which a specific medium had a small chemical capacity), the likelihood of failing one or more of the residual analyses tests increased. The first test for model validity is to answer the questions. (1) Does the model make sense? (2) Does it have a theoretical basis? The model makes sense when it states that the number of bed volumes treated before chemical breakthrough is a function of both the flow rate (and hence contact time), the rate constant for sorption/ion-exchange/removal, and the initial concentration of the pollutant. The primary test of the models is in the residual analysis.

6) Development of the Model Equations into an Algorithm

It was expected that the model equations proposed as a result of this research can be combined to provide an algorithm for predicting the removal efficiency of a stormwater filter. The first equation is based on work by Urbonas (1999), who tested several sand filters for stormwater treatment. Urbonas' equation states that the flow rate (measured as a loading rate with units of length/time) is a function of the suspended solids loading. The power equation of Urbonas was not found to explain the decrease in flow as a result of suspended solids loading for the fibrous media. Therefore, the data were re-examined using a linear relationship between suspended solids loading and flow rate.

The second equation of the algorithm is based on the equation proposed by Bohart and Adams (1920) for the removal of chlorine by charcoal and is termed the BDST analysis of fixed-bed adsorbers. As it is traditionally applied, the BDST model relates the breakthrough time of the filter to the depth of the filter bed. The integrated equation proposed by Bohart and Adams (1920) was modified and simplified based upon the known characteristics of the stormwater filter media into an equation that relates the breakthrough curve (shown as the ratio of the effluent concentration to the influent concentration) to the number of bed volumes treated. However, for the peat-sand media because of its capacity for divalent cations (including the heavy metals), the full BDST model equation should be used. The applicable equations are Equations 4.7-1 and 4.7-2. The results of these analyses are shown in Table 4-70 for the Urbonas model equation and in Table 4-71 for the modified BDST model.

$$u = ki \bullet Lm\text{-}c \tag{4.7-1}$$

where u = unit flow/loading rate (m/day), ki = flow rate reduction rate constant (day/m), Lm = suspended solids loading, and c = power of the flow rate reduction with suspended solids loading.

$$\ln\left(\frac{C_0}{C_e}-1\right) = -1x10^6\,kx\sum_{j=1}^{n}\left(\frac{C_0(BV)}{u}\right) \tag{4.7-2}$$

where C0 = influent concentration (mg/L), Ce = effluent concentration (mg/L), k = rate constant predicted from laboratory breakthrough curves (day-1), x = bed depth (m), u = unit flow/loading rate (m/day), and BV = bed volumes treated.

Table 4-70. Model Equations for the Effect of Solids Loading on Flow Rate

Filtration Media	Equation for Effect of Suspended Solids Loading on Flow Rate
Sand	$u = 44500 \bullet L_{m,sand}^{-1.02}$
Carbon-Sand	$u = 14800 \bullet L_{m,carbon}^{-0.77}$
Peat-Sand	$u = 2000 \bullet L_{m,peat}^{-0.71}$
Compost-Sand	$u = 1.6 \times 10^{13} \bullet L_{m,compost}^{-4.09}$
Zeolite-Sand	$u = 60 \bullet L_{m,zeolite}^{-0.23}$
Agrofiber-Sand	$u = 205 - 0.09 \bullet L_{m,agrofiber}$
Cotton-Sand	$u = 106 - 0.01 \bullet L_{m,cotton}$

Table 4-71. Design Criteria for Modified BDST Equation (capacity and rate constants) – Statistically Significant Models Only

Pollutant	Media	Capacity (mg/g)	k (day^{-1})
Hardness	Carbon-Sand	20	4.96×10^{-10}
	Peat-Sand	20	2.30×10^{-9}
	Compost-Sand	1.2	1.20×10^{-9}
	Sand	1.2	4.13×10^{-10}
Calcium (Ca)	Carbon-Sand	0.6	1.12×10^{-8}
	Peat-Sand	>10	4.91×10^{-9}
	Compost-Sand		
	Sand		
Magnesium (Mg)	Carbon-Sand	0.2	2.14×10^{-8}
	Peat-Sand	>1	2.95×10^{-8}
	Compost-Sand	0.4	3.06×10^{-8}
	Sand		
Zinc (Zn)	Carbon-Sand		8.17×10^{-9}
	Peat-Sand		
	Compost-Sand		
	Sand		2.91×10^{-8}
Sodium (Na)	Carbon-Sand		
	Peat-Sand	0.001	4.37×10^{-8}
	Compost-Sand		
	Sand		
Potassium (K)	Carbon-Sand	0.53	1.99×10^{-8}
	Peat-Sand		
	Compost-Sand		
	Sand		

7) Pilot-Scale Testing Using a Detention Pond in a Residential Area

Pilot-scale tests were then run to compare the results obtained in the laboratory-scale columns with filters more closely sized to those that would be used in a full-scale design. For some parameters and some media, the predictions were excellent. The lack of ability to predict removal for all pollutants may be due to the inability of the media to remove a sufficient fraction of the pollutants in the laboratory-scale breakthrough experiments such that relatively accurate rate constants could be calculated. It also may be due to the extremely low influent concentrations of some pollutants to the pilot-scale filters. The influent concentration of the pollutants may already be at the point where an equilibrium adsorption-desorption activity controls the removal ability of the filter. However, it can be seen that the results of the batch testing of capacity and kinetics are not applicable to the operation of a filter in the field.

8) Filtration as a Stormwater Control

Filter media apparently clogs before the media can experience chemical break-though when using stormwater. It is not yet clear if depth filtering media will be a cost-effective stormwater control, considering the pre-treatment needed to prevent this clogging. The necessary pretreatment alone may provide adequate control, without the additional filtration cost. Large-scale filtration installations (especially sand) have been shown to perform well for extended periods of time with minimal problems. The use of supplemental materials (such as organic compounds) should increase their performance for soluble compounds. The use of upflow filtration is also expected to increase the life of filters before clogging, for some media types (not for peat/sand combinations).

9) Overall Summary

The confirmation of the modeling equations for a few pollutants for some media indicate that this modeling approach has the potential to provide an estimate of the life of the filter, i.e., the time until media replacement is needed, for applications where the influent concentration is not quite as low as it was in the two detention ponds used in this pilot-scale study. The best application for these filters may then be to further treat the effluent from a detention pond in critical source areas, such as scrap metal recyclers, rather than as a polisher for effluent from a detention pond in a residential area. During these field pilot-scale tests and related full-scale tests, the minimum attainable concentration for these metals using media filtration appears to be approximately 5 to 10 μg/L for copper and lead, and approximately 10 to 20 μg/L for zinc, irrespective of media. These tests do confirm the utility of different filtration arrangements, especially concerning the problems associated with clogging and reduced filtration performance.

CHAPTER 5.0

SWALE STUDIES

5.1 Swale Hydraulic Study Objectives

- Develop a method by which the transport mechanisms and removal rates of metals can be predicted when swale geometry, flow rate and grass type / condition are specified
- Develop a method by which the hydraulic characteristics of shallow open channel flow in a swale can be predicted on the basis of flow rate, cross sectional geometry and slope, and grass type / condition contaminates in urban runoff.

5.1.1 Background / Literature Review

A vegetated swale is a broad, shallow channel with a dense stand of vegetation covering the side slopes and bottom. Swales can be natural or man-made, and are designed to trap particle pollutants (suspended solids and trace metals), promote infiltration, and reduce the flow velocity of storm water runoff (EPA 1999a). Swales were selected for further investigation in this project because of their known effectiveness, low operational and maintenance costs, and application flexibility. There is a fairly large body of literature dealing with the use of swales and filter strips for the removal of metals and other contaminants in stormwater runoff, but they do not provide much useful information for swale design (Kercher, et al. 1983; Wigington, et al. 1986; Dillaha, et al. 1988; Lee, et al. 1989; Kruzic and Schroeder 1990; Nair and Schnoor 1992; Sansalone and Buchberger 1995; Woodard and Rock 1995; Newberry and Yonge 1996; and Pratt 1996). Quantitative data on fate of runoff pollutants through swales is scarce or nonexistent. With regard to hydraulic design, there are in existence a number of sources and guidelines (Smith 1942; Stillwater Outdoor Hydraulic Laboratory 1947; Cox and Palmer 1948; Ree 1949 and 1958; Ree and Palmer 1949; Fredenhagan and Doll 1954; Cowan 1956; Chow 1959; Federal Highway Administration 1988). These documents, however, have addressed the design of channels (swales) where the depth of flow and hydraulic radius are relatively large in comparison to the height of the grasses. For optimal metals removal, one needs quite shallow depths of flow relative to the grass height, and for these conditions the roughness characteristics can be significantly different than those for the case of greater flow depths (Dillaha, et al. 1988).

The following sections will detail previous research performed to determine the hydraulic characteristics of swales, the mechanisms of metal removal, and the existing guidelines for construction of urban vegetated swales.

5.1.1.1 Hydraulic Characteristics

Historically the design of vegetated swales has been based on principles of flow conveyance and soil conservation in the swale rather than for pollution control. The United States Department of Agriculture funded extensive research on the topic (erosion control) during the late 1930's to the mid 1950's. Two of the most predominant research efforts were made at the Spartanburg Outdoor Hydraulic Laboratory (USDA 1940) and at the Stillwater Outdoor Hydraulic Laboratory (USDA 1954). These facilities examined the relationship between the hydraulic resistance of various grasses and the depths of flow conducted over the vegetated swales.

The most important hydraulic characteristic of vegetation is the resistance it offers to flowing water. A common measure of this resistance is Manning's "n", which quantitatively expresses the degree of retardation to flow (Ree 1949; Cowan 1956). Disturbances in the fluid are produced and maintained largely by roughness and irregularities of the bed and sometimes by the severe whipping and oscillation of stems and leaves of vegetation by the flow. The more violent the disturbance, the greater is the retardance to the flow and the greater are the forces acting to scour the bed (USDA 1954). Experiments showed that as greater depths and velocities of flow were obtained, more force is exerted on the plant until the leaves are pushed aside. Thus, there is a larger opening for the flow and reduced turbulence because the blades of grass no longer project across the stream. The result is a reduction in the flow-retarding properties of the cover, which is reflected in lower n values. Since the rearrangement of plants is largely due to velocity, the greater the velocity, the greater the reduction in "n" (Ree 1958). The highest "n" values generally occur when the vegetation is partially bent over and is oscillating in the flow (Palmer 1946). The most striking thing observed about values of Manning's "n", was their great range. Values ranged from 0.02 to 0.6. One type of vegetation in a channel could show nearly this much variation, depending on the flow profiles (Ree 1949).

The hydraulics of flow over a vegetated slope involve the relationship of depth and velocity of flow to the height and physical structure of the vegetation. The Stillwater Outdoor Hydraulic Laboratory observed that resistance to flow is a function of the degree of flattening of the vegetation, which is influenced directly by the velocity and depth of flow (Ree and Palmer 1949). Based on these observations, Manning's "n" values were established as a function of the product of velocity and hydraulic radius (determined by channel cross-section, slope, and depth), when the velocities were great enough to displace the vegetation (Ree 1958). When the vegetation remained upright and was not disturbed by the flow, the "n" values bore no consistent relationship to the product of VR (Ree 1960). Grasses including Centipede, Bluegrass, and Zoysia are continuously agitated by overland sheet flow and thus maintain a relationship between "n" and VR, regardless of the depth of flow. The product of mean velocity (V) and hydraulic radius (R) was ultimately determined to be a satisfactory criterion for estimating "n" (Cox and Palmer 1948). It was additionally found that if "n" values were plotted against the product of VR instead of depth, approximately the same curve would be obtained for a given type of vegetation, regardless of the shape or slope of a channel. Plants of similar physical characteristics display similar n-VR curves. This discovery led to a retardance classification of vegetation types based on their n-VR curves (Ree 1949). The flows of very small depth encountered a maximum resistance as indicated by the large "n" values. The "n" values then start to decrease rapidly with increases in flow depth. When bending and submerging are complete, the n curve starts to level out, approaching a constant value (Ree 1949).

The flow of water in a thin layer over soil surfaces has been variously termed "sheet flow", "shallow flow", "overland flow", and "low flow". A flow ceases to be classified as small when its depth and velocity are such that bending and complete inundation of vegetation occurs. Thus, "low flow" varies in depth and velocity according to land slope and physical structure of the vegetation (Palmer 1946). For optimal metals removal, one needs quite shallow depths of flow relative to the grass height, and for these conditions the roughness characteristics can be significantly different than those for the case of greater flow depths (Dillaha, et al. 1988). It is in these low flow regimes that a vegetated swale's various metal removal pathways (sedimentation, infiltration and bioaccumulation) are most effective.

The ideal cross section for a laboratory study is a rectangular one with vertical sidewalls being made of some frictionless material. A test channel, or flume, built in this manner would yield the relationship between Manning's "n" and depth of flow unaffected by so-called corner effects (Ree 1958). Using depth of flow as the hydraulic radius has the effect of considering only the bed as the wetted perimeter (Cox and Palmer 1948). For any channel, regardless of its actual cross-sectional shape, the depth and hydraulic radius of a flow are practically equal when the depth is small compared to the width of the flow.

5.1.1.2 Mechanisms of Metal Removal

Grass swales control heavy metal contamination through a combination of sedimentation, infiltration through the soil, and biological uptake. While the removal pathways have been identified in previous research, case studies that have quantitative data on fate of runoff pollutants through swales are scarce or nonexistent.

Sedimentation. Runoff events following an extensive dry period are considered to have the highest concentrations of pollutants due to a large amount of sediment and contaminants that could be deposited on the roadway. The first removal pathway, sedimentation, is especially important because a majority of the metals in runoff occur as particle bound species. Sediments having the highest concentrations of heavy metals are associated with street and parking lot runoff (Pitt 1996). Sedimentation occurs as velocity is decreased by vegetal retardance within a grassed swale. The EPA (1999b) stated that a properly designed vegetated swale may achieve 25 to 50 percent reduction in particulate pollutants, including sediment and sediment attached phosphorus, metals, and bacteria. Other studies have documented, for sediment and sediment bound contaminant, capture rates between 50 and 90 percent (Walsh, et al. 1997; Wang 1981; and Newberry and Yonge 1996). Schueler (1994b) documented an exemplary swale study in Florida with sediment removal rates of (98%), organic carbon (64%), and nitrogen (45 – 48%). This swale was ideally constructed with a low slope, high infiltration rate (sandy soil), and dense grass-cover; thus accounting for its superior effectiveness. The largest portions of the metals retained were within the initial 1 m of grass strip and 0.4 inches in depth. These analytical findings were supported by visual observations that indicated that the stormwater sediment was retained in the upper 1 m of the grass strip (Newberry and Young 1996). In the study of Kao, et al. (1975), grassed areas were alternated with areas of bare soil, and this arrangement not only yielded high sediment removal efficiencies, but the sediments also tended to accumulate in the bare soil areas where they can be (relatively) easily removed once they have accumulated. Sedimentation occurs in regions of the swale where the flow velocity is reduced, so while the actual removal process occurs in the grassed sections, the visible results materialize within the bare soil sections. In wet and humid climates where vegetation readily establishes itself, the use of bare soil areas may not be a practical alternative, but the use of such areas (or perhaps xeriscaping) in more arid climates could be quite feasible. While numerous studies have discussed the use of swales and filter strips for the removal of metals and other contaminants in stormwater runoff, these studies had conflicting results on the swales' capabilities to remove metals and other contaminant. These studies, however, did mutually conclude that the effectiveness of these removal methods decreases with time as sediment accumulates within the swales or filter strips (Kao, et al. 1975; Kercher, et al. 1983; Wigington, et al. 1986; Dillaha, et al. 1988; Lee, et al. 1989; Kruzic and Schroeder 1990; Nair and Schnoor 1992; Sansalone and Buchberger 1995; Woodard and Rock 1995; Newberry and Yonge 1996; and Pratt 1996).

Infiltration. Schueler (1994c) stated that the bulk of the observed pollutant removal in swales could be accounted for by simple infiltration of runoff through the bottom of the swale. This implies that the major pollutant removal mechanism in swales is an underground one (infiltration) and not necessarily a surface one (filtering and adsorption). Most metals can be sorbed to soil (Wang 1981). While it might appear at first that intentional infiltration of metal-bearing stormwater might represent a threat to groundwater resources, it has been found that this is generally not the case (Wigington, et al. 1986; Pitt, et al. 1996; Mikkelson, et al. 1996a, b). Heavy metals have an affinity for soil, and tend to be sorbed onto the soil particles in the first several centimeters as infiltrating water percolates downward. Schueler (1994c) indicated that most trace metals were indeed trapped in the upper five centimeters of swale soils, and did not migrate into nearby ground water. Additional studies have found high concentrations of metals, including 1000-6600 ug/kg lead and 490-2400 ug/kg iron, in the top 15 cm of soil underneath a highway swale, but nearby ground water was unaffected (Walsh and Barrett 1997). Wang (1981) conducted a study of highway filter strips, similar to grassed swales, and determined that lead levels in soil samples obtained farther than about 30 m from the roadway were indistinguishable from background levels, with metal concentrations decreasing rapidly within the first 9 m.

While sedimentation and infiltration occur during every storm, adsorption requires that the metal be present in runoff as a positively charged cation that can be adsorbed to a negatively charged particle in the soil or organic layer (Schueler 1994).

5.1.1.3 Channel Design

The ultimate goal of this part of the project was to develop a set of guidelines that will enable optimization of vegetated swales for the removal of metals and other contaminants in urban runoff. However, like all other hydraulic problems, economic optimization of vegetative swales can only be achieved through a cost / benefit relationship, involving six key parameters. These parameters include: permissible velocity, slope, infiltration rate (of soil), grass selection, channel shape / size, and maintenance efforts (EPA 1999a). All of these parameters can have direct influences on the others (velocity is a function of channel slope, grass density can dampen velocity, etc.), so a careful balance must be maintained during the design of any vegetated swale. Optimizing pollutant removal in a vegetated swale is a process of creating a design that will maximize the hydraulic residence time, while ensuring sedimentation, infiltration and biological uptake pathways are preserved.

Velocity. The first parameter, velocity, is of great importance due to its influence on sedimentation rates and the vegetal retardance coefficient. Grassed waterways are designed for velocities of flow that are safe for the channel to support. If the vegetation cover fails, erosion may occur, because bare ground erodes under such velocities of flow (Fredenhagan and Doll 1954). It follows that a uniform, sod forming vegetation having a dense, relatively deep root system will offer the greatest protection against scour. (USDA 1954). Vegetative swales are typically ineffective during and are vulnerable to large storms because of the high-velocity flows which may result. A typical design storm used for sizing swales has a six-month frequency (EPA 1999a). Swales are generally not used where the maximum flow rate exceeds 5 feet per second and are commonly used to treat small drainage areas of less than 5 acres (EPA 1999a). Prime factors in determining permissible velocities are: physical nature of vegetation (type and distribution of root growth and top growth and physical condition), erodibility of soil, uniformity of cover, and bed slope (USDA 1954).

Slope. The difficulty of providing adequate soil protection increases with increases in both velocity and bed slope. If possible, higher flows should be related to the shallowest slope possible to allow for the longest hydraulic retention times and, subsequently, the retention time for metals (Newberry and Yonge 1996). Grass swales should be used on sites with relatively flat slopes of less than 4 percent (EPA 1999b). Slope will typically be mandated by the topography available at the construction site.

Infiltration Rate. Soil infiltration rates should be greater than 0.2 mm/sec (0.5 in/hr) to allow for metal removal and to provide drainage for swales at low slopes (EPA 1999a). The use of permeable soils is expected to increase the retention potential by allowing greater filtration and increasing the preference of the metals to sorb to the soil and biota of the grass strip (Newberry and Young 1996; Pitt 1999). Infiltration rates can be maximized by locating the swales on sandy soils, using soil amendments, or by installation of perforated underdrains (Schueler 1994c).

Grass Selection. There are a number of factors to be considered in the establishment of an adequate vegetative cover. The selection of grass for a channel lining depends mainly on the climate and soil in which the plant will grow and survive under given conditions (Chow 1959). Local sod/seed suppliers will provide a list of appropriate grasses for your region. The soil must be suitable for the growth of vegetation, or must be properly conditioned. A well-prepared seedbed for the vegetated waterway is one that is fine, firm, moist, and free of growing weeds and objectionable debris. If a seedbed is not properly prepared, the resulting sod is likely to be rough, spotty, thin, and weedy. It will be difficult to maintain, or the grass may even fail before becoming established (Fredenhagan and Doll 1954).

Channel Shape. The Handbook for Channel Design for Soil and Water Conservation (USDA 1954) contains a rather complete set of diagrams for determining the dimensions of trapezoidal and triangular channels with many different side slopes, and for dealing with parabolic channels. In order to optimize a swale for metals removal, the shape must ensure that the height of the flow is less than or equal to the height of the vegetative cover (Dillaha, et al. 1988). In addition to the cross-sectional shape, the length of the swale is equally important. Long swales increase the hydraulic retention time, allow greater infiltration, and increase the metal removal capabilities. While no recommendation has been made for the optimal swale length, Wang (1981) conducted a study of highway filter strips, similar to grassed swales, and determined that lead levels in soil samples obtained farther than about 30 m from the roadway were indistinguishable from background levels, with metal concentrations decreasing rapidly within the first 9 m. The optimal swale length would depend on the flows expected and soil infiltration rates.

Maintenance. The useful life of a vegetated swale system is directly related to its maintenance frequency. Grass should never be cut shorter than the design flow depth. Timely and proper mowing increases the density and vigor of the turf, retards weed growth, prevents weed seed formation and develops a tougher sod. Swale performance can vary with season, based on grass condition and density (EPA 1999a).

Advantages of Grassed Swales. The advantages of a grass swale may include reduced peak flows, increased pollution removal, and low capital costs. Grass swale systems are inexpensive to construct, an estimated $6,900 per acre drained. Swales also require a small percentage of land (Kercher and Landon 1983).

Vegetated swales can serve as part of a storm water drainage system and can replace curbs, gutters and storm sewer systems. Therefore, swales are best suited for residential, industrial, and commercial areas with small flows and small populations (EPA 1999a).

While limited quantitative performance data exist for vegetated swales, it is known that check dams, small slopes, permeable soils, dense grass cover, increased contact time, and small storm events all contribute to successful pollutant removal by the swale system.

5.1.2 Experimental Methods

The following sections describe the physical characteristics of the indoor / outdoor swales and the experimental methods used for data collection in this research.

These experiments include:

- 5.1.2.1 Indoor Swales
  - Determination of Hydraulic Characteristics
- 5.1.2.2 Outdoor Swales (1&2)
  - Determination of Hydraulic Characteristics
  - Surface Runoff Sample Collection
- 5.1.2.3 Amended Swale (3)
  - Surface Runoff and Subsurface Leachate Collection
- 5.1.2.4 Rainwater Collection
  - Rainfall Monitoring
  - Rainfall Collection
- 5.1.2.5 Survey
- 5.1.2.6 Additional Experiments
  - Soil Analysis
  - Determination of Infiltration Rate
  - Determination of Hydraulic Retention Time

5.1.2.1 Indoor Swales

The indoor laboratory swales were constructed to accomplish two goals 1) to develop multiple flow profiles for analysis of the grass swales' hydraulic characteristics, specifically Manning's "n", and 2) to examine the ability of common grasses to absorb / retain heavy metals.

General Description. The indoor swales are comprised of three channels that are two feet wide, eight feet long, and six inches deep. The channels are rectangular in shape with nearly frictionless sidewalls. This channel shape yields a relationship between Manning's "n" and depth, unaffected by corner effects (Ree 1958). The frame was built of wood and sealed with non-reactive marine epoxy paint (Figure C-1). Drains were installed along the length of the swale, in two-foot increments, to aid in the recovery of infiltrated water (Figure C-2 and C-3). Water is supplied to the swales via a 1.5 horsepower pump and piping system (Figure C-4). Sheet flow is established over the swales by allowing the water to pass through a gravel filter, prior to introduction (Figure C-5). Once the water supply system and frame construction was complete, the three swales were filled with a loamy sand soil. This type of soil allows for high infiltration rates while supplying essential nutrients to the experimental grasses. A mixture of seventy percent topsoil and thirty percent sand (by weight) yielded a sand / silt / clay ratio typical of loamy sands. Local topsoil was used in the mixture to provide a reference from which to compare the indoor and outdoor swale sets. Once mixed, the soil was placed in the three indoor swales and planted with the various grass types (Figure C-6). The grasses selected for indoor experimentation, based on their common use and physical characteristics, were Centipede, Zoysia, and Kentucky Bluegrass.

Determination of Hydraulic Characteristics. The determination of hydraulic characteristics is accomplished by the manipulation of swale parameters to produce multiple flow profiles. Detailed hydraulic evaluation of the resulting profiles yields an average Manning's "n" value for each configuration. Swale parameters include slope, flow, and grass type / length. Utilizing a removable section of pipe in the water supply system and elevating the head works, the slope can be altered (Figure C-7 and C-8). The experiments utilize seven preset slopes including: 0.1%, 0.2%, 0.5%, 1.0%, 2.0%, 5.0%, and 10.0%. Flow rate (discharge) is controlled by a series of rotary valves (Figure C-9). Manipulation of those valves can vary the flow between 50 and 150 L / min with less than 2 mm of water surface oscillation over the testing period of one hour. Grass length is controlled by trimming to the desired height.

Experimental Procedure:

1. Set slope to desired level
2. Cut grass to desired length
3. Close all drains, to collect infiltrating water.
4. Turn on pump and adjust to desired flow (inline flow meter is used as a guide)
5. Let water circulate over swale until soil is saturated (approximately 10 min)
6. Sample soil depth and water surface depth according to grid pattern (See Figure C-10)
7. Once all measurements have been recorded, the pump is shut off and all drains opened to release infiltrated water.
8. Measurements are transferred into a project spreadsheet for calculation of the various hydraulic characteristics.

All measurements are taken using a specialized pin gauge and recorded to the closest millimeter. These values represent the distance from the reference bar (Figure C-11) to the surface of interest (soil or water surface). The profile data are transferred into project spreadsheets, in which calculations of the various hydraulic characteristics are performed.

5.1.2.2 Outdoor Swales (#1 and #2)

The outdoor swales were constructed to examine the swales' abilities to remove metal concentrations from urban runoff. Experiments were conducted to determine the hydraulic characteristics of the grasses, measure metal removal rates (via sedimentation and biological uptake), and to monitor metal accumulations in the underlying soils (infiltration).

General Description. Three sets of outdoor swales were constructed on the University of Alabama, Tuscaloosa campus. The first, Swale Set #1, is located adjacent to a small student parking lot and drains an area of 0.042 acres or 1800 ft^2. The swale set is comprised of three radiating channels, which are connected by means of a concrete flow splitter (Figure C-12). This flow splitter ensures an even distribution of stormwater, during rainfall events, to each of the three channels. Each swale is two feet wide, ten feet long, six inches in depth and rectangular in shape. The cross sectional geometry was selected to ease construction / maintenance efforts, while assuring results that vwere comparable to laboratory findings. The first channel was planted with Zoysia at a slope of 23.6 %, the second channel contains Centipede grass at a slope of 11.2 %, and the final channel (bare earth), at a slope of 2.1%, serves as a control. Sharp crested weirs were installed in the entrance of each channel, to provide a means of calculating flow.

Swale Set #2 receives runoff from a larger student parking lot, draining an area of 0.278 acres or 12,100 ft^2. This swale set is very similar to swale set #1, with the exception of an additional ten-foot long entrance channel (Figure C-13). This channel, lined with concrete, was required to transport stormwater to topography suitable for the experimental channels' construction. The swale set is comprised of three radiating channels, which are connected by means of a concrete flow splitter. The flow splitter ensures an even distribution of stormwater between the channels. Each swale is two feet wide, ten feet long, six inches in depth and rectangular in shape. The first channel was planted with Zoysia at a slope of 4.5 %, the second channel contains Centipede grass at a slope of 11.4 %, and the final channel (bare earth), at a slope of 8.6%, serves as a control. Sharp crested weirs were installed in the entrance of each channel, to provide a means of calculating flow.

Test Channel Slopes:	Swale 1	Swale 2
Control (Bare Soil)	2.1%	8.6%
Centipede	11.2%	11.4%
Zoysia	23.6%	4.5%

The Zoysia and Control channels allowed channel slopes to be investigated, while the Centipede channels allowed duplication of results.

The final swale set, the "Amended Swale", will be described in the following section (5.1.2.3).

Determination of Hydraulic Characteristics The determination of hydraulic characteristics in the outdoor swales was accomplished by an examination of the flow profiles resulting from rainfall events. The swale parameters of slope and grass type / height were established prior to the storm. However, flow in the outdoor swales can be quite unstable. Flow rates are dependent on the intensity / duration of the storm event and the time of concentration of the drainage area. All efforts were made to collect profile data as quickly as possible, to limit the errors which unstable flow may introduce.

Experimental Procedure:

1. Select outdoor swale for measurement
2. Clean out any surface trash that may have accumulated in the swale.
3. Let stormwater circulate over swale until soil is saturated (approximately 10 min).
4. Sample soil depth and water surface depth according to grid pattern
5. Once all measurements have been recorded, select new swale for measurement
6. Continue steps 1-5 until storm event ends or flow is insufficient for measurements.
7. Completed measurements of the flow profiles are transferred into a project spreadsheet for calculation of the various hydraulic characteristics.

All measurements are taken using a specialized pin gauge and recorded to the closest millimeter. These values represent the distance from the reference bar (Figure C-11) to the surface of interest (soil or water surface). The data are analyzed in the same manner as the indoor laboratory swales.

Collection of Surface Runoff Samples. Surface (stormwater) runoff samples were taken periodically to monitor the swales' metal removal capabilities. Individual grab samples of 100 mL were collected in twenty-minute intervals to create composites for the entire storm event. Two composites were made for each of the six channels in Outdoor Swale Sets #1 and #2 (six channels = twelve composites). The sampling locations were as follows: 1) as water entered the swale over the weir and 2) as water exits the swale. These composite samples are analyzed in the laboratory for TSS, TDS, total metals, and total filtered metals. The resulting data are then compared (entrance vs. exit) to determine the metal removal ability of each swale.

5.1.2.3 Amended Swale (#3)

The amended outdoor swale (Swale Set #3) was designed to compare the infiltration rate and metal retention capacity between an amended soil and a natural soil. Field observations were used to compare the infiltration rates, while metal accumulation rates were monitored in the lab.

General Description. The amended swale set (Swale Set #3) receives stormwater runoff from Campus Drive, a high traffic road adjacent to the swale entrance, draining an area of 0.289 acres or 12,600 ft^2. The head of the channel test section was located approximately 25 feet from where the roadway runoff entered the swale. The experimental channel exists on a 0.49% slope. The swale set was comprised of two, 24 feet long, parallel channels separated by a $1/8^{th}$ inch

Plexiglas divider (Figure C-14). The divider extends ten inches below and rises six inches above the soil surface, ensuring flow (surface and subsurface) between the two channels is separated. The natural soil channel was undisturbed, apart from the installation of two pan lysimeters (Figure C-15). The amended channel was constructed by excavating the existing soil eight inches deep, nine inches wide, and twenty-four feet long; yielding a trench volume of twelve cubic feet. The channel was lined with filter fabric, covered in stone, and fit with two pan lysimeters, prior to the addition of amended soil. The amended soil was prepared by mixing fifty percent sand with fifty percent peat moss by volume (weight ratio = 7.7:1 (sand: moss)). The sand and peat moss were placed separately in five gallon buckets and lightly packed to verify that no voids existed in the containers. The ingredients were mixed in a concrete mixer to ensure homogeneity. The final "amended" soil was compacted into the trench and bordering vegetation was encouraged to grow over the surface. Finally, a concrete flow splitter was installed at the swale entrance to evenly distribute the arriving stormwater.

Surface Runoff and Subsurface Leachate Collection. The following procedure details how surface and subsurface water samples are taken for the determination of the metal retention capabilities of the amended swale set.

Sampling Procedure:

1. Clean out any surface trash that may have accumulated in the swale.
2. Collect grab samples at each sampling location (300 mL).
3. Measure flow depth at each sampling location (See Figure C-16).
4. Measure flow velocity at each sampling location (See Figure C-16).
5. Measure infiltration rate.
6. Form composite samples for all sampled locations.

Note: Steps 2-5 are performed at each sampling location, before the next sampling location is addressed.

Individual grab samples of 300 mL (100 mL for metal analysis and 200 mL for dissolved solids / total suspended solids) were collected in 20-minute intervals to create composite samples for the entire storm event. The final composite was separated and preserved as needed (HNO_3 and refrigeration for metals; refrigeration for TSS, TDS). The approximate velocity was determined by floating a marker to observe distance over time (V= D/T). Flow depth is recorded during each sampling. Utilizing the top of the Plexiglas divider as a reference, the distance to ground surface (Ds) and distance to water surface (Dw) was measured. Accurate water depth (WD) was acquired from the comparison of these values (WD = Ds – Dw). Lysimeter saturation time was determined by visual inspection during storm events. Depth measurements (inside lysimeter) vs. time are used to estimate the infiltration rate. Double ring infiltration experiments were conducted before and after selected storm events (See 5.1.2.6 Additional Experiments).

5.1.2.4 Rainwater Collection / Monitoring

Rainfall Monitoring. Rainfall was monitored by a Qualimetrics tipping bucket rain gauge and recorded on an Onset HOBO event logger (Figure C-17 and C-18). The rainfall gauge was installed on top of the Civil and Environmental Engineering Building (MIB). This location was selected because of its close proximity to experimental swales and relative security. Close attention was paid to the manufacturer's setup guide to ensure optimal performance.

Rainfall data recorded on the HOBO event logger were downloaded twice per month, utilizing Onset's Boxcar 3.6 software. The data indicated the date and time for each 0.01 inch of rainfall. The collected information was then transferred into Microsoft Excel where graphs are prepared for each month. These graphs display storm events that occurred during the month, and can be analyzed to determine the duration and intensity of each event.

Storm duration was determined by the length of time between the first data pulse (0.01 inch of rainfall) to the final data pulse in any storm event. The final data pulse is defined as the point at which no subsequent data are recorded within thirty minutes. Storm intensity is calculated based on the elapsed time between data pulses.

Rainwater Collection. The use of actual stormwater was essential for the various experiments throughout the project. During rainfall events, stormwater was collected from a culvert located behind the Civil and Environmental Engineering Building (Figure C-19). This catchment is situated so that it receives runoff from both a student parking lot and adjoining rooftops. Utilizing a sump pump, vinyl hoses, and multiple transport (twenty gallon) containers, large volumes of stormwater were quickly obtained. The transport containers were ultimately dumped into two 150 gallon, high-density plastic tanks (Figure C-20) until needed. Baseline tests were conducted to ensure that the collection process did not influence metal concentrations observed in the stormwater. Additional tests were performed to examine the impact that holding time had on the reserved stormwater. Data for these experiments are available in the data appendix.

5.1.2.5 Survey

Surveying was conducted utilizing a WILD Heerbrugg scale-reading theodolite (Model T-16). The Stadia (distance; leveling) method was used to determine the drainage area and channel slopes for each of the project swales. Boundaries of the drainage area were determined by observation of flow conditions during actual rainfall events. Distance and elevation measurements were tabulated in Microsoft Excel for analysis. Angles were rounded to the nearest 0.5-degree to permit manual drafting of the drainage basins. The drainage areas were plotted on engineering graph paper at one of two scales (1inch = 20ft or 1 inch = 40ft). Drainage areas were determined by counting squares. The Stadia method is accurate to +- 1ft for distance and +-0.01 ft for elevation. While care was taken to ensure accurate results in area estimates some errors could exist. Worst-case scenarios would indicate an error of one graphic square: +-25 ft^2 for Swale #1 and +-100 ft^2 for Swales 2 and 3.

5.1.2.6 Additional Experiments

Several experiments / procedures were conducted to determine the characteristics of different soils being utilized in the project. The information acquired from these tests was then used to assess the soil's influence on the hydraulic characteristics data collection. Below is a list of the standard ASTM methods employed, including any alterations of those methods which were made.

ASTM Methods Used:

- ASTM C 136-96a: Standard Test Method for Sieve Analysis of Fine and Coarse Aggregates.
- ASTM C 702-93: Standard Practice for Reducing Samples of Aggregate to Testing Size.
- ASTM D 653-96: Standard Terminology Relating to Soil, Rock, and Contained Fluids.
- ASTM D 2216-92: Standard Test Method for Laboratory Determination of Water (Moisture) Content of Soil and Rock.
- ASTM D 2488-93: Standard Practice for Description and Identification of Soils (Visual-Manual Procedure)
- ASTM D 4220-95: Standard Practice for Preserving and Transporting Soil Samples.

Modified ASTM Methods Used:

- ASTM D 3385-94: Standard Test Method for Infiltration Rate of Soils in Field Using Double-Ring Infiltrometer.

  This method was modified by utilizing three "Turf Tec International - Field Double Ring Infiltrometers" simultaneously per site, instead of the single (large) double ring infiltrometer, as recommended in the ASTM method. This alternative approach uses the mean of the three observed infiltration rates to represent the overall infiltration rate of the selected area. The smaller field infiltrometers allowed for rapid determination of the infiltration rate and are more time/cost effective.

Additional Experiments:

- Estimation of Surface Contact Time

  Estimation of surface contact time for the swales was performed in the following manner.

  1. An estimate of Q (flow) was developed from previous hydraulic characteristic / storm data.
  2. Velocity was calculated as a function of Q and cross sectional area (V=Q/A).
  3. Travel time was calculated for each section based on reach distance and the calculated velocity.
  4. The sum of travel times over the entire swale was assumed equal to the hydraulic residence time or surface contact time.
  5. Results indicate that the surface contact time, for a 10 foot rectangular swale, ranged from 28 to 97 sec. These times are based on the data available at the time of analysis.

5.1.3 Results / Discussion

5.1.3.1 Hydraulic Characterization Results

The goal of the indoor swale set was to investigate the hydraulic characteristics, specifically Manning's "n", for three grass species (Bluegrass, Centipede, and Zoysia) where the depth of flow was quite shallow relative to the grass's blade height. Shallow depths of flow are essential to optimizing metal removal rates in vegetated swales. The experimental methods / calculations were modeled after the Stillwater Hydraulic Laboratory in Stillwater Oklahoma

(USDA 1954). Their research, as discussed in 5.1.1.1, led to the development of standard retardance curves, or a series of graphs, which display the relationship between Manning's "n" and the product of velocity and hydraulic radius (VR) for different types of vegetation (See Figure 5-1). Each curve on the graph represents a class of vegetation with similar physical characteristics. Grasses including Centipede and Kentucky Bluegrass are contained within "Retardance Class C."

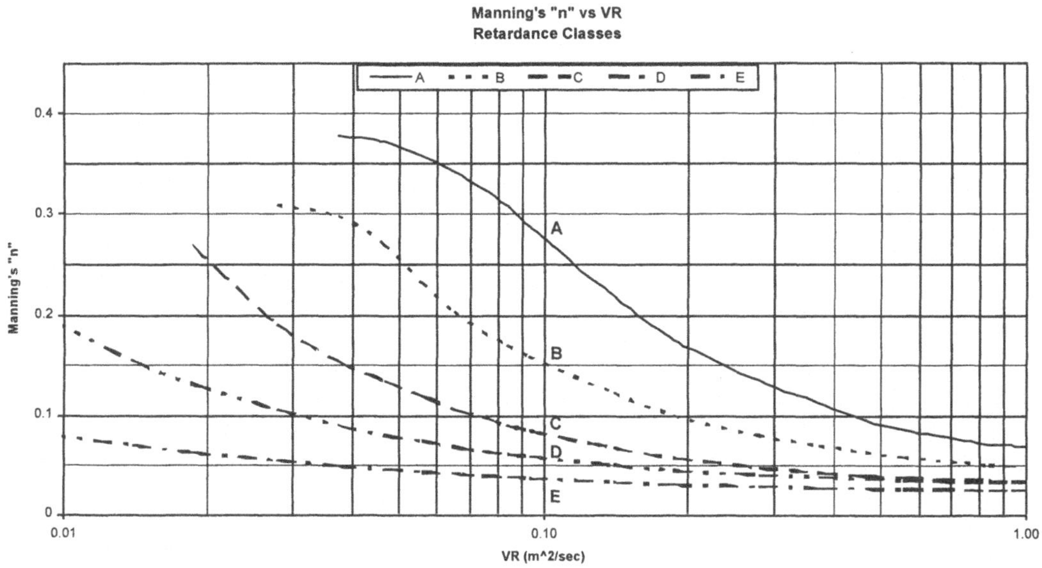

Figure 5-1: Original Stillwater Laboratory Retardance Curves

Figure 5-2, illustrates a typical set of raw data that were used to develop the final retardance curves from the Stillwater experiments. Data collected with VR values greater than 1.00 ft^2/sec are in close proximity to the final curve. However, as VR decreases, a general dispersion of the raw data can be seen. The highest variation in the calculated values of Manning's "n", (0.39 to 0.21 +-0.18), was observed at a VR of 0.02 ft^2/sec (0.002m^2/sec). It is at this point, VR equal to or less than 0.002m^2/sec, where observations for the indoor laboratory swales in this research began.

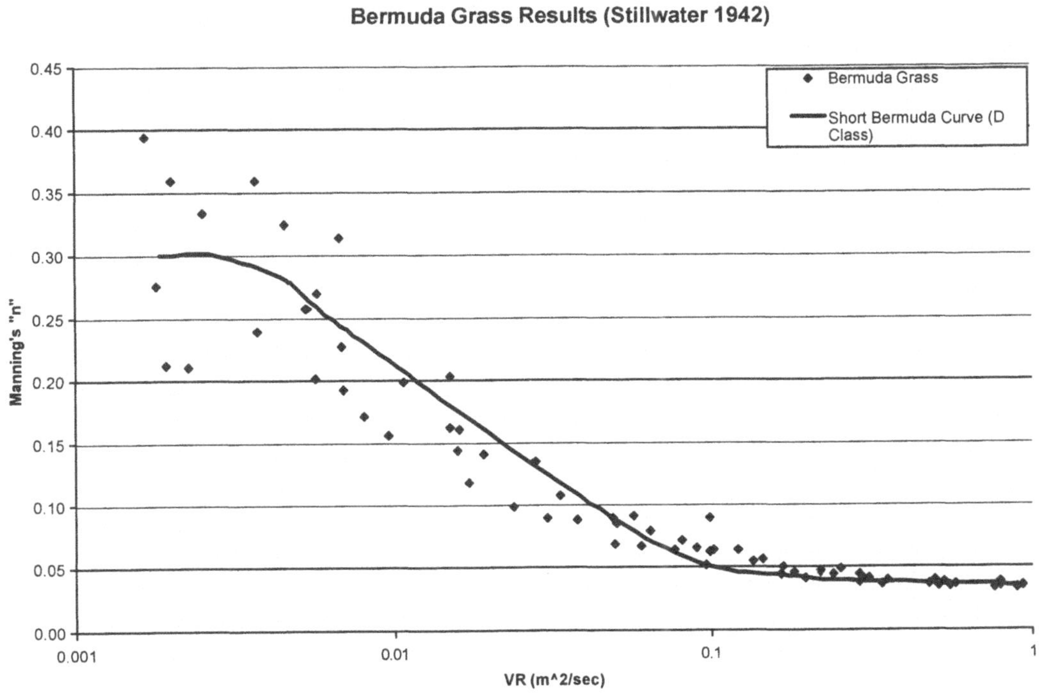

Figure 5-2: Bermuda Grass Results (Spreading Trend as VR Decreases)

The following figures (Figure 5-3, Figure 5-4, and Figure 5-5) show the relationship between Manning's "n" and the product of velocity and hydraulic radius (VR) that were observed in the various grasses contained within the "Indoor Laboratory Swales". While the raw data were slightly dispersed, the spread was no greater than that observed in the Stillwater experiments, even at much lower values of VR.

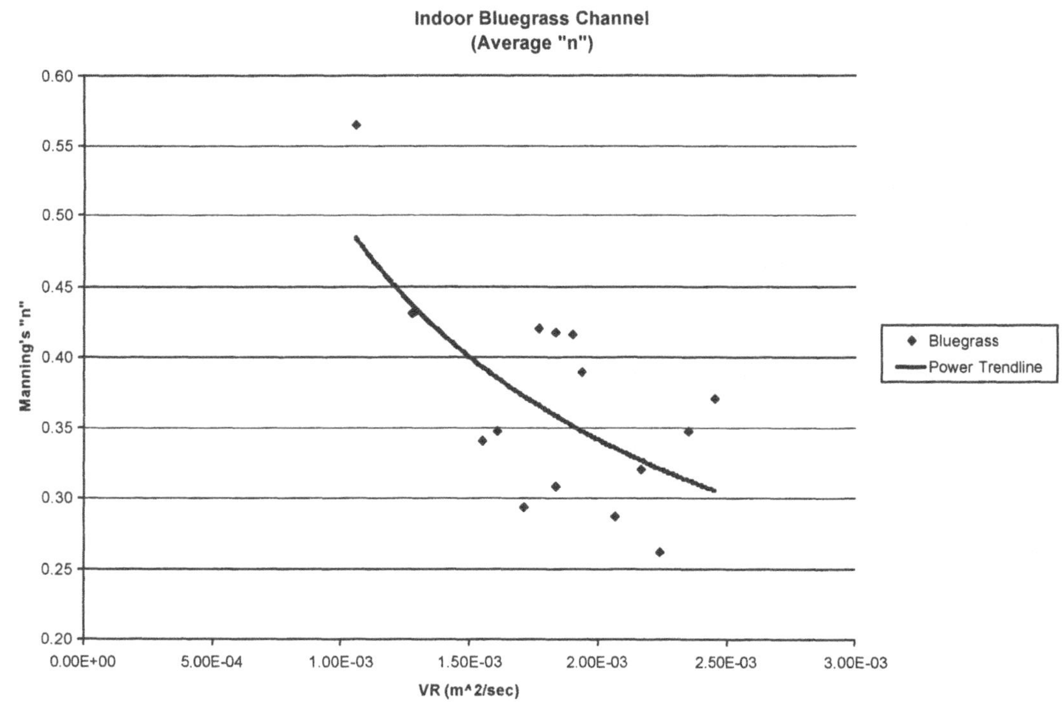

Figure 5-3: Indoor Bluegrass Results

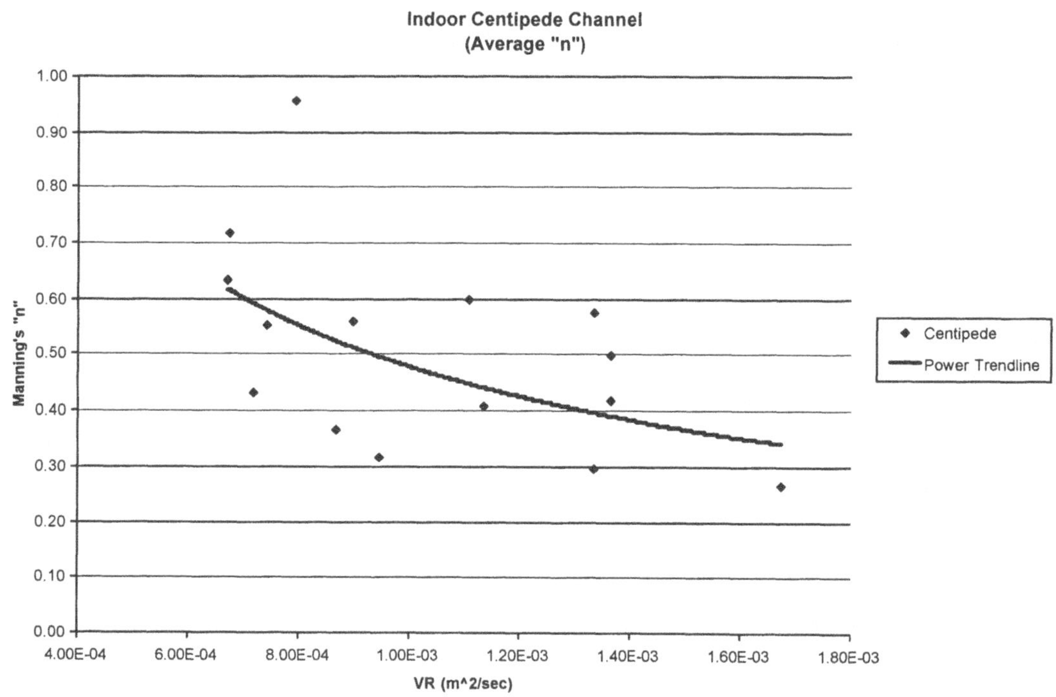

Figure 5-4: Indoor Centipede Results

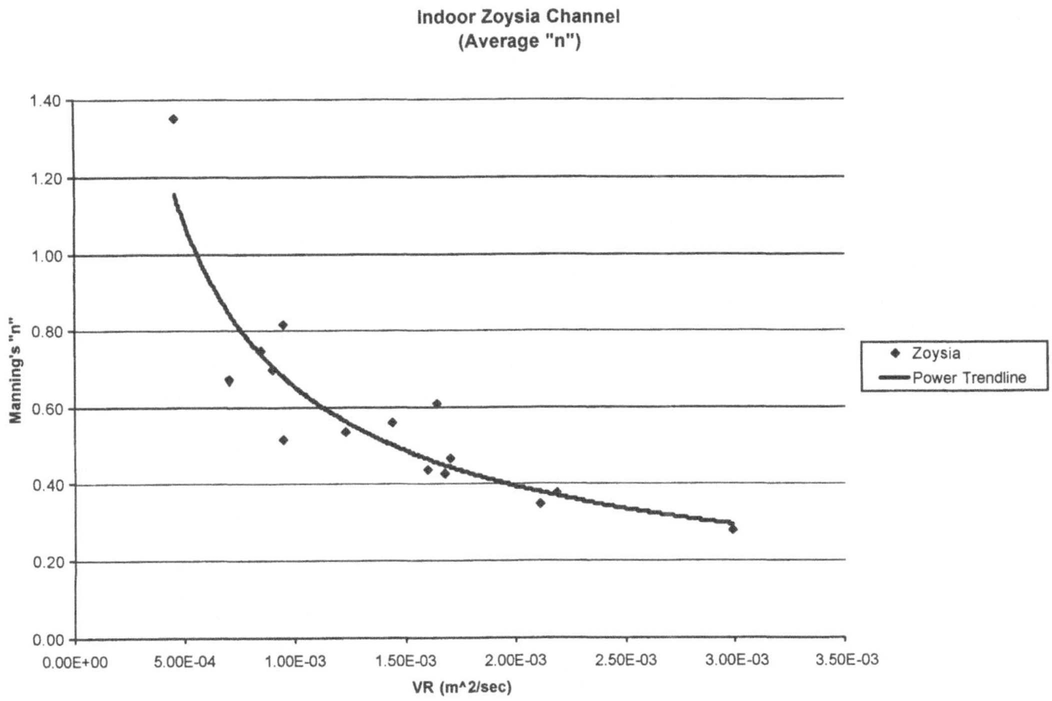

Figure 5-5: Indoor Zoysia Results

In addition to the hydraulic characterization experiments conducted on the indoor swales, several measurements were taken at the outdoor swales (#1 and # 2) for comparison. Completing the required measurements for the determination of Manning's "n" at the outdoor swales, which rely on large storm events, was very difficult. While more than twenty storm events were evaluated, only four dates provided enough information to plot. These data values may be seen in Figure 5-6. This graph shows three distinct regions 1) the indoor swale data, 2) the outdoor swale data and 3) the historical Stillwater retardance curves. While the three regions do not align perfectly, they are quite close. Reasons for possible discrepancies include: condition of grass at time of sampling (height, density, etc.), underlying soil properties, slope (outdoor slopes were higher than lab tests), variations in flow (volume / velocity) during the outdoor experiments, and differences between sheet flow and channelized flow.

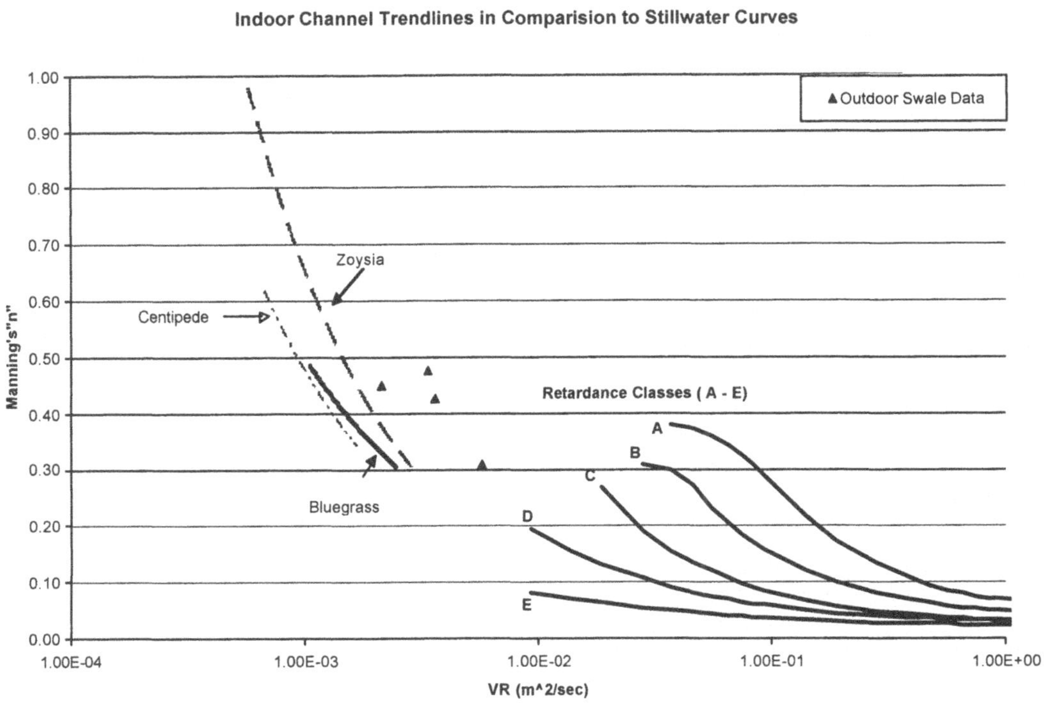

Figure 5-6: Indoor Grass "n" vs. VR Curves / Stillwater "n" vs. VR Curves

The current experimental grass channels appear to best fit the Retardance Class 'D' curve. A vegetation's retardance class is based primarily on its physical characteristics, including blade height and stand density. Experiments at the Stillwater Laboratory indicated that Centipede and Kentucky Bluegrass were of Retardance Class 'C', however, these grasses were allowed to grow in excess of six inches during testing. The current (WERF) experimental channels were only grown to approximately three inches in height. This reduction in grass height justifies a decrease in the observed retardance class (USDA 1954). Bermuda grass, for example, can range from Retardance Class 'B' to 'E' based solely on grass height.

5.1.3.2 Outdoor Swale (#1 and #2) Results

In addition to the hydraulic characterization experiments conducted at the outdoor swales (Swales #1 and # 2 discussed above), a series of samples were taken to examine the swales' abilities to remove heavy metals from surface flow. Stormwater samples were captured at the entrance / exit of each channel (Centipede, Control, and Zoysia). Composite samples, for an entire storm event, were formed by taking 100ml grabs in 20min intervals. The collected stormwater was then analyzed for TSS and heavy metal concentrations (Cu, Cr, Pb, Zn, Cd, Fe, Na, Mg, Ca, and K). Metal concentrations below the ICP's level of detection (LOD) were omitted from the final analysis. Metal concentrations (mg/L of metal) were divided by the TSS (total suspended solids in kg/L) to yield values of mg metal / kg TSS. Finally, the effluent concentrations were divided by the influent concentrations to calculate a relative percent change in concentration:

[1- (effluent/influent)*100 = Percent Change]

Results for the outdoor swales are shown in Figure 5-7 below. Positive values represent a removal of pollutants from the stormwater, while negative values indicate possible scour of metals from the swale.

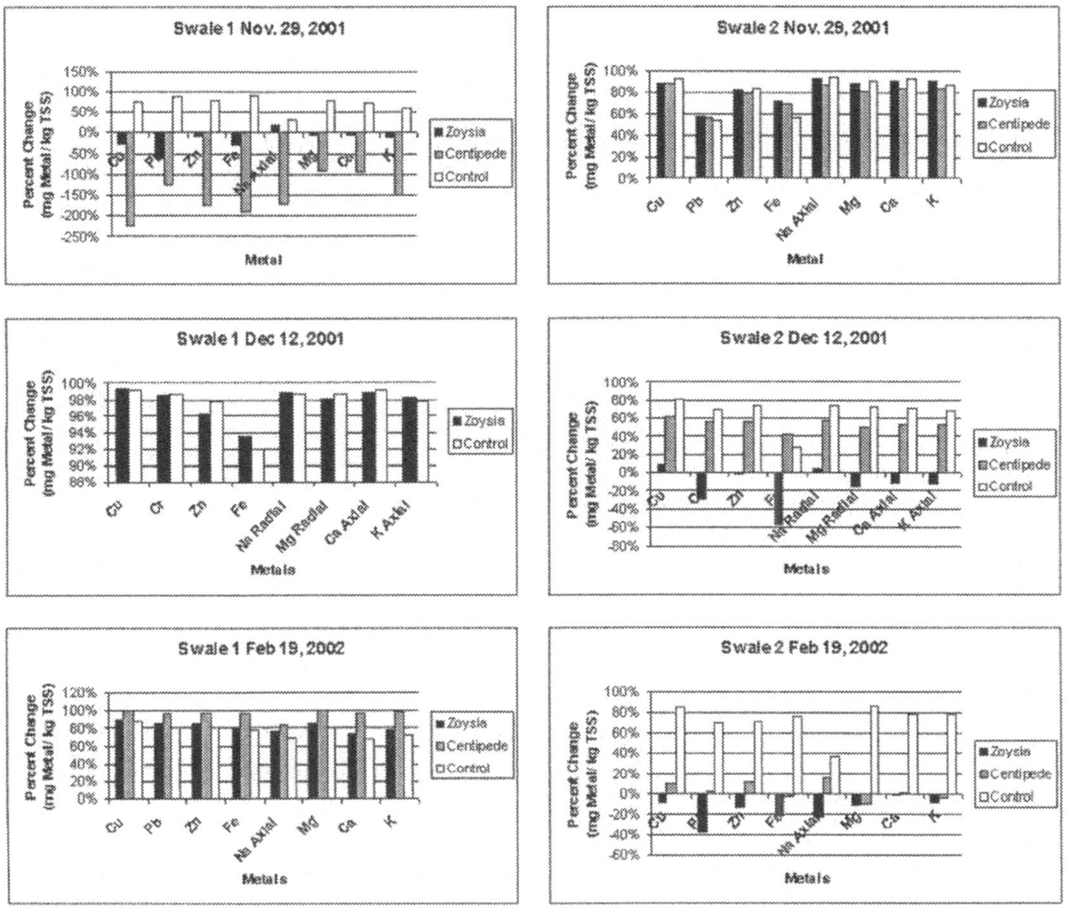

Figure 5-7: Percent Change (mg Metal / kg TSS) observed in Outdoor Swales 1 & 2

The figure above illustrates that grass swales can reduce metal concentrations in stormwater (Swale # 2 11/29/01, Swale #1 12/12/01, and Swale #1 2/19/02), but may possibly deteriorate the overall water quality under some conditions (Swale #1 11/29/01, Swale #2 Zoysia 12/12/01, and Swale #2 2/19/02). When metals are removed, the percent change is almost 100 %, while during other events the concentrations may remain the same or double. Great variability was observed, during the monitored storm events, for the swales' overall metal removal performance. Variations in metal removal rates between the different grasses and the control were less pronounced. The percent change for all metals (withen a given channel) were very similar.

Swale hydraulics appear to be more crucial in improving stormwater quality than the grass species, during single storm events. Field observations made during multiple rainfall events indicate that deposition of sediment and erosion of underlying clean soil occurred, and both erosion and deposition may occur simultaneously. Variations in the amount of suspended solids in the influent and effluent could significantly effect the percent change. Figure 5-8 shows a comparison between the maximum allowable shear stress and results of hydraulic characterization tests conducted at the outdoor swales.

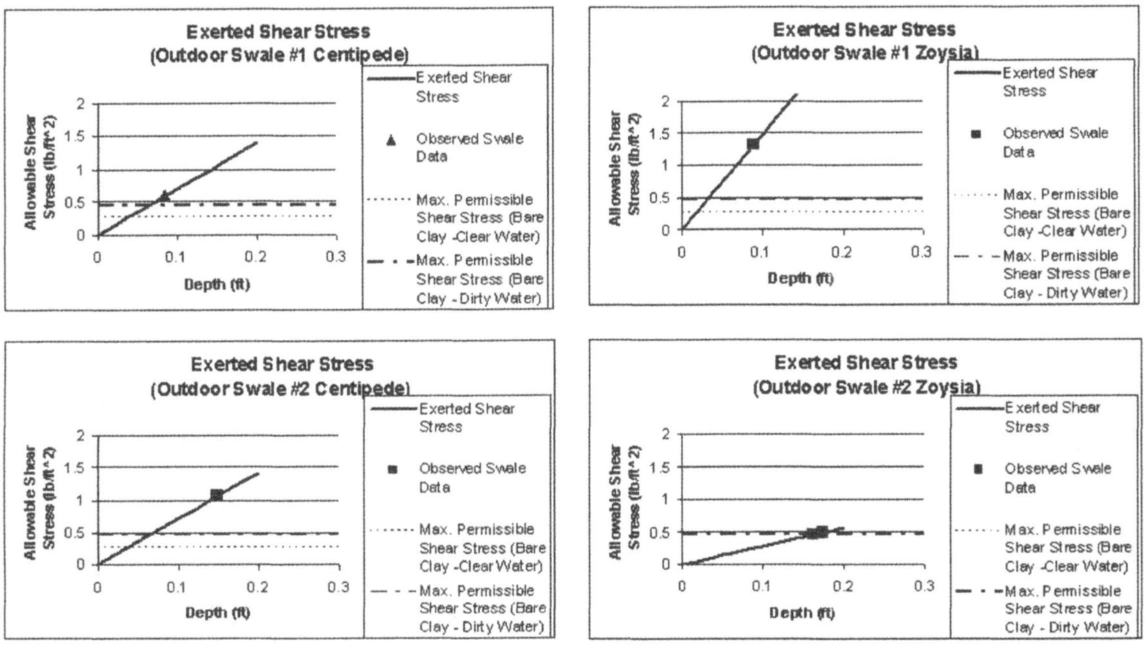

Figure 5-8: Shear Stress Observations in Outdoor Swales 1 & 2

Data points above the horizontal lines indicate possible erosion of bare clay from the swale channel. While the grasses offer increased protection to the underlying soil, the graphs above indicate a strong potential for erosion to occur in the vegetated swales, especially where there may be voids in the grass cover.

Figure 5-9, demonstrates that as slope increases the potential for erosion will also increase. Other factors that may influence the metal removal rates, during single storm events, include: storm intensity, temporal variations in storm intensity, flow velocity, grass condition (density, height, etc), and underlying soil properties.

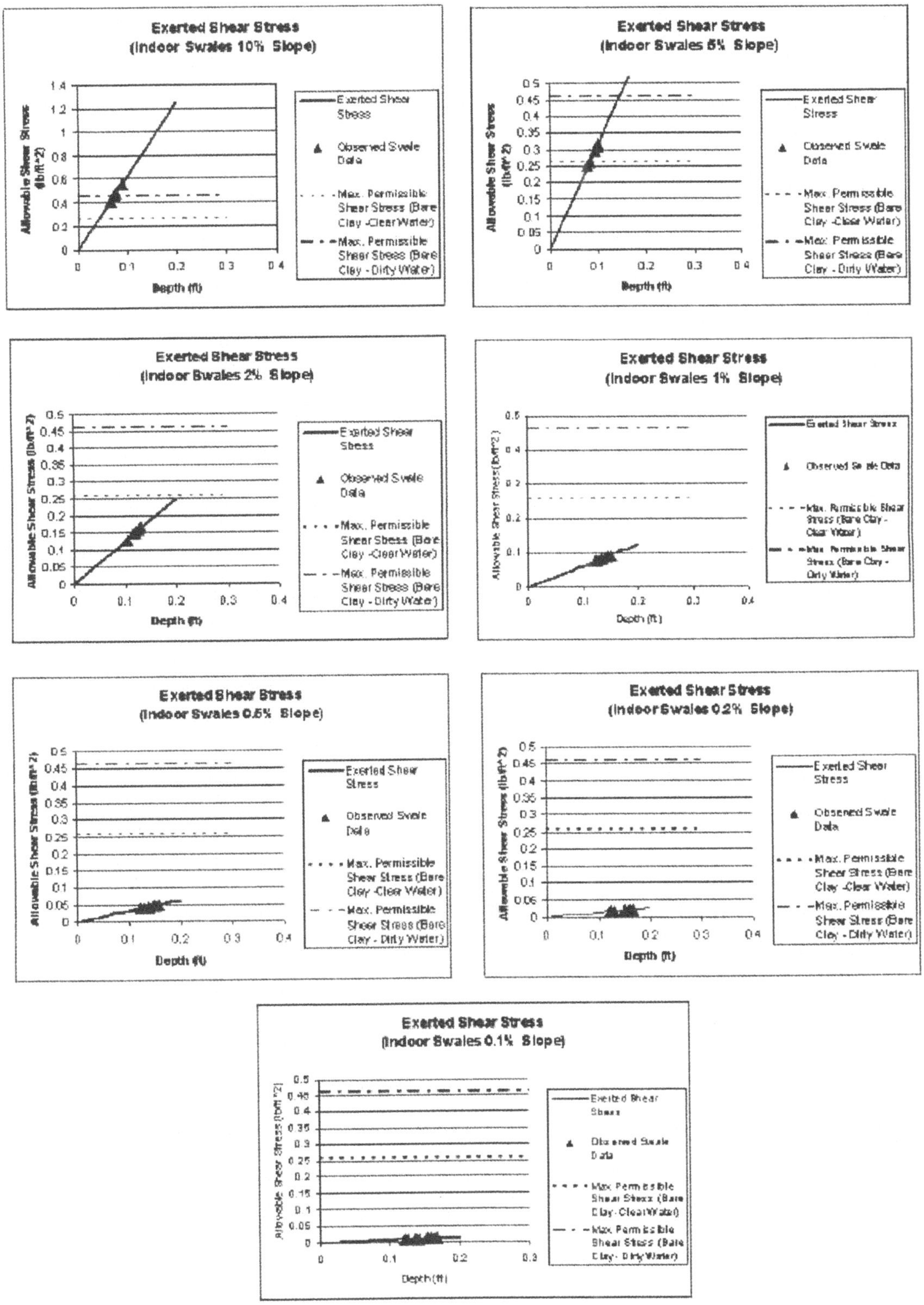

Figure 5-9: Shear Stress Observations in Indoor Swales

5.1.3.3 Amended Swale (#3) Results

The amended swale (outdoor swale # 3) was constructed to compare the infiltration rate and metal retention capacity between an amended soil (50% peat moss/ 50% Soil by volume) and a natural soil. This swale was described in section 5.1.2.3 and appendix photo C-16, seen below.

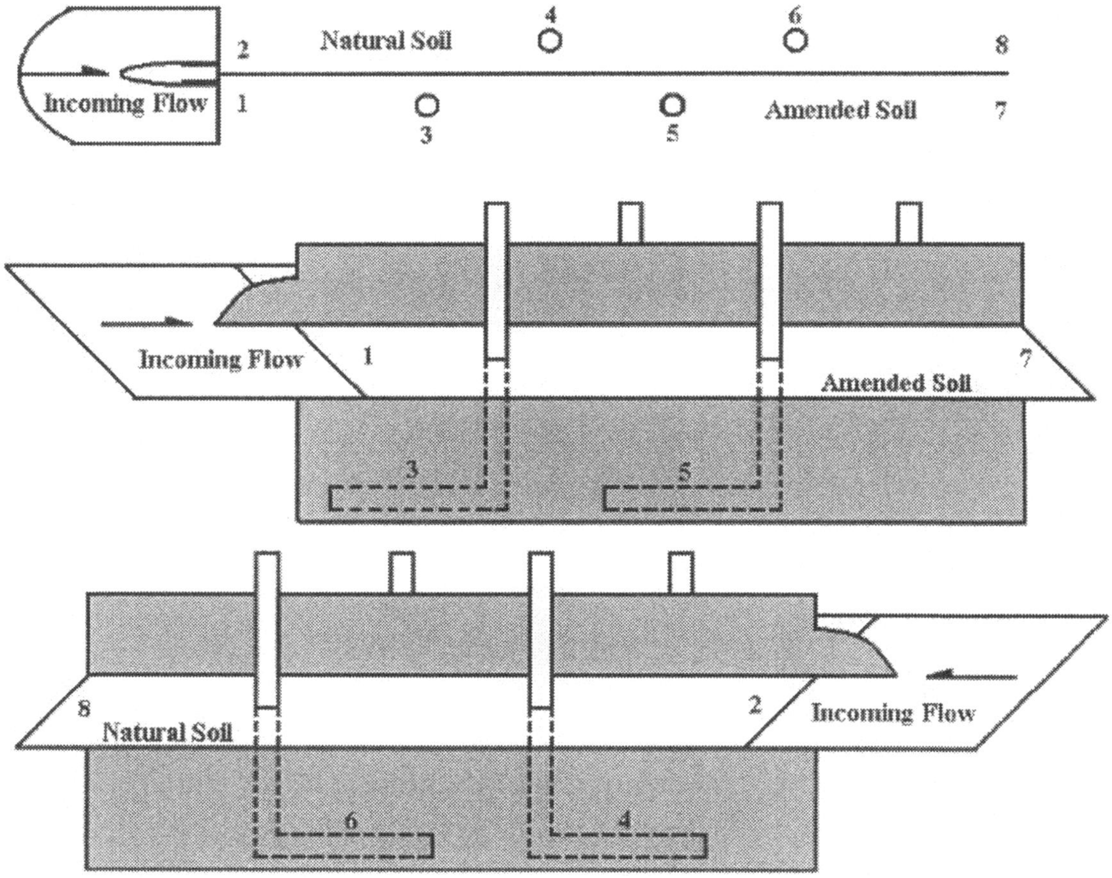

Appendix Photo C-16

The samples collected were analyzed for total metals (Cu, Cr, Pb, Zn, Cd, Fe, Na, Mg, Ca, and K), total suspended solids (TSS), and total dissolved solids (TDS).

Table 5-1, shows that a majority of total solids were as dissolved solids at all sampling locations for most storm events.

Table 5-1: Percent of Total Solids (TS) as Dissolved Solids (TDS)

	8/6/2001	9/3/2001	11/29/2001	2/19/2002
SWALE 3 Location #1	78%	61%	49%	71%
SWALE 3 Location #2	54%	30%	42%	70%
SWALE 3 Location #3	72%	88%	55%	67%
SWALE 3 Location #4	94%	69%	60%	59%
SWALE 3 Location #5	80%	87%	49%	65%
SWALE 3 Location #6	98%	68%	44%	42%
SWALE 3 Location #7	74%	72%	46%	71%
SWALE 3 Location #8	82%	46%	41%	71%

Metal concentrations were determined in the laboratory through ICP analysis. Occasionally, metal concentrations were reported as below the ICP's level of detection (LOD). This indicates that metals concentration detected, by the ICP, are statistically identical to zero. These values were therefore treated as null data during final data analysis.

The first step was to analyze the various metals for both the amended and natural soils. Table 5-2, shows the median and standard deviation for all metals (in mg/L) at the swale inlet / outlet and for the swale's surface and subsurface sampling locations. Tables 5-3 and 5-4, show the relative percent change between the various sampling locations. Percent change was calculated using the following formula:

Comparison: X (entrance) vs. Y (exit)

Percent Change for Storm event # n = 1- (Y/ X) *100

Percent Change = Median of (Percent Change for all storm events)

The percent change in metal concentrations was calculated for each metal during each storm event. The median of those values is the percent change, for the specified comparison. A positive value represents a reduction in metal concentrations in Y compared to X. A negative value indicates location Y has higher metal concentrations than location X.

Many of the median metal concentrations in the surface runoff downstream of the swale, were lower than the median initial concentrations, especially for the natural soil (8 of 10 constituents). However, the concentration reductions were generally small (5 to 35%), with typical reductions of about 20%. Only Cadmium was observed to decrease in the surface flows for the amended soil swale. The concentration increases were likely associated with scoured sediment from previous storms. Only four events were monitored at each swale, while at least thirty rains likely occurred during this period. As previously indicated by the calculated shear stress values, depositions and scour processes are dependent on hydraulic conditions that varied greatly during tests.

The concentration changes between the surface and subsurface sampling locations showed fewer differences. The observed concentration differences were low for these tests due to the unexpectedly low inlet metal concentrations at these sites, and large flows which reduced the residence and contact times.

Table 5-2: Metal Concentrations Observed in Swale # 3

Copper	Amended	#1	Natural	#2
	Median	Std. Dev.	Median	Std. Dev.
Inlet	0.010	0.063	0.008	0.050
	Amended	#7	Natural	#8
	Median	Std. Dev.	Median	Std. Dev.
Exit	0.006	0.038	0.010	0.041

Copper	Amended	#1 & #7	Natural	#2 & #8
	Median	Std. Dev.	Median	Std. Dev.
Surface	0.008	0.050	0.009	0.043
	Amended	#3 & #5	Natural	#4 & #6
	Median	Std. Dev.	Median	Std. Dev.
Subsurface	0.010	0.036	0.009	0.032

Chromium	Amended	#1	Natural	#2
	Median	Std. Dev.	Median	Std. Dev.
Inlet	0.006	0.004	0.007	0.004
	Amended	#7	Natural	#8
	Median	Std. Dev.	Median	Std. Dev.
Exit	0.005	0.006	0.005	0.003

Chromium	Amended	#1 & #7	Natural	#2 & #8
	Median	Std. Dev.	Median	Std. Dev.
Surface	0.006	0.005	0.007	0.004
	Amended	#3 & #5	Natural	#4 & #6
	Median	Std. Dev.	Median	Std. Dev.
Subsurface	0.007	0.004	0.006	0.006

Lead	Amended	#1	Natural	#2
	Median	Std. Dev.	Median	Std. Dev.
Inlet	0.009	0.011	0.010	0.004
	Amended	#7	Natural	#8
	Median	Std. Dev.	Median	Std. Dev.
Exit	0.010	0.008	0.008	0.006

Lead	Amended	#1 & #7	Natural	#2 & #8
	Median	Std. Dev.	Median	Std. Dev.
Surface	0.009	0.009	0.010	0.005
	Amended	#3 & #5	Natural	#4 & #6
	Median	Std. Dev.	Median	Std. Dev.
Subsurface	0.008	0.007	0.013	0.007

Zinc	Amended	#1	Natural	#2
	Median	Std. Dev.	Median	Std. Dev.
Inlet	0.032	0.032	0.043	0.018
	Amended	#7	Natural	#8
	Median	Std. Dev.	Median	Std. Dev.
Exit	0.032	0.011	0.029	0.008

Zinc	Amended	#1 & #7	Natural	#2 & #8
	Median	Std. Dev.	Median	Std. Dev.
Surface	0.032	0.023	0.037	0.017
	Amended	#3 & #5	Natural	#4 & #6
	Median	Std. Dev.	Median	Std. Dev.
Subsurface	0.022	0.017	0.026	0.013

Cadmium	Amended	#1	Natural	#2
	Median	Std. Dev.	Median	Std. Dev.
Inlet	0.003	1 value	0.003	1 value
	Amended	#7	Natural	#8
	Median	Std. Dev.	Median	Std. Dev.
Exit	0.002	1 value	0.002	1 value

Cadmium	Amended	#1 & #7	Natural	#2 & #8
	Median	Std. Dev.	Median	Std. Dev.
Surface	0.002	0.000	0.003	0.001
	Amended	#3 & #5	Natural	#4 & #6
	Median	Std. Dev.	Median	Std. Dev.
Subsurface	0.003	0.001	0.002	0.000

Iron	Amended	#1	Natural	#2
	Median	Std. Dev.	Median	Std. Dev.
Inlet	0.766	0.498	1.687	1.041
	Amended	#7	Natural	#8
	Median	Std. Dev.	Median	Std. Dev.
Exit	0.907	0.665	1.068	0.479

Iron	Amended	#1 & #7	Natural	#2 & #8
	Median	Std. Dev.	Median	Std. Dev.
Surface	0.853	0.562	1.268	0.849
	Amended	#3 & #5	Natural	#4 & #6
	Median	Std. Dev.	Median	Std. Dev.
Subsurface	0.994	0.680	2.614	1.352

Sodium	Amended	#1	Natural	#2
	Median	Std. Dev.	Median	Std. Dev.
Inlet	0.354	0.350	0.472	0.376
	Amended	#7	Natural	#8
	Median	Std. Dev.	Median	Std. Dev.
Exit	0.423	0.425	0.570	0.582

Sodium	Amended	#1 & #7	Natural	#2 & #8
	Median	Std. Dev.	Median	Std. Dev.
Surface	0.415	0.372	0.472	0.472
	Amended	#3 & #5	Natural	#4 & #6
	Median	Std. Dev.	Median	Std. Dev.
Subsurface	0.541	0.166	0.561	0.237

Magnesium	Amended	#1	Natural	#2
	Median	Std. Dev.	Median	Std. Dev.
Inlet	1.079	0.173	1.359	0.194
	Amended	#7	Natural	#8
	Median	Std. Dev.	Median	Std. Dev.
Exit	1.101	0.163	1.201	0.257

Magnesium	Amended	#1 & #7	Natural	#2 & #8
	Median	Std. Dev.	Median	Std. Dev.
Surface	1.101	0.165	1.347	0.236
	Amended	#3 & #5	Natural	#4 & #6
	Median	Std. Dev.	Median	Std. Dev.
Subsurface	1.088	0.586	1.464	0.304

Calcium	Amended	#1	Natural	#2
	Median	Std. Dev.	Median	Std. Dev.
Inlet	5.800	1.236	6.484	2.147
	Amended	#7	Natural	#8
	Median	Std. Dev.	Median	Std. Dev.
Exit	5.915	0.957	6.127	2.573

Calcium	Amended	#1 & #7	Natural	#2 & #8
	Median	Std. Dev.	Median	Std. Dev.
Surface	5.915	1.090	6.484	2.294
	Amended	#3 & #5	Natural	#4 & #6
	Median	Std. Dev.	Median	Std. Dev.
Subsurface	4.392	2.220	10.138	4.510

Potassium	Amended	#1	Natural	#2
	Median	Std. Dev.	Median	Std. Dev.
Inlet	0.955	0.444	1.650	0.852
	Amended	#7	Natural	#8
	Median	Std. Dev.	Median	Std. Dev.
Exit	1.491	0.546	1.365	0.761

Potassium	Amended	#1 & #7	Natural	#2 & #8
	Median	Std. Dev.	Median	Std. Dev.
Surface	1.397	0.517	1.624	0.824
	Amended	#3 & #5	Natural	#4 & #6
	Median	Std. Dev.	Median	Std. Dev.
Subsurface	2.271	2.769	2.051	1.739

Table 5-3: Inlet & Outlet Surface Metal Comparisons (Swale # 3)

	Inlet & Outlet Surface Metal Comparisons	ID #'s	Percent Change	Alpha
Copper Cu	**Inlet & Outlet Surface Metal Comparisons**	**ID #'s**	**Percent Change**	**Alpha**
	Natural Inlet vs. Natural Outlet	2 vs 8	18%	0.26
	Amended Inlet vs. Amended Outlet	1 vs 7	45%	0.18
	Amended Inlet vs. Natural Inlet	1 vs 2	19%	0.20
	Amended Outlet vs. Natural Outlet	7 vs 8	-13%	0.09
Cromium Cr	**Inlet & Outlet Surface Metal Comparisons**	**ID #'s**	**Percent Change**	**Alpha**
	Natural Inlet vs. Natural Outlet	2 vs 8	23%	0.08
	Amended Inlet vs. Amended Outlet	1 vs 7	14%	0.38
	Amended Inlet vs. Natural Inlet	1 vs 2	-8%	0.13
	Amended Outlet vs. Natural Outlet	7 vs 8	-5%	0.76
Lead Pb	**Inlet & Outlet Surface Metal Comparisons**	**ID #'s**	**Percent Change**	**Alpha**
	Natural Inlet vs. Natural Outlet	2 vs 8	29%	0.09
	Amended Inlet vs. Amended Outlet	1 vs 7	7%	0.46
	Amended Inlet vs. Natural Inlet	1 vs 2	-18%	0.70
	Amended Outlet vs. Natural Outlet	7 vs 8	20%	0.17
Zinc Zn	**Inlet & Outlet Surface Metal Comparisons**	**ID #'s**	**Percent Change**	**Alpha**
	Natural Inlet vs. Natural Outlet	2 vs 8	38%	0.07
	Amended Inlet vs. Amended Outlet	1 vs 7	-9%	0.48
	Amended Inlet vs. Natural Inlet	1 vs 2	-50%	0.26
	Amended Outlet vs. Natural Outlet	7 vs 8	11%	0.73
Cadmium Cd	**Inlet & Outlet Surface Metal Comparisons**	**ID #'s**	**Percent Change**	**Alpha**
	Natural Inlet vs. Natural Outlet	2 vs 8	40%	N/A
	Amended Inlet vs. Amended Outlet	1 vs 7	15%	N/A
	Amended Inlet vs. Natural Inlet	1 vs 2	-34%	N/A
	Amended Outlet vs. Natural Outlet	7 vs 8	5%	N/A
Iron Fe	**Inlet & Outlet Surface Metal Comparisons**	**ID #'s**	**Percent Change**	**Alpha**
	Natural Inlet vs. Natural Outlet	2 vs 8	26%	0.08
	Amended Inlet vs. Amended Outlet	1 vs 7	11%	0.64
	Amended Inlet vs. Natural Inlet	1 vs 2	-76%	0.10
	Amended Outlet vs. Natural Outlet	7 vs 8	-7%	0.66
Sodium Na	**Inlet & Outlet Surface Metal Comparisons**	**ID #'s**	**Percent Change**	**Alpha**
	Natural Inlet vs. Natural Outlet	2 vs 8	15%	0.56
	Amended Inlet vs. Amended Outlet	1 vs 7	-5%	0.31
	Amended Inlet vs. Natural Inlet	1 vs 2	-21%	0.14
	Amended Outlet vs. Natural Outlet	7 vs 8	9%	0.54
Magnesium Mg	**Inlet & Outlet Surface Metal Comparisons**	**ID #'s**	**Percent Change**	**Alpha**
	Natural Inlet vs. Natural Outlet	2 vs 8	17%	0.08
	Amended Inlet vs. Amended Outlet	1 vs 7	0%	0.34
	Amended Inlet vs. Natural Inlet	1 vs 2	-32%	**0.01**
	Amended Outlet vs. Natural Outlet	7 vs 8	-3%	0.59
Calcium Ca	**Inlet & Outlet Surface Metal Comparisons**	**ID #'s**	**Percent Change**	**Alpha**
	Natural Inlet vs. Natural Outlet	2 vs 8	17%	0.13
	Amended Inlet vs. Amended Outlet	1 vs 7	-12%	0.30
	Amended Inlet vs. Natural Inlet	1 vs 2	-32%	**0.04**
	Amended Outlet vs. Natural Outlet	7 vs 8	11%	0.70
Potassium K	**Inlet & Outlet Surface Metal Comparisons**	**ID #'s**	**Percent Change**	**Alpha**
	Natural Inlet vs. Natural Outlet	2 vs 8	38%	0.22
	Amended Inlet vs. Amended Outlet	1 vs 7	-44%	0.08
	Amended Inlet vs. Natural Inlet	1 vs 2	-71%	**0.04**
	Amended Outlet vs. Natural Outlet	7 vs 8	30%	0.95

Table 5-4: Surface & Subsurface Metal Comparisons (Swale # 3)

Copper	Surface & Subsurface Metal Comparisons	ID #'s	Percent Change	Alpha 1	Alpha 2
Cu	Natural Surface vs. Natural Subsurface	2, 8 vs 4, 6	4%	0.14	0.19
	Amended Surface vs. Amended Subsurface	1, 7 vs 3, 5	27%	0.12	0.37
	Amended Surface vs. Natural Surface	1, 7 vs 2, 8	-8%	0.35	0.58
	Amended Subsurface vs. Natural Subsurface	3, 5 vs 4, 6	7%	0.58	0.70

Cromium	Surface & Subsurface Metal Comparisons	ID #'s	Percent Change	Alpha 1	Alpha 2
Cr	Natural Surface vs. Natural Subsurface	2, 8 vs 4, 6	-10%	0.69	0.67
	Amended Surface vs. Amended Subsurface	1, 7 vs 3, 5	-2%	0.59	0.43
	Amended Surface vs. Natural Surface	1, 7 vs 2, 8	-3%	0.83	0.89
	Amended Subsurface vs. Natural Subsurface	3, 5 vs 4, 6	12%	0.96	0.96

Lead	Surface & Subsurface Metal Comparisons	ID #'s	Percent Change	Alpha 1	Alpha 2
Pb	Natural Surface vs. Natural Subsurface	2, 8 vs 4, 6	-38%	**0.05**	**0.02**
	Amended Surface vs. Amended Subsurface	1, 7 vs 3, 5	21%	0.06	0.12
	Amended Surface vs. Natural Surface	1, 7 vs 2, 8	5%	0.31	0.31
	Amended Subsurface vs. Natural Subsurface	3, 5 vs 4, 6	-52%	**0.00**	**0.00**

Zinc	Surface & Subsurface Metal Comparisons	ID #'s	Percent Change	Alpha 1	Alpha 2
Zn	Natural Surface vs. Natural Subsurface	2, 8 vs 4, 6	23%	0.13	0.12
	Amended Surface vs. Amended Subsurface	1, 7 vs 3, 5	33%	**0.03**	0.28
	Amended Surface vs. Natural Surface	1, 7 vs 2, 8	-8%	0.46	0.76
	Amended Subsurface vs. Natural Subsurface	3, 5 vs 4, 6	-35%	0.57	0.64

Cadmium	Surface & Subsurface Metal Comparisons	ID #'s	Percent Change	Alpha 1	Alpha 2
Cd	Natural Surface vs. Natural Subsurface	2, 8 vs 4, 6	32%	0.47	0.36
	Amended Surface vs. Amended Subsurface	1, 7 vs 3, 5	-27%	0.66	0.53
	Amended Surface vs. Natural Surface	1, 7 vs 2, 8	-16%	0.59	0.74
	Amended Subsurface vs. Natural Subsurface	3, 5 vs 4, 6	38%	0.38	0.46

Iron	Surface & Subsurface Metal Comparisons	ID #'s	Percent Change	Alpha 1	Alpha 2
Fe	Natural Surface vs. Natural Subsurface	2, 8 vs 4, 6	-94%	0.06	**0.05**
	Amended Surface vs. Amended Subsurface	1, 7 vs 3, 5	-26%	0.29	0.28
	Amended Surface vs. Natural Surface	1, 7 vs 2, 8	-15%	0.09	0.08
	Amended Subsurface vs. Natural Subsurface	3, 5 vs 4, 6	-144%	**0.03**	**0.03**

Sodium	Surface & Subsurface Metal Comparisons	ID #'s	Percent Change	Alpha 1	Alpha 2
Na	Natural Surface vs. Natural Subsurface	2, 8 vs 4, 6	-1%	0.61	0.61
	Amended Surface vs. Amended Subsurface	1, 7 vs 3, 5	-47%	0.50	0.50
	Amended Surface vs. Natural Surface	1, 7 vs 2, 8	-1%	0.32	0.35
	Amended Subsurface vs. Natural Subsurface	3, 5 vs 4, 6	5%	0.99	0.98

Magnesium	Surface & Subsurface Metal Comparisons	ID #'s	Percent Change	Alpha 1	Alpha 2
Mg	Natural Surface vs. Natural Subsurface	2, 8 vs 4, 6	-19%	0.09	0.16
	Amended Surface vs. Amended Subsurface	1, 7 vs 3, 5	-9%	0.37	0.35
	Amended Surface vs. Natural Surface	1, 7 vs 2, 8	-13%	**0.03**	**0.01**
	Amended Subsurface vs. Natural Subsurface	3, 5 vs 4, 6	-26%	0.28	0.35

Calcium	Surface & Subsurface Metal Comparisons	ID #'s	Percent Change	Alpha 1	Alpha 2
Ca	Natural Surface vs. Natural Subsurface	2, 8 vs 4, 6	-62%	**0.00**	**0.00**
	Amended Surface vs. Amended Subsurface	1, 7 vs 3, 5	18%	0.26	0.15
	Amended Surface vs. Natural Surface	1, 7 vs 2, 8	-8%	0.10	**0.05**
	Amended Subsurface vs. Natural Subsurface	3, 5 vs 4, 6	-128%	**0.00**	**0.00**

Potassium	Surface & Subsurface Metal Comparisons	ID #'s	Percent Change	Alpha 1	Alpha 2
K	Natural Surface vs. Natural Subsurface	2, 8 vs 4, 6	-31%	**0.03**	0.06
	Amended Surface vs. Amended Subsurface	1, 7 vs 3, 5	-66%	**0.02**	**0.03**
	Amended Surface vs. Natural Surface	1, 7 vs 2, 8	-6%	0.14	0.06
	Amended Subsurface vs. Natural Subsurface	3, 5 vs 4, 6	16%	0.22	0.12

In addition to the percent change analysis, the Student t-test was run to analyze whether two samples were likely to have come from two underlying populations having the same mean. This investigation was conducted in Excel, utilizing the TTEST function with two tails and a paired analysis. An Alpha value (see Table 5-3, Table 5-4) of 0.05 signifies that there is a 5% probability (or less) that the two samples analyzed camefrom populations with identical means. Alpha values smaller than or equal to 0.05 indicate that the population means are significantly different from one another. Values equal to or less than 0.05 are bold in the above tables (Table 5-3, Table 5-4).

5.1.4 Conclusions / Benefits

Indoor laboratory swale experiments have produced retardance curves (Manning's "n" vs. VR graphs, Figure 5-3 thru 5-6) for Centipede, Kentucky Bluegrass, and Zoysia species for flow depths that will optimize heavy metal removal.

The hydraulic characteristics of grass swales may be more important than type of vegetation for removing heavy metals from stormwater during single storm events. Because of the potentials for both sediment deposition and scouring, swales can improve or deteriorate water quality during storm events.

The outdoor swale tests showed highly variable results. Many of the concentration reductions were quite large, but some "negative removals", possibly associated with scour of previously deposited materials, were also noted.

5.2 Swale Phytoremediation Study Objective

To develop guidelines for the design of vegetated swales for the removal of metals and other contaminants in urban rumoff

5.2.1 Background

Phytoremediation is an innovative technology that is gaining strength in the environmental field for treatment of contaminated water. Phytoremediation involves the use of plants to clean up sites by removing pollutants from soil and water. Plants can degrade organic contaminants and can stabilize metal contaminants by acting as filters or traps. At sites contaminated with metals, plants either stabilize or remove the metals from the soil and water through two mechanisms: phytoextraction and rhizofiltration. Phytoextraction refers to the uptake of metals by plant roots into plant stems and leaves; plants can then be harvested and the metal removed from the system; harvested metal-loaded biomass can be incinerated or composted to recycle the metals. Rhizofiltration is defined as the use of plant roots to absorb, concentrate and precipitate heavy metals from polluted effluents (Raskin *et al.* 1994). Several plant species have been tested in recent years for their potential use in phytoremediation. Crops such as *Brassica juncea* (L.) Czern. (Indian mustard) were shown to possess great potential for phytoextraction of various metals from soils (Blaylock *et al.* 1997; Ebbs *et al.* 1997; Huang *et al.* 1997 ; Raskin *et al.* 1994). Other species like corn (*Zea mays* L.) or pea (*Pisum sativum* L.) were also found to efficiently extract Pb from contaminated soils (Huang *et al.* 1997). Grasses like oat (*Avena sativa*) or barley (*Hordeum vulgare*) were also very proficient in extracting Cu, Cd and particularly Zn from amended soils (Ebbs and Kochian 1998). Similarly, several other grasses and dicotyledonous crops (Indian mustard, sunflower, wild cabbage, tobacco) efficiently removed Cu, Cd, Cr^{6+}, Ni,

Pb and Zn from aqueous solutions by tissue absorption and root-mediated precipitation (in the case of Pb) (Dushenkov *et al.* 1995). Rhizofiltration by Indian mustard and sunflower (*Heliantus annuus* L.) plants was effectively employed to eliminate Uranium from groundwater in greenhouse experiments and in a pilot-scale rhizofiltration system (Dushenkov *et al.* 1997). An additional benefit of the use of plants for remediation purposes is that plant growth favors the development of active microbial communities in soils and sediments, and microorganisms can significantly contribute to the metal remediation process through sorption and precipitation mechanisms (Avery 1995; Beveridge 1986; Garnham 1997; Kapoor and Viraraghavan 1997; Macaskie and Basnakova 1998; Urrutia 1997). These and other studies indicate that phytoextraction/ rhizofiltration may provide a cost-effective method to treat pollutants at concentrations that are too low for efficient removal by conventional methods while too high to allow discharge to the environment (Dushenkov *et al.* 1997). Furthermore, since this methodology is also useful for treating a wide variety of environmental contaminants, phytoremediation may be an appropriate technology for treatment of stormwaters runoff.

Few studies have followed the potential of grass varieties in metal accumulation and uptake. Newberry and Yonge (1996) studied the effect of grass strips in retardation of heavy metals from stormwater in highway runoff. However, no specific variety of grass used is indicated in this study. More recently, the ability of several grass species (Annual Ryegrass, Kentucky Bluegrass and Tall Fescue) to uptake Pb from soil was compared (Gum 1998), showing removal percentages of 18% Pb for Annual Ryegrass, and 16% for Tall Fescue, which are then considered good candidates for phytoremediation of Pb metal from soil.

In this project, we studied the effect of phytoextraction by turf grass varieties in swale systems for removal of metals from stormwater runoff. Centipede (*Eremochloa ophiuroides* (munro) Hack) and Zoysia (*Zoysis japonica* Steud.) were the species tested because they are the varieties most commonly used in the South in home lawns and yards. However, since other species, particularly Kentucky Bluegrass (*Poa pratensis* L.), are more frequently used in northern areas of the country in which most of the existing literature on swale systems was conducted, this variety was tested as a comparison to the southern varieties in the controlled environment of the indoor swales. Previous literature data on Pb retention by these species showed that Kentucky bluegrass (*Poa pratensis* L.) was able to accumulate Pb to a 16.5 % of its dry weight, while Centipede grass (*Eremochloa ophiuroides* (munro) Hack) accumulated up to a 13%, and the Japanese lawngrass (*Zoysis japonica* Steud.) showed an important but slightly lower (6%) retention ability (Dushenkov *et al.* 1995).

5.2.2 Experimental Methods

5.2.2.1 Sample collection from swales (mesocosm and field)

A) Indoor swales. The indoor swales were seeded with three grass varieties: Zoysia (*Zoysia japonica* Steud.), Centipede (*Eremochloa ophiuroides* (munro) Hack) and Kentucky bluegrass (*Poa pratensis* L.). Once the grasses were well established, they were sampled for time zero (before treatment with stormwater) analysis (sampling date: 6/23/00). Duplicate samples were collected with a soil corer (2" x 2"). The soil was separated and retained for water content and ATP determinations. The grasses were washed with distilled water to eliminate any soil residue. The rooting systems were not completely developed in some of the varieties, so plant mass was determined for whole plants, (no separation between shoots and roots). These were dried processed as indicated below. ATP content (as a measure of microbial activity) was determined by the procedures described below.

Determination of swale systems' maximum capacity to retain metals. Conducted through introduction of 'spiked' stormwater to indoor swale systems. Stormwater is collected and stored in 150 Ga. Containers. Then Cu, Pb, Cd and Cr(III) are added in concentrations that represent 100 times greater values than the average content in stormwater; and Zn is added in a concentration approximately 10 times greater than the average Zn content.

Metal	Final concentration mg L^{-1}	Added as
Zn	5	$ZnSO_4.7H_2O$
Cu	5	$CuCl_2.2H_2O$
Pb	1	$Pb(NO_3)_2$
Cd	0.25	$Cd(NO_3)_2.4H_2O$
Cr(III)	0.25	$CrCl_3$

The stormwater was stirred for 2 hours and metals allowed to equilibrate with particulates within the stormwater for 24 hours. Then the stormwater was run through the swale seeded with Kentucky Bluegrass for 2 hours. Runoff water was collected every 2 feet down the swale slope (4 samples); the final runoff was also collected both the first flush and after saturation (2 samples). The treated stormwater is maintained within the swale to allow for metal retention by soil and plants. The treated swale was sampled until the soil run dry. Bluegrass was sampled at times zero, 2, 6 and 9 days. All spiked-stormwater and runoff samples are analyzed for pH, TOC and metals following the methods included in Table 5-5 below. Grass samples were analyzed for metal content as explained above and soils for pH, and cation exchange capacity by the methods indicated below in Table 5.5. Similar procedures were repeated on the Zoysia and Centipede indoor swales as natural stormwater was available (by rainfall events) during the fall of 2001. Zoysia and Centipede were sampled at 2, 6, 9 and 13 days after addition of the spiked-stormwater.

Table 5-5. Summary of analytical methods for swales task no 6 (phytoextraction studies)

Measurement		Type	Method	Instrumentation
Waters	pH	Critical	EPA 150	pH meter
	Metals (Pb, Cu, Cr, Cd, Zn)	Critical	S.M. 3500	ICP/MS
	pH in water	Critical	EPA 9045C	pH meter
Soils	CEC	Critical	EPA 9080	ICP/MS
	ATP	Non-critical	PA extraction	ATP luminometer
	Dry mass	Critical		Balance
Grasses	Metals and cations	Critical	S.M. 3500	ICP/MS

B) Outdoor swales. Outdoor swales were sampled on the following dates:

- Time zero (**t_0**): July 2000 for Outdoor swale #1; October 2000 for Outdoor swale #2.
- Time two (**t_2**): June 2001
- Time three (**t_3**): November 2001

5.2.2.2 Metal analysis on plant tissues

Known volumes of plant and root material were collected using 2" x 2" AMS soil core samplers with butyrate plastic liners. Soil and plant material were separated, the soil collected for moisture, ATP, and CEC determinations. Plant roots and shoots were also separated after washing off the remaining soil particles with distilled water, rinsed with ultra pure (18 ?) distilled water and dried at 80° C for 2 days for determination of dry weight (root or shoot) biomass. Dried tissue was then calcinated at 500° C for 18 h. The ash was dissolved in HNO_3 acid (Optima Ultra pure), diluted to a 2% concentration, filtered by a 0.2 µm sterile membrane filter, and metals analyzed by ICP.

5.2.2.3 Microbial activity in soils

Microbial activity and metabolism in all swale plots was evaluated by following production of ATP in the soil throughout the duration of the experiment. Two methods were used:

A) Phosphate-sulfuric acid method of Eiland (Eiland 1983).

Reagents

Buffer A: Tris buffer (250 mM, pH 7.5) with 4 mM EDTA
Buffer B: Tris buffer (25 mM, pH 7.5) with 0.4 mM EDTA, sterilized by autoclaving for 20 min at 121 C
Extractant A: 27.3 ml of 95% H_2SO_4 and 44.5 g $Na_2HPO_4.2H_2O$ in 1000 ml distilled water
Extractant B: 0.005% solution of alkyldemethyl-benzylammonium chloride in buffer B

ATP standards: disodium ATP in buffer B, 0.02-0.2 ug mL^{-1}

Enzyme solution: luciferin-luciferase enzyme mixture (from Sigma) made up in 5 ml of bacteria-free distilled water.

Extraction of ATP. Add 1 g of soil/sediment sample to a 50 ml plastic centrifuge tube containing 10 ml of extractant A; make three replicates for each sample. Shake tubes in an ice bath on a reciprocating shaker table (200 rpm) for 15 min. Homogenize the suspensions by vortexing for 5 sec and immediately pipet a 50 µl portion of suspension into 1.5 ml of buffer A in a microcentrifuge tube. Store frozen until ATP analysis.

Preparation of internal ATP standard. Add 1 g of soil/sediment sample to each of 3 tubes containing 9.5 ml of extractant A and 0.5 ml of standard ATP solution (10 µg ml^{-1}). Shake tubes in an ice batch on a reciprocating shaker table for 15 min. Homogenize the suspensions by vortexing for 5 sec and immediately pipet triplicate 50 µl portions of suspension into 1.5 ml of buffer A in a microcentrifuge tube. Store frozen until ATP analysis.

Measurement of ATP (equal procedure for samples and standards)

Thaw the frozen samples. Homogenize the contents of the microcentrifuge tubes by vortexing, and then pipet 150 µl into a photometer cuvette containing 50 µl of extractant B. Place the cuvette into the photometer chamber, and then add 100 µl of the luciferin-luciferase enzyme using the special autopipettor which fits into the photometer. Light emission is measured over a 10 sec integration period. If maximal light intensity is not reached during the first ten seconds, light emission should be measured over a 30 sec integration period.

Calculations. Plot the concentration of ATP in the standards against the corresponding photometer readings and calculate best-fit straight line by linear regression analysis. Use the regression equation to calculate the ATP content of the extraction mixtures, and then correct these values for sample dilutions and soil/sediment water content and express them in µg ATP per ml of wet soil/sediment. Convert µg ATP ml^{-1} to biomass-C ml^{-1} assuming a C:ATP mass ratio of 250.

B) The PA extraction procedure (Horwath & Paul 1994).

The components of the PA mixture are designed to quench metabolic processes by destabilizing cell compartmentalization and solubilizing ATP. The H_3PO_4 extracts ATP, inactivates ATPasas and, saturates phosphate-binding sites. EDTA chelates metal ions and prevents inhibition of the luciferin-luciferase reaction. Adenosine saturates ATP-binding sites. Urea quenches metabolic reactions by denaturing enzymes that catalyze the metabolism of nucleotides. Dimethylsulfoxide (DMSO) and detergents remove and lyse cells from soil surfaces.

Reagents

BRIJ 35 detergent (Across organics), 0.5 g of in 34 ml of water at 45° C.

10 N phosphoric acid, 228 ml per liter of water

10 M Urea, 600 g per liter of water

DMSO

Adenosine, 5 g per liter

1 M EDTA, 45.22 g per 100 ml

Water used to make the above solutions is purified (18 mΩ), filtered through 0.2 µm, and autoclaved.

Reagents are combined in the following order:

34 ml BRIJ 35

20 ml of 10 N phosphoric acid

20 ml of 10 M Urea

20 ml DMSO

4 ml adenosine
2 ml of 1 M EDTA

1. M Tricine buffer, pH 11.2

25 mM Tricine buffer at pH 7.8 with 5 mM $MgSO_4$, 1 mM EDTA and 1 mM dithiothreitol.
Procedure. Soil samples (1-2 g wet soil) are combined with 10 ml of the extractant in a sterile centrifuge tube and shaken vigorously for 30 min. The samples are then centrifuged (30,000 x g, 20 min). The supernatant is then diluted to 100 ml with 0.1 M Tricine buffer at pH 11.2 to yield the desired pH between 7-8.

ATP is measured in a reaction vessel containing 50 μl sample extract, 50 μl Tricine buffer pH 7.8 and 100 μl fire fly luciferase.

In our samples, light production was measured in an ATP photometer with 10 sec integration time.

5.2.3 Results and Discussion

5.2.3.1 Optimization of ATP extraction procedure for Fe-rich local soils.

Important effort during this period was placed on optimizing the procedure for extraction of ATP from the local soils used on the experimental swales (both indoor and outdoor). The procedure used initially (Eiland 1983) was providing low recoveries for internal standards, lower than those reported in the literature for this procedure. This is most certainly the result of the abundance of Fe(III) oxides in the local soils, since these minerals are well known sorbents for organic and inorganic components. In fact, Eilan's procedure uses a combination of sulfuric acid and phosphate to extract ATP from soils in an attempt to compensate for possible adsorption; the phosphate is added to saturate sorption sites that could bind the ATP extracted from the microbial cells. However, since our local soils are extremely rich in Fe oxides, it is possible that this amount of phosphate is not enough to prevent adsorption of the ATP extracted from microbial cells into the oxides, which is, therefore, not quantified. To evaluate this possibility, experiments were conducted with our swale soils using Eiland's extractant to which increasing concentrations of phosphate were added, up to a (4x) concentration from the amount proposed in the original method. The efficiency of these modified extractants was evaluated by using internal ATP standards, that is, natural soils to which known concentrations of ATP were added experimentally. Results from these trials offered once again very low recoveries for the internal standards (less than 10%), even at the highest phosphate concentrations (4 times that in the original procedure). It was then concluded that this method for ATP extraction was not appropriate for our Fe-rich soils due to adsorption of the ATP molecules liberated from the microbial cells to the soil matrix, which results in severe underestimation of the soil ATP content and microbial activity.

In consequence, we searched for an alternative extraction procedure for soil ATP. Among several possibilities, the PA extraction method (Webster *et al.* 1985; Horwath & Paul 1994) was chosen because it uses a complex mixture of reagents specifically designed to saturate ATP-binding sites in the soil matrix and complex metal ions that interfere with the luciferin-luciferase reaction. The procedure was optimized for use with our samples. Results for the soils of the outdoor swales #2 were much more satisfactory, with good internal standard recoveries (Table 5-6). ATP extractions for the previously sampled swales soils (outdoor #1 and indoor) were then repeated using the PA extraction method.

Table 5-6. Dry weight (g) and ATP (μg/g dry soil) determinations for time zero samples in outdoor swales #2. Values indicate means ± standard deviation for the mean

Sample	Grasses Dry weight (g)	% of total biomass	Soil water g water/g soil	μg ATP/g dry soil[&]
Centipede 1 Shoots	2.86	62.25	0.134 ± 0.005	0.543 ± 0.145
Roots	1.73	37.75		
Centipede 2 Shoots	2.69	69.71	0.128 ± 0.043	0.717 ± 0.00*
Roots	1.17	30.29		
Zoysia 1 Shoots	1.07	25.72	0.115 ± 0.016	0.298 ± 0.048
Roots	3.11	74.28		
Zoysia 2 Shoots	2.02	49.76	0.158 ± 0.003	0.536 ± 0.095
Roots	2.04	50.24		

[&] determined using the PA extraction procedure (Webster *et al.* 1984)

5.2.3.2 Metal content in plant tissues

A) Outdoor swales. The metal content of the grass samples collected from the outdoor swales is presented below for all three sampling times, for both top and bottom of swales.
Chromium. Cr content in grass shoots (Figure 5-10) is relatively low throughout the sampling period and somewhat variable over time. In general, both species seem to accumulate similar amounts, ranging from 5 to 25 mg Cr per Kg of dry shoot material.

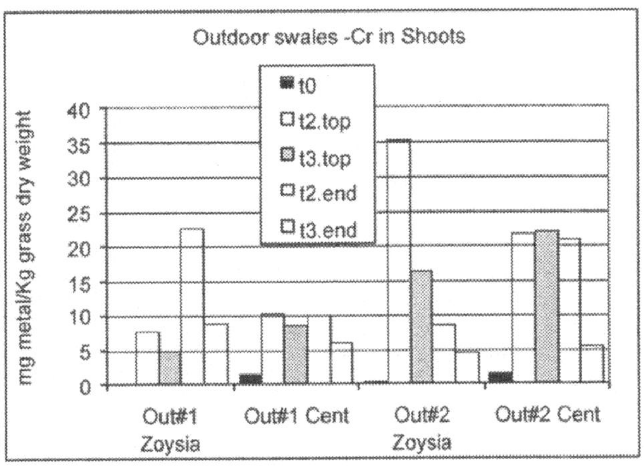

Figure 5-10 Cr content in shoot tissues of grass samples collected from the Outdoor swales at t0, t2 and t3 sampling dates.

Cr content in grass roots (Figure 5-11) is very low (1 to 10 mg Cr/Kg dry weight) and essentially similar for both grass varieties and locations.

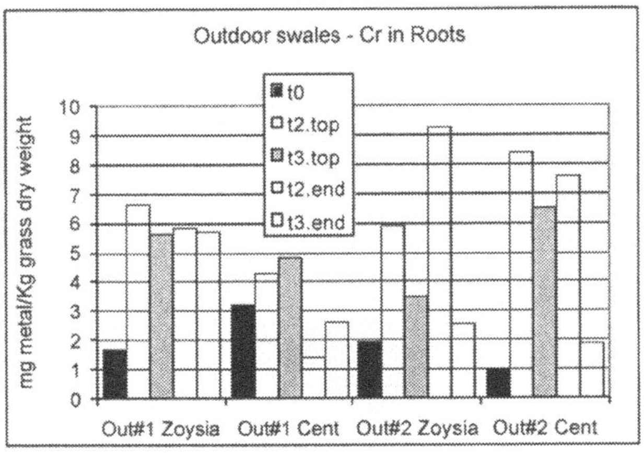

Figure 5-11 Cr content in root tissues of grass samples collected from the Outdoor swales at t0, t2 and t3 sampling dates.

The location of the Outdoor #2 system showed greater metal contents during the summer of 2001 (t2) than during the fall sampling (t3) (Table 5-7). Greater dryness in the soil is indicative of lack of rainfall, in this case, during the end of summer and fall of 2001.

Table 5-7. Water content in soils (%) at each sampling time.

Outdoor#1	**July 2000 – t0**	**June 2001 – t2**	**Nov 2001 – t3**
Zoysia, 1.1	11.7	16.2	5.1
1.2	12.5	18.0	5.6
2.1	13.4	16.8	7.0
2.2	13.6	17.2	6.0
Centipede 1.1	10.2	19.5	6.4
1.2	10.4	17.3	7.5
2.1	12.0	18.3	7.0
2.2	12.1	17.9	7.8
Bare soil 1.1	10.2	15.2	6.1
2.1	9.6	14.8	4.5
Outdoor#2	**Oct 2000 – t0**	**June 2001 – t2**	**Nov 2001 – t3**
Zoysia, 1.1	13.0	15.6	4.5
1.2	13.7	17.1	4.6
2.1	9.7	16.5	3.1
2.2	15.8	16.9	3.6
Centipede 1.1	10.3	16.5	1.4
1.2	12.6	20.6	8.3
2.1	15.6	19.6	6.8
2.2	16.0	18.3	8.9
Bare soil 1.1	10.3	16.0	7.3
2.1	11.1	15.2	8.8

Lead. Pb contents in grass shoots for the outdoor swales (Figure 5-12) was always below 90 mg Pb/Kg dry weight, which is characteristic of non-metalloaccumulators (ref). Particularly greater concentrations were found at the top of the Zoysia swale for the location #2, indicating a possible greater source of contaminant metal in this location than in nearby location #1 in which concentrations of Pb in shoot tissues were always below 40 mg/Kg.

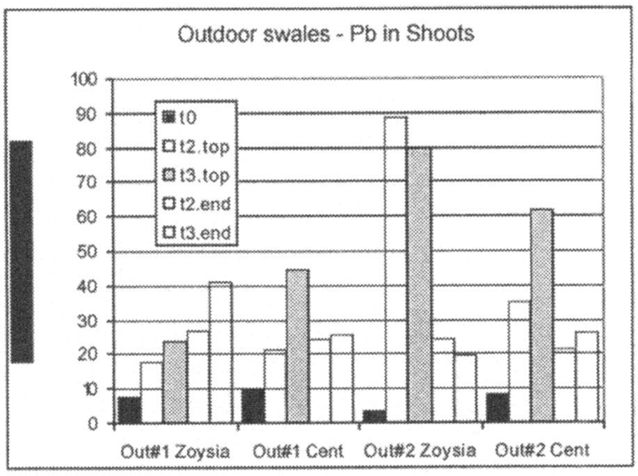

Figure 5-12 Pb content in grass shoots of samples collected from the Outdoor swales at t0, t2, and t3 sampling dates.

No substantial differences in Pb accumulation are apparent between both varieties of grass within location #1 but Zoysia appears to be able to translocation Pb to its aerial tissues better than Centipede in location #2 (Figure 5-13). No clear reason for this difference is currently apparent.

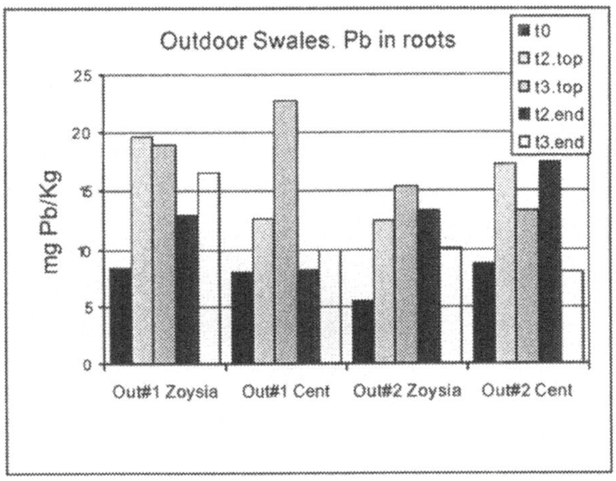

Figure 5-13 Pb contents in roots of grass samples collected from the Outdoor swales during the t0, t2 ad t3 sampling times.

Pb contents in root tissues are lower than those found in aerial parts of the plant, ranging from 5 to 23 mg/Kg (Figure 5-13). Pb accumulation in roots was relatively similar within locations and for both varieties and, although there is a certain train to greater accumulations on those samples collected at the top of the swales (that first receive the stormwater influx) this is not a significant difference.

The data obtained on Pb content by these grasses indicates a greater degree of translocation of Pb from roots to aerial tissues, which is a relatively rare phenomenon in plant species (Baker *et al.* 2000). This is due to the chemical characteristics of lead, which easily precipitates at circumneutral pH with ions like phosphate or sulfate, which are readily abundant in plant roots. Despite this, there are plant varieties which act as Pb hyperaccumulators (e.g., *Thlaspi rotundifolium, Alyssum wulfenianum*) (Baker *et al.* 2000). Even though grasses are not particularly hyperaccumulator, previous data was obtained in which grass varieties showed greater accumulation of Pb in aerial parts than expected (Deram and Petit 1997; Williams *et al.* 1977; Barry and Clark 1978), in agreement with our results herein. The grass *Arrhenatherum elatius* was reported to hyperaccumulate Pb, being the first report of such phenomena in the *Poaceaei (*Deram and Petit 1997). Large values of Pb contents in aerial tissues were also reported for *Festuca ovina* (130 to 11750 mg/Kg) (Barry and Clark 1978) and pasture species growing on mining waste in the UK (mean shoot Pb concentration of 13488 mg/Kg) (Williams *et al.* 1977). In comparison with these numbers from the literature, our results show a low degree of Pb accumulation in the plant shoots (below 100 mg/Kg); however, the fact that Pb contents are greater in shoots than in the roots of the grasses suggest an accumulation nevertheless, and therefore emphasize that common lawn grass varieties may still be useful for alleviating low levels of Pb input associated with stormwater runoff.

Copper. Cu content in plant shoots is presented in Figure 5-14. For this metal, there is a clear effect due to the location of the swale. Location #2 shows greater Cu content in the plant shoots than location #1, for both grass varieties.

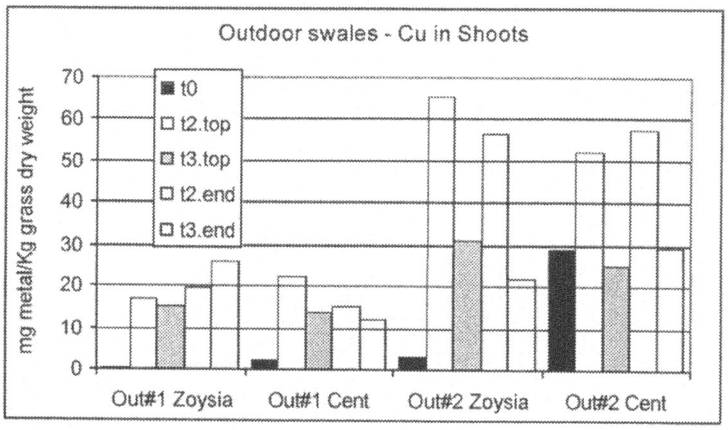

Figure 5-14Copper content in shoots of grass samples collected from the outdoor swales at t0, t2 and t3 sampling dates.

Cu ranged from 1 to 28 mg/Kg at the Location #1 for both Zoysia and Centipede, while ranged from 2 to 65 mg/Kg for location #2. The top of the swales at location #2 presented approximately double Cu contents in plant material than those plants located at the bottom of the swale. Since the input of stormwater takes place mainly through the top (head) of the swale system, these results indicate an effective retention mechanism of the metal by soil/plants at the initial reception point that then reduces the amount of metal available for those plants at the bottom of the system. Both grass varieties seem to present similar responses (Figure 5-14).

Parallel differences are observed in root metal content (Figure 5-15). Cu contents range from 2 to 25 mg/Kg for location #1 and 6 to 60 mg/Kg in location #2. Again, similar responses were obtained for both grass varieties at each location, which indicates differences in Cu input in the stormwater

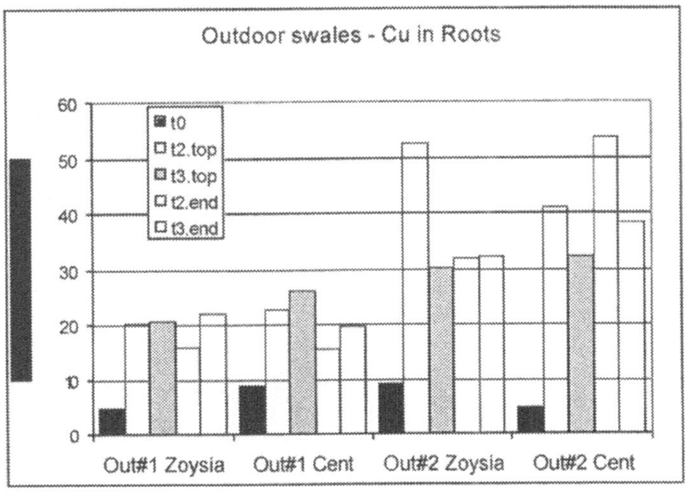

Figure 5-15 Cu content in roots of grass varieties collected from the outdoor swales at t0, t2 and t3 sampling dates.

Unlike Pb, Cu contents were approximately equal in grass root and shoot systems. Again, both grass varieties seem equally apt to accumulate Cu from stormwater runoff sources.

Cadmium. Cd content in plant tissues in outdoor swales was extremely low, always below 1 mg/Kg, both in shoots (Figure 5-16) and roots (Figure 5-17).

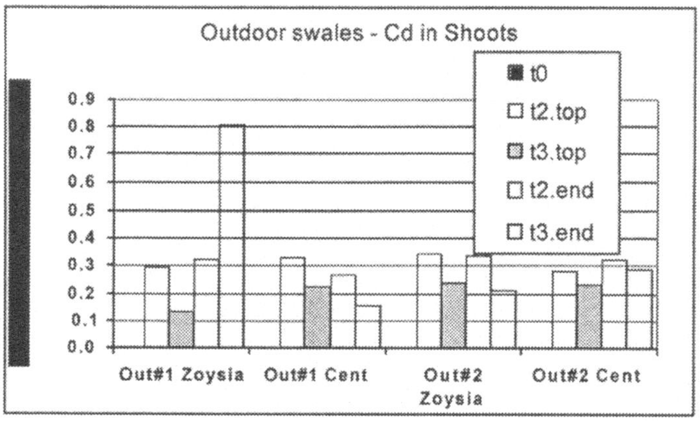

Figure 5-16 Cd content in shoots of grass samples collected from the outdoor swales on t0, t2 and t3 sampling dates.

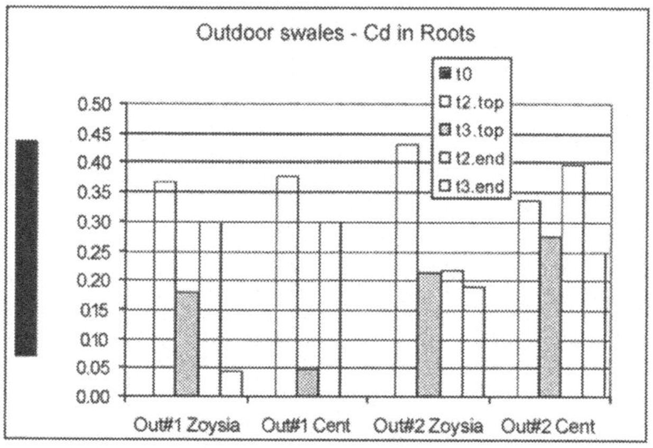

Figure 5-17 Cd content in roots of grass samples collected from the outdoor swales on t0, t2 and t3 sampling dates.

Overall, Cd content in roots has mean values between 0.2-0.3 mg/Kg while mean values in shoots range from 0.15 to 0.25 mg/Kg. Values greater than 100 mg/Kg are considered hyperaccumulation while contents between 10 to 100 mg/Kg are common in plants grown in lead/zinc mineralized soils (Baker *et al.* 2000). In sight of these numbers, the grass varieties here studied present no signs of Cd accumulation or contamination.

B) Indoor swales. Experiments with the indoor swales were conducted with the intention of obtaining maximum metal retention abilities by the systems, both by sorption to the soil and by uptake/accumulation into the plant tissues. All three grass varieties, Bluegrass, Centipede and Zoysia were used, as described in the methods section.

Chemical composition of the runoff waters after passing through the swales in comparison with the composition of the initial spiked stormwater show great variation in the efficiency of metal retention within the swale systems depending on the metal in question, for all three grass varieties (Table 5-8) Pb was the metal retained with the best efficiency (57 to 65%), followed by

Cu (~40% retention), Cd (21-28% retention), Zn (13-26%) and Cr (14-16% for Zoysia and Centipede, 37% for Bluegrass).

Table 5-8. Percentage of soluble metal in the stormwater retained by the indoor swales during the indoor spike experiments. These values were calculated from the concentrations of metals in the input stormwater and final runoff, after 24 hours contact time.

Metal retained (%)	Cu	Cr	Pb	Zn	Cd
Zoysia	39.8	16.3	64.8	13.3	21.1
Centipede	39.0	13.8	57.1	20.4	27.9
Bluegrass	40.3	36.8	66.7	26.1	25.2

While the concentration of metals decreased on passing through the swale systems, the concentrations of Fe and basic cations (Ca, Mg, Na, K) in the runoff waters was substantially increased, as a clear evidence of cation exchange processes taking place within the soil to counteract the addition of heavy metals. Cation exchange is one of the well known processes taking place in soils that counteract external inputs of elements and contribute to soil pH buffering. Increases in Fe and basic cations for the three grass varieties are included in Table 5-9.

Table 5-9. Increase in basic cations and Fe in runoff waters from swale systems after addition of heavy metal-spiked stormwater. Values are expressed as % from the initial element content in each particular stormwater.

Increased concentration (%)	Fe	Na	Mg	Ca	K
Zoysia	6.0	23.3	17.4	12.4	75.8
Centipede	44.5	61.7	86.9	44.4	124.6
Bluegrass	338.5	77.2	51.9	17.2	23.4

For Zoysia (Figure 5-18) the metal retained in a greater extent was Pb, with a 65% extraction efficiency overall. These numbers are even higher at the top of the swale systems (drain 1) in which metals were almost completely retained within the swale. Cr retention was relatively small.

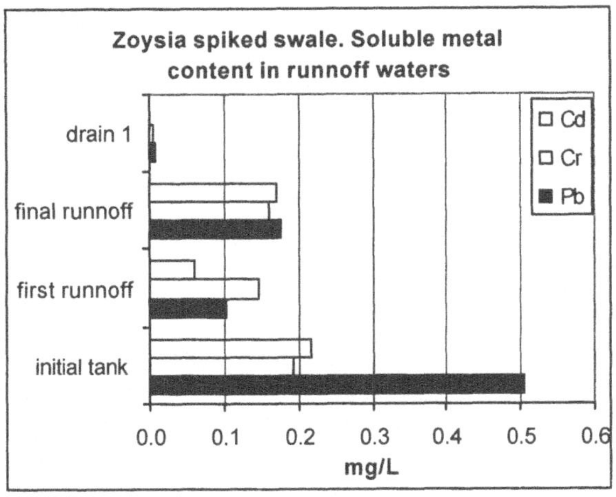

Figure 5-18 Soluble fraction of Pb, Cd and Cr in spiked-stormwater (initial tank) and runoff waters from the Zoysia indoor swale.

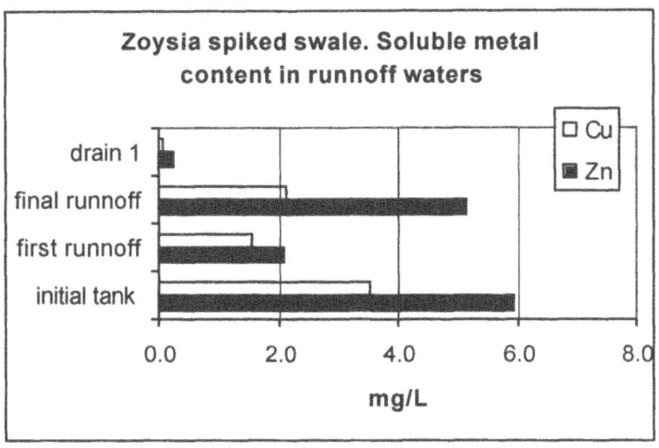

Figure 5-19 Soluble fraction of Cu and Zn in spiked-stormwater (initial tank) and runoff waters from the Zoysia indoor swale.

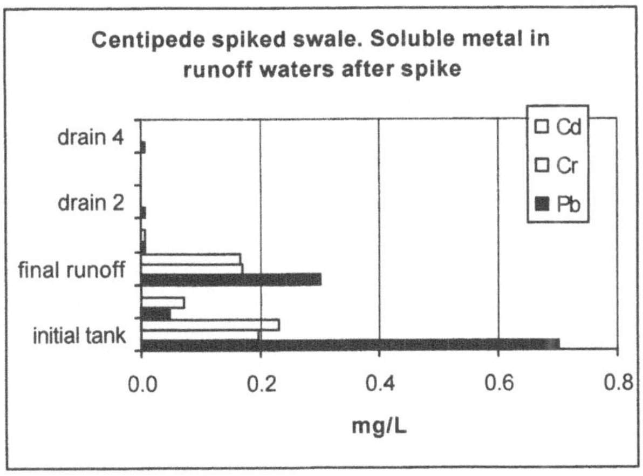

Figure 5-20 Soluble Pb, Cd and Cr remaining in runoff waters of the Centipede indoor swale after spike with metal-enriched stormwater (initial tank).

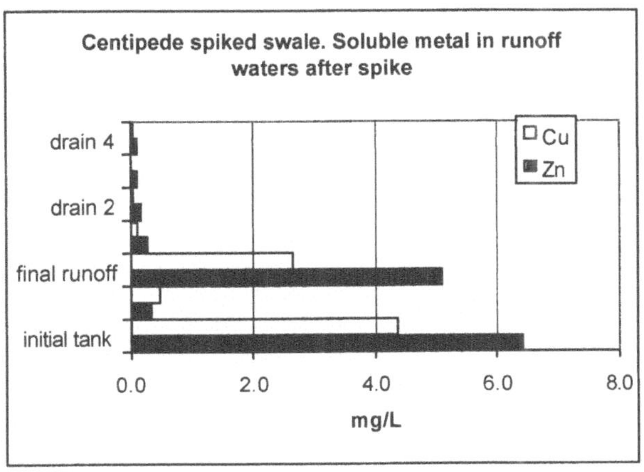

Figure 5-21 Soluble Cu and Zn remaining in runoff waters of the Centipede indoor swale after spike with metal-enriched stormwater (initial tank).

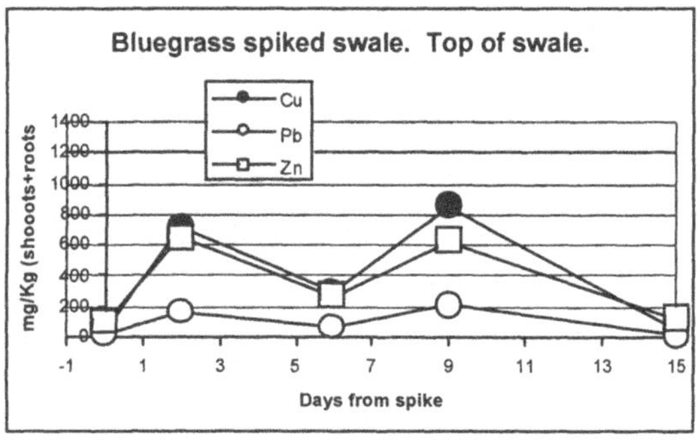

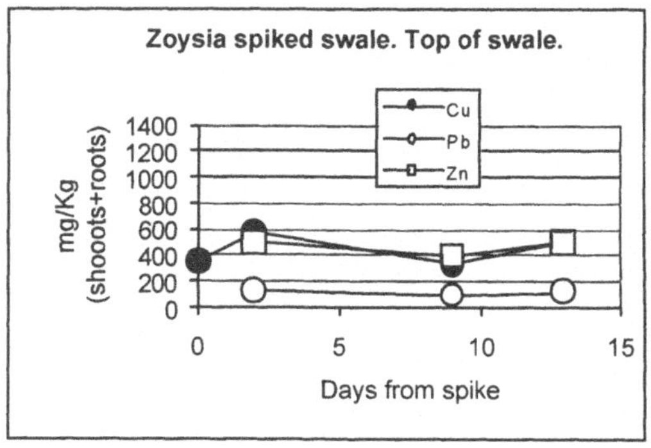

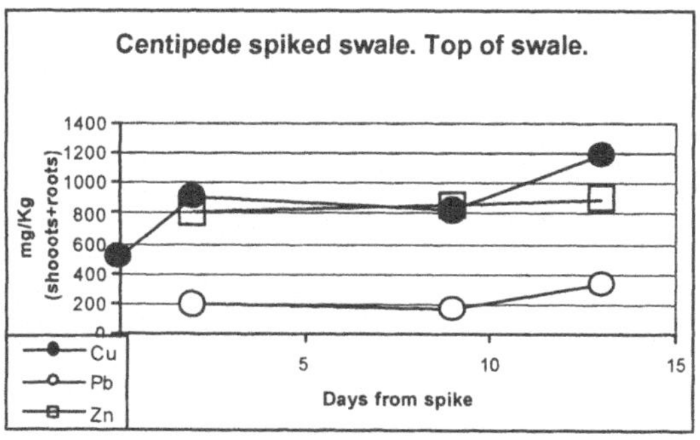

Figure 5-22 Cu, Pb and Zn content in plant tissues (shoots and roots) for the three grass varieties over time after the addition of the metal-spiked stormwater.

These results indicated that the Centipede variety was able to uptake more Cu and Zn than Zoysia or Kentucky Bluegrass, and slightly more Pb; Cd contents were similar for all three varieties (~ 30 mg/Kg); the ability of accumulate Cr was extremely low in all three cases (~ 10 mg/Kg).

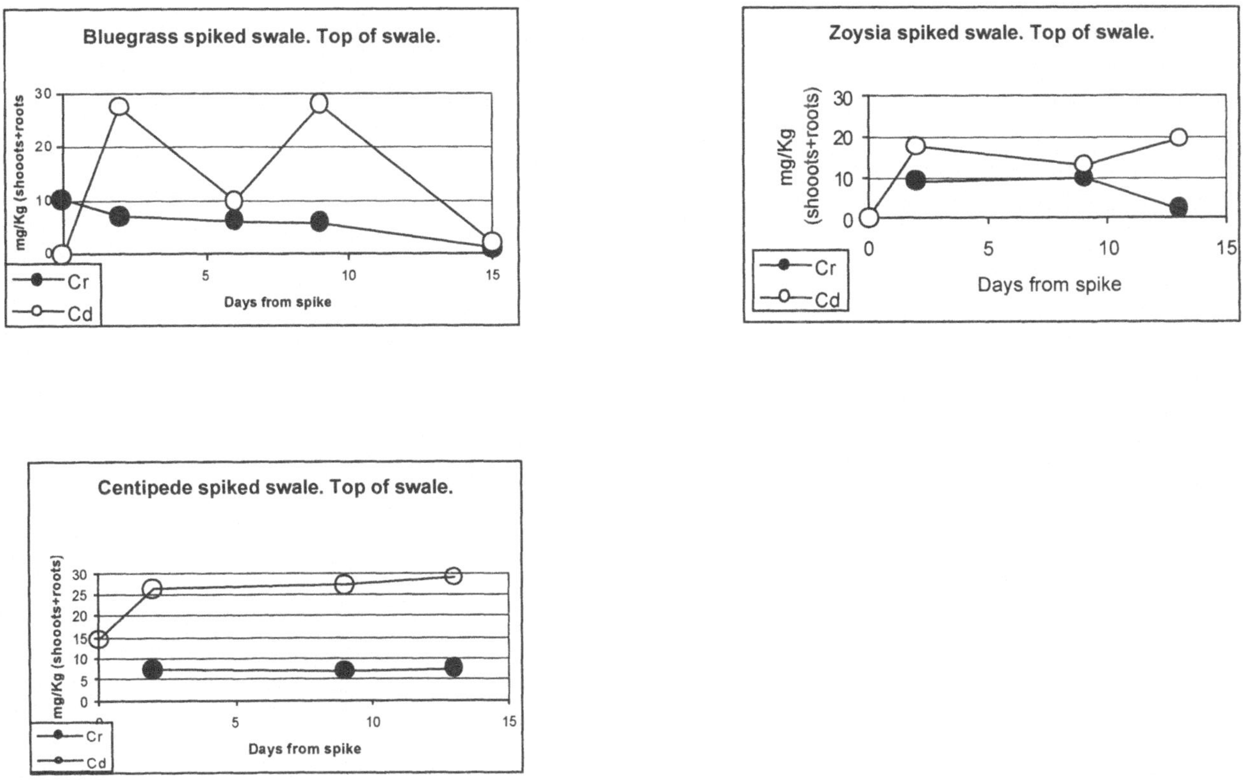

Figure 5-23 Cr and Cd content in plant tissues (shoots and roots) for the three grass varieties over time after the addition of the metal-spiked stormwater.

5.2.4 Conclusions

Both Centipede and Zoysia grass varieties, common in many lawns in the Southeastern US have certain capability to uptake metals from soil that receives stormwater. Experiments with metal spiked stormwater, however, suggested that Centipede had a greater maximum ability to accumulate Cu and Zn than Zoysia and Kentucky Bluegrass, most used in northern areas of the country. Centipede seemed to also be able to accumulate more Pb. Cd and Cr accumulated in all three at similar rate and in very small amounts (~30 mg/Kg and 10 mg/Kg, respectively). Zoysia accumulated the least Pb, Cu and Zn of all three varieties in these studies.

Even though the data obtained in the outdoor swales after the first year seemed to indicate that Zoysia was a more efficient metal accumulator, these final results show that this is not the case;

Zoysia grass accumulated the metal faster than Centipede (see t2 vs t3 results in outdoor swales figures), but by the second year this difference was not apparent anymore.

In conclusion, this study presents preliminary data on the role of common grass varieties in phytoremediaton of metals in contaminated soils and waters, particularly in the low metal concentration environment of stormwater runoff. This is of interest because little information is still available on the role of common plant species in this process, since the field of phytoremediation is a relatively new one. The data obtained suggests that there is a relatively similar behavior among the different grass species. In other words, any grass added to the surface of a swale system would represent a positive influence on metal up take and purification of the runoff water. However, if a recommendation should be made for systems designed to have a long life as filters for metal accumulation from stormwater, Centipede could possibly be the best choice for its resilience to drought, its nutritional frugality, and its greater ability to accumulate key contaminant metals such as Cu, Zn and Pb.

CHAPTER 6.0

TREATABILITY OF HEAVY METALS

6.1 Measurements of Heavy Metal Associations with Particulates

The objectives of these tests were to determine the associations of heavy metals (along with some major constituents and nutrients) with different-sized particulates in stormwater. The binding strengths of these metals to the particulates were also examined by using a sequential extraction procedure using different acids and bases under several pH values. Also, experiments examining the characteristics of the filterable (<0.45 μm) portion of the heavy metals were also conducted. Appendix H summarizes much historical information concerning the characteristics of stormwater that affects its treatability. That appendix is an excerpt from an on-going EPA research project that is examining advanced filtration techniques for stormwater treatment.

6.1.1 Introduction

Many studies have identified metals in urban runoff as a major contributor to the degradation of urban streams and rivers. Metals of most concern in urban runoff are copper, cadmium, chromium, lead, mercury, and zinc. Of these metals, copper and zinc are currently the most prevalent in terms of occurrence and concentration in urban runoff with trace quantities of lead, nickel, cadmium and chromium commonly found in all areas of the United States. Metals in urban runoff can occur as dissolved, colloidal and particulate-bound species. Therefore, it is important to measure all forms of heavy metals, especially the particulate and filterable fractions, when determining their fate and effects. If possible, associations of the metals with different particle sizes should also be determined. Finally, to obtain the most meaningful data on either bioavailability or toxicity, it is important that chemical speciation techniques be applied (Florence and Bately 1980). Chemical speciation may be defined as the determination of the individual concentrations of the various chemical forms of an element when together make up the total concentration of that element in a sample. Speciation of metals is dependent upon chemical and physical parameters such as pH, temperature, and the presence of ligands and particulates. Depending upon the chemical form of the metal, a water with a high total metal concentration may be less toxic than another water with a lower total metal concentration (Florence and Bately 1980).

The threat from metals to humans and aquatic life is due to their toxicity, persistence and bioaccumulation. It is important to determine the speciation of a metal because the toxicity of many metals is related to their speciation and valence state. Most metals are essential nutrients for living cells, but only in small quantities. When metals are present in excess, they can become cumulative toxins. Other metals, such as lead and mercury, have no nutritional value and are always considered dangerous, even at relatively small concentrations. Associations with fine particulates also affects heavy metal toxicity to many forms of aquatic life.

The development of analytical techniques which can reliably measure the concentration of the various chemical forms of a trace metal in a water sample is a challenging problem. The analytical schemes developed by Florence and Batley (1980), Florence (1977) and Bott (1995) enable associations of metals to different categories to be measured. Their schemes do not provide a detailed species distribution, but rather an estimation of the fraction of the metal present in toxic forms. Florence and Bately (1980) used Anodic Stripping Voltammetry (ASV) with a chelating resin separation and UV irradiation to examine the dissolved (<0.45 micron) metal fractions, as shown in Figure 6-1. A chelating resin (Chelex-100) was used to separate ionic metals from metals strongly bound in metal-ligand complexes, or strongly adsorbed to colloidal particles. The UV-irradiation was applied to the water at a pH of 4.8 to cause dissociation of the metals from organic complexes, or colloids. The use of a chelating resin such as Chelex-100, in combination with ASV, has been utilized by many researchers, and various forms of the resin have been characterized (Figura and McDuffle 1977, 1979, 1980; Yousef, *et al.* 1985). The methods used during this project were modified, based on the Florence and Bately scheme.

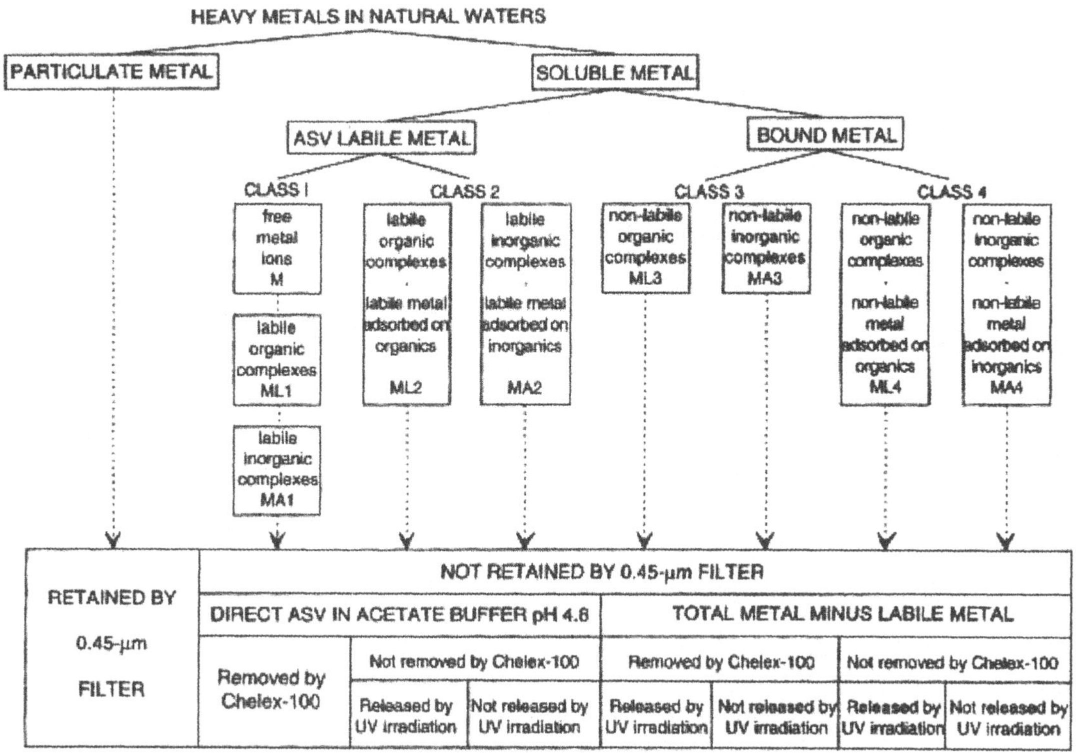

Figure 6-1. Diagram of sequential extraction (Florence and Batley 1980), as presented by Bott (1995).

The main purpose of treating stormwater is to reduce its adverse impacts on receiving waters. One of the most important characteristics affecting the treatability and fate of heavy metals in stormwater is their association with different particle sizes. Most stormwater treatment methods includes the physical removal of particulates. Since many metals of interest are associated with particulates in stormwater, knowledge of the critical size of these particulates is crucial for the design of stormwater treatment. Basic particle size distributions of stormwater need to be known as well as the associations of the metals with the different sizes. Finally, the strength of the associations of the heavy metals with the particulates under varying pH conditions were examined as part of these tests. In a related series of tests during the filtration evaluations, other measurements were made examining the retention of heavy metals onto filtration media under both aerobic and anaerobic conditions.

6.1.2 Methods

6.1.2.1 Sample Collection and Rain Conditions

Stormwater samples were collected from sites in the Birmingham and Tuscaloosa, AL, areas, as shown in Table 6-1. Eight stormwater samples were collected during June and July of 2001, and are listed as sample numbers 1 to 8 in Table 6-1. Collection sites for these first eight samples included an apartment complex parking lot, an industrial area, shopping center parking lot and a campus parking lot. Highly mixed sheet flow runoff was collected into 10L plastic containers and kept stored under refrigeration until analysis. All samples were taken at, or near, the beginning of the rains which were all of moderate to high intensity. A second set of samples were collected in June and July of 2002 and are listed as samples 11 to 18 in Table 6-1. All samples from this second set were collected from storm drain outlets from either commercial or residential drainage areas. All samples were collected during heavy rain events. Samples were collected in 10L plastic containers and stored under refrigeration until analyzed.

Table 6-1. Sample Collection Information

Sample No.	Sampling date	Site Description/Drainage Area	Type of Sample	Rain History
1	June 22, 2001	Apartment building parking lot, Center Point area of Birmingham.	Sheetflow runoff in front of storm drain.	No rain for approx. 2 weeks. Sample taken at beginning of storm.
2	June 28, 2001	University parking lot, Birmingham.	Sheetflow runoff before storm drain.	Recent rain within few days. Sample taken at beginning of storm.
3	July 5, 2001	Semi-industrial area, Trussville, AL (east of Birmingham).	Sheetflow runoff from street in front of storm drain.	Rain previous day. Sample taken approx. 1 hr after start of rain.
4	July 5, 2001	Shopping center parking lot, Trussville.	Sheetflow runoff in front of storm drain.	Rain previous day. Sample taken approx. 1 hr after start of rain.
5	July 20, 2001	Home improvement shopping center parking lot near wood storage area, Homewood, AL .	Sheetflow runoff in front of storm drain.	No rain for approx. 2 weeks. Sample taken at beginning of storm.
6	July 20, 2001	Shopping center parking lot, Homewood.	Sheet flow runoff.	No rain for approx. 2 weeks. Sample taken approx . 10 min after start of rain.
7	July 20, 2001	Residential neighborhood, Birmingham.	Sheet flow runoff from street.	No rain for approx. 2 weeks. Sample taken approx. 30min after start of rain.
8	July 20, 2001	University parking lot, Birmingham.	Sheet flow runoff in front of storm drain.	No rain for approx. 2 weeks. Sample taken approx. 35min after start of rain.
9	June 26, 2002	Residential/Commercial Tuscaloosa, AL	Storm Drain Outlet	No rain for approx. two days. First sample taken 45 min after it started raining. All samples taken while still raining.
10	June 26, 2002	Commercial Tuscaloosa, AL	Storm Drain Outlet	
11	June 26, 2002	Commercial Tuscaloosa, AL	Storm Drain Outlet	
12	June 26, 2002	Residential Tuscaloosa, AL	Storm Drain Outlet	
13	June 26, 2002	Commercial Tuscaloosa, AL	Storm Drain Outlet	
14	July 11, 2002	Commercial Tuscaloosa, AL	Storm Drain Outlet	No rain for approx. a day and a half. The first sample was taken 15 min after it started raining. Rain lasted for an hour after the last sam ple was taken.
15	July 11, 2002	Commercial Tuscaloosa, AL	Storm Drain Outlet	
16	July 11, 2002	Residential/Commercial Tuscaloosa, AL	Storm Drain Outlet	
17	July 11, 2002	Commercial Tuscaloosa, AL	Storm Drain Outlet	
18	July 11, 2002	Residential	Storm Drain Outlet	

2: Sample collection information

6.1.2.2. Stormwater Analyses

Each of the first eight stormwater samples was analyzed in four separate experiments, according to the flowchart shown in Figure 6-2:

- The initial characterization of the stormwater sample.
- The sample is sieved/filtered and a portion of each filtrate is preserved for further analysis.
- The metal-binding strength of the particles in the sample is investigated by exposing the water to various pH levels using a weak acid, strong acid, and a strong base.

- ◆ Ionic and colloidal metal fractions in the water samples are separated by exposure to both ultraviolet irradiation and a chelating resin.

The second set of samples were not tested for metal-binding strength nor were the ionic and colloidal metal fractions separated by exposure to ultraviolet irradiation and a chelating resin. The following sections describe these tests and the results

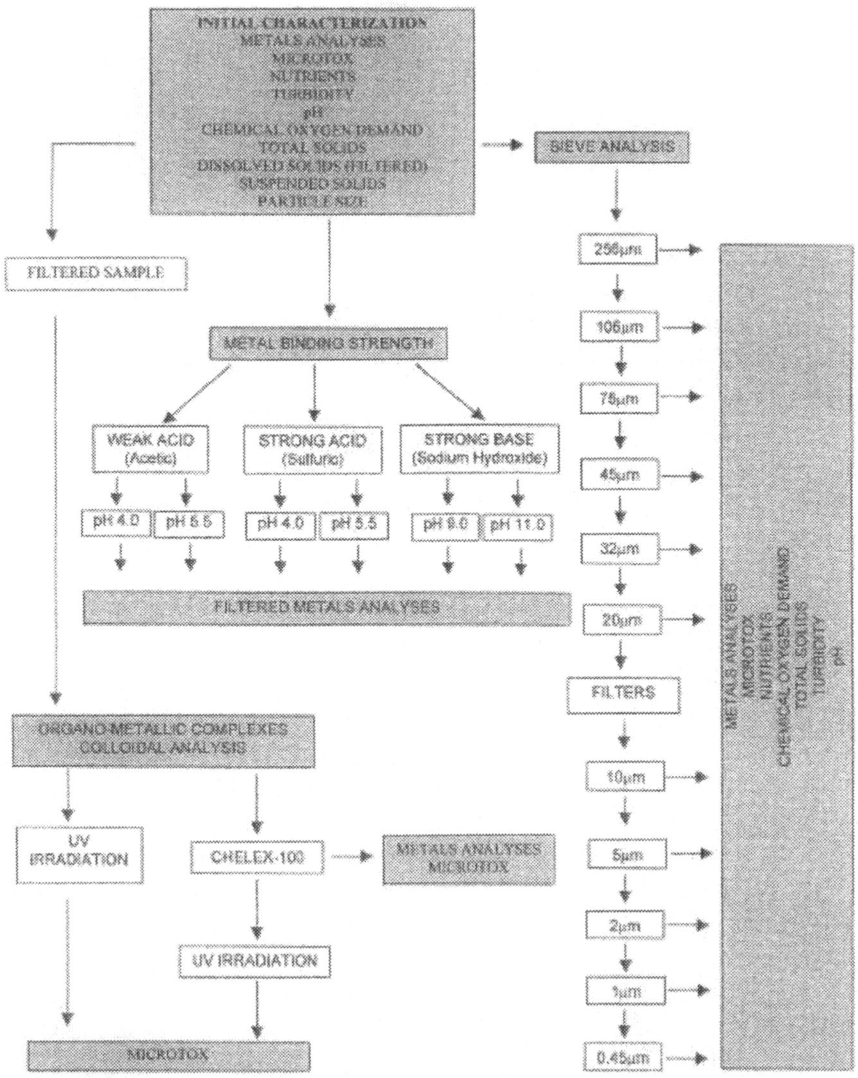

Figure 6-2. Flow chart of original analysis scheme

Initial Characterization. Each water sample was analyzed for metals, toxicity, turbidity, pH, solids (total, dissolved, suspended), particle size, chemical oxygen demand, and nutrients (ortho-phosphate, total phosphorus, nitrate, and ammonia). The analytical methods used for the analyses are listed in Table 6-2.

Table 6-2: Analytical Techniques

Analytical Parameter	Analysis Method (EPA Method number shown if available; equivalent Standard Method in parentheses, if available).
pH	EPA Method 150 (Standard Methods 4500-H^+.B.)
Turbidity	EPA Method 180.1 (Standard Methods 2130.B.)
Ortho-Phosphate	EPA Method 365.2 (Standard Methods 4500-P.E.)
Total Phosphorus	EPA Method 365.2 (Standard Methods 4500-P B, 5 and P.E.)
Nitrate	EPA Method 353.3 (Standard Methods 4500-NO_3^-.E.)
Ammonia	EPA Method 350.2
Toxicity	Microtox Rapid Toxicity Screening (UAB Laboratory SOP)
Chemical Oxygen Demand	EPA Method 410.4 (Standard Methods 5220.D.)
Solids (Total, Dissolved, Suspended)	EPA Method 160.1 and 160.2 (Standard Methods 2540.B.C.D.)
Particle Size Distribution	Standard Methods 2560.B. (Coulter Counter, UAB Laboratory SOP)
Acid digestion for Metals Analysis	Microwave Digestion using CEM Microwave MDS 2100 (UAB Laboratory SOP)
Metals Analysis	Standard Methods 3120. (Inductively Coupled Plasma, ICP)

Sieve Analysis. For the first set of samples, approximately 5 liters of each water sample was processed by sequentially passing it through six standard 3 inch diameter stainless steel Tyler sieves, ranging in size from 250 to 20 μm (see Figure 6-2). The water was then sequentially passed through five precision membrane filters ranging from 10 to 0.45 μm in size (Osmonics Inc. polycarbonate, except for the 0.45 μm which were MicronSep cellulosic). After each sieve/filter, 400 mL of water was put aside for metals, nutrients, COD, total solids, turbidity and pH analyses, and another 40 mL was placed in a small glass vial for toxicity analysis. The second set of samples were passed through three standard 3 inch diameter stainless steel Tyler sieves ranging in size from 250 to 45 μm and then through four precision membranes ranging from 10 to 0.45 μm in size.

Metal-Binding Strength. In order to evaluate the effect of pH on the metal-binding strength on the particles in the stormwater, each of the first eight water sample was exposed to varying pH levels using acetic acid (a weak acid), sulfuric acid (a strong acid) and sodium hydroxide (a base). The samples were lowered to pH levels of approximately 4.0 and 5.5 using the acids, and they were also raised to pH levels of approximately 9.0 and 11.0 using the base. An unmodified sample was also treated the same to act as a control sample, along with ultrapure water samples that have been adjusted with the acids and base to act as reagent blanks. The pH adjustments were made by using 50mL samples (in triplicate) and titrating the sample to the specified pH using either 0.05- 0.2M of the specific acid or 0.05- 0.1M of NaOH and placing the sample on a shaking table for three days. Afterwards, the samples were filtered through a 0.45 μm filter, digested, and analyzed for heavy metals using an ICP. The results can be compared to the initial unfiltered and filtered (through 0.45 μm membrane filters) heavy metal results.

Organo-Metallic Complexes/Colloidal Analysis. To separate the ionic and colloidal fractions, each of the first eight samples was exposed for six hours to ultraviolet light in order to cause dissociation of metals from organic complexes or colloids. A portion was then put aside for toxicity analysis. A sample was also exposed to Chelex-100 chelating resin in order to separate ionic metals from metals strongly bound to metal-ligand complexes, or those strongly adsorbed to colloidal particles. For each 100mL of 0.45-μm filtered water, 5g of Chelex-100 resin was added and the water sample was gently shaken for one hour. After Chelex resin exposure, the water sample was again filtered, and portions were put aside for toxicity and heavy metal analyses. After Chelex resin exposure, another portion of the same sample was UV irradiated for 6 hours and put aside for toxicity analysis.

Quality Assurance/Quality Control. All laboratory tests were also run using a method blank of ultrapure (18mΩ) water and instrument performance was checked with external standards when applicable. For most of the analyses performed, few, if any, problems were detected and the method blank gave expected results. But when testing the method blanks for total nitrogen, a few samples came out higher than expected. Other problems with the total nitrogen method was its' relatively insensitivity compared to the concentrations found in the samples. Appendix F contains tables for all tests run on each stormwater sample and includes the results from the method blanks. Figures 6- 3 and 6-4 and plots of additional QA/QC tests using different laboratory waters for copper and zinc, showing little change in initial sample dissolved metal concentrations during the sieving and filtration steps. There appeared to be some small zinc contamination (about 5 μg/L) for one of the samples associated with the filter holders that were used for the membrane filters, possibly associated with the sealed magnetic elements used to secure the filter arrangements (there was also some minor iron contamination noted for these filtered samples when using the laboratory water). All new plastic filter holders were used in the second series of tests to eliminate this small problem.

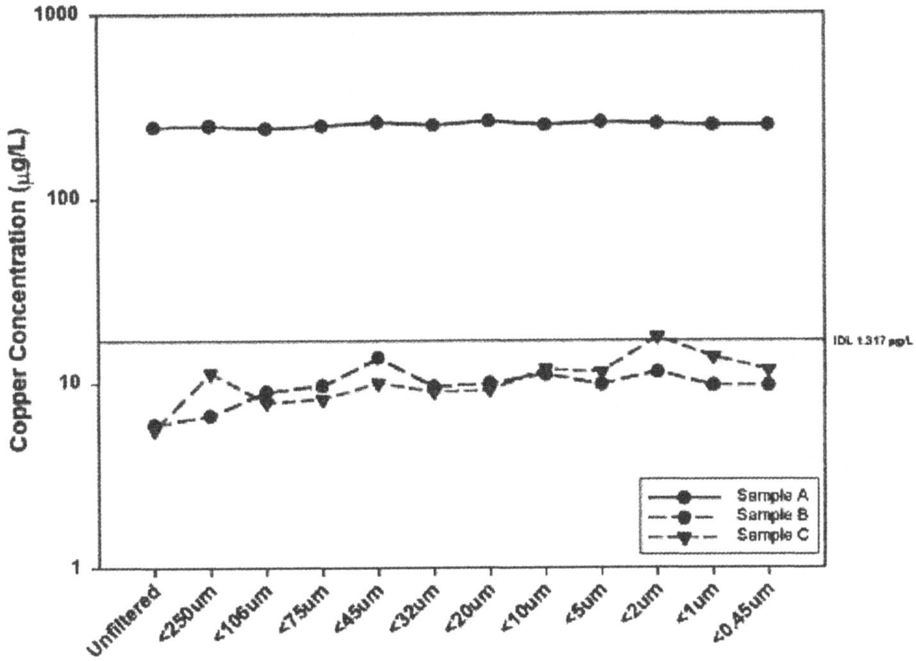

Figure 6-3. QA/QC analyses for copper.

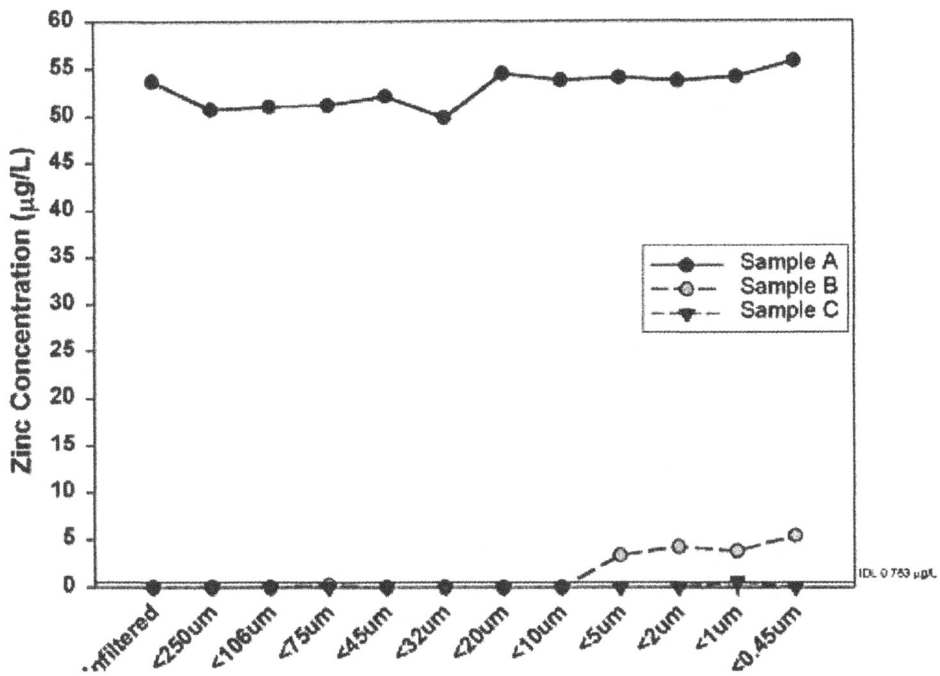

Figure 6-4. QA/QC analyses for zinc.

6.1.3 Observations of Associations of Stormwater Pollutants with Different Sized Particulates

Initially, eight samples were collected and analyzed for metals, nutrients, toxicity, turbidity, pH, COD, particle size, and solids. An additional ten samples (samples 9 to 18) were collected and analyzed in the summer of 2002.

Figures 6-5 through 6-32 are plots showing the concentrations of the samples, after filtering the water through various sieves and membrane filters. Appendix F shows the numerical data for the majority of these tests, including a summary table that lists the total and filterable concentrations for each sample, for comparison to the test results. Figures 6-5 and 6-6 show the measured changes in total solids after sieving and filtering through different sizes apertures. As expected, the general trend is a lower concentration of total solids as the particle size decreases. All samples were also divided into three portions for particle size analyses: >106 μm, <0.45 μm, and the intermediate range from 0.45 to 106 μm. The Coulter Counter results for the intermediate size range were compiled with results from the other two size ranges to obtain the complete particle size distribution. Appendix E shows the calculations for these particle size analyses for the first eight samples. Figure 6-33 is a plot of the Coulter Counter results for the second set of samples (9 to 18).

Obviously, concentrations should all decrease with filtration. However, there were periodic jumps in concentrations for some conditions, reflecting variability in the analytical method and the sample handling. Also, the samples behaved differently from each other, with some, notably samples #5 and #8, generally having the highest concentrations and the most dramatic changes with filtration. Sample #5 was from a commercial parking area, near a home improvement store storage area, while sample #8 was from a university parking lot. Both sites had about a 2 week inter-event dry period.

Constituents that did not change significantly with filtration included nitrate and sodium, as expected. Other analytes (COD and cadmium) also had little change, except for a single sample in each case. Most of the analytes had large decreases with filtration, especially for the more contaminated samples (turbidity, phosphorus, phosphate, magnesium, chromium, copper, iron, lead, and zinc). Total solids and COD had much smaller changes with filtration, with substantial fractions associated with the filterable (<0.45 μm) fraction.

Table 6-3 summarizes the calculated "potency" factors for several size ranges for the heavy metals during these tests. These are the masses of these constituents associated with the particulates, expressed as mg constituent per kg of suspended solids. As expected, these values increase with decreasing particle size, except for iron which is seen to generally remain constant. The iron measurements show that these stormwater particulates were about 1.5 to 3% iron, by mass. There were also large zinc components for the smallest sizes. Typical soil values for iron are about 50,000 mg Fe per kg solids, somewhat higher than these observations, while copper, lead, and zinc are all generally less than typical soil values of about 100 mg/kg (much less than shown here) (Lindsay 1979).

Typical street dirt (mostly for the smallest <43 μm size particles) contents of these heavy metals are about 100 to 500 mg/kg for copper, about 500 to 7500 mg/kg for lead, and about 250 to 1200 mg/kg for zinc (Bannerman, *et al.* 1983, Pitt 1979, Pitt 1985, Pitt and McLean 1986, Pitt and Sutherland 1982, and Terstrip, *et al.* 1982). The lead values found during this research are much less than the earlier street dirt values, due to the decreased use of lead in gasoline. However, the copper and zinc values observed are much greater than the street dirt values.

In Toronto, Pitt and McLean (1986) found copper values of about 30 to 200 mg/kg for most residential and commercial source area particulates (<125 μm), but from 100 to over 1000 mg/kg at industrial areas. Similar values for residential and commercial area lead soil concentrations were generally from 200 to 1,000 mg/kg and slightly higher in industrial areas. Zinc ranged from 50 to 2,000 mg/kg in all land use areas. Again, the observed copper and zinc values were much greater than these previous measurements, while the lead values were much less.

Table 6-3. Summary Table Showing Heavy Metal Associations for Different Particle Sizes

	Copper		Iron		Lead		Zinc	
particle size (μm)	mg Cu/kg SS	COV	mg Fe/kg SS	COV	mg Pb/kg SS	COV	mg Zn/kg SS	COV
>250	50	na	28604	1.50	117	0.58	266	0.88
106 to 250	2137	1.45	21730	0.85	375	1.03	3486	0.79
45 to 106	1312	1.16	14615	0.72	226	0.85	2076	0.88
10 to 45	735	0.97	26221	0.54	229	0.50	1559	0.74
2 to 10	4668	1.60	18508	1.16	868	0.78	13641	1.88
0.45 to 2	2894	1.21	29267	1.31	199	1.40	13540	1.56

6.1.3.1 Colloidal Analysis

As noted above, Chelex-100 resin was added to portions of the filtered (0.45 μm filter) stormwater samples for the first eight samples. After shaking for one hour, the samples were again filtered (again using a 0.45 μm filter). Table 6-4 shows the changes in major constituent and heavy metal concentrations before and after exposure to the Chelex. Ionic forms of the metals will bond to the Chelex, while colloidal bound metals and metals strongly bound to ligands will remain in solution and be filtered. Therefore, the "after" Chelex exposed concentrations reflects these bound metals, while the differences in the "before" and "after" exposure concentrations reflects the portions associated with free ionic forms.

More than 90% of the filterable forms of calcium, magnesium, potassium, iron, and zinc were in ionic forms, with very little colloidal, or other bound forms. More than 80% of the filterable chromium and lead were also ionic, while only about 50% of the filterable copper and 30% of the filterable cadmium were ionic.

Table 6-4. Effects of exposure to Chelex-100 on filtered concentrations

Analyte	Sample 1		Sample 2		Sample 3		Sample 4		Sample 5	
	Before	After	Before	After	Before	After	Before	After	Before	After
Calcium	29.3	0.2	19.4	0.4	11.0	0.2	8.1	0.2	30.3	0.3
Magnesium	2.4	0.1	2.8	0.1	1.2	0.1	2.1	0.1	6.0	0.1
Potassium	1.7	0.1	1.1	0.1	2.1	0.2	1.9	0.2	5.2	0.6
Cadmium	**5.6**[1]	6.4	6.5	**5.8**	5.8	**5.8**	5.6	6.0	9.1	**1.4**
Chromium	**8.5**	3.3	12.9	3.4	5.4	**2.0**	3.2	**2.4**	77.7	5.6
Copper	26.0	16.6	25.5	18.4	20.6	15.2	12.5	15.1	87.4	13.4
Iron	1934	64.9	1520	47.1	1270	129	248	48.8	6020	150
Lead	**11.3**	2.4	35.1	12.0	5.8	**1.3**	3.8	**1.6**	52.3	**0.6**
Zinc	210	9.3	103	14.9	50.6	9.9	19.0	7.6	960	13.3

	Sample 6		Sample 7		Sample 8		Average		St. Dev.		COV	
	Before	After	Before	After	Before	After	Before	After	Before	After	Before	After
Calcium	3.4	**0.1**	3.6	**0.1**	5.0	**0.1**	110	1.6	11.2	0.1	0.1	0.1
Magnesium	0.8	**0.0**	0.8	**0.0**	2.2	**0.0**	18.2	0.4	1.7	0.0	0.1	0.1
Potassium	0.6	**0.1**	1.3	**0.1**	3.7	0.3	17.7	1.8	1.5	0.2	0.1	0.1
Cadmium	5.9	**0.9**	0.4	**0.7**	0.2	**0.6**	39	28	3.0	2.7	0.1	0.1
Chromium	2.4	**1.5**	2.2	**1.7**	11.2	**1.3**	124	21	25.5	1.4	0.2	0.1
Copper	12.5	**4.2**	4.6	5.6	7.6	**4.0**	197	93	26.6	6.0	0.1	0.1
Lead	2.3	**0.6**	4.6	**0.5**	10.3	**2.2**	126	21	18.1	3.9	0.1	0.2
Iron	365	82.0	1410	156	9900	349	22700	1030	3380	99	0.1	0.1
Zinc	22.3	7.9	21.1	19.7	51.4	7.9	1430	91	320.9	4.3	0.2	0.0

[1] bold values are slightly lower than the calculated limit of detection, but are still useful indicators for the trends (although there is slightly reduced confidence in their measured values)

As indicated in Figure 6-2, UV exposure experiments were also conducted on the first eight samples. Because the ICP is only sensitive to total forms of the metals (after proper digestion), only toxicity tests were meaningful for these tests. These results were more variable than anticipated, partially due to relatively low toxicity for the initial samples. Five of the eight samples had increased toxicity after UV exposure, compared to toxicity values before UV exposure (indicating disassociation of organo-metallic complexes), while three had reduced toxicity (unknown reasons). These results were not statistically significant. Since almost all of the dissolved chromium, lead, iron, and zinc, and about half of the copper, were in free ionic forms, relatively little amounts would have been associated with organo-metallic complexes available for UV destruction in any case.

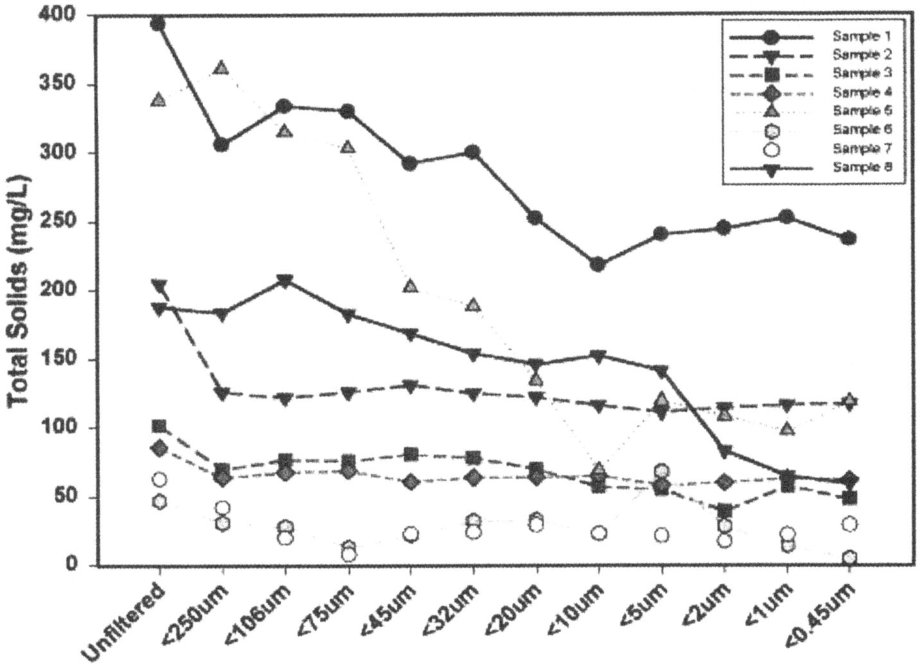

Figure 6-5. Total solids associations with particle sizes (all analyses with sieves) for samples 1 to 8.

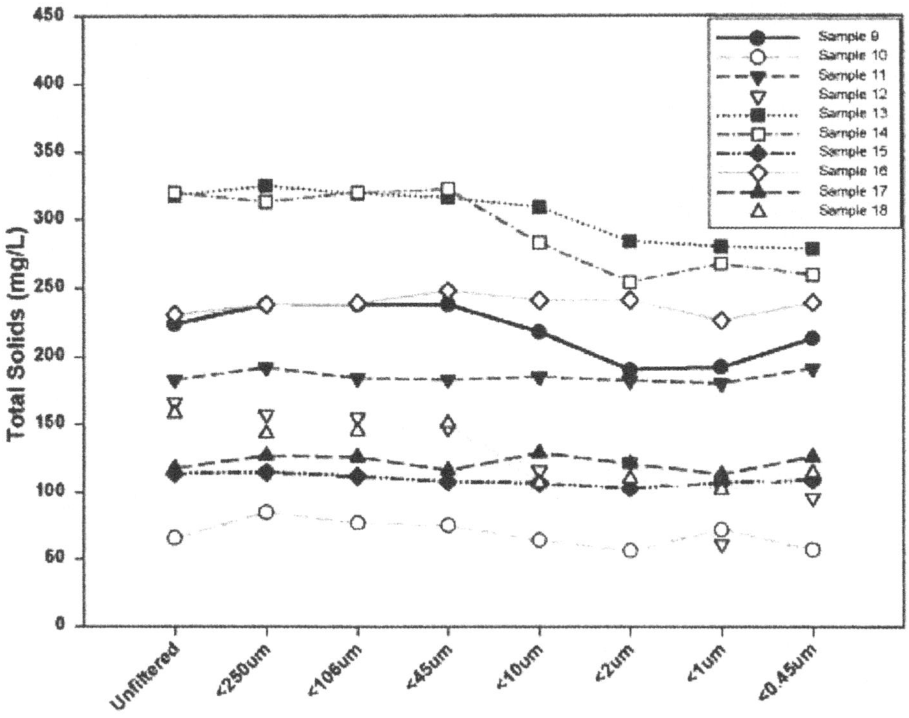

Figure 6-6. Total solids associations with particle sizes (all analyses with sieves) for samples 9 to 18.

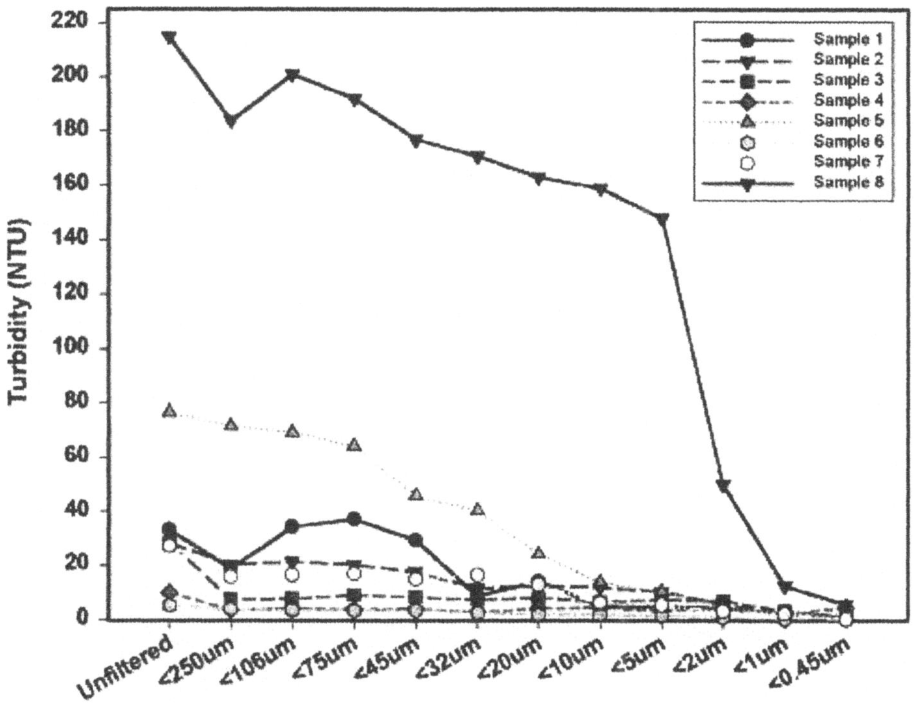

Figure 6-7. Turbidity associations with particle sizes (all analyses with sieves) for samples 1 to 8.

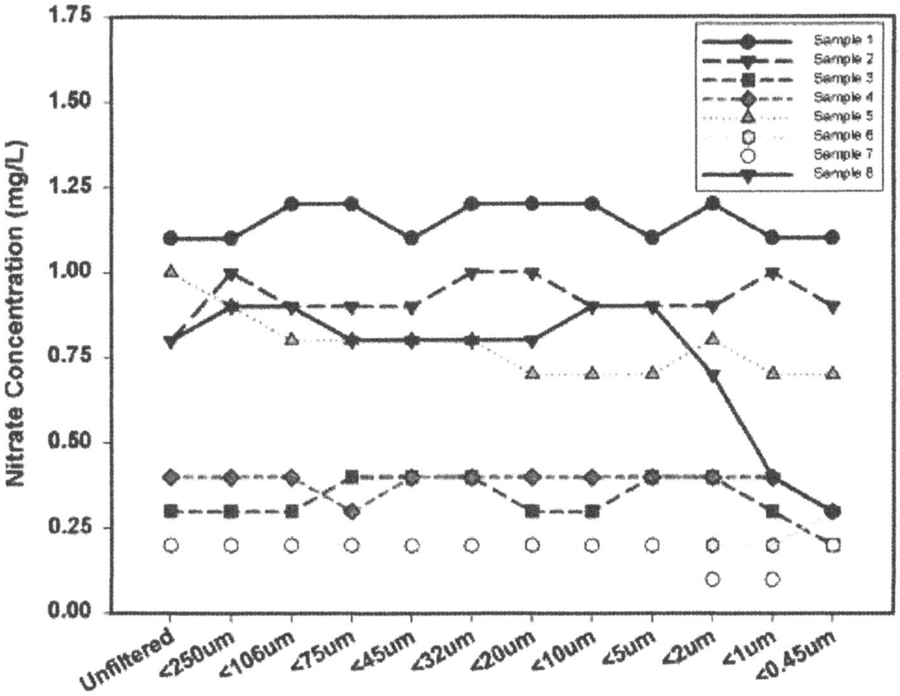

Figure 6-8. Nitrate associations with particle sizes (separations with sieves and membrane filters) for first 8 samples.

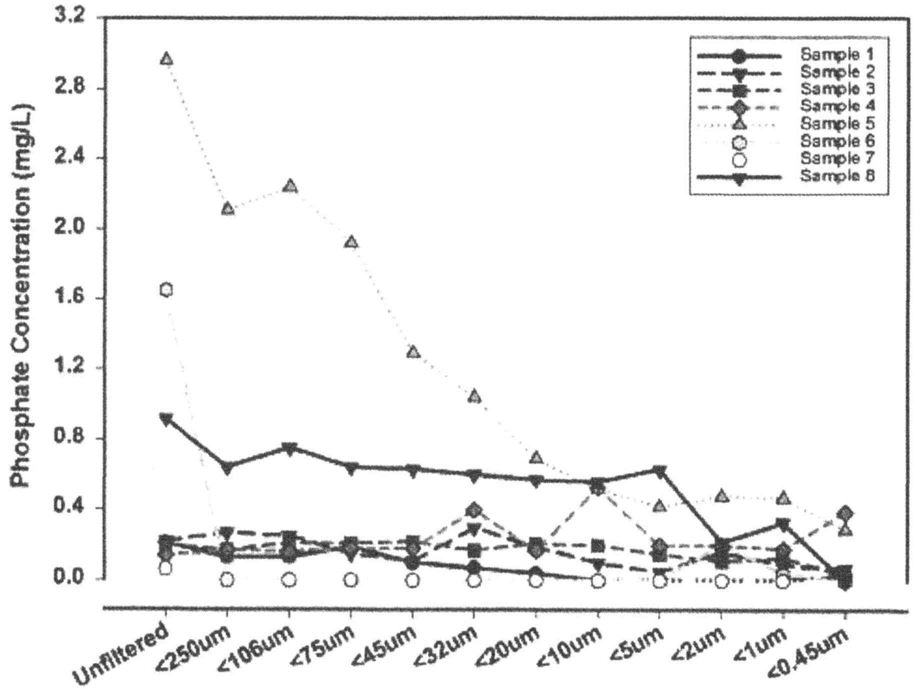

Figure 6-9. Phosphate associations with particle sizes (separations with sieves and membrane filters) for first 8 samples.

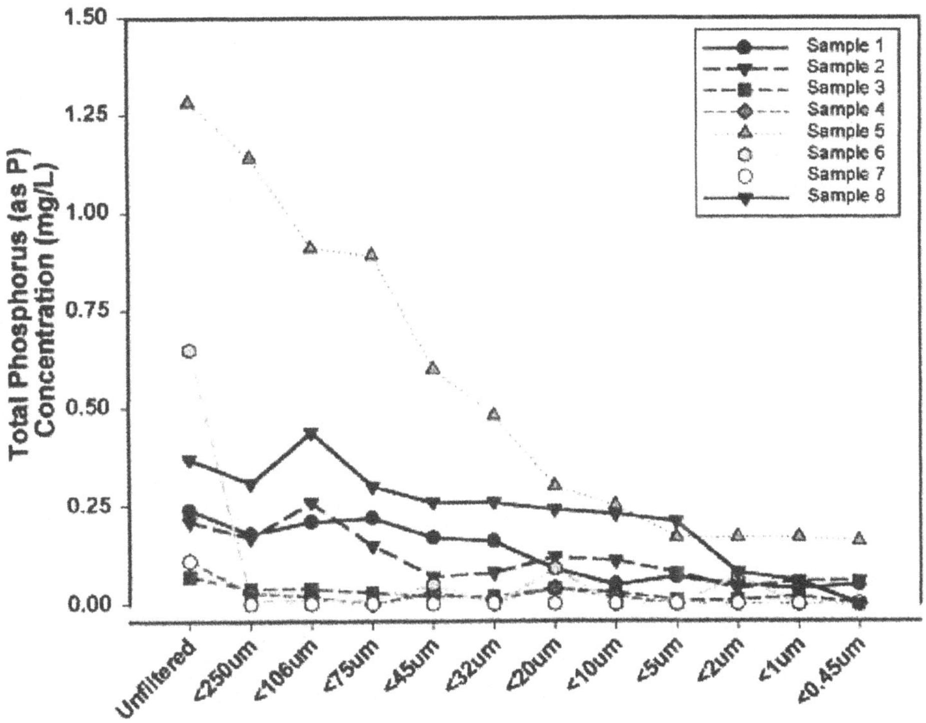

Figure 6-10. Total phosphorus associations with particle sizes (separations with sieves and membrane filters) for first 8 samples.

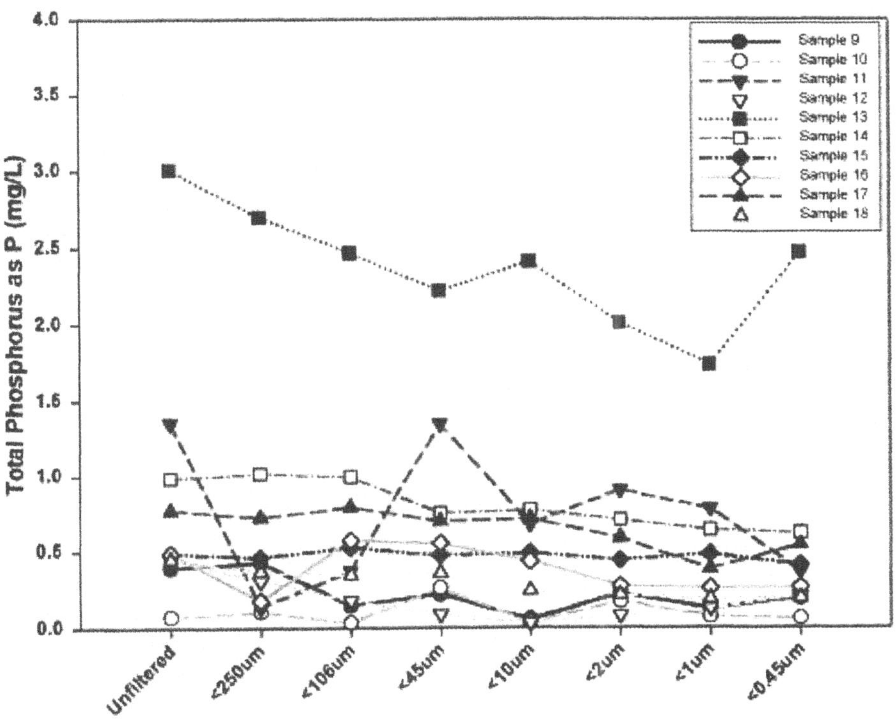

Figure 6-11. Total phosphorus associations with particle sizes (separations with sieves and membrane filters) for samples 9 to 18.

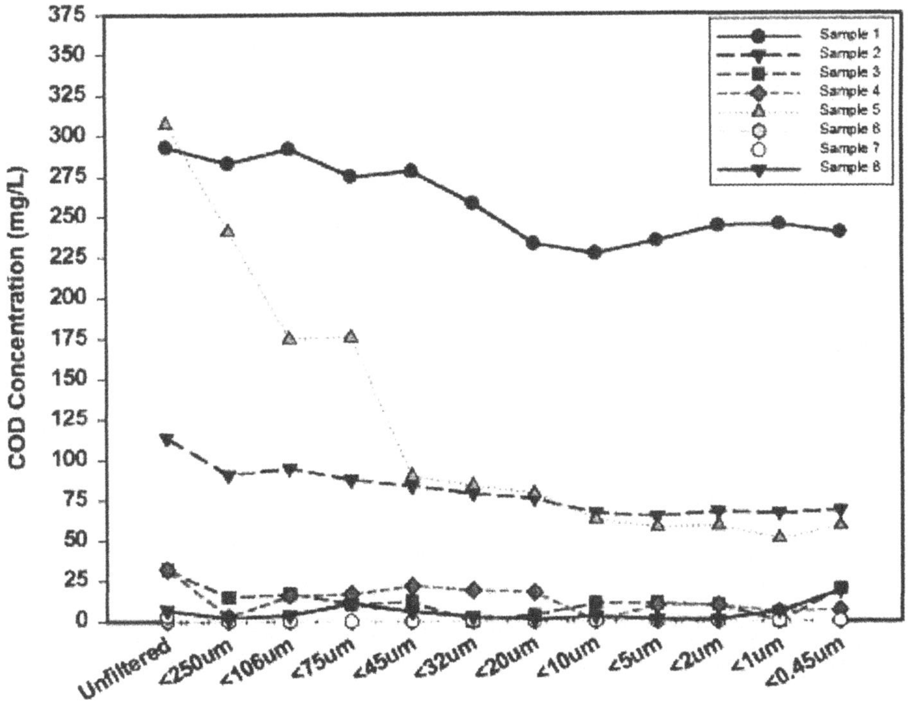

Figure 6-12. COD associations with particle sizes (separations with sieves and membrane filters) for first 8 samples

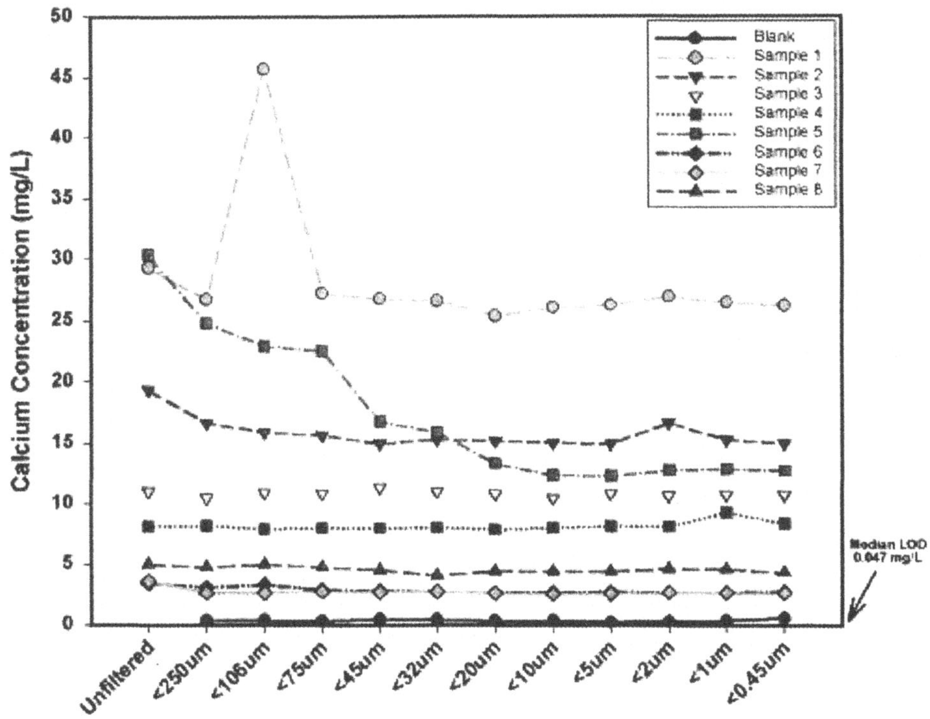

Figure 6-13. Associations of calcium with different sized particulates for first 8 samples.

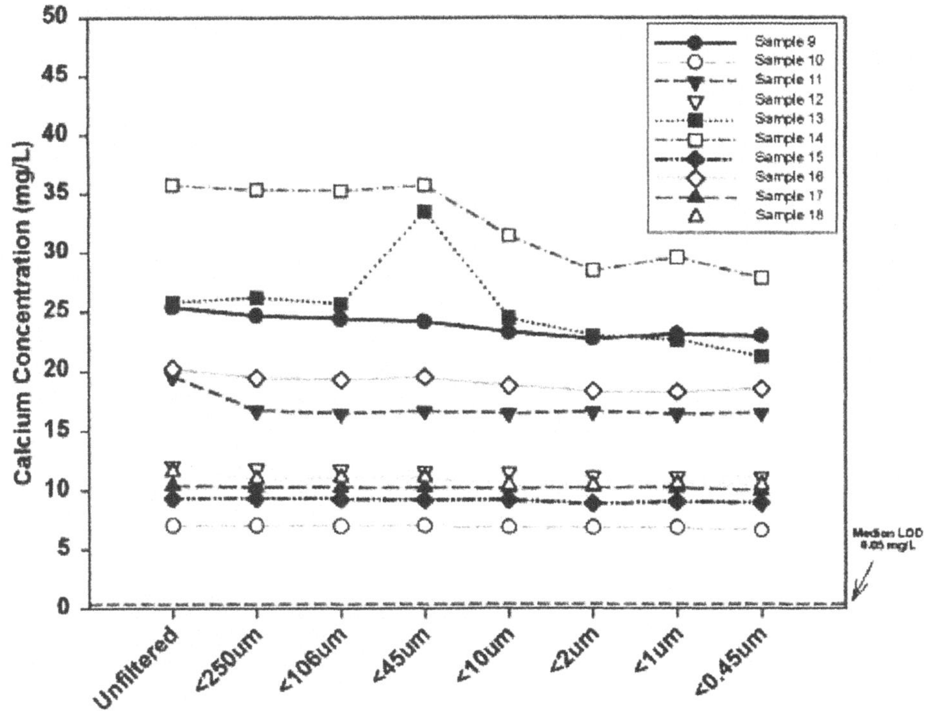

Figure 6-14. Associations of calcium with different sized particulates for samples 9 to 18.

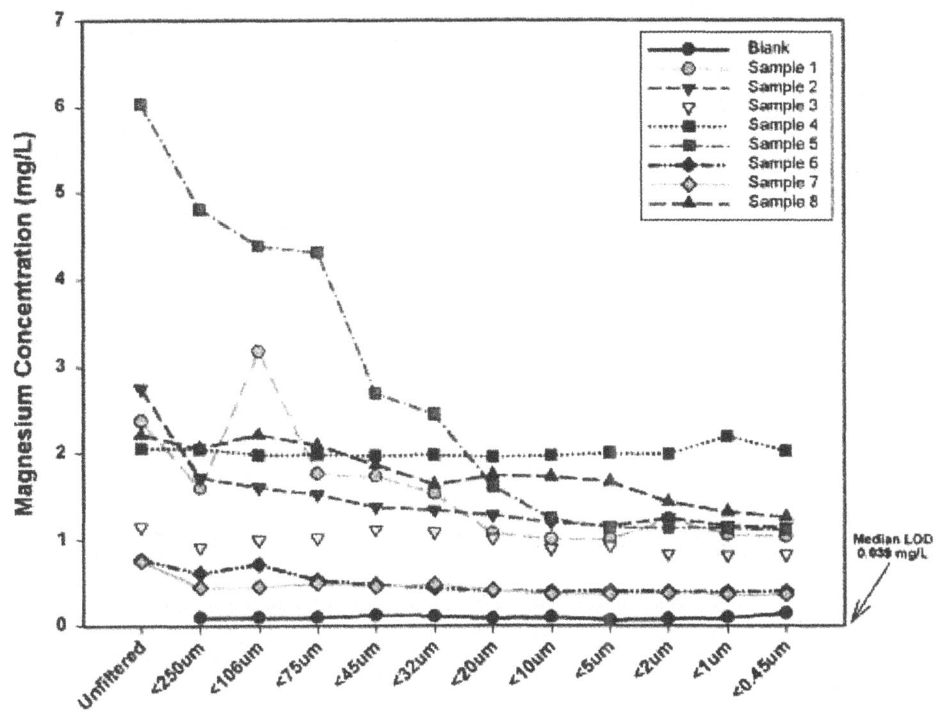

Figure 6-15. Associations of magnesium with different sized particulates for first 8 samples.

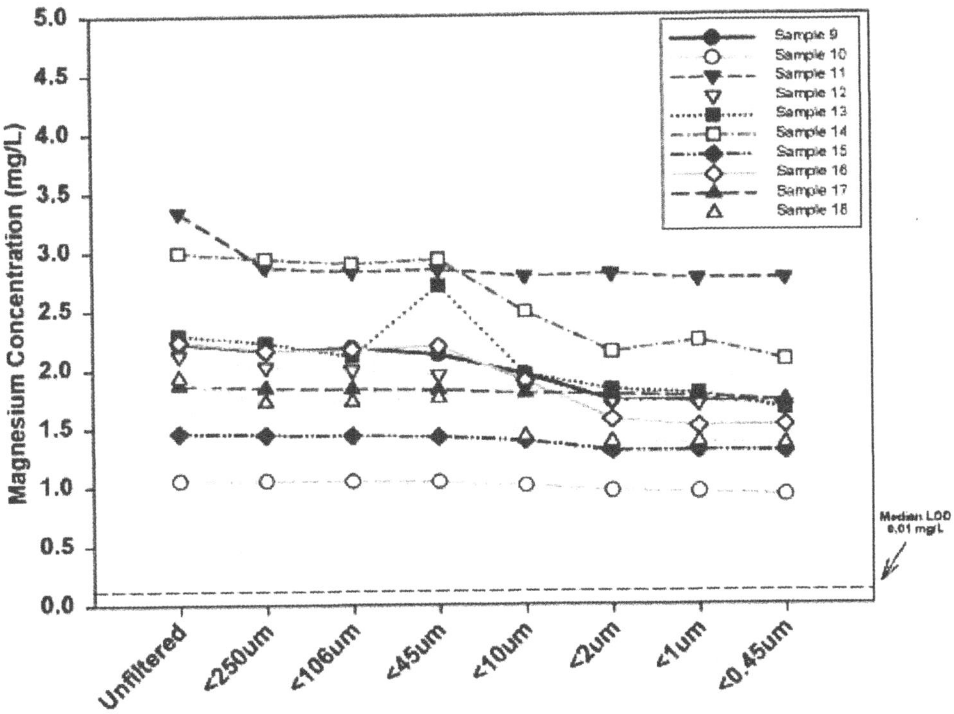

Figure 6-16. Associations of magnesium with different sized particulates for samples 9 to 18.

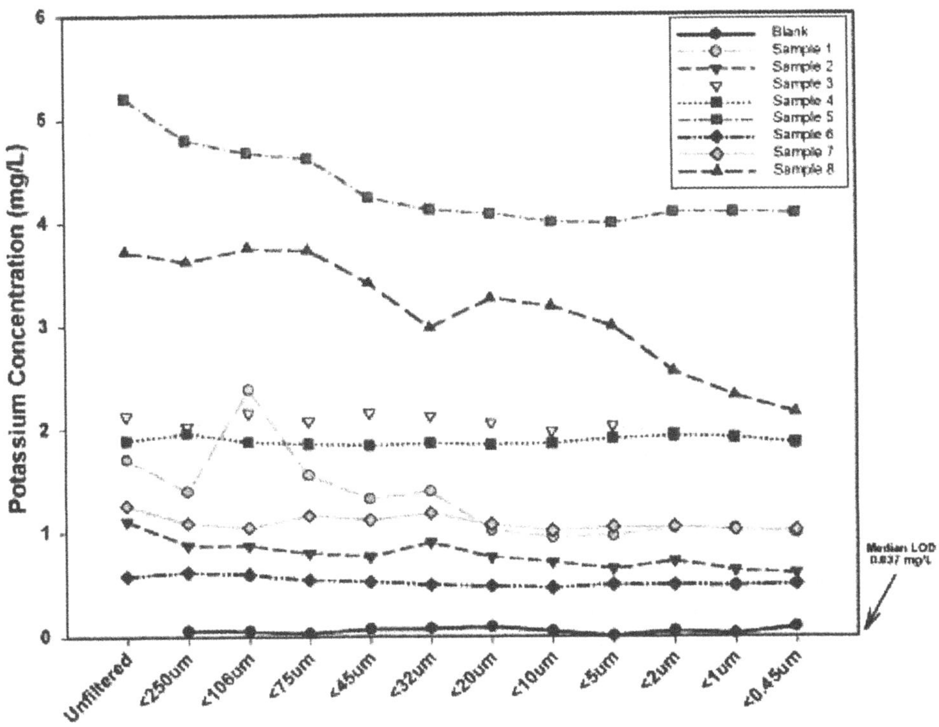

Figure 6-17. Associations of potassium with different sized particulates for first 8 samples.

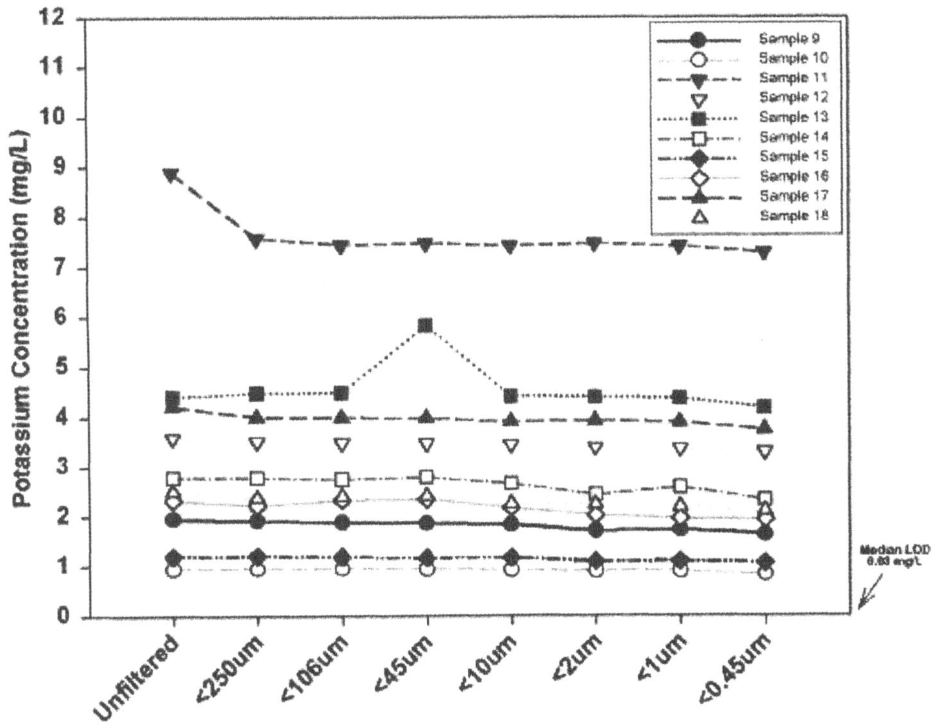

Figure 6-18. Associations of potassium with different sized particulates for samples 9 to 18.

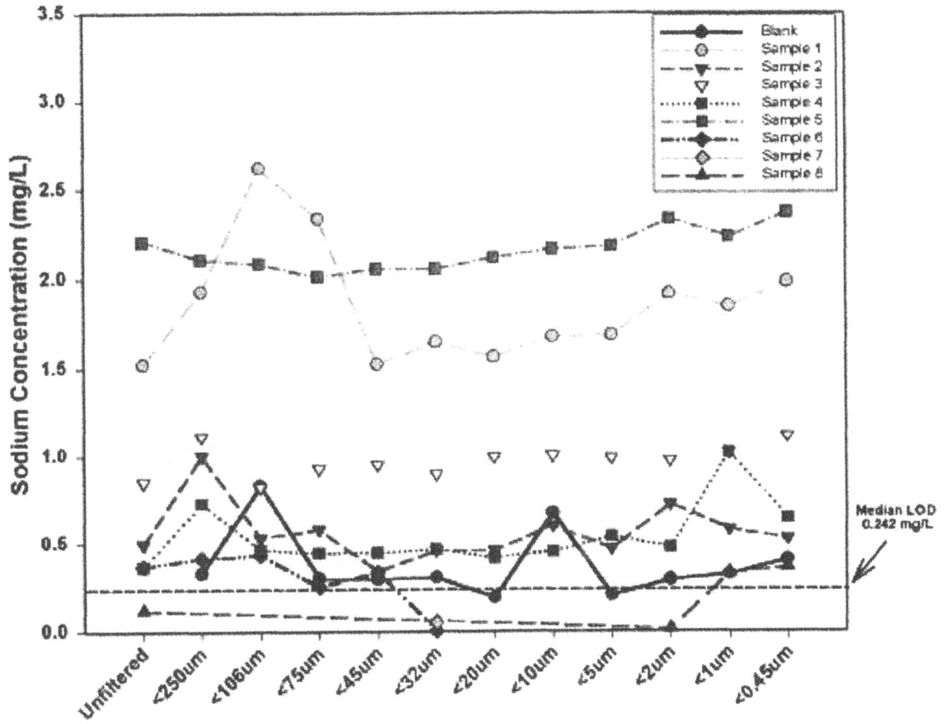

Figure 6-19. Associations of sodium with different sized particulates for first 8 samples.

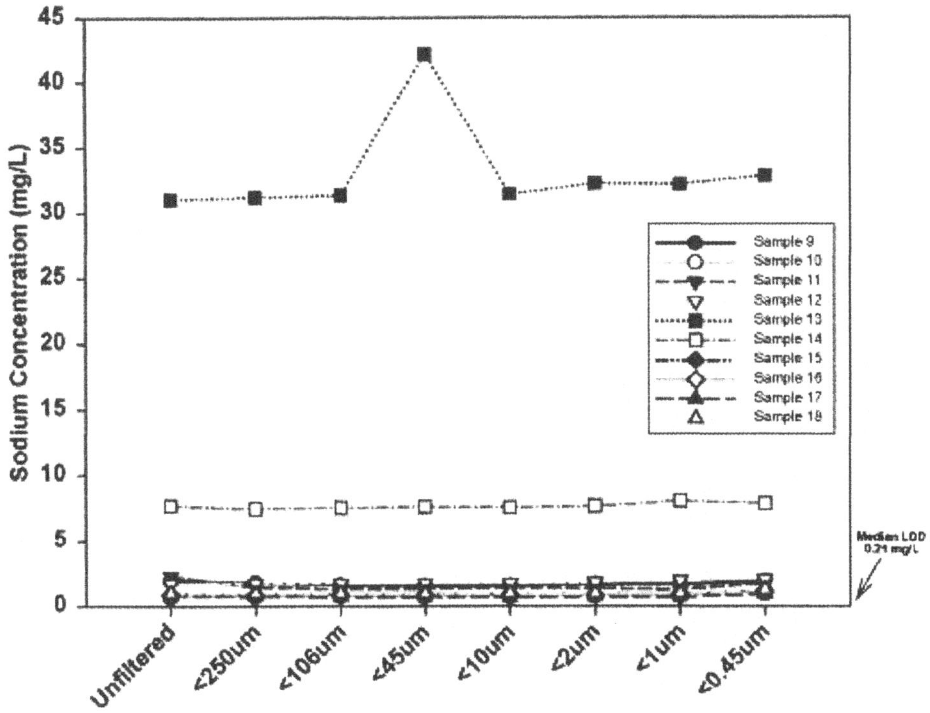

Figure 6-20. Associations of sodium with different sized particulates for samples 9 to 18.

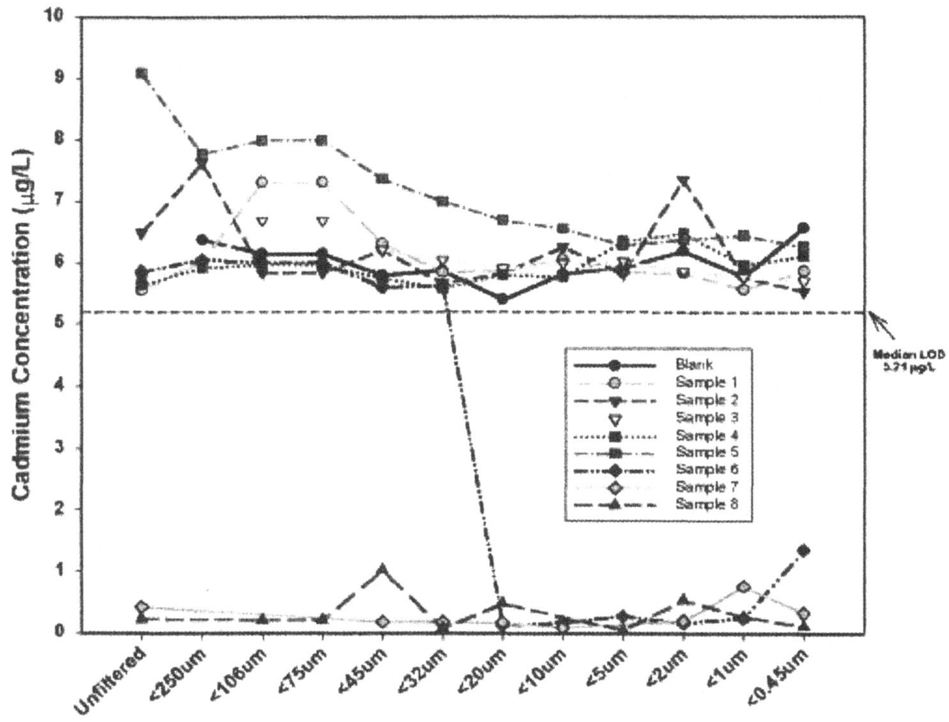

Figure 6-21. Associations of cadmium with different sized particulates for first 8 samples.

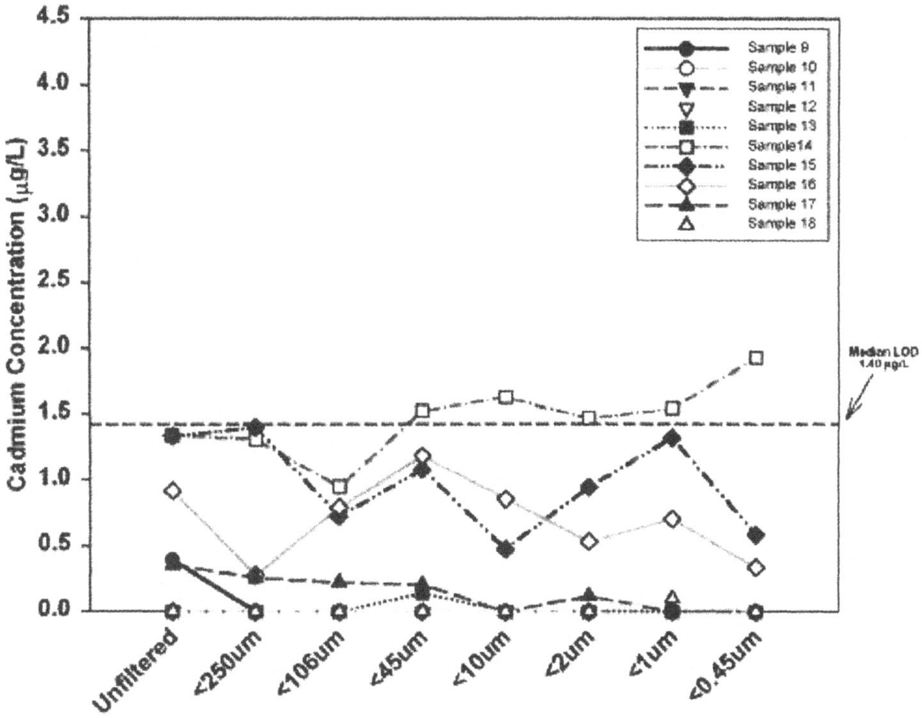

Figure 6-22. Associations of cadmium with different sized particulates for samples 9-18.

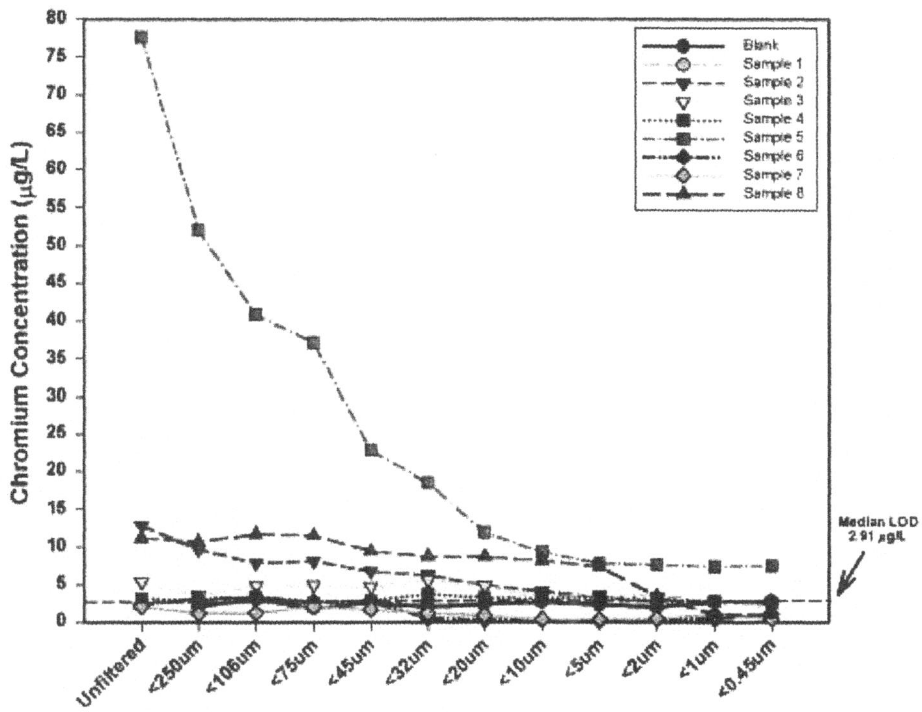

Figure 6-23. Associations of chromium with different sized particulates for first eight samples.

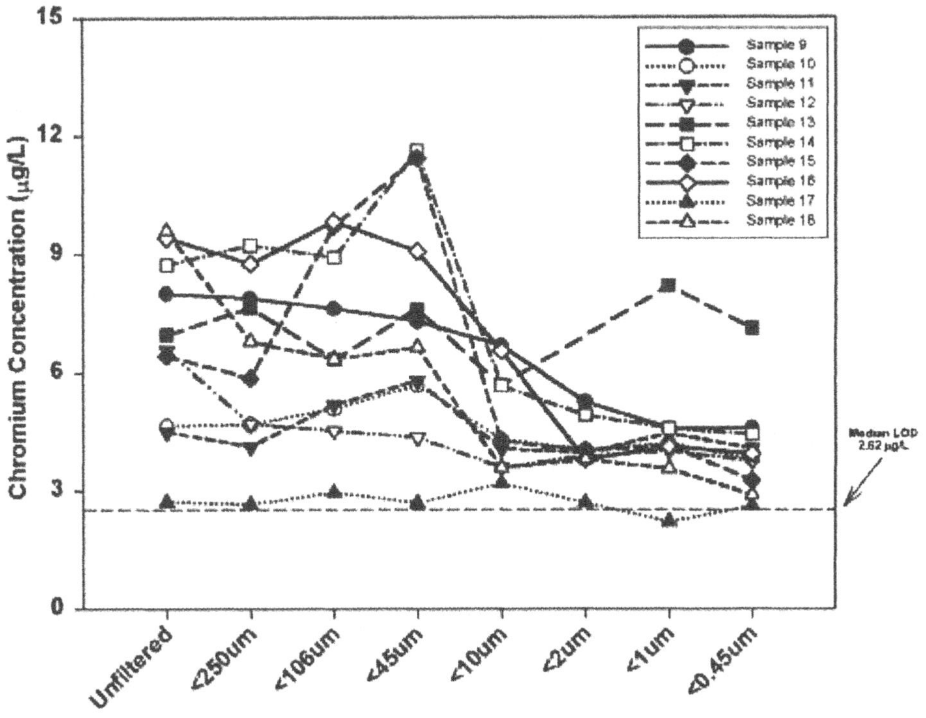

Figure 6-24. Associations of chromium with different sized particulates for samples 9-18.

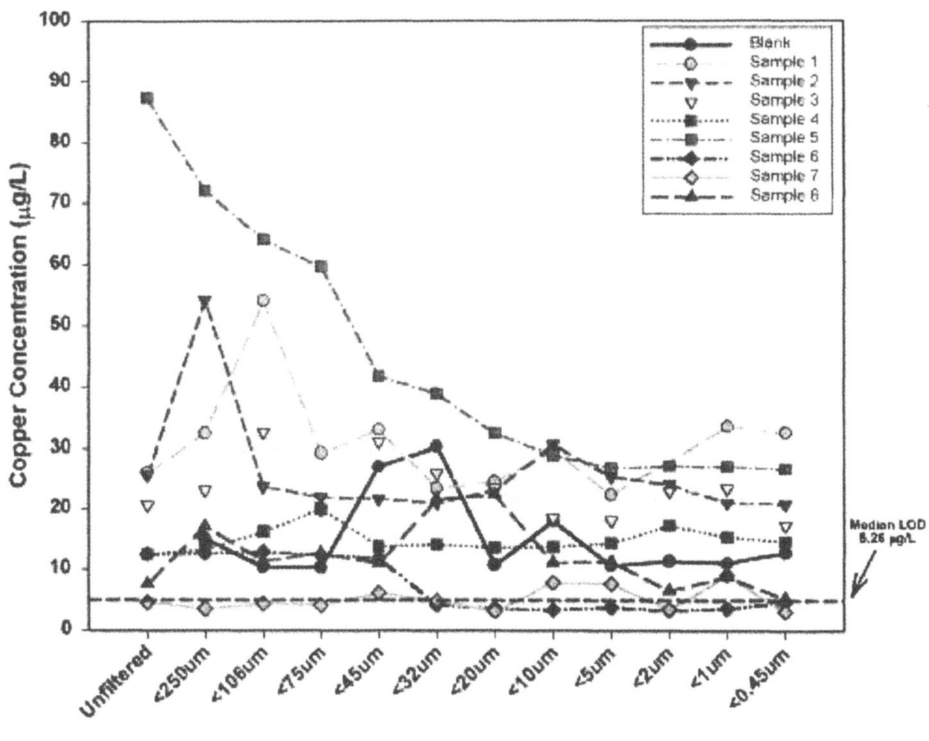

Figure 6-25. Associations of copper with different sized particulates for first eight samples.

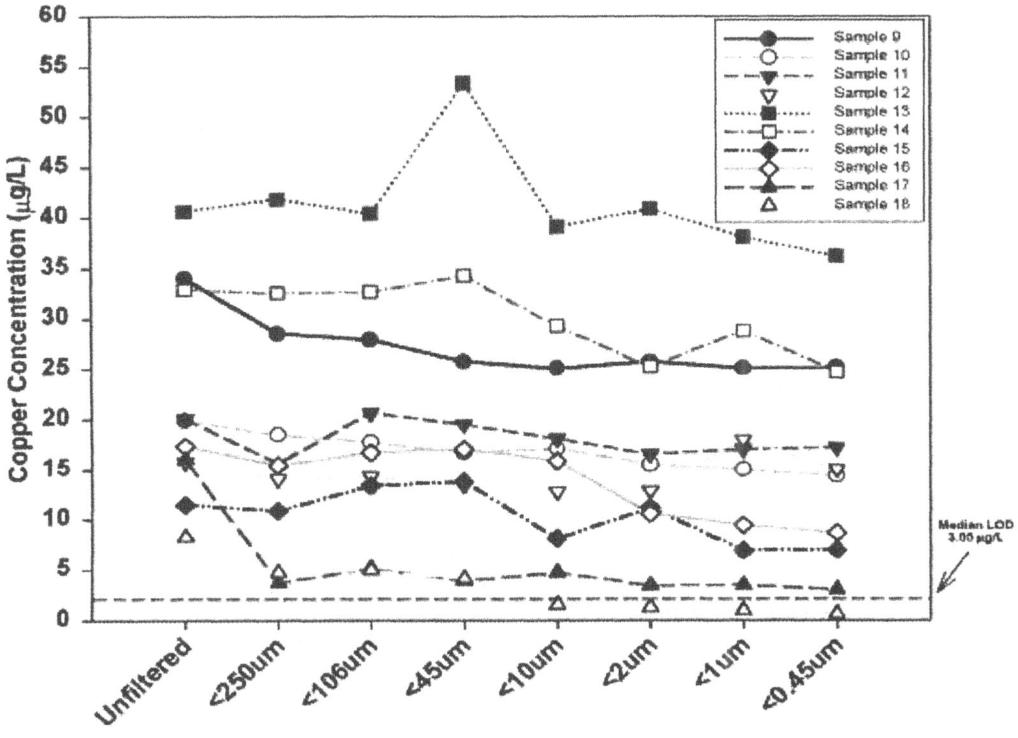

Figure 6-26. Associations of copper with different sized particulates for samples 9 to 18.

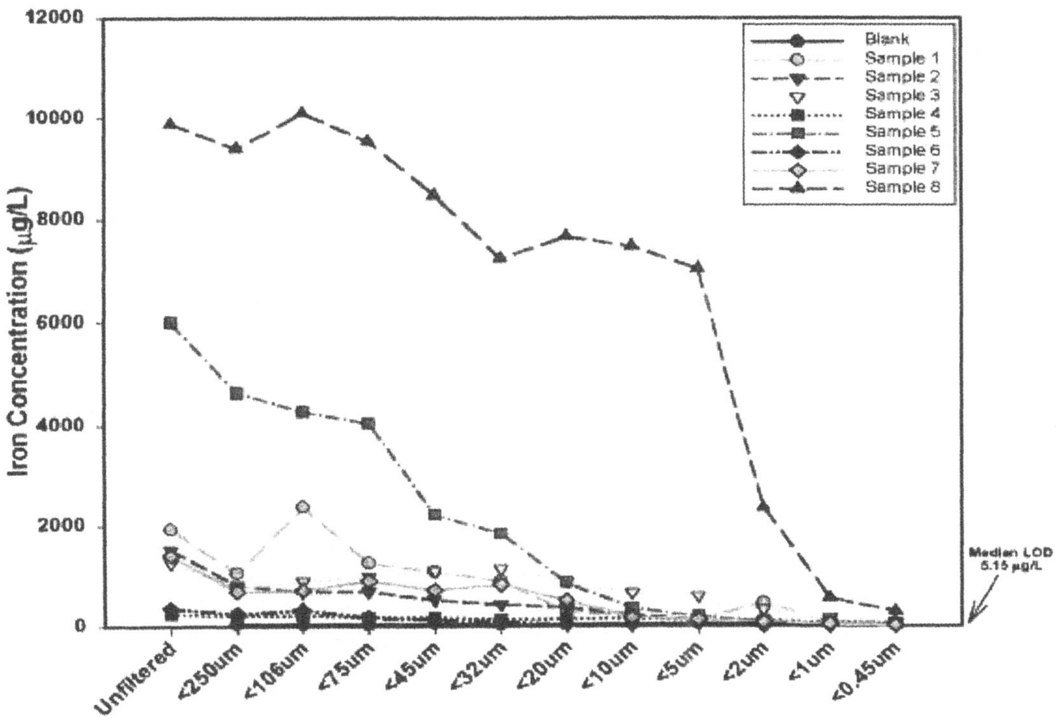

Figure 6-27. Associations of iron with different sized particulates for first 8 samples.

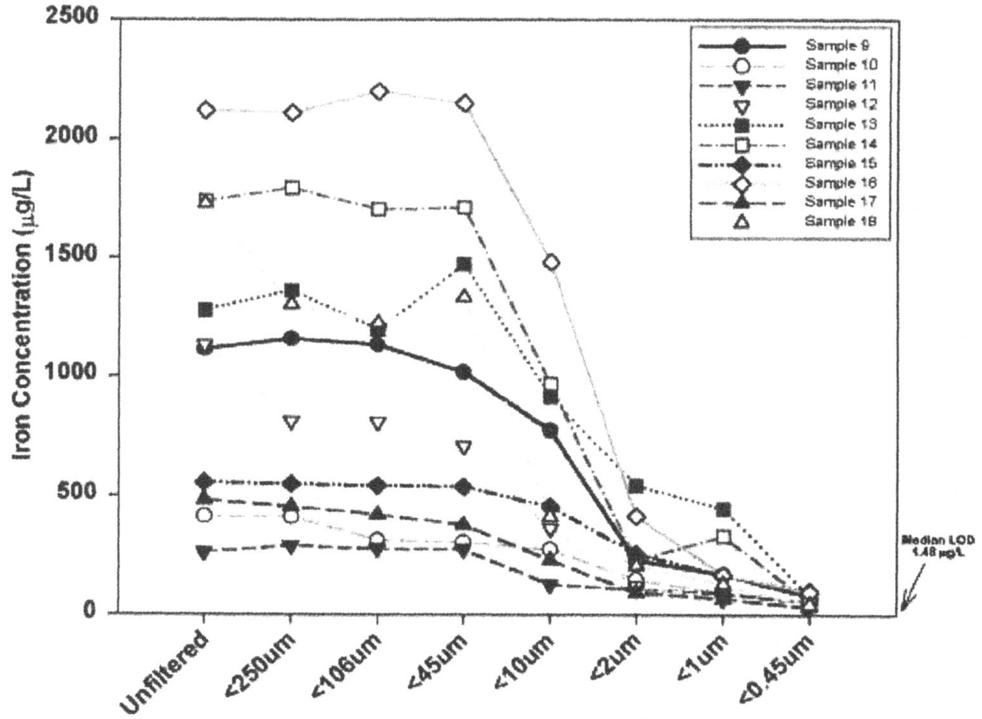

Figure 6-28. Association of iron with different sized particulates for samples 9 to 18.

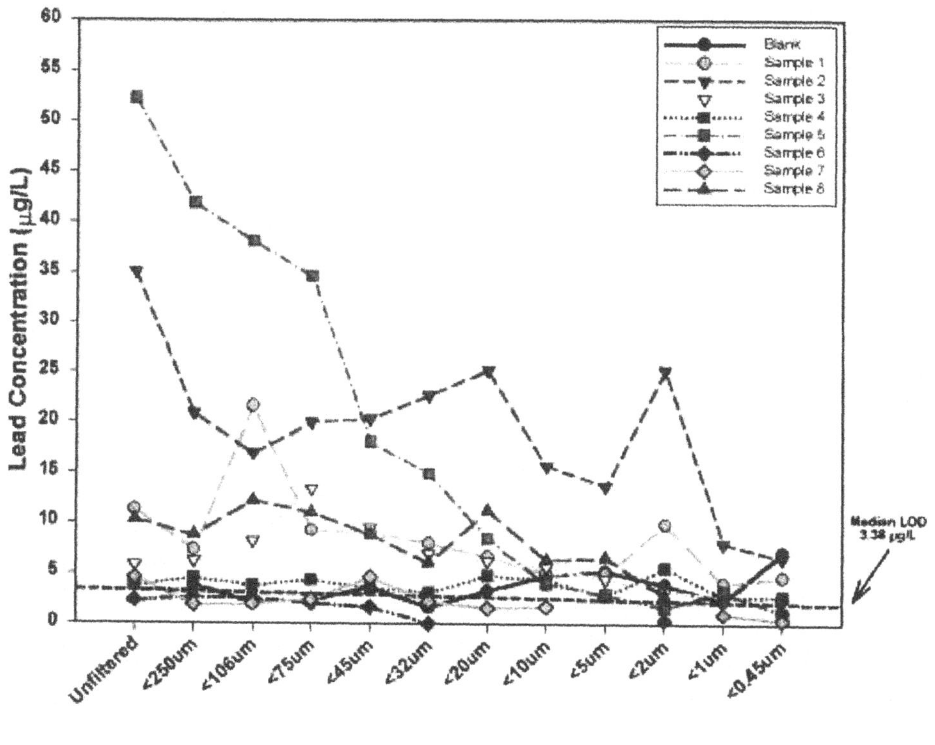

Figure 6-29. Associations of lead with different sized particulates for first 8 samples.

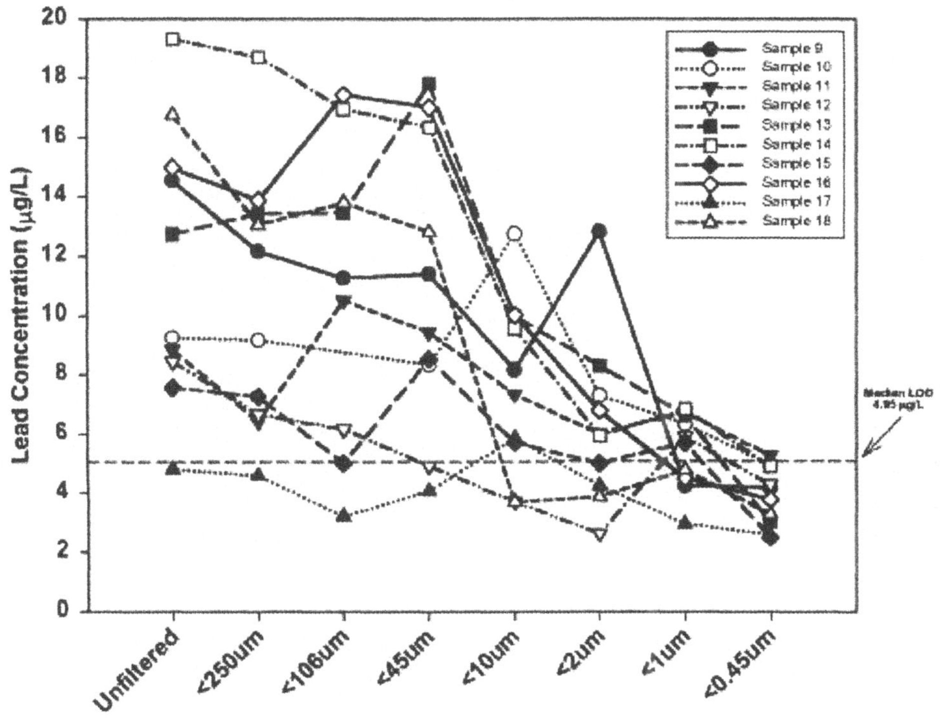

Figure 6-30. Associations of lead with different sized particulates for samples 9 to 18.

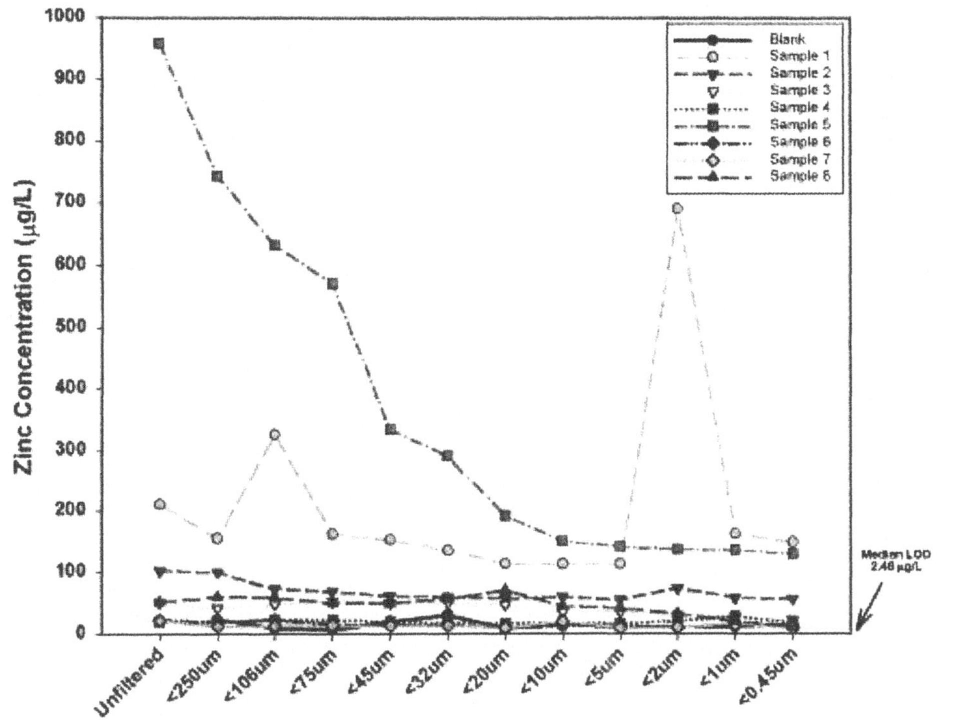

Figure 6-31. Associations of zinc with different sized particulates for first 8 samples.

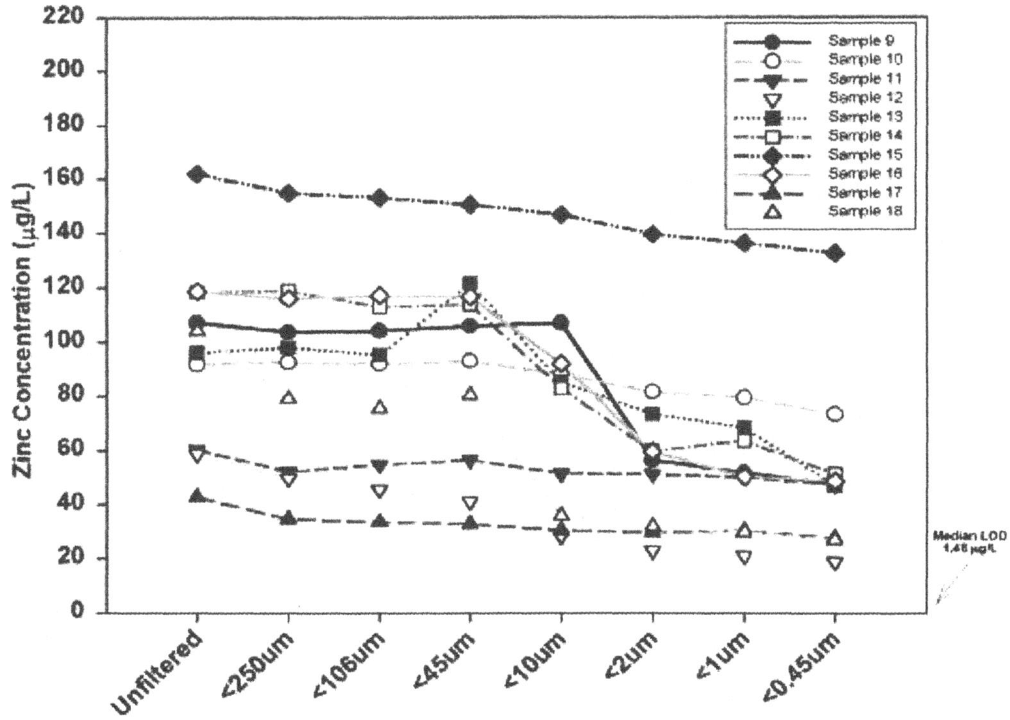

Figure 6-32. Associations of zinc with different sized particulates for samples 9-18.

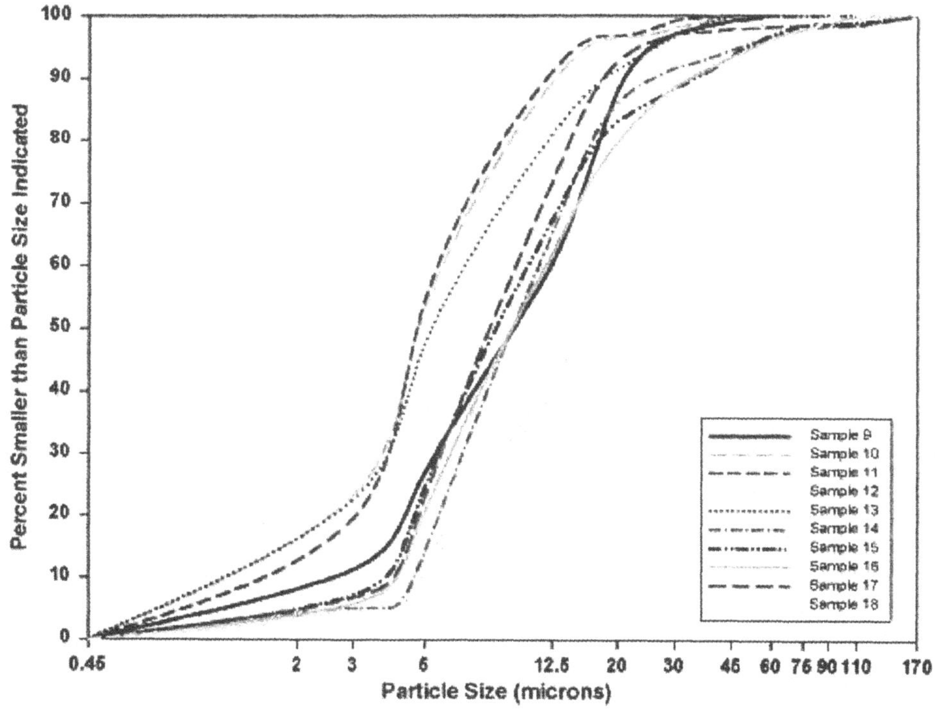

Figure 6-33. Particle size distribution for samples 9-18.

6.1.4 Strengths of Major Constituent and Heavy Metal Associations with Particulates

Figures 6-34 through 6-42 show plots of the resulting filterable concentrations after the first eight stormwater samples were subjected to varying pH conditions, and Appendix G shows the numerical data for these tests. Appendix table blanks indicate concentrations that were too low compared to the instrument detection limits. Appendix G also lists the total and filterable concentrations for each sample, for comparison to the test results.

It was expected that the filtered concentrations after exposure would be intermediate between the original filterable (filtered through 0.45 µm filters and digested) and the initial total (unfiltered and digested) concentrations. If the concentrations under all pH conditions were similar to the concentrations of the original filtered samples, then the particulate bound metals are strongly associated with the particulates and are unlikely to enter the overlying water column. However, if the concentrations approached the total metal concentrations, and were much higher than the filtered metal concentrations, then the metals are weakly associated with the particulates and my become disassociated during normal environmental conditions, with potentially adverse water quality consequences.

The major constituents (calcium, potassium and sodium) had very little change for any pH exposure condition; all concentrations were very similar, with the exceptions of samples #2 and #5 for calcium which had increased retention under basic conditions. Magnesium also had several samples that had better retention at high pH, compared to low pH conditions. Cadmium had three samples that were strongly associated with the particulates, but the other five samples were weakly bound, or were mostly associated with the original filterable fraction. Chromium and copper both had one sample (#5) that had a very high totally digested concentration (78 and 87 µg/L) that were strongly bound under all pH conditions, but the other samples all indicated weakly bound chromium and copper. Only small fractions of the iron were released under the pH conditions of the exposure tests for all samples, indicating that iron was always strongly bound. Most of the exposed lead concentrations were closer to the original dissolved lead concentrations than the total lead concentration, indicating strong particulate associations for all pHconditions. Several of the zinc samples indicated weak associations with particulates at pH 4, but were more strongly associated at higher pH conditions.

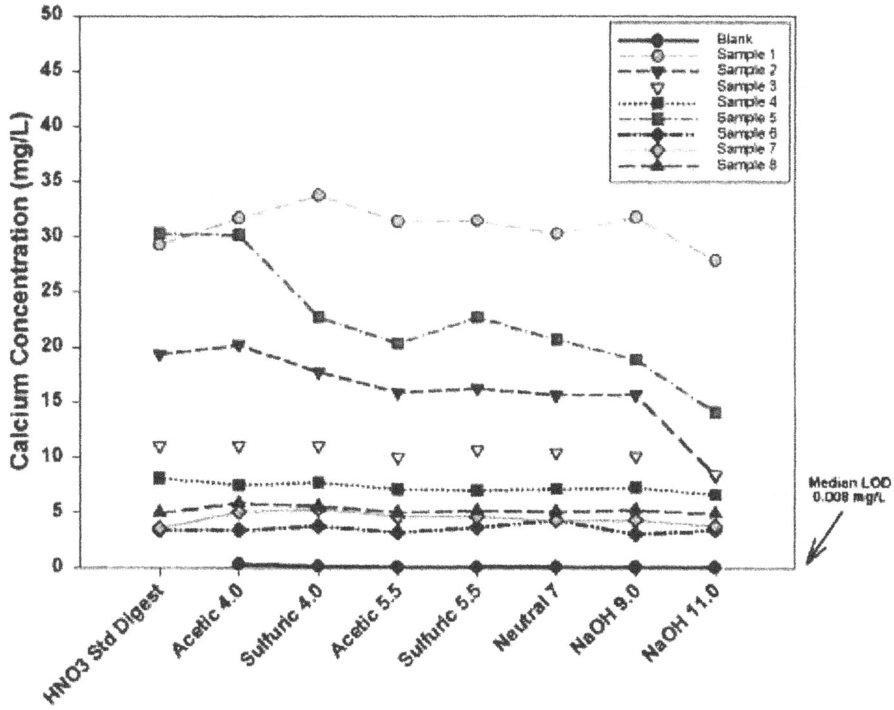

Figure 6-34. Leachability of calcium under different pH conditions.

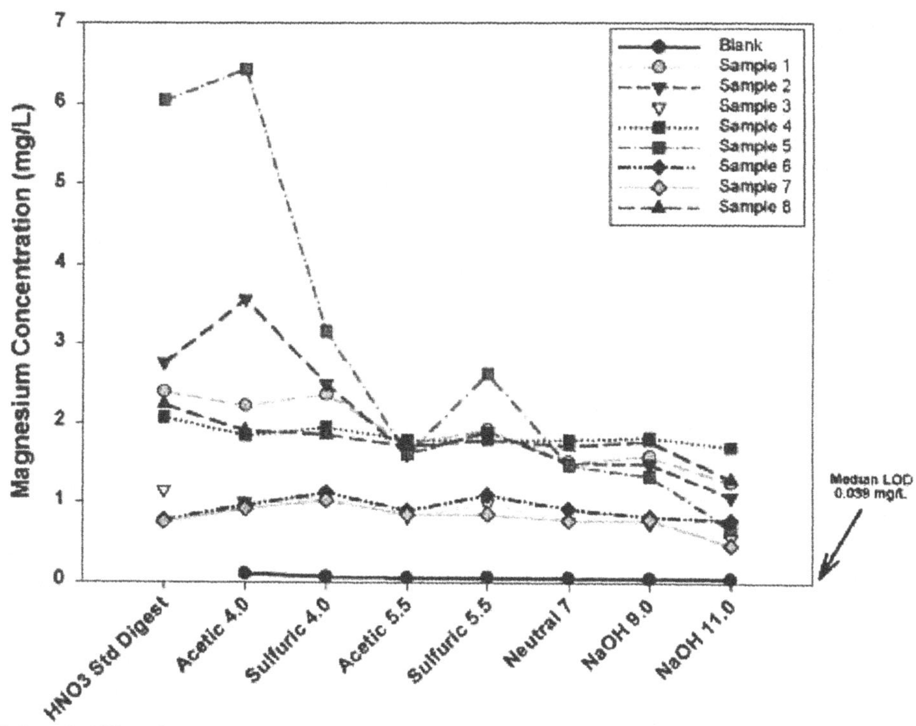

Figure 6-35. Leachability of magnesium under different pH conditions.

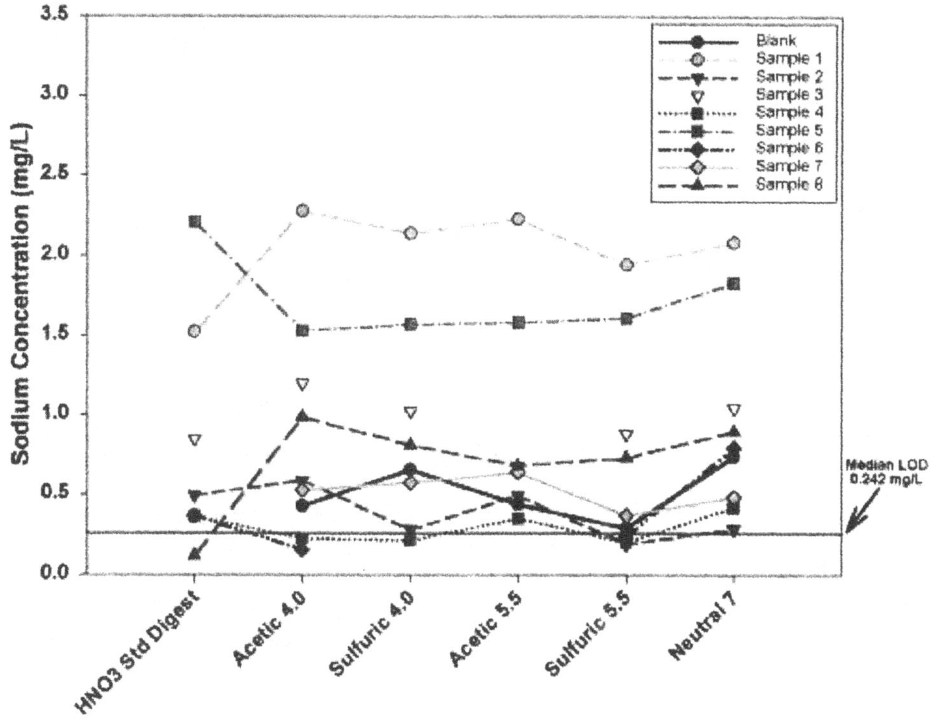

Figure 6-36. Leachability of sodium under different pH conditions.

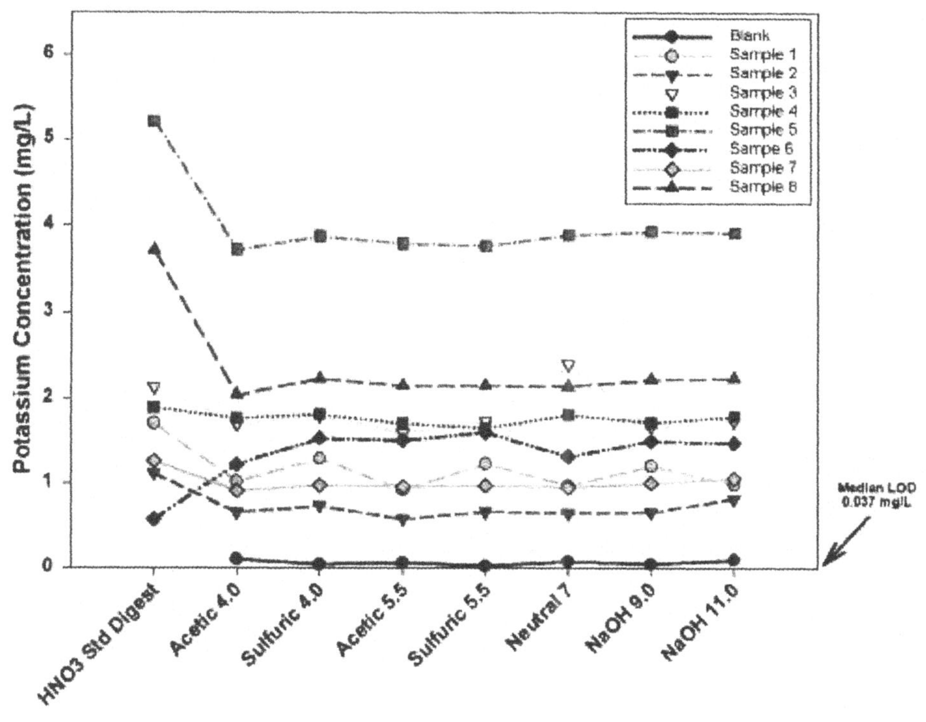

Figure 6-37. Leachability of potassium under different pH conditions.

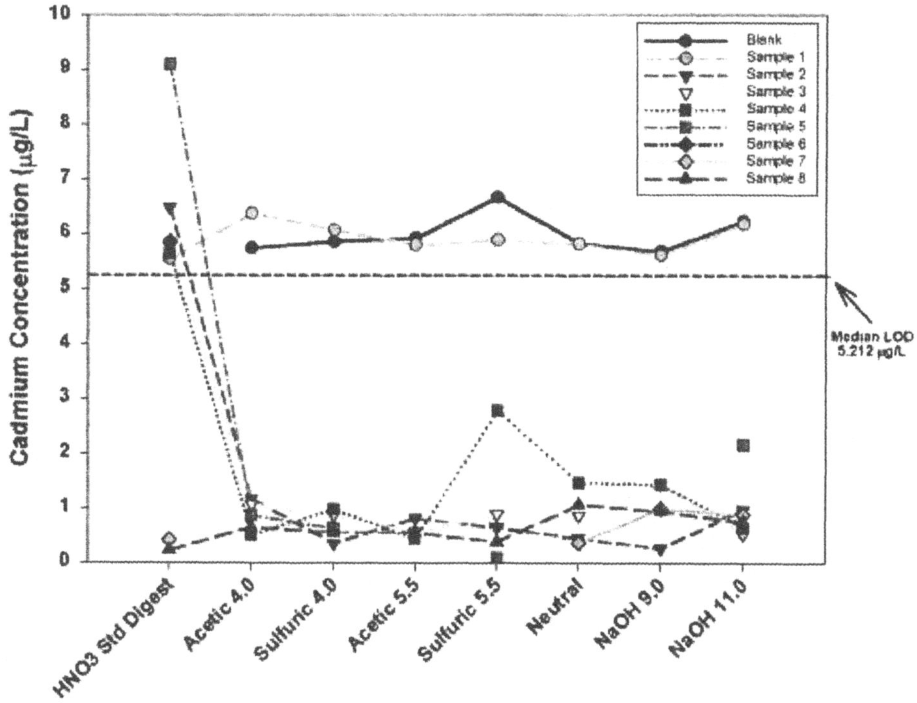

Figure 6-38. Leachability of cadmium under different pH conditions.

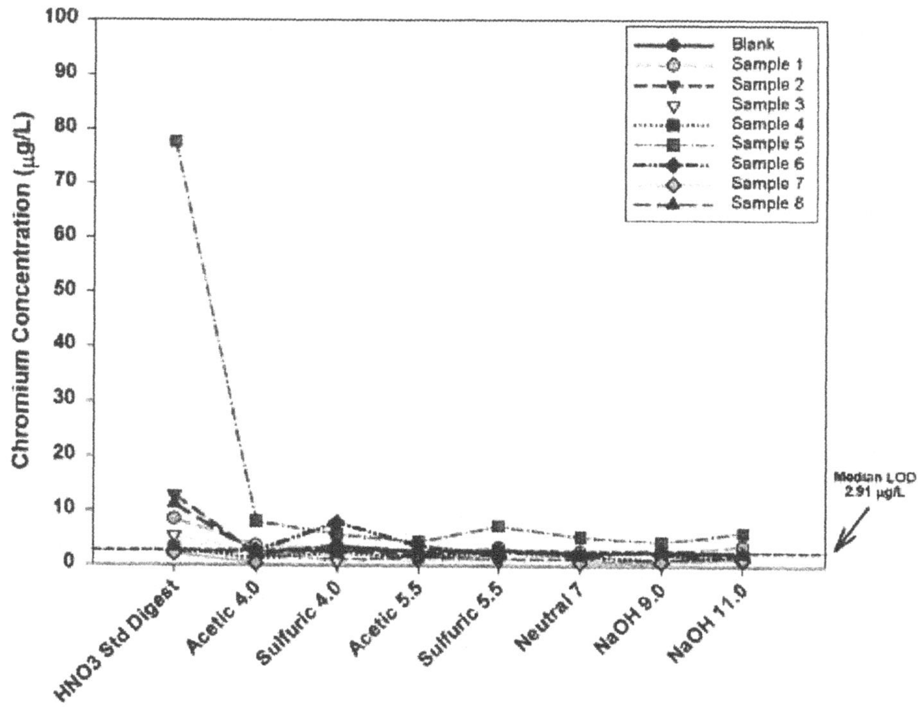

Figure 6-39. Leachability of chromium under different pH conditions.

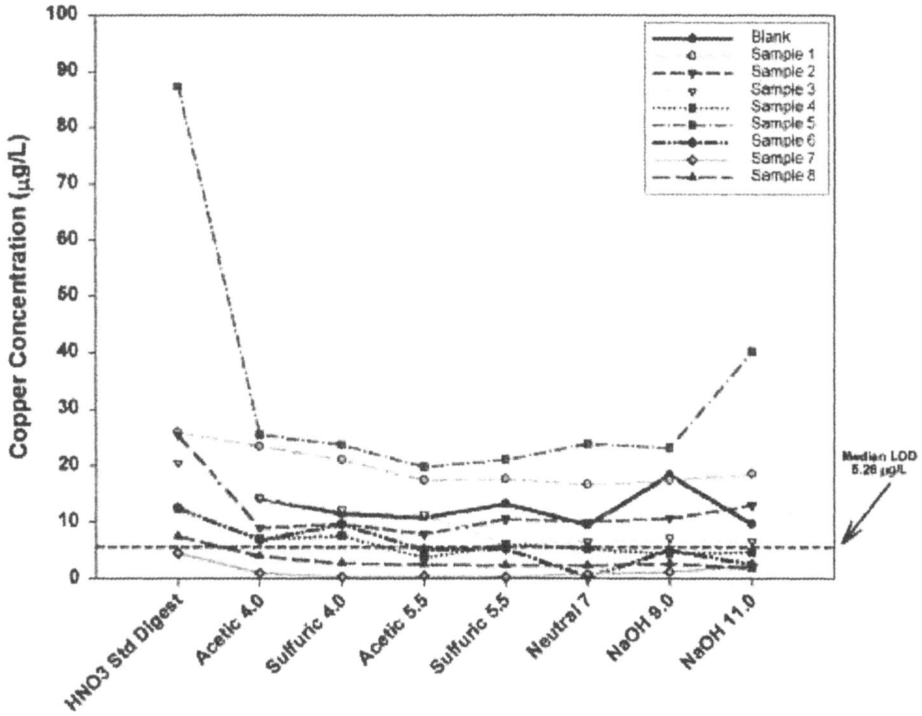

Figure 6-40. Leachability of copper under different pH conditions.

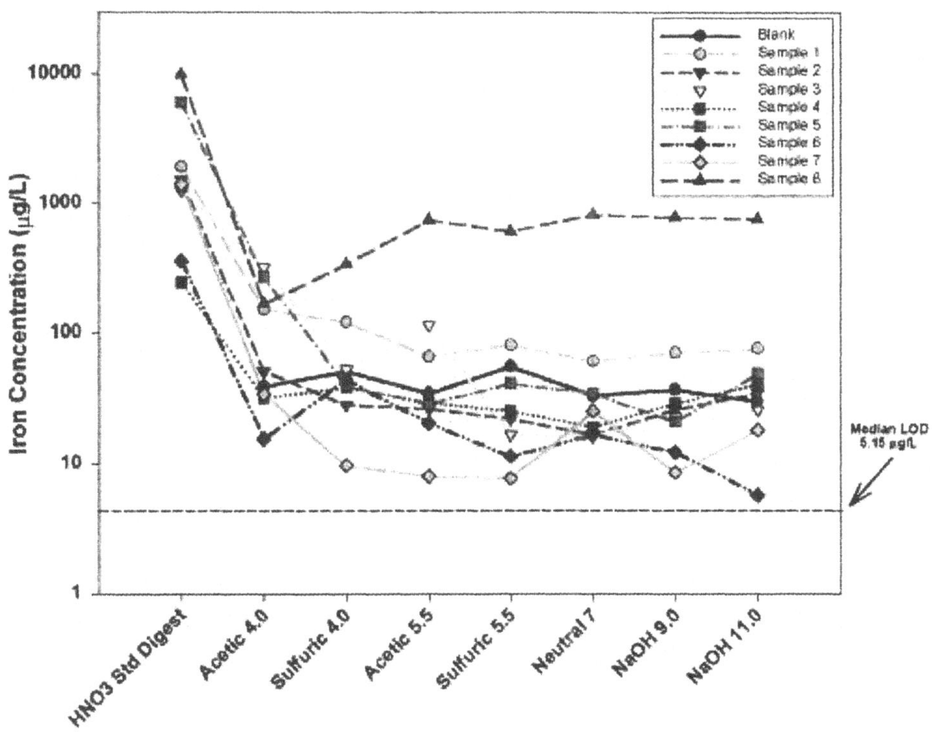

Figure 6-41. Leachability of iron under different pH conditions.

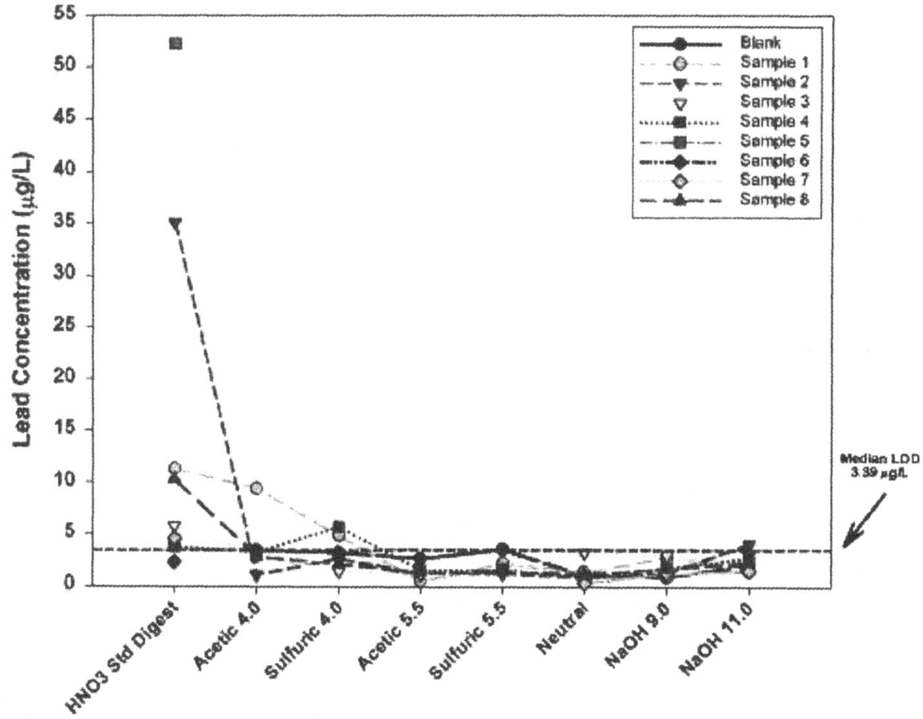

Figure 6-42. Leachability of lead under different pH conditions.

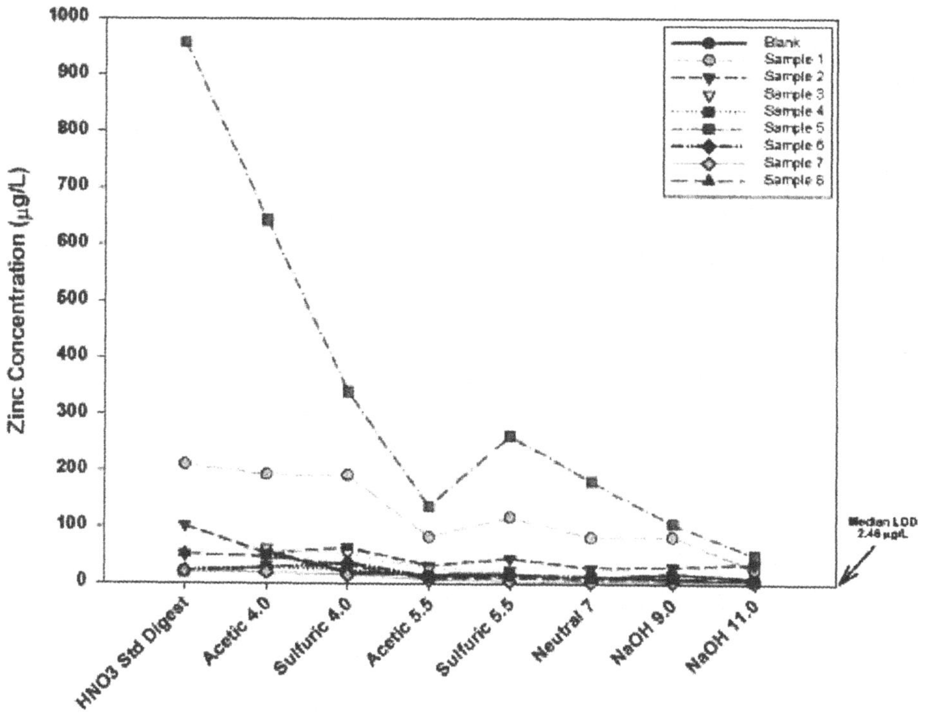

Figure 6-43. Leachability of zinc under different pH conditions.

6.2 Stormwater Heavy Metal Treatability

6.2.1 Associations with Particle Sizes

Most stormwater treatment efforts involve the physical removal of particulates. In order to better design sedimentation stormwater treatment devices, it is important to understand which pollutants are associated with different sized particulates and how they may be controlled during the removal of the particulates. Using data from the first eight samples, Table 6-5 lists the remaining concentrations for several monitored pollutants after controlling for particle sizes ranging from 20 to 0.45 μm. The corresponding percentage reductions are shown in Table 6-6. Some pollutants can be reduced by a reduction in particulates, such as suspended solids, total phosphorus and most heavy metals. Other pollutants, such as nitrates, are reduced much less, even after filtration down to 0.45 μm.

Table 6-5. Average Remaining Concentrations after Controlling for Different Sized Particles

		Residual Concentration after Removing all Particulates Greater than Size Shown (mg/L, unless otherwise noted)			
	Unfiltered	20 μm	5 μm	1 μm	0.45 μm
Total Solids	178	106	102	86	84
Suspended Solids	94	22	18	2	0
Turbidity (NTU)	53	30	24	4	2
Total-P	0.38	0.12	0.07	0.04	0.03
Total-N	2.13	1.50	1.25	1.38	1.63
Nitrate	0.6	0.60	0.60	0.53	0.50
Phosphate	0.8	0.23	0.18	0.15	0.10
COD	99	51	48	47	52
Ammonia	0.26	0.17	0.14	0.12	0.11
pH	7.35	7.45	7.66	7.39	7.53
Cadmium (μg/L)	4.9	3.9	3.8	3.8	3.8
Chromium (μg/L)	15.4	4.7	3.8	2.8	2.5
Copper (μg/L)	24.6	18.3	16.2	17.8	15.6
Iron (μg/L)	2830	1350	1050	140	80
Lead (μg/L)	15.7	9.2	6.0	3.7	2.8
Zinc (μg/L)	179	64	53	53	51

Table 6-6. Average Percentage Reduction in Pollutants after Controlling for Different Particle Sizes

	Percent Pollutant Reduction after Removing all Particulates Greater than Size Shown			
	20 μm	5 μm	1 μm	0.45 μm
Total Solids	40%	43%	52%	53%
Suspended Solids	76	81	98	100
Turbidity	43	55	92	96
Total-P	68	82	89	92
Total-N	30	41	35	23
Nitrate	0	0	12	17
Phosphate	71	78	81	88
COD	48	52	52	47
Ammonia	35	46	54	58
Cadmium	20	22	22	22
Chromium	69	81	82	84
Copper	26	34	34	37
Iron	52	63	95	97
Lead	41	62	76	82
Zinc	64	70	70	72

Most well designed wet detention ponds remove most particulates down to about 1 to 5 μm, depending on the rain conditions and drainage area. Smaller ponds may only be able to remove the particulates down to about 20 μm, while no pond can remove the filterable fraction by physical removal processes alone. The following list shows which constituents can be controlled to different levels and the associated particulate removal targets:

>90%
- suspended solids (0.45 to 1 μm)
- turbidity (0.45 to 1 μm)
- total phosphorus (0.45 μm)
- iron (0.45 to 1 μm)

80 to 89%
- suspended solids (5 μm)
- turbidity (2 μm?)
- total phosphorus (1 to 5 μm)
- phosphate (0.45 to 1 μm)
- chromium (0.45 to 5 μm)
- lead (0.45 μm)

70 to 79%
- suspended solids (20 μm)
- turbidity (3 μm?)
- phosphate (5 to 20 μm)
- lead (1 μm)
- zinc (0.45 to 5 μm)

all <50%
- total nitrogen
- nitrate
- COD
- cadmium
- copper

Of course, samples for other locations and situations would likely result in different levels of control. However, the predicted level of control associated with particulate control at least to 1 to 5 μm are close to what is expected for a well-designed wet detention pond designed for the control of these particulates, with the exception of copper which would normally be better controlled than indicated with this data. These data enable us to specify the level of treatment of particulates to obtain specific pollutant treatment goals.

The Chelex-100 tests indicated that almost all of the filterable (through 0.45 μm filters) heavy metals (and major ions) were not associated with colloids and organo-metallic complexes, but occurred as free ions. The main exception was cadmium, with only about 30% of the filterable cadmium in ionic forms, and copper with about half of the filterable copper in ionic forms. The UV exposure tests used to indicate likely organo-metallic complexes were quite uncertain due to the low concentrations of all complexed metals in the filterable sample fraction and the imprecision of the toxicity tests at such low values.

6.2.2 Strength of Bonding with Particulates

As noted above and in the literature review, most of the heavy metals in stormwater are associated with the particulates larger than 0.45 μm in size, although some exceptions exist (usually zinc, and sometimes copper, can predominantly be associated with the filterable fraction of stormwater). The fate of these particulate-bound heavy metals is of interest when designing stormwater controls. Specifically, do these metals stay bound to the particulates under typical environmental conditions of pH and Eh? This is of special interest when considering sediment in urban streams, and captured sediment in stormwater controls. These experiments were therefore conducted to examine the likelihood of the metals disassociating from the particulates under pH conditions ranging from about 4 to 11. In addition, tests were conducted with both weak and strong acids. Related tests were conducted as part of the filter media evaluation task of this research to measure the disassociation potential of heavy metals (and nutrients) under aerobic and anaerobic conditions having extreme Eh values.

Table 6-7 summarizes the results of these tests, indicating the ease with which the major constituents and heavy metals can be disassociated from the particulates under different pH conditions. As expected, most of the calcium and sodium were in dissolved forms during these tests and remained so for the 3 day duration under all exposure conditions. However, one sample (#5, from the home improvement center storage area) did have significant particulate forms of calcium. Both magnesium and zinc were weakly bound at low pH, but were strongly bound at high pH conditions. In fact, the zinc likely precipitated under the high pH conditions, as the filterable concentrations after exposure were actually less than the original sample filterable concentrations. The potassium and all metals were strongly bound to the particulates under all conditions. Most of the cadmium concentrations were well below the limit of detection after the exposure, and were also less than the original filterable cadmium concentrations, indicating possible precipitation during the test. A couple of the potassium and cadmium samples remained dissolved during the exposure, while some of the dissolved copper and lead precipitated at 5.5 and neutral conditions. There was no apparent difference between exposure to the strong or the weak acids at the two low pH conditions for any constituent.

Table 6-7. Summary of Strengths of Associations of Major Constituents and Heavy Metals with Particulates under Varying pH Exposure Conditions

	pH exposure conditions (after 3 days on shaking table)				
	pH = 4	pH = 5.5	neutral	pH = 9	pH = 11
Calcium	dissolved[1]	dissolved	dissolved	dissolved	dissolved
Magnesium	weak[2]	moderate[3]	moderate	moderate	strong[4]
Potassium	strong	strong	strong	strong	strong
Sodium	dissolved	dissolved	dissolved	dissolved	dissolved
Cadmium	precipitated[5]	precipitated	precipitated	precipitated	precipitated
Chromium	strong	strong	strong	strong	strong
Copper	strong	strong	strong	strong	strong
Iron	strong	strong	strong	strong	strong
Lead	strong	strong	strong	strong	strong
Zinc	weak	strong	strong	precipitated	precipitated

[1] dissolved constituents occur then the exposed concentrations were close to the original filterable concentration, and most of the constituent was in the dissolved (filterable) fraction.
[2] weak associations occur when the exposed concentrations are close to the original total concentration which was substantially larger than the dissolved concentration.
[3] moderate associations occur when the exposed concentrations are intermediate between the original total and filterable concentrations.
[4] strong associations occur when the exposed concentrations remain close to the original filterable concentration which is substantially smaller than the total concentration.
[5] precipitation may have occurred when the exposed concentrations are less than the original filterable concentration.

These tests indicate that the heavy metals of concern remain strongly bound to the particulates during long exposures at the extreme pH conditions likely to occur in receiving water sediments. They will also likely remain strongly bound to the particulates in stormwater control device sumps or detention pond sediments where particulate-bound metals are captured. The associated tests examining metal binding to filtration media under aerobic and anaerobic conditions also found that the heavy metals will likely remain strongly associated with a variety of organic and inorganic media under varying Eh conditions. However, those other tests did indicate that captured nutrients are likely to come off the media under anaerobic conditions.

6.2.3 Coprecipitation as a Stormwater Heavy Metal Removal Mechanism

A preliminary investigation of stormwater metal co-precipitation, using biogenic iron and sulfur minerals, was planned during this project. After the initial literature review, it was concluded that this process, if occurring at all in stormwater treatment systems, would be extremely rare due to the requirement for strictly anaerobic conditions. Stormwater conveyance and treatment systems may periodically go anaerobic, but only for short periods between runoff events when treatment is of little importance. However, stream sediments in urban receiving waters may be anaerobic for extended periods of time. Therefore, co-precipitation of heavy metals may be an important fate mechanism for stormwater heavy metals, but it is unlikely to be a significant removal mechanism that can be exploited for stormwater treatment.

Most dissimilatory iron-reducing bacteria are strictly anaerobic organisms (*Geobacteriaceae*). Although organisms of the genus *Shewanella* are alternate anaerobes (i.e., can live under aerobic conditions), it is only during anaerobic metabolism that the formation of magnetite (and therefore the possibility of metal co-precipitation) takes place. This in itself may limit the feasibility of co-precipitation of metals with biogenic magnetite as a result of dissimilatory Fe(III) reduction in stormwater treatment systems, which typically sustain anaerobic conditions only periodically. It should also be noted that magnetite is not always an end-product of dissimilatory Fe(III)-reduction, but that magnetite formation is in itself a function of the particular growth conditions (Fredrickson, *et al.* 1998; Cooper, *et al.* 2000).

More extensive tests of filter media metal retention behavior under extended aerobic and anaerobic conditions were therefore conducted as part of this research, instead of the originally planned co-precipitation experiments. Studies on the effect of anaerobiosis on metal retention by filter systems (Clark, *et al.* 2001 and Clark 2000) as part of this project, indicated that heavy metals are not mobilized from filter systems under anaerobic conditions. It was found that metal retention within the filters was not different from what was observed in oxygenated environments. However, it is plausible that under certain, specific environmental conditions, co-precipitation of metals by iron- and sulfate- reducing bacteria may take place in stormwater treatment systems.

CHAPTER 7.0

DESIGN GUIDELINES FOR THE TREATMENT OF STORMWATER HEAVY METALS

7.1 Introduction

Much information has been collected during this WERF-sponsored research project. This chapter summarizes the project information that can be directly used for the design of stormwater controls for the reduction of heavy metals. The first subsection (Treatability of Stormwater Heavy Metals) includes a brief overview of the associations of metals with different stormwater fractions (particle sizes, colloids, and ionic forms), from Chapter 6. This section also describes the results of tests that examined the strength of those associations. As an example, some metals may be weakly associated with particulates under certain pH or oxidation conditions. If these particulate-bound metals accumulate in sediment where these adverse conditions may occur, then the metals may be disassociated with the particulates and affect the water column and biota. These associations are also useful in predicting the performance of sedimentation controls in removing heavy metals. This information can be used to predict the heavy metal control associated with a wet detention pond that is designed to remove particles of a specific size, and larger, for example. The next subsection (Design of Filters for the Control of Stormwater Heavy Metals) summarizes material from Chapters 3 and 4 and includes several topics including the selection of filtration media, the clogging rates under different conditions, the metal retention capacity of the different media, etc. Special tests were also conducted during this research to investigate several important issues associated with the possible use of upflow filters. These filters hold promise in being much more cost-effective than the traditional slow filters that are most commonly used for the treatment of stormwater, but the behavior of the media under anaerobic conditions needed to be examined. A current, on-going EPA research project (a SBIR project being conducted by USInfrastructure, Inc.) is further investigating upflow filter designs, and some of the literature information on stormwater treatability in Appendix 6D is from that on-going study. Additional tests were also conducted during this WERF project that investigated the clogging behavior of the different media under upflow conditions. This subsection also outlines a simple approach in the selection of filtration media and the sizing of stormwater filters. The last subsection (Design of Grass-Lined Swales for the Control of Stormwater Heavy Metals) summarizes material mostly from Chapter 5. This subsection examines both the hydraulic and the water quality objectives of a swale design. It includes information on the selection of the grass species and other channel parameters. This subsection also shows how the important project information on the determination of channel roughness is used in the design of channels.

7.2 Review of Treatability of Stormwater Heavy Metals

This subsection contains information from Chapter 6 describing certain aspects of the treatability of heavy metals in stormwater, specifically the associations of the metals with different particle sizes and the filtered fraction.

7.2.1 Associations of Stormwater Heavy Metals with Different Particle Sizes and the Filterable Fraction

Table 7-1 summarizes the measured particle size associations during this research. As noted in Chapter 6, these data are similar to that found elsewhere.

Table 7-1. Summary Table Showing Heavy Metal Associations for Different Particle Sizes

	Copper		Iron		Lead		Zinc	
particle size (µm)	mg Cu/kg SS	COV	mg Fe/kg SS	COV	mg Pb/kg SS	COV	mg Zn/kg SS	COV
>250	50	na	28604	1.50	117	0.58	266	0.88
106 to 250	2137	1.45	21730	0.85	375	1.03	3486	0.79
45 to 106	1312	1.16	14615	0.72	226	0.85	2076	0.88
10 to 45	735	0.97	26221	0.54	229	0.50	1559	0.74
2 to 10	4668	1.60	18508	1.16	868	0.78	13641	1.88
0.45 to 2	2894	1.21	29267	1.31	199	1.40	13540	1.56

The association tests using the Celex-100 resin found that more than 90% of the filterable forms of calcium, magnesium, potassium, iron, and zinc were in ionic forms, with very little colloidal, or other bound forms. Also, more than 80% of the filterable chromium and lead were also ionic, while only about 50% of the filterable copper and 30% of the filterable cadmium were ionic.

This data can be used to estimate the level of control that may be associated with different designs of particle trapping devices. Some pollutants can be significantly reduced by a reduction in particulates, such as suspended solids, total phosphorus and most heavy metals. Other pollutants, such as nitrates, are reduced much less, even after filtration down to 0.45 µm. As an example, most well designed wet detention ponds remove most particulates down to about 1 to 5 µm, depending on the rain conditions and drainage area. Smaller ponds may only be able to remove the particulates down to about 20 µm, while no pond can remove the filterable fraction by physical removal processes alone. The following list, previously shown in Chapter 6, shows which constituents can be controlled to different levels and the associated particulate removal targets:

>90%

suspended solids (0.45 to 1 µm)

turbidity (0.45 to 1 μm)
total phosphorus (0.45 μm)
iron (0.45 to 1 μm)

80 to 89%
suspended solids (5 μm)
turbidity (2 μm?)
total phosphorus (1 to 5 μm)
phosphate (0.45 to 1 μm)
chromium (0.45 to 5 μm)
lead (0.45 μm)

70 to 79%
suspended solids (20 μm)
turbidity (3 μm?)
phosphate (5 to 20 μm)
lead (1 μm)
zinc (0.45 to 5 μm)

all <50%
total nitrogen
nitrate
COD
cadmium
copper

The predicted levels of control associated with particulate control at least to the 1 to 5 μm particle sizes are close to what is expected for a well-designed wet detention pond designed for the control of these particulates. An exception is copper, which would normally be better controlled than indicated from these data. These data can be used to specify the level of treatment of particulates to obtain specific pollutant treatment goals for other stormwater pollutants.

7.2.2 Strengths of Associations of Stormwater Metals to Particulates

As noted in Chapter 6 and in the literature review shown in Appendix A6, most of the heavy metals in stormwater are associated with the particulates larger than 0.45 μm in size, although some exceptions exist (usually zinc, and sometimes copper, can predominantly be associated with the filterable fraction of stormwater). The fate of these particulate-bound heavy metals is of interest when designing stormwater controls. Specifically, do these metals stay bound to the particulates under typical environmental conditions of pH and Eh? This is of special interest when considering sediment in urban streams, and captured sediment in stormwater controls. Special experiments were therefore conducted to examine the likelihood of the metals disassociating from the particulates under pH conditions ranging from about 4 to 11. In addition, tests were conducted with both weak and strong acids. Related tests were conducted as part of the filter media evaluation task of this research to measure the disassociation potential of heavy metals (and nutrients) under aerobic and anaerobic conditions having extreme Eh values (summarized in the next subsection).

Table 7-2 summarizes the results of these tests, indicating the likelihood that the major constituents and heavy metals can be disassociated from the particulates under different pH conditions. As expected, most of the calcium and sodium were in dissolved forms during these tests and remained so for the duration under all exposure conditions. Both magnesium and zinc were weakly bound at low pH, but were strongly bound at high pH conditions. In fact, the zinc likely precipitated under the high pH conditions, as the filterable concentrations after exposure were actually less than the original sample filterable concentrations. The potassium and all metals were strongly bound to the particulates under almost all conditions. Only small fractions of the iron were released under the pH conditions of the exposure tests for all samples, indicating that iron was always strongly bound. Most of the exposed lead concentrations were closer to the original dissolved lead concentrations than the total lead concentration, indicating strong particulate associations for all pH conditions. Several of the zinc samples indicated weak associations with particulates at pH 4, but were more strongly associated at higher pH conditions. There was no apparent difference between exposure to the strong or the weak acids at the two low pH conditions for any constituent.

Table 7-2. Summary of Strengths of Associations of Major Constituents and Heavy Metals with Particulates under Varying pH Exposure Conditions

	pH exposure conditions (after 3 days on shaking table)				
	pH = 4	pH = 5.5	neutral	pH = 9	pH = 11
Calcium	dissolved[1]	dissolved	dissolved	dissolved	dissolved
Magnesium	weak[2]	moderate[3]	moderate	moderate	strong[4]
Potassium	strong	strong	strong	strong	strong
Sodium	dissolved	dissolved	dissolved	dissolved	dissolved
Cadmium	precipitated[5]	precipitated	precipitated	precipitated	precipitated
Chromium	strong	strong	strong	strong	strong
Copper	strong	strong	strong	strong	strong
Iron	strong	strong	strong	strong	strong
Lead	strong	strong	strong	strong	strong
Zinc	weak	strong	strong	precipitated	precipitated

[1] dissolved constituents occur then the exposed concentrations were close to the original filterable concentration, and most of the constituent was in the dissolved (filterable) fraction.
[2] weak associations occur when the exposed concentrations are close to the original total concentration which was substantially larger than the dissolved concentration.
[3] moderate associations occur when the exposed concentrations are intermediate between the original total and filterable concentrations.
[4] strong associations occur when the exposed concentrations remain close to the original filterable concentration which is substantially smaller than the total concentration.
[5] precipitation may have occurred when the exposed concentrations are less than the original filterable concentration.

These tests indicate that the heavy metals of concern remain strongly bound to the particulates during long exposures at the extreme pH conditions likely to occur in receiving water sediments (the only exception was for zinc under the lowest pH condition tested). They will also likely remain strongly bound to the particulates in stormwater control device sumps or detention pond sediments where particulate-bound metals are captured. The associated filter media tests that examined metal binding to filtration media under aerobic and anaerobic conditions also found that the heavy metals will likely remain strongly associated with a variety of organic and inorganic media under varying Eh conditions. However, those other tests did indicate that captured nutrients are likely to come off the media under anaerobic conditions.

7.3 Design of Filters for the Control of Stormwater Heavy Metals

7.3.1 Selection of Filtration Media

The selection of a filtration media for removal of heavy metals from stormwater is dependent on many issues. The following discussion summarizes some of the results from tests conducted during this research and reported in Chapters 3 and 4 that affect the selection of filtration media. The most important issues are related to pollutant removal, pollutant retention capacity, clogging, and leaching of pollutants from the media. Costs and disposal considerations are also of considerable importance.

7.3.1.1 Pollutant Removal

The ability of a filtration media to remove the targeted heavy metal is probably the most important consideration in the design of a stormwater filter. The performance of the different tested media varied on the concentration of the metals and on the specific metal being investigated. Much prior laboratory work on filtration used very large concentrations of single metal solutions. These types of analyses were not found to be suitable indicators of performance under typical stormwater conditions.

Generally, the peat-sand mix, the St. Cloud zeolite, compost, and iron-oxide coated sand showed the best overall performance, even though the other media sometimes out-performed them for certain metals at different concentrations. Some media caused some degradation of the water. For example, kudzu and agrofiber caused an increase in the dissolved concentrations of several metals and major ions. Other media that released noticeable levels of major ions included TP 207 resin, peanut hulls and compost. During detailed tests using compost, zeolite, and peat-sand, the peat-sand mix was also found to perform best at removing dissolved metals.

All the media removed total suspended solids well, but only peat showed any removal of total dissolved solids. Compost appeared to increase the total dissolved solids concentration. Some improvement in total suspended solids and particulate bound metal removal efficiency was noted for the compost as residence time increased. The 0.5 and 1 minute residence times had slightly worse removal efficiencies than the 3, 5, and 10 minute residence times for compost; there was no observable improvement in particulate bound removal efficiency as residence time increased passed 3 minutes. Copper and chromium typically had lower removal efficiencies than the other metals. Peat had the best removal efficiencies for particulate bound metals. Removal efficiencies of compost and zeolite were approximately the same.

The peat-sand mix had the most dramatic impact on pH, dropping the pH from around 7 down to approximately 4 during the initial tests. No relationship between pH change and residence time was noted for peat, but the pH drop was less after large volumes of stormwater were treated.

Compost caused an increase in color in the treated stormwater, with increasing color with increasing column residence time. In several runs with a long residence time, peat-sand also slightly increased color, while in some of the runs with short residence times, peat-sand slightly decreased color.

Performance Affected by Stormwater Characteristics. The results of this project emphasize the importance of characterizing the wastewater to be treated before selecting a media. As an example, the TP 207 commercial ion-exchange resin had far better removal of zinc than any of the other media at high initial zinc concentrations, but had lower removal than many of the media at the low concentrations typically found in stormwater.

Behavior of Filter Media during Different pH Conditions. Sorption capacity increased for copper for all media as the pH neared its ideal sorption conditions at a pH of 5.0. For lead, the greater sorption capacity was observed at a lower pH (2.0). Zinc showed variable behavior. However, the capacity of compost for zinc decreased as the pH increased. As expected, once the metals were adsorbed onto the media, only negligible removal occurred during rinsing and subsequent exposure. The sorption onto the peat and compost appeared to be permanent under all conditions, likely due to the formation of complexes with the organic compounds on the surface of these materials. These complexes are likely to be 'permanent,' as shown by the lack of desorption under the significantly reducing conditions occurring in the compost.

Performance Differences Found During Batch vs. Flow Measurements. The sorption capacities determined during batch and continuous flow (breakthrough) studies varied greatly. In general, the batch-study capacities are at least equal to, if not greater than, the breakthrough capacities. The use of batch studies is not very applicable for determining capacities under actual field conditions.

Small Scale vs. Pilot-Scale Measurements. The results from the pilot-scale tests were compared to the prior results obtained during smaller laboratory-scale columns. For some parameters and some media, the predictions were excellent. The lack of ability to predict removal for all pollutants may be due to the inability of the media to remove a sufficient fraction of the pollutants in the laboratory-scale breakthrough experiments such that relatively accurate rate constants could be calculated. It also may be due to the extremely low influent concentrations of some pollutants to the pilot-scale filters. However, as noted above, the results from the batch testing of capacity and kinetics are not applicable to the operation of a filter in the field.

Microbial Growth Effects. Results of special column tests (comparing sterilized and regular columns) indicated that microbial growth had no effect on metals adsorption onto the media.

7.3.1.2 Conventional Slow Rate Filtration Clogging Tests

The clogging of the filters during conventional down flow, slow rate filtration can be predicted using the equations presented in Table 7-3 (based on work by Urbonas).

Table 7-3. Model Equations for the Effect of Solids Loading on Flow Rate

Filtration Media	Equation for Effect of Suspended Solids Loading on Flow Rate*
Sand	$u = 44500 \bullet L_{m,sand}^{-1.02}$
Carbon-Sand	$u = 14800 \bullet L_{m,carbon}^{-0.77}$
Peat-Sand	$u = 2000 \bullet L_{m,peat}^{-0.71}$
Compost-Sand	$u = 1.6 \times 10^{13} \bullet L_{m,compost}^{-4.09}$
Zeolite-Sand	$u = 60 \bullet L_{m,zeolite}^{-0.23}$
Agrofiber-Sand	$u = 205 - 0.09 \bullet L_{m,agrofiber}$
Cotton-Sand	$u = 106 - 0.01 \bullet L_{m,cotton}$

*Statistically significant coefficients ($\alpha = 0.05$).

7.3.2 Potential Benefits and Problems Associated with Upflow Filtration

Filter media apparently clogs before the media can experience chemical break-though when using stormwater. It is not yet clear if filter media that rely on "depth filtering" (where the sediment penetrates relatively far into the media) will be a cost-effective stormwater control, considering the pre-treatment needed to prevent this clogging. The necessary pretreatment alone may provide adequate control, without the additional filtration cost (Pitt, *et al.* 1999). Large-scale filtration installations (especially sand) have been shown to perform well for extended periods of time with minimal problems. The use of supplemental materials (such as organic compounds) should increase their performance for soluble compounds. The use of upflow filtration is also expected to increase the life of filters before clogging, for some media types.

7.3.2.1 Clogging Tests during Upflow Conditions

Laboratory tests were conducted to measure the sediment clogging problems associated with upflow filtration. Table 7-4 presents the results of these tests. These are all much greater than usually obtained during conventional gravity filtration where the captured sediment is retained on top of (or in) the filter media.

Table 7-4. Cumulative Suspended Solids Loading and Flow at Breakthrough (Failure) in Upflow Filtration

Filter Media	Cumulative Suspended Solids at Breakthrough/Failure (g/m²)	Cumulative Water Loading at Breakthrough/Failure (m)
Sand	4000	2.7
Activated Carbon-Sand	7000	4.7
Peat-Sand	13000	8.3
Compost-Sand	7500	4.7
Cotton-Sand	4500	2.9
Agrofiber-Sand	15000	9.6

Based upon the behavior observed during these experiments, the media was classified into the following three categories:

1. Sudden breakthrough with no decrease in flow rate (agrofiber-sand, cotton-sand, and carbon-sand): The clay front moved up through the media until it reached the top of the column and breakthrough occurred, i.e., true depth filtration. Taller filters should have a longer life.

2. Gradual breakthrough with decreasing flow rate (compost-sand and sand): Much of the clay was trapped at the bottom of the column and did not penetrate the column, i.e., surface filtration. Flow rate was reduced due to the blocking of the pores at the bottom by the trapped clay. No large clay front was observed. A larger sump to trap the non-penetrating clay particles off-line may be necessary. The behaviors of these media are similar to their behaviors during downflow filtration in that the predominant mechanism occurring is surface filtration. These media may be amenable to modeling using the physical clogging equation used for downflow filtration.

3. Gapping media (peat-sand): The media failed when a gap developed between the sand underlayer and the mixed media layer when the fluidized bed velocity was exceeded or when the media floated in water. The media failed when the pressure at the bottom of the filter pushed the peat-sand layer up to an unacceptable height and added peat to the effluent water. Gapping media may not be effective in upflow filtration unless well contained.

7.3.2.2 Head Loss during Upflow Filtration Conditions

In these tests, the upflow column design functioned well. Pumping the water allowed contact time to be controlled. At residence times of three to ten minutes, the compost and zeolite columns showed little sign of increased headloss during the runs, even when influent suspended solids concentrations were around 400 mg/L. At these lower flow rates, most of the suspended solids appeared to settle out in the sump area of the columns.

Peat-sand had the greatest headloss and the greatest change in headloss over the course of each test. Zeolite and compost had considerably less headloss, and typically only showed an increase in headloss for the two shortest residence times (1 and 0.5 minutes). The compost performed best (shown in Table 7-5). The degree of change in headloss increased as the total suspended solids concentration of the influent increased.

Table 7-5. Initial and final headloss (ft) across compost column.

Series	Actual Residence Time minutes	Flow Rate, m/day	Headloss in feet		
			Start	End	Increase during run
1	9.77	11.12	0.12	0.12	0.00
	4.23	25.68	0.12	0.12	0.00
	3.04	35.76	0.46	0.46	0.00
	1.19	90.99	0.81	0.92	0.12
	0.48	224.95	1.62	1.62	0.00
2	15.09	7.20	0.12	0.12	0.00
	5.16	21.05	0.23	0.23	0.00
	3.01	36.10	0.12	0.12	0.00
	0.94	115.25	0.69	0.69	0.00
	0.50	215.59	1.50	1.85	0.35
3	14.33	7.58	0.12	0.12	0.00
	5.30	20.52	0.12	0.12	0.00
	3.17	34.33	0.12	0.12	0.00
	1.04	104.64	0.81	0.92	0.12
	0.53	205.99	1.27	2.54	1.27

7.3.3.3 Residence/Contact Time in Columns

There was an observable relationship between dissolved metal removal and residence time. An increase in dissolved metal removal was observed as residence time increased. However, the impact of residence time on removal varied depending on the metal and the media. Lead, chromium, and copper had the most notable increase in dissolved metal removal as residence time increased. Zinc and cadmium, which had the greatest percent removal for all the media (above 90% initially), showed no improvement in removal efficiency as residence time increased.

7.3.3.4 Loss of Captured Materials from Media

Tests were conducted to measure the amounts of previously captured material lost during long interevent periods that may cause the media to be exposed to anaerobic conditions. As noted in Chapter 4, the heavy metals were strongly attached to the media for all conditions, while noticeable losses of some nutrients were observed.

7.3.3 Example Design Approach

The information obtained during earlier EPA sponsored research (Clark and Pitt 1999; Clark 2000) can be used to develop design guidelines for stormwater filtration, especially in conjunction with reported information in the literature and the new information obtained during this research. The design of a stormwater filter needs to be divided into two phases. The first phase is the selection of the media to achieve the desired pollutant removal goals. The second phase is the sizing of the filter to achieve the desired run time before replacement of the media.

The main objective of filter media research is usually to monitor a variety of filtration media to determine their pollutant removal capabilities. However, it soon becomes apparent that the filters are more limited by clogging caused by suspended solids in the stormwater, long before reductions in their pollutant removal capabilities can be identified. Therefore, measurements in filter run times, including flow rates and clogging parameters, need to be considered in filter design. Small-scale filter set-ups commonly used for pollutant removal measurements probably under-predict the actual run times that can be achieved under full-scale applications. Even with the increased filter depth utilization and better drying between storms that may be achieved with full-scale applications, pretreatment of stormwater so the suspended solids content is about 10 mg/L, or less, is probably necessary in order to take greater advantage of the pollutant retention capabilities of most of the media. This level of pretreatment, however, may make further stormwater control unnecessary, except for unusual conditions, as many other pollutants will also typically be reduced to low levels. Of course, it may be more cost-effective to consider shortened filter run times, without pretreatment, and not utilize all of the pollutant retention capabilities of the media.

7.3.3.1 Selection of Filtration Media for Pollutant Removal Capabilities

The selection of the filter media needs to be based on the desired pollutant removal performance and the associated conditions. If based on suspended solids alone for untreated stormwater (a likely common and useful criteria), then the filtration media would be ranked according to the following, according to Clark and Pitt (1999):

1) >90% control of suspended solids: compost/sand, act. carbon/sand, Zeolite/sand, Enretech/sand
2) 80 - 90% control of suspended solids: sand, peat/sand
3) very little control of suspended solids: filter fabrics

If based on a wider range of pollutants for untreated stormwater, then the ranking would be as follows:

1) sand, act. carbon/sand, Enretech/sand (no pollutant degradation, but sand by itself may not offer "permanent" pollutant retention until aged and has biological growths and/or deposition of silts and oils)
2) Zeolite/sand (no degradation)
3) compost/sand (color degradation)
4) peat moss/sand (turbidity and pH degradation)
5) filter fabrics alone (very little pollutant removal benefit)

Pre-settling of the stormwater was conducted to reduce the solids loadings on the filters to increase the run times before clogging in order to take better advantage of the pollutant retention capabilities of the filters. Settling reduced the stormwater suspended solids to about 10 mg/L, with about 90% of the particles (by volume) less than 10 μm in size. The untreated stormwater had a suspended solids concentration of about 30 to 50 mg/L, but many of the particles were larger, with about 90% of the particles being less than 50 μm. The pre-settling also reduced the other stormwater pollutants (color and turbidity by about 50%, and COD by about 90%, for example). This pre-settling was similar to what would occur with a well designed and operated wet detention pond. This pre-settling had a significant effect on the filter performance and the rankings would be as follows, considering a wide range of stormwater pollutants (suspended solids removal by itself would not be a suitable criteria, as it is not likely to be reduced any further by the filters after the pre-settling):

1) peat moss/sand (with degradation in color, turbidity, and pH)
2) activated carbon/sand (no degradation, but fewer benefits)

3) Enretech/sand, forest/sand, sand (few changes, either good or bad)
4) compost/sand (many negative changes)

Obviously, knowing the stormwater control objectives and options will significantly affect the selection of the treatment media. This is most evident with the compost material. If suspended solids removal is the sole criterion, with minimal stormwater pre-treatment, then it is the recommended choice (if one can live with a slight color increase in the stormwater, which is probably not too serious). However, if a filter is to be used after significant pre-treatment in order to have a longer filter life, a compost filter would be the last choice (not considering economics).

The following list summarizes the likely significant reductions in concentrations observed for different filter media (Clark and Pitt 1999):

- Sand: Medium to high levels of control for most pollutants, if the stormwater is not pre-treated. These levels of control are associated with retention of suspended solids and the associated particulate fractions of the pollutants. Can relatively easily flush previously captured pollutants. With pretreatment, has little additional benefit. Likely minimum effluent concentrations: 10 mg/L for suspended solids, 50 HACH color units, 10 NTU for turbidity.
- Peat moss/sand: Medium to high levels of control for most pollutants, for both untreated and pre-settled stormwater. Largest range and number of pollutants benefited under pre-settled conditions. Caused increases in color and turbidity, and reductions in pH (by about one pH unit). Likely minimum effluent concentrations: 5 mg/L for suspended solids, 85 HACH color units, 10 - 25 NTU for turbidity.
- Activated carbon/sand: Very good control for most pollutants, especially if the stormwater is not pre-treated. Also large number of benefited pollutants under pre-settled conditions. Caused no adverse changes for any pollutant. Likely minimum effluent concentrations: 5 mg/L for suspended solids, 25 HACH color units, 5 NTU for turbidity.
- Zeolite/sand: Medium to high levels of control for many pollutants for untreated stormwater, but no likely benefits for pre-settled stormwater. Caused increased color and turbidity on pre-settled stormwater. Likely minimum effluent concentrations: 10 mg/L for suspended solids, 75 HACH color units, 15 NTU for turbidity.
- Compost/sand: Medium to very high levels of control for many pollutants for untreated stormwater, but worsened water quality for many pollutants if pre-settled. Increased color under all conditions and had increased phosphate and potassium in effluent. Likely minimum effluent concentrations: 10 mg/L for suspended solids, 100 HACH color units, 10 NTU for turbidity.
- Enretech/sand: Medium to high levels of control for many pollutants for untreated stormwater, but had little effect on pre-settled stormwater. Likely minimum effluent concentrations: 10 mg/L for suspended solids, 80 HACH color units, 10 NTU for turbidity.
- Filter fabrics: No significant and/or important reductions for any pollutants using either untreated or pre-settled stormwater.

7.3.3.2 Design of Filters for Specified Filtration Durations

The filtration durations measured during these tests can be used to develop preliminary filter designs. It is recommended that allowable suspended solids loadings be used as the primary controlling factor in filtration design. Clogging is assumed to occur when the filtration rate becomes less than about 1 m/day. Obviously, the filter would still function at smaller filtration flow rates, especially for the smallest rains in arid areas, but an excessive amount of filter by-passing would likely occur for moderate rains in humid areas. Tables 7-6 and 7-7 summarize the observed filtration capacities of the different media tested by Clark and Pitt (1999). The wide ranges in filter run times as a function of water are mostly dependent on the suspended solids content of the water, especially when the water is pre-treated. Therefore, the suspended solids loading capacities are recommended for design purposes.

Table 7-6. Filtration Capacity as a Function of Suspended Solids Loadings (Clark and Pitt 1999)

Filtration Media	Capacity to 20 m/day	Capacity to 10 m/day	Capacity to <1 m/day
Sand	150-450 gSS/m²	400->2000 gSS/m²	1200-4000 gSS/m²
Peat/sand	100-300	150-1000	200-1700
Peat	?	?	200
Leaves	?	?	2100
Activated carbon/sand	150-900	200-1100	500->2000
Zeolite/sand	200-700	800-1500	1200->2000
Compost/sand	100-700	200-750	350-800
Enretech/sand	75-300	125-350	400-1500

Table 7-7. Filtration Capacity as a Function of Pre-Treated Water (generally <10 mg SS /L) Loading (Clark and Pitt 1999)

Filtration Media	Capacity to 20 m/day	Capacity to 10 m/day	Capacity to <1 m/day
Sand	6-20 m	8->25 m	13->40 m
Peat/sand	3-17	4-22	7-30
Activated carbon/sand	5-25	6->25	15->40
Zeolite/sand	7-25	8->25	14->40
Compost/sand	3-20	4-30	6->30
Enretech/sand	3-11	4-25	15->30

The most restrictive materials (the Enretech and Forest Products media) are very fibrous and still show compaction, even when mixed with sand. The most granular media (activated carbon and the Zeolite) are relatively uniform in shape and size, but have sand interspersed to fill the voids to slow the water to increase the contact time for better pollutant removal. The sand has the highest filtration rates because it has the most uniform shape and size.

The test observations indicated that only about 2.5 cm of the filter columns (about 10%) were actually used for solids retention during these tests. A full-scale filter could utilize about 5 times these depths for solids retention, if care was taken to allow selective piping to deeper depths, while not providing short-circuiting through the complete filter column. This could be most easily accomplished by placing a turf grass layer on top of the media (as in the peat-sand filter designs of the Metropolitan Washington (D.C.) Council of Governments). It is recommended that the roots of the grass used in the cover layer do not extend below about one-half of the filtration depth (the root depth should therefore be up to about 12 cm). Mechanical removal of the clogged layer to recover filter flow rates was not found to be very satisfactory by Clark and Pitt (1999), but has been used successfully during full-scale operations. Great care must be taken when removing this layer, as loosening the media, besides increasing the flow capacity of the filter, will also enable trapped pollutants (associated with the suspended solids) to be easily flushed from the media.

The flow rates through filters that have thoroughly dried between filter runs significantly increases. Small-scale tests restrict complete drying during normal inter-event periods which may occur more commonly with full-scale filters. Wetting and drying of filters (especially peat) has been known to produce solution channels through the media that significantly increases the flow. If these solution channels extend too far through the filter, they would cause short-circuiting and would therefore reduce pollutant retention. Adequate filter depths will minimize this problem. Table 7-8 shows the observed increases in filter flow rates for saturated (and partially clogged filters) and the associated flow capacity recovery for filters that have been thoroughly dried and then re-wetted. The filter fabrics did not indicate any flow rate improvements with wetting and drying, while the peat moss/sand filter had the greatest improvement in flow capacity (by about ten times), as expected. The other media showed much more modest improvements (but still about two to three times).

Table 7-8. Filter Flow Rates (m/day) for Saturated (and Partially Clogged) Filters and Recovered Filtration Capacity after Through Drying (Clark and Pitt 1999)

Filter Condition	sand	peat moss/ sand	act. carbon/ sand	Zeolite/ sand	compost/ sand	Enretech/ sand	forest/ sand	Emcon™ fabric	Gunder-boom™ fabric
Saturated/ partially clogged	13	4.0	17	17	13	8.4	8.4	850	200
Recovered flow capacity after drying	40	42	33	39	32	24	17	850	200
Increase in flow (multiple)	3.1X	11X	1.9X	2.3X	2.5X	2.9X	2.0X	1.0X	1.0X

The above filter capacity ranges are associated with varying test conditions and may be further grouped into approximate categories, as shown on Table 7-9, which are multiplied by 5 to account for an anticipated greater filter flow capacity associated with full-scale applications.

Table 7-9. Expected Full-Scale Media Flow Capacities (Clark and Pitt 1999)

Capacity to <1 m/day	Capacity to 10 m/day	Filtration Media Category
5,000 gSS/m²	1,250 gSS/m²	Enretech/sand; Forest/sand
5,000	2,500	Compost/sand; Peat/sand
10,000	5,000	Zeolite/sand; Act. Carbon/sand
15,000	7,500	Sand

Filter designs can be made based on the predicted annual discharge of suspended solids to the filtration device and the desired filter replacement interval. As an example, Table 7-10 shows the volumetric runoff coefficients (Rv) that can be used to approximate the fraction of the annual rainfall that would occur as runoff for various land uses and surface conditions, based on small-storm hydrology concepts (Pitt 1987). In addition, Table 7-11 summarizes likely suspended solids concentrations associated with different urban areas and waters.

Table 7-10. Volumetric Runoff Coefficients (Rv) for Different Urban Areas (Pitt 1987)

Area	Volumetric Runoff Coefficient (Rv)
Low density residential land use	0.15
Medium density residential land use	0.3
High density residential land use	0.5
Commercial land use	0.8
Industrial land use	0.6
Paved areas	0.85
Sandy soils	0.1
Clayey soils	0.3

Table 7-11. Typical Suspended Solids Concentrations in Runoff from Various Urban Surfaces (Pitt, et al. 1995)

Source Area	Suspended Solids Concentration (mg/L)
Roof runoff	10
Paved parking, storage, driveway, streets, and walk areas	50
Unpaved parking and storage areas	250
Landscaped areas	500
Construction site runoff	10,000
Combined sewer overflows	100
Detention pond water	20
Mixed stormwater	150
Effluent after high level of pre-treatment of stormwater	5

Using the information in the above two tables and the local annual rain depth, it is possible to estimate the annual suspended solids loading from an area. The following three examples illustrate these simple calculations.

1) A 1.0 ha paved parking area, in an area receiving 1.0 m of rain per year:

(50 mg SS/L) (0.85) (1 m/yr) (1 ha) (10,000 m^2/ha) (1,000 L/m^3) (g/1,000 mg) = 425,000 g SS/yr

Therefore, if a peat/sand filter is to be used having an expected suspended solids capacity of 5,000 g/m^2 before clogging, then 85 m^2 of this filter will be needed for each year of desired operation for this 1.0 ha site. This is about 0.9% of the paved area per year of operation. If this water is pre-treated so the effluent has about 5 mg/L suspended solids, then only about 0.2% of the contributing paved area would be needed for the filter. A sand filter would only be about 1/3 of this size.

2) A 100 ha medium density residential area having 1.0 m of rain per year:

(150 mg SS/L) (0.3) (1 m/yr) (100ha) (10,000 m^2/ha) (1,000 L/m^3) (g/1,000 mg) = 45,000,000 g SS/yr

The unit area loading of suspended solids for this residential area (450 kg SS/ha-yr) is about the same as in the previous example (425 kg SS/ha-yr), requiring about the same area (on a percentage basis) dedicated for the filter. The reduced amount of runoff is balanced by the increased suspended solids concentration.

3) A 1.0 ha rooftop in an area having 1.0 m of rain per year:
(10 mg SS/L) (0.85) (1 m/yr) (1 ha) (10,000 m^2/ha) (1,000 L/m^3) (g/1,000 mg) = 85,000 g SS/yr

The unit area loading of suspended solids from this area (85 kg SS/ha-yr) is much less than for the other areas and would only require a filter about 0.2% of the roofed drainage area per year of operation.

It is recommended that the filter media be about 50 cm in depth and that a surface grass cover be used, with roots not extending beyond half of the filter depth. This should enable a filtration life of about five times the basic life observed during these tests.

Normally, it is highly recommended that significant pre-treatment of the stormwater be used to reduce the suspended solids concentrations to about 10 mg/L before filtration for pollutant removal. This would not be needed for the rooftop runoff example, as the SS in that water is typically similar to this value. This pre-treatment can be accomplished using grass filters, wet detention ponds, or other specialized treatment (such as the sedimentation chamber in the multi-chambered treatment train, MCTT, applicable for critical source areas). The selection of the specific filtration media should be based on the desired pollutant reductions, but should in all cases include amendments to plain sand if immediate and permanent pollutant reductions are desired.

7.4 Design of Grass-Lined Swales for the Control of Stormwater Heavy Metals

7.4.1 Heavy Metal Concentration Reductions during Outdoor Swales Tests

Grass swales were found to reduce metal concentrations from stormwater. However, under some conditions, the water leaving a swale may be degraded. Field observations made during multiple rainfall events indicated that deposition of sediment and erosion of underlying soil occurred, and both deposition and erosion may occur simultaneously. During conditions favoring metal capture, the removal rates were close to 100%. During other rains, some scour appeared to resuspend previously captured material. This large variability in the metal removal performance was due to changes in the hydraulic conditions. Much less variability was found between the different grasses, while very little variability was found between the different metals. The swale hydraulics are therefore much more critical when reducing metal concentrations than the grass species.

7.4.1.1 Amended Soils Results

Many of the median concentrations in the surface runoff after the swale were less than the median initial concentrations, especially for the natural soil (8 of 10 constituents), however, the concentration reductions were generally small (5 to 35%, with typical reductions of about 20%). Only cadmium was observed to decrease in the surface flows for the amended soil swale. The concentration increases were likely associated with scoured sediment from previous storms.

The concentration changes between the surface and subsurface sampling locations showed fewer differences. The observed concentration differences were low for these tests due to the unexpectedly low metal concentrations at these sites, and the large flows that reduced the residence and contact times. Volume and mass reductions were not directly measured, but the amended soils had much greater infiltration rates than the non-amended natural soils, resulting in significantly reduced mass discharges of the heavy metals.

7.4.1.2 Grass Types and Metal Uptake

Both Centipede and Zoysia grass varieties, common in many lawns in the Southeastern US, can uptake metals from soils that receive stormwater. Experiments with stormwater found that Centipede had a greater maximum ability to accumulate Cu and Zn than Zoysia and Kentucky Bluegrass. Centipede seemed to also be able to accumulate more Pb. Cd and Cr accumulated in all three at similar rates and in very small amounts (~30 mg/Kg and 10 mg/Kg, respectively). Zoysia accumulated the least Pb, Cu and Zn of all three varieties in these studies.

Even though the data obtained in the outdoor swales after the first year seemed to indicate that Zoysia was a more efficient metal accumulator, the final results found that this is not the case; initially, Zoysia grass accumulated the metals faster than Centipede, but by the second year, this difference was no longer apparent.

Either grass variety seems to be able to accumulate metal from stormwater runoff. However, Centipede may possibly be the best choice for its resilience to drought, its nutritional frugality, and its greater ability to accumulate key contaminant metals such as Cu, Zn and Pb.

7.4.2 Grass-Lined Channel Design Approach

The ultimate goal of the grass-lined swale monitoring task of this project was to develop a set of guidelines that can optimize vegetated swales for the removal of metals and other contaminates in urban runoff. However, like all other hydraulic problems, optimization of vegetative swales can only be achieved thru a cost/benefit relationship, between six key parameters. These parameters include: permissible velocity, slope, infiltration rate (of soil), grass selection, channel shape/size, and maintenance efforts (EPA 1999a). All of these parameters can have a direct influence on the other (velocity is a function of channel slope, grass density can dampen velocity, etc.), so a careful balance must be maintained during the design of any vegetated swale. Optimization of a vegetated swale is therefore creating a design that will maximize the hydraulic residence time, while insuring sedimentation, infiltration and biological uptake pathways are preserved.

7.4.2.1 Velocity

The first parameter, velocity, is of great importance due to its influence on sedimentation rates and the vegetal retardance coefficient. Grassed waterways are designed for velocities of flow that are safe for the cover the site will support. If the vegetation cover fails, erosion may occur, because bare ground erodes under such velocities of flow (Fredenhagan and Doll 1954). It follows that a uniform, sod forming vegetation having a dense, relatively deep root system will offer the greatest protection against scour (USDA 1954). Vegetative swales are typically ineffective in and are vulnerable during large storms, because of the high-velocity flows which may result. A typical design storm used for sizing swales is a six-month frequency. Swales are generally not used where the maximum flow rate exceeds 5 cubic feet per second, and are commonly used to treat small drainage areas of less than 5 acres (EPA 1999a). Prime factors in determining permissible velocities are: physical nature of vegetation (type and distribution of root growth and top growth and physical condition), erodibility of soil, uniformity of cover, and bed slope. (USDA 1954)

7.4.2.2 Slope

The difficulty of providing adequate soil protection increases with increases in both velocity and bed slope. If possible, higher flows should be related to the shallowest slope possible to allow for the longest hydraulic retention times and, subsequently, the retention time for metals. (Newberry and Young 1996). Grass swales should be used on sites with relatively flat slopes of less than 4 percent (EPA 1999b). Slope will typically be mandated by the topography available at the site.

7.4.2.3 Infiltration Rate

Soil infiltration rates should be greater than one-half inch per hour to allow for metal removal and to provide drainage for swales at low slopes (EPA 1999a). The use of permeable soils is expected to increase the retention potential by allowing greater filtration and increasing the preference of the metals to sorb to the soil and biota of the grass strip (Newberry and Young 1996; Pitt, et al. 1999). Infiltration rates can be optimized by locating the swales on sandy soils, the use of soil amendments, or installation of perforated underdrains (Schueler 1994c).

7.4.2.4 Grass Selection

There are a number of factors to be considered in the establishment of an adequate vegetative cover. The soil must be suitable for the growth of vegetation, or it must be properly conditioned. A well-prepared seedbed for the vegetated waterway is one that is fine, firm, moist, and free of growing weeds and objectionable debris. If a seedbed is not properly prepared, the sod obtained is likely to be rough, spotty, thin, and weedy. It will be difficult to maintain, or the grass may even fail before becoming established. (Fredenhagan and Doll 1954).

7.4.2.5 Channel Shape

The *Handbook for Channel Design for Soil and Water Conservation* (USDA 1954) contains a rather complete set of diagrams for determining the dimensions of trapezoidal and triangular channels with many different side slopes, and for solving parabolic channels. In order to optimize a swale for metals removal, the shape must ensure that the height of the flow is less than or equal to the height of the vegetative cover (Dillaha, et al. 1988). In addition to the cross-sectional shape, the length of the swale is equally important. Longer swales increase hydraulic retention time, and therefore increase metal removal capabilities. Wang (1981) conducted a study of highway filter strips, similar to grassed swales, and determined that lead levels in soil samples obtained farther than about 30 m from the roadway were indistinguishable from background levels, with metal concentrations decreasing rapidly within the first 9 m.

7.4.2.6 Maintenance

The useful life of a vegetated swale system is directly proportional to its maintenance frequency. Grass should never be cut shorter than the design flow depth. Timely and proper mowing increases density and vigor of the turf, retards weed growth, prevents weed seed formation and develops a tougher sod. Swale performance can vary with season, based on grass condition and density (EPA 1999a).

7.4.2.7 Advantages of Grassed Swales

The advantages of a grass swale compared to a storm sewer system may include reduced peak flows, increased pollution removal, and lower capital costs. Grass swale systems are less expensive to construct, an estimated $6,900 per acre drained. The swales also require a smaller percentage of land (Kercher and Landon 1983).

Vegetated swales can serve as part of a storm water drainage system and can replace curbs, gutters and storm sewer systems. Therefore, swales are best suited for residential, industrial, and commercial areas with low flow and smaller populations. (EPA 1999a).

7.4.3 Design of Stable Grass-Lined Channels

The following discussion is summarized from Temple, *et al.* (1987) in *Stability Design of Grass-Lined Open Channels*, USDA Agricultural Handbook # 667. The stability limits of concern when designing a grass-lined channel are those related to the prevention of channel erosion. Since significant bed load transport, along with its associated detachment and redeposition, is incompatible with the maintenance of a quality grass cover, further design considerations may be needed to limit particle or aggregate detachment processes. This limitation results in the logical dominant parameter being the boundary stress effective in generating a tractive force on detachable particles or aggregates.

For the soils most often encountered in grass-lined channel design, particle detachment begins at levels of total stress low enough to be withstood by the vegetation without significant damage to the individual vegetal elements. When this occurs, the vegetation is undercut and the weaker vegetation is removed. This removal decreases the density and uniformity of the cover, which in turn leads to greater stresses at the soil-water interface, resulting in an increased erosion rate. Supercritical flow causes a more severe problem by the tendency for slight boundary or cover discontinuities to cause flow and stress concentrations to develop during these flow conditions.

For very erosion-resistant soils, the vegetal elements may sustain damage before the effective stress at the soil-water interface becomes large enough to detach soil particles or aggregates. Although the limiting condition in this case is the stress on the vegetal elements, failure progresses in much the same fashion. Damage to the vegetal cover results in an increase in effective stress on the soil boundary until conditions critical to erosion are exceeded. The ensuing erosion further weakens the cover, and unraveling occurs.

The potential for rapid unraveling of a channel lining once a weak point has developed, combined with the variability of vegetative covers, forces design criteria to be conservative. Very dense and uniform covers may withstand stresses substantially larger than those specified for short periods without significant damage. Reducing of the stability limits is not advised, however, unless a high level of maintenance guarantees that an unusually dense uniform cover will always exist. Also, unusually poor maintenance practices or nonuniform boundary conditions should be reflected in the design.

7.4.3.1 Selection of Channel Lining and Grass Systems based on Expected Roughness Conditions

The value of Manning's "n" in grasses is a function of grass type, and the product of velocity and hydraulic radius (VR). Grasses are divided into retardance classes based on their physical characteristics (height, width, density, etc.). Most sod forming grasses are classified as type C. These grasses can have "n" values ranging from 0.03 - 0.3 depending on VR, with a typical value of 0.03 in open channels. Figure 7-1 an example of a VR-n curve from the tests conducted during this research, super-imposed on early data from the Stillwater, OK, USDA field tests. This type of figure has long been used to design stable channels that are grass lined.

Tests were conducted under both controlled indoor locations and at outdoor swales during actual rains to obtain flow retardance curves needed for the design of grass-lined channels. Figure 7-1 shows three distinct regions 1) the indoor swale data 2) the outdoor swale data and 3) the historical retardance curves.

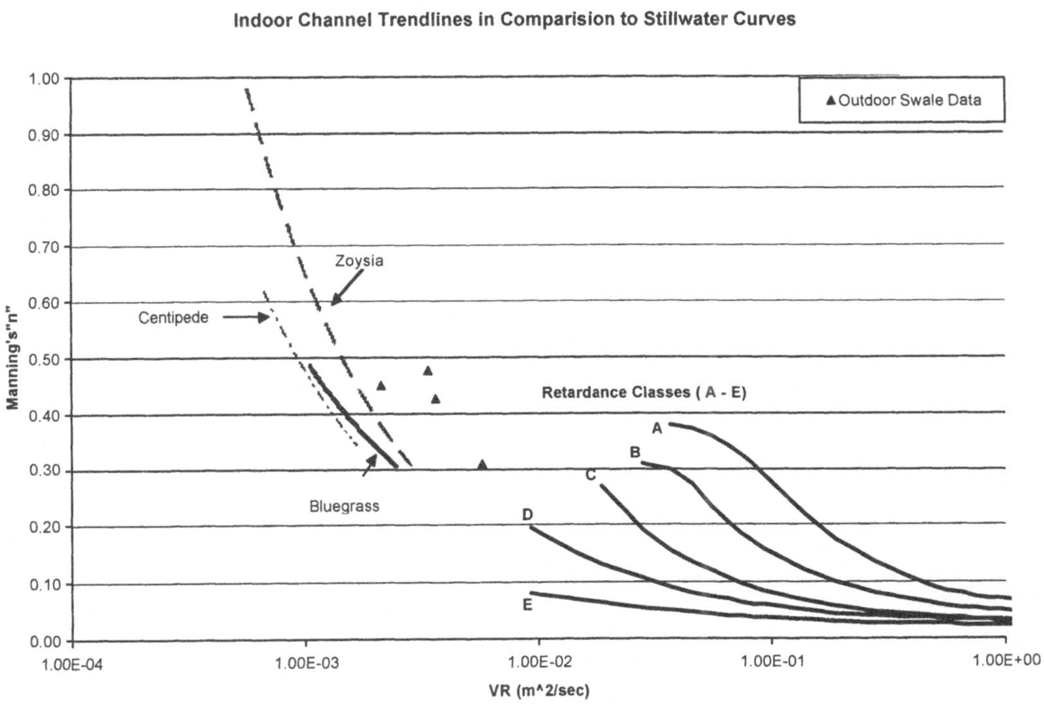

Figure 7-1: Indoor Grass "n" vs. VR Curves / Stillwater "n" vs. VR Curves

7.4.3.2 Example Problem

The appropriate Manning's "n" to use varies on the time frame: bare soil retention and vegetation establishment (short-term) and for fully grassed conditions (long-term) (Chow 1959). The following example illustrates how the VR-n curve can be used to select a stable channel system.

A channel slope is 20:1 (S=0.05), with a 4 foot (1.2 m) bottom width and 1:1 side slopes. The grass cover is expected to be in retardance group D. The two-year design storm produces a short-term peak flow of 45 cfs (1.27 cms). Assume that the limiting shear stress is 4.5 lb/ft^2 (250 N/m^2) (protected with an erosion control blanket having an n = 0.03).

Determine the depth of flow (Figure 7-2, from Chow 1959):

$$AR^{\frac{2}{3}} = \frac{nQ}{1.49S^{0.5}} = \frac{0.03(45cfs)}{1.49(0.05)^{0.5}} = 4.05$$

and $b^{8/3} = (4\text{ ft})^{8/3} = 40.32$

therefore $AR^{2/3}/b^{8/3} = 4.05/40.32 = 0.10$

With a 1:1 side slope trapezoidal channel, the ratio of y/b from Figure 7-2 is 0.25, and the depth is therefore: 4(0.25) = 1.0 ft.

The shear stress is therefore:

?yS= (62.4 lb/ft^3) (1.0 ft) 0.05 ft/ft) = 3.1 lb/ft^2

This calculated shear stress is less than the critical value of 4.5 lb/ft^2 and the short-term channel stability under bare soil conditions protected with an erosion control mat, but before establishment of vegetation, is assumed to be satisfactory.

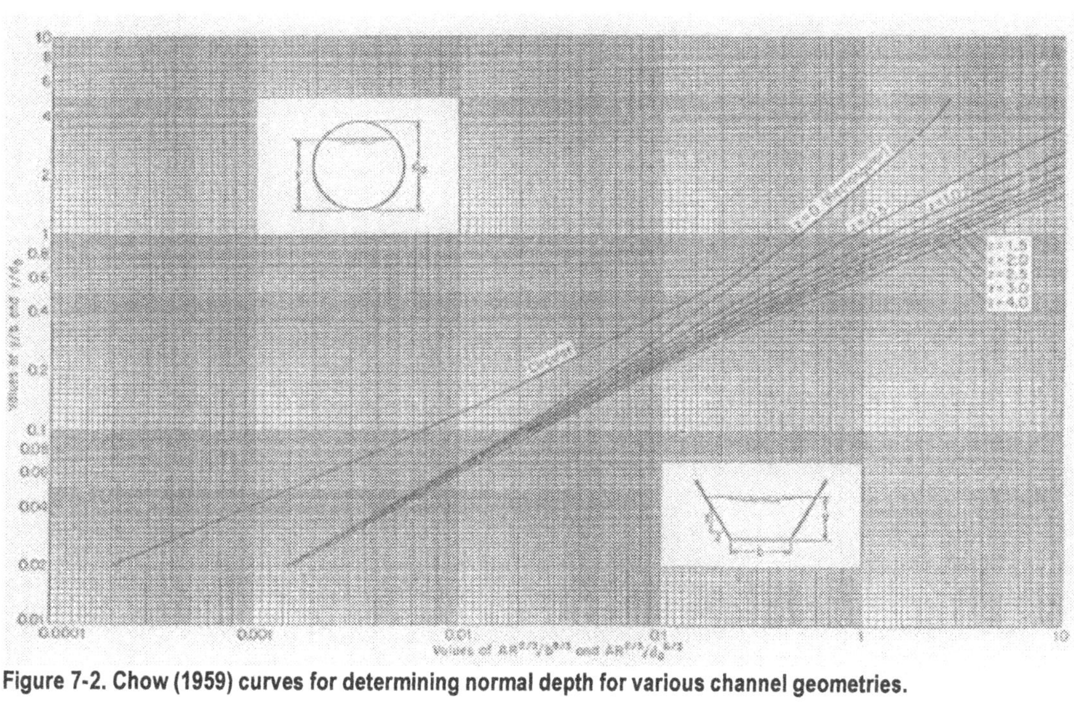

Figure 7-2. Chow (1959) curves for determining normal depth for various channel geometries.

During mature conditions, assume a $Q_{peak} = Q_{10year} = 70\ ft^3/s$ ($2\ m^3/s$). The limiting shear stress is 6 lb/ft^2 ($300\ N/m^2$). Assume an initial $n_{vegetated} = 0.05$.

The first step again is to determine the depth of flow (easiest to use Figure 7-2 again, from Chow 1959):

$$AR^{\frac{2}{3}} = \frac{nQ}{1.49S^{0.5}} = \frac{0.05(70cfs)}{1.49(0.05)^{0.5}} = 10.51$$

and $b^{8/3} = (4\ ft)^{8/3} = 40.32$

therefore $AR^{2/3}/b^{8/3} = 10.51/40.32 = 0.26$

With a 1:1 side slope trapezoidal channel, the ratio of y/b from Figure 7-1 is 0.43, and the depth is therefore: 4(0.43) = 1.7 ft.

From geometry, the area is therefore 9.7 ft^2, the velocity is ($70\ ft^3/sec$)/($9.7\ ft^2$) = 7.2 ft/sec, P is 8.8 ft, and R is 9.7/8.8 = 1.1 ft. VR is therefore (7.2 ft/sec)(1.1 ft) = 7.9 ft^2/sec. From Figure 7-1, the estimated new value for n is therefore 0.032, using a retardance class of D. The depth must therefore be recalculated, using this new value for n:

$$AR^{\frac{2}{3}} = \frac{nQ}{1.49S^{0.5}} = \frac{0.032(70cfs)}{1.49(0.05)^{0.5}} = 6.72$$

and $b^{8/3} = (4\ ft)^{8/3} = 40.32$

therefore $AR^{2/3}/b^{8/3} = 6.72/40.32 = 0.17$

With a 1:1 side slope trapezoidal channel, the ratio of y/b from Figure 7-2 is 0.34, and the depth is

therefore: 4(0.34) = 1.4 ft.

From geometry, the area is 7.6 ft^2, the velocity is 70/7.6 = 9.2 ft/sec, P is 8.0 ft, and R is 7.6/8.0 = 0.95 ft. The revised VR is therefore (9.2 ft/sec)(0.95 ft) = 8.7 ft^2/sec. Re-examining Figure 7-1, the revised value of n is still close to 0.032.

The shear stress is therefore:

?yS= ($62.4\ lb/ft^3$) (1.4 ft) 0.05 ft/ft) = 4.4 lb/ft^2

This calculated shear stress is less than the critical value of 6 lb/ft^2 and the long-term channel stability is assumed to be satisfactory.

7.4.4 Water Quality Analyses of Grass-Lined Channels

The expected water quality benefits associated with grass lined swales, having natural soil or amended soil, are shown in Tables 7-13 through 7-15. These calculations were made using WinSLAMM, the Source Loading and Management Model (Pitt and Voorhees 1995 and

2002). This model was calibrated using local Birmingham area conditions and set up to examine the volume and mass reductions associated with two swale designs. The area modeled was a typical medium density residential area. The swale was 4 ft wide, and 350 ft of swale length was available per acre for this area. These tables show how the swale, even with only 0.25 in/hr infiltration rates, was able to totally infiltrate the flows from the small rains, up to about 0.25 inch in depth. The swale performance decreased with larger rains, as expected. The 112 rains during the 1976 rain year were also examined (a typical rain year for local conditions).

Table 7-12. Calculated Runoff Volume Reductions Associated with Swales (ft^3/100 acres)

		0.25 in/hr (loam)		1.25 in/hr (loamy sand)	
Rain Depth (in)	Before Swale	After Swale	% reduction	After Swale	% reduction
0.01	4.75	0	100	0	100
0.05	1332	0	100	0	100
0.1	4147	0	100	0	100
0.25	20540	0	100	0	100
0.5	51670	14760	71	0	100
0.75	85620	43000	50	0	100
1	118600	70260	41	0	100
1.5	198360	144300	27	0	100
2	292250	232500	20	0	100
2.5	396690	325500	18	40710	90
3	503370	415030	18	61670	88
4	754990	649500	14	227570	70
1976 rain year	7189000	4454000	38	627700	91

Table 7-13. Calculated Particulate Solids Reductions Associated with Swales (lbs/100 acres)

		0.25 in/hr (loam)		1.25 in/hr (loamy sand)	
Rain Depth (in)	Before Swale	After Swale	% reduction	After Swale	% reduction
0.01	0	0	n/a	0	n/a
0.05	35	0	100	0	100
0.1	67	0	100	0	100
0.25	880	0	100	0	100
0.5	1260	75	94	0	100
0.75	1590	290	82	0	100
1	2250	770	66	0	100
1.5	3590	2240	38	0	100
2	5240	4090	22	0	100
2.5	7240	5940	18	740	90
3	9260	7640	17	1140	88
4	14520	12490	14	4380	70
1976 rain year	144029	71730	50	10840	92

Table 7-14. Calculated Total Copper Reductions Associated with Swales (lbs/100 acres)

		0.25 in/hr (loam)		1.25 in/hr (loamy sand)	
Rain Depth (in)	Before Swale	After Swale	% reduction	After Swale	% reduction
0.01	0	0	n/a	0	n/a
0.05	0	0	n/a	0	n/a
0.1	0.01	0	100	0	100
0.25	0.03	0	100	0	100
0.5	0.06	0.01	83	0.01	83
0.75	0.09	0.03	67	0.01	89
1	0.28	0.11	61	0.01	96
1.5	0.38	0.27	29	0.03	92
2	0.21	0.17	19	0.08	62
2.5	0.36	0.28	22	0.1	72
3	0.37	0.33	11	0.08	78
4	0.66	0.6	9	0.29	56
1976 rain year	8.3	4.3	48	1.5	82

Table 7-15. Calculated Total Zinc Reductions Associated with Swales (lbs/100 acres)

		0.25 in/hr (loam)		1.25 in/hr (loamy sand)	
Rain Depth (in)	Before Swale	After Swale	% reduction	After Swale	% reduction
0.01	0	0	n/a	0	n/a
0.05	0.02	0	n/a	0	n/a
0.1	0.07	0.03	57	0.02	71
0.25	0.55	0.15	73	0.14	75
0.5	0.95	0.66	31	0.33	65
0.75	1.21	0.88	27	0.37	69
1	2	0.96	52	0.4	80
1.5	7.2	5.9	18	1.1	85
2	11	9.2	16	1.4	87
2.5	5.9	5.5	7	1	83
3	16	15	6	3.3	79
4	59	55	7	13	78
1976 rain year	334	230	31	87	74

The particulate solids (Total Suspended Solids) had greater control than the runoff volume because of filtration that occurs during the smaller rains. Total copper removals were somewhat less than the TSS removals due to the presence of filterable forms of copper. Zinc removals were less due to the greater presence of filterable zinc forms.

Even with the small infiltration rate, the annual pollutant reductions are estimated to be between about 30 and 50%. With amended soils, bringing the infiltration rate to 1.25 in/hr, the pollutant removals greatly increased to between about 75 to 90%. Concentration reductions are much less, as indicated during the field measurements, and as shown in the example on Table 7-16. In this case, high TSS reductions are evident for the small rains, with about 50% reductions expected for all events smaller than about 0.75 inches. However, the reductions decrease dramatically as the rains increase. During the typical rain year, the overall TSS reductions would likely be low. These removal rates are similar to what has been reported during field observations.

Table 7-16. Particulate Solids (mg/L) Changes

		0.25 in/hr (loam)	
Rain Depth (in)	Before Swale	After Swale	% reduction
0.01	1.27	0	100
0.05	423	0	100
0.1	260	0	100
0.25	690	0	100
0.5	390	82	79
0.75	300	109	64
1	300	175	42
1.5	290	260	10
2	290	280	3
2.5	290	290	0
3	300	295	2
4	310	310	0

REFERENCES

Ab Razak, I.A., and Christensen, E.R. 2001. Water Quality Before and After Deep Tunnel Operation in Milwaukee, Wisconsin. *Water Res* 35:2683.

Abuzald, N.S., and Nakhla, G.F. 1994. Dissolved oxygen effects on equilibrium and kinetics of phenolics adsorption by activated carbon. *Environmental Science & Techonology* 28(2):216–221.

Adler, J. 1981. Automatic device for emptying a detention pond at a constant low flow rate and velocity. *1981 International Symposium on Urban Hydrology, Hydraulics, and Sediment Control,* University of Kentucky, Lexington.

Aires, N., and Tabuchi, J.P. 1995. Hydrocarbon separators and stormwater treatment. [French]. *TSM, Special Issue: Stormwater,* No. 11:862–864.

Akeley, R.P. 1980. Retention ponds for control of urban stormwater quality. In: *Proceedings National Conference on Urban Erosion and Sediment Control: Institutions and Technology,* EPA-905/9-80-002, Chicago.

Akeley, R.P. 1980. Retention ponds for control of urban stormwater quality. In *Proceedings National Conference on Urban Erosion and Sediment Control: Institutions and Technology,* EPA-905/9-80-002, Chicago.

Al-Haj Ali, A., and El-Bishtawi, R. 1997. Removal of lead and nickel ions using zeolite tuff. *Journal of Chemical Technology and Biotechnology* 69:27–34.

Allard, L.A., Graham, E., Platz, W., Carr, R., and Wheeler, J. 1996. The StormTreat™ system used as a storm water best management practice. In: *Proceedings of Watershed'96, A National Conference on Watershed Management.* Baltimore, MD, June 8–12. Tetra Tech, Inc. ISBN: 1-57278-028-2. pp. 463–465.

Allen, H.E. (editor) 1996. *Metal Contaminated Aquatic Sediments.* Chelsea, MI: Ann Arbor Press. 350 pp.

Allen, S., McKay, G., and Khader, K. 1988. Multi-component Sorption Isotherms of Basic Dyes onto Peat. *Environmental Pollution* 52:39–55.

Allen, S., Brown, P., McKay, G., and Flynn, O. 1992. An evaluation of single resistance transfer models in the sorption of metals ions by peat. *Journal of Chemical Technology and Biotechnology* 54:271–276.

Allen, S.J. 1981. *Mathematical Modeling of the Adsorption of Dyes from Effluent Using Peat.* Ph.D dissertation, Queens University, Belfast.

Alley, W.M. 1981. Estimation of Impervious-Area Washoff Parameters. *Water Resources Research* 17(4):1161–1166.

Alley, W.M. 1980. Determination of the Decay Coefficient in the Exponential Washoff Equation. *International Symposium on Urban Runoff,* University of Kentucky, Lexington.

American Public Health Association. 1995. *Standard Methods for the Examination of Water and Wastewater,* 18th edition. American Public Health Association, American Water Works Association, Water Pollution Control Federation, Alexandria, VA. 900 pages.

American Public Works Association. 1969. *Water Pollution Aspects of Urban Runoff. Water* Pollution Control Research Series WP-20-15, Federal Water Pollution Control Administration.

Ammon, D.C. 1979. *Urban Pollutant Buildup and Washoff Relationships.* Masters of Engineering Thesis, University of Florida.

Anderson, B.C., Brown, A., Watt, W.E., and Marsalek, J. 1998. Biological Leaching of Trace Metals from Stormwater Sediments: Influential Variables and Continuous Reactor Operation. *IAWQ 19th Biennial Int. Conf.,* Vancouver, Can. 8:72.

Anderson, M.A., and Rubin, A.J. 1981. *Adsorption of Inorganics at Solid-Liquid Interfaces.* Ann Arbor, MI: Ann Arbor Science Publishers, Inc. 260 pp.

Andral, M.C., Roger, S., Montrejaud-Vignoles, M., and Herremans, L. 1999. Particle Size Distribution and Hydrodynamic Characteristics of Solid Matter Carried by Runoff from Motorways. *Water Environ. Res.* 71(4):398.

Appel, P.L., and Hudak, P.F. 2001. Automated Sampling of Stormwater Runoff in an Urban Watershed, North-Central Texas. *J. Environ. Sci. Health Part A—Toxic/Hazardous Substances & Environmental Engineering* 36:897.

Aquatic Eco-Systems. 1999. Catalog. Aquatic Eco-Systems, Inc., Apopka, FL.

Armstrong, David E., and Llena, Reynaldo. 1992. *Stormwater Infiltration: Potential for Pollutant Removal.* Report prepared for the Wisconsin Department of Natural Resources (Madison) and the U.S. Environmental Protection Agency, Chicago.

Aronson, D.A., and Seaburn, G.E. 1974. *Appraisal of Operating Efficiency of Recharge Basins on Long Island, New York.* U.S. Geological Survey Water-Supply Paper 2001-D. USGS, Washington, DC.

Aronson, G., Watson, D., and Pisano, W. 1983. *Evaluation of Catchbasin Performance for Urban Stormwater Pollution Control.* Municipal Environmental Research Laboratory: Office of Research and Development, EPA-600/2-83-043, Cincinnati, OH.

American Society of Civil Engineers. 1975. *Sedimentation Engineering.* Ed. by Vito A. Vanoni. Manuals and Reports on Engineering Practice, No. 54.

Atasi, K., Fujita, G., Le Platte, G., Hufnagel, C., Keeler, G., Graney, J., and Chen, T. 2001. Impact of Atmospheric Deposition on the Headworks of a Wastewater Treatment Plant—a Case Study. *Water Sci. Tech.* 43:223a.

Atasi, K.Z., Fugita, G., LePlatte, G., Hufnagel, C., Keeler, G., Graney, J., and Chen, T. 1999. Impact of Atmospheric Deposition on the Headworks of Wastewater Treatment Plant—A Case Study. *Proc.Water Environ. Fed. 72nd Annu.Conf. Exposition,* [CD-ROM], New Orleans, LA.

Atasi, K.Z., Hufnagel, C., and Chen, T. 2000. Impact of Atmospheric Deposition on Surface Water Runoff of Toxic Chemicals in Urban Environment. In: *Watershed 2000 Management Conference,* Vancouver, BC, July. Water Environment Federation, CD-ROM. 2000.

Avery, S. V. 1995. *J. Chem. Technol. Biotechnol.* 62:3–16.

Ayyoubi, A. *Physical Treatment of Urban Storm Water Runoff Toxicants.* 1993. Master's Thesis. The University of Alabama at Birmingham Department of Civil Engineering, Birmingham, AL.

Bailey, H.C., Elphick, J.R., Potter, A., and Zak, B. 1999. Zinc Toxicity in Stormwater Runoff from Sawmills in British Columbia. *Water Res.* (G.B.) 33(11):2721.

Baker, A.J.M., McGrath, S.P., Reeves, R.D., Smith, J.A.C. 2000. Metal hyperaccumulator plants: a review of the ecology and physiology of a biological resource for phytoremediation of metal-polluted soils. In: N. Terry and G. Bañuelos (eds.) *Phytoremediation of contaminated soil and water,* pp. 85–107. Boca Raton, FL: Lewis Pub. CRC Press.

Baker, L.A., ed. 1994. *Environmental Chemistry of Lakes and Reservoirs.* American Chemical Society, Washington, DC. 615 pp.

Baladès, J.D., Bourgogne, P., and Gourioux. 1995a. A reservoir structure pavement for the urban storm drainage of a supermarket area of 120.000 m^2 in Bordeaux (in French). *International Conference NOVATECH 95,* Lyon. pp. 329–336.

Baladès, J.D., Legret, M., and Madiec, H. 1995b. Porous pavements as tools to control the pollutant loads on urban streets (in French). *International Conference NOVATECH 95,* Lyon. pp. 345–352.

Ball, J.E., Jenks, R., and Aubourg, D. 1998. An Assessment of the Availability of Pollutant Constituents on Road Surfaces. *Sci. Total Environ.* 209(2,3):243.

Ball, J.E. 2000. Runoff From Road Surfaces—How Contaminated is It? In: *Proc. Hydro 2000—Hydrology and Water Resources Symposium,* Perth, WA, Australia.

Baloch, S.J., Meyer, M.L., and Johnson, D.K. 1997. Mercury and Suspended Sediment Loadings in the Lower Minnesota River. *Environ. Sci. Technol.* (U.K.) 31(1):198.

Bannerman, R., Owens, D.W., Dodds, R.B., and Hornewer, N.J. 1993. Sources of pollutants in Wisconsin stormwater. *Water Science and Technology* 28(3–5):241–259.

Bannerman, R., Konrad, J.G., Becker, D., Simsiman, G.V., Chesters, G., Goodrich-Mahoney, J. and Abrams, S. 1979. *The IJC Menomonee River Watershed Study.* U.S. Environmental Protection Agency, EPA-905/4-79-029-C, Chicago, IL.

Bannerman, R., Baun, K., Bohn, M., Hughes, P.E., and Graczyk, D.A. 1983. *Evaluation of Urban Nonpoint Source Pollution Management in Milwaukee County, Wisconsin.* U.S. Environmental Protection Agency, PB 84-114164, Chicago, IL.

Bannerman, R.T., Legg, A.D., and Greb, S.R. 1996. Quality of Wisconsin Stormwater 1989–94. USGS Open-File Report 96-458.

Bao-rui, Yan. 1988. Investigation into Mechanisms of Microbial Effects on Iron and Manganese Transformations in Artificially Recharged Groundwater. *Water Science and Technology* 20(3):47–53.

Barbosa, A.E., and Hvitved-Jacobsen, T. 1999. Highway Runoff and Potential for Removal of Heavy Metals in an Infiltration Pond in Portugal. *Sco. Total Environ.* 235(1–3):151.

Barbosa, A.E., and Hvitved-Jacobsen, T. 2001. Infiltration Pond Design for Highway Runoff. Treatment in Semiarid Climates. *J. Environ. Eng.* 127:1014.

Barfield, B. 1986. Analysis of the effects of existing and alternate design criteria on the performance of sediment detention ponds. *1986 Sediment and Stormwater Conference,* sponsored by the Maryland Water Resources Administration, Sediment and Stormwater Division, Salisbury State College, MD, July–August.

Barkdoll, M.P., Overton, D.E., and Beton, R.P. 1977. Some Effects of Dustfall on Urban Stormwater Quality. *Water Pollution Control Federation* 49(9):1976–84.

Barrett, M.E., and Borroum, S. 2001. A Preliminary Assessment of the Cost, Maintenance Requirements and Performance of Sand Filters. In: *Proc. ASCE EWRI Conf.—Bridging the Gap: Meeting the World's Water and Environmental Resources Challenges.* CD-ROM.

Barry, S.A.S., and Clark, S.C. 1978. Problems of interpreting the relationship between the amounts of lead and zinc in plants and soil on metalliferous wastes. *New Phytol.* 81:773–783.

Barry, S.C., Taylor, S.E., and Birch, G.F. 1999. Heavy Metals in Urban Stormwater Canals Entering Port Jackson, Australia and Their Impact on the Estuarine Environment. In: *Proc. the Eighth International Conference on Urban Storm Drainage.* August 30–September 3, Sydney, Australia, ed. by I.B. Joliffe and J.E. Ball. The Institution of Engineers Australia, The International Association for Hydraulic Research, and The International Association on Water Quality, 1825.

Bartone, D.M., and Uchrin, C.G. 1999. Comparison of Pollutant Removal Efficiency for Two Residential Storm Water Basins. *J. Environ. Eng.-ASCE* 125(7):674.

Bartosova, A., and Novotny, V. 1999. Model of Spring Runoff Quantity and Quality for Urban Watersheds. *Water Sci. and Technol.* (G.B.) 39(12):249.

Becker, F.A., Hedges, P.D., and Smisson, R.P.M. 1995. The relationship between pollutants and specific sewage settling velocity fractions. *NOVATECH 95, 2nd International Conference on Innovative Technologies in Urban Storm Drainage,* May 30–June 1, Lyon, France. pp. 607–610. Organized by Eurydice 92 and Graie.

Bedner, R.E., and Fluke, D.J. 1980. *Demonstration of Debris Basin Effectiveness in Sediment Control.* Environmental Protection Agency, Industrial Environmental Research Laboratory–Cincinnati, EPA-600/7-80-154. Cincinnati, OH.

Benke, A.C., Willeke, G.E., Parrish, F.K., and Stites, D.L. 1981. *Effects of Urbanization on Stream Ecosystems.* School of Biology, Environmental Resources Center, Report No. ERC 07-81, Georgia Institute of Technology, Atlanta, GA.

Benoit, G., Rozan, T.F., Patton, P.C., and Arnold, C.L. 1999a. Trace Metals and Radionuclides Reveal Sediment Sources And Accumulation Rates in Jordan Cove, Connecticut. *Estuaries* 22(1):65.

Benoit, G., Wang, E.X., Nieder, W.C., Levandowsky, M., and Breslin, V.T. 1999b. Sources and History of Heavy Metal Contamination and Sediment Deposition in Tivoli South Bay, Hudson River, New York. *Estuaries* 22(2A):167.

Bernard, C., Herviou, P., Poujol, T., and Zobrist, C. 1996. Procedures for treatment of stormwater and quality objectives: two practical representative examples. In: *Proceedings: Urban Wet Weather Pollution: Controlling Sewer Overflows and Stormwater Runoff.* Quebec City, Quebec. June 16–19. Water Environment Federation, Alexandria, VA. pp. 1559–1568.

Berthouex, P.M., and Brown, L.C. 1994. *Statistics for Environmental Engineers.* Boca Raton, FL: Lewis Publishers. 200 pp.

Betson, R.P. 1978. Precipitation and Streamflow Quality Relationships in an Urban Area. *Water Resources Research* 14(6):1165–1169.

Beveridge, T.J. 1986. The immobilization of soluble metals by bacterial walls. *Biotechnol. Bioengineer. Symp.* 16:127–139.

Beveridge, T.J. 1989. Metal ions and bacteria. In: *Metal Ions and Bacteria,* T.J. Beveridge and R.J. Doyle, eds. New York: John Wiley and Sons. pp. 1–29.

Birch, G., and Taylor, S. 1999. Source of Heavy Metals in Sediments of the Port Jackson Estuary, Australia. *Sco. Total Environ.* 227(2–3):123.

Birch, G.F. 2000. Marine Pollution in Australia, with Special Emphasis on Central New South Wales Estuaries and Adjacent Continental Margin. *Internat. J. Environ. Pollut.* 13:573.

Birch, G.F., Scollen, A., Snowdon, R., and Suh, J. 1999. Sources of Heavy Metals in Stormwater Draining into Port Jackson, Sydney, Australia. In: *Proc. the Eighth International Conference on Urban Storm Drainage.* August 30–September 3, Sydney, Australia, ed. by I.B. Joliffe and J.E. Ball. The Institution of Engineers Australia, The International Association for Hydraulic Research, and The International Association on Water Quality. 2202.

Biskner, C.D., Baron, C.A., and Millcan, J.H. 1978. EPA's goal for suspended solids is not met with media filtration. *Pulp & Paper:* 126–128.

Blanchard, G., Maunaye, M., and Martin, G. 1984. Removal of heavy metals from waters by means of natural zeolites. *Water Research* 18(12):1501–1507.

Blaylock, M.J., Salt, D.E., and Dushenkov, S. 1997. Enhanced accumulation of Pb in Indian mustard by soil-applied chelating agents. *Environ. Sci. Technol.* 31:860–865.

Blocki, S.W. 1993. Hydrophobic zeolite adsorbent: a proven advancement in separation technology. *Environmental Progress* 12(3):226–230.

Boggess, D.H. 1975. *Effects of a Landfill On Ground-Water Quality.* United States Department of the Interior Geological Survey, Open File Report 75-594. Prepared in cooperation with the City of Fort Myers. U.S. Government Printing Office, Washington, DC.

Bohart, G., and Adams, E. 1920. Some aspects of the behavior of charcoal with respect to chlorine. *Journal of the American Chemical Society* 42:523–545.

Boller, M. 1997. Tracking Heavy Metals Reveals Sustainability Deficits Of Urban Drainage. System. *Water Sci. Technol.* 35(9):77.

Bondurant, J.A., Brockway, C.E., and Brown, M.J. 1975. Some aspects of sedimentation pond design. In: *Proceedings National Symposium on Urban Hydrology and Sediment Control,* University of Kentucky, Lexington.

Bonzongo, J-C., Butler, K.F., Rowls, E.A., and Lyons, W.B. 1999. Impacts of Land Use and Physicochemical Settings on Methyl Mercury Levels in the Mobile-Alabama River System (MARS). In: *Proceedings: AWRA's 1999 Annual Water Resources Conference—Watershed Management to Protect Declining Species,* December, Seattle, WA. American Water Resources Association. 525.

Booth, D.B., and Jackson, C.R. 1997. Urbanization of aquatic systems: Degradation thresholds, stormwater detection, and the limits of mitigation. *Journal of the American Water Resources Association* 33(5):1077–1090.

Bott, A.W. 2001. Voltametric determination of trace concentrations of metals in the environment. *Current Separations* 14(1):24–30. 1995.

Boudries, H., Broguet, C., Mouchel, J-M., and Thévenot, D.R. 1996. Urban Runoff Impact on Composition and Concentration of Hydrocarbons in River Seine Suspended Solids. In: *Proc. 7th Int. Conf. on Urban Storm Drainage,* Hannover, Germany, IAHR/IAWQ Joint Committee on Urban Storm Drainage. 569.

Box, G.E.P., Hunter, W.G., and Hunter, J.S. 1978. *Statistics for Experimenters.* New York: John Wiley and Sons, Inc.

Brady, D., Rose, P.D., and Duncan, J.R. 1994. The use of hollow fiber cross-flow microfiltration in bioaccumulation and continuous removal of heavy metals from solution by *Saccharomyces cerevisiae. Biotechnology and Bioengineering* 44:1362–1366.

Bremer, P.J., and Geesey, G.G. 1993. Interactions of bacteria with metals in the aquatic environment. In: *Particulate Matter and the Aquatic Environment,* S.S. Rao, ed. Boca Raton, FL: Lewis Publishers. pp. 41–63.

Brierly, C.L., Brierly, J.A., and Davidson, M.S. 1989. Applied microbial processes for metals recovery and removal from wastewater. In: *Metal Ions and Bacteria,* T.J. Beveridge and R.J. Doyle, eds. New York: John Wiley & Sons. pp. 359–382.

Brown, C., Placchi, C., and Gersberg, R. 1998. Modeling the Impacts of Surface Water Hydrology and Land Use on Water Quality in the Tijuana River Watershed, *Proc. Watershed Manage.—Moving from Theory to Implementation*, Denver, CO. WEF (CP3804). 477.

Brown, David. 1982. *Effects of Effluent Spray Irrigation on Groundwater at a Test Site near Tarpon Springs, Florida.* United States Department of the Interior Geological Survey Open File Report 81-1197. Prepared in cooperation with Pinellas County, FL. USGS, Denver, CO.

Brown, J. 2000. Kudzu as a Medium for Adsorption of Heavy Metals in Dilute Aqueous Wastestreams. *Practice Periodical Hazard., Toxic, Radioactive Waste Manage.* 4:82.

Brown, P.A., Gill, S.A., and Allen, S.J. 2000a. Metal removal from wastewater using peat. *Water Research* 34(16):3907–3916.

Brown, P., Jefcoat, I.A., Parrish, D., Gill, S., and Graham, E. 2000b. Evaluation of the adsorptive capacity of peanut hull pellets for heavy metals in solution. *Advances in Environmental Research* 4(1):19–29.

Brown, P.A. 1993. The application of peat and lignite to the removal of heavy metals from industrial wastewater. Ph.D dissertation, Queens University, Belfast.

Brown, P.A., Brown, J.M. and Allen, S.J. 2001. The Application of Kudzu as a Medium for the Adsorption of Heavy Metals From Dilute Aqueous Wastestreams. *Bioresource Tech.* 78:195.

Brydges, T., and Robinson, G. 1986. Two examples of urban stormwater impoundment for aesthetics and for protection of receiving waters. In: *Restoration of Lakes and Inland Waters: International Symposium on Inland Waters and Lake Restoration,* Portland, ME, September 1980. Proceedings published by the U.S. Environmental Protection Agency, Washington, DC.

Buffleben, M.S., Zayeed, K., Kimbrough, D., Stenstrom, M.K., and Suffet I.H. 2001. Evaluation of Urban Non-Point Source Runoff of Hazardous Metals that Enters Santa Monica Bay, California. *5th International Conf.: Diffuse/Nonpoint Pollution and Watershed Management.* CD-ROM.

Burton, G.A., Jr., and Pitt, R. 2001. *Stormwater Effects Handbook: A Tool Box for Watershed Managers, Scientists, and Engineers.* Boca Raton, FL: CRC Press, Inc. 1085 pp.

Burton, G.A., and Pitt, R. 2002. *Handbook for Evaluating Stormwater Runoff Effects, A Tool Box of Procedures and Methods to Assist Watershed Managers.* New York: CRC/Lewis Publishers.

Burton, G.A., Jr. 1989. Evaluation of seven sediment toxicity tests and their relationships to stream parameters. *Toxicity Assess.* 4:149–159.

Burton, G.A., Jr., Stemmer, B.L., and Winks, K.L. 1989. A multitrophic level evaluation of sediment toxicity in Waukegan and Indiana Harbors. *Environ. Toxicol. Chem.* 8:1057–1066.

Burton, G.A., Jr. 1991. Assessing freshwater sediment toxicity. *Environ. Toxicol. Chem.* 10(12).

Burton, G.A., Jr., and Stemmer, B.L. 1988. Evaluation of surrogate tests in toxicant impact assessments. *Toxicity Assessment: An International Journal* 3:255–269.

Burton, G.A., Jr., and Scott, J. 1992. Sediment toxicity evaluations. *Environ. Sci. Technol.* 25.

Burton, G.A., Jr., ed. 1992. *Sediment Toxicity Assessment.* Boca Raton, FL: Lewis Publishers.

Burton, G.A., Jr., Pitt, R., and Clark. S. 2000. The role of whole effluent toxicity test methods in assessing stormwater and sediment contamination. *CRC Critical Reviews in Environmental Science & Technology* 30:413–447.

Butcher, J.B., Pollman, C., and Thirolle, E.A. 2000. Mercury TMDLs for Arivaca and Penna Blanca Lakes, AZ. *Watershed 2000 Management Conference*, July, Vancouver, BC. Water Environment Federation, CD-ROM.

Butler, Kent S. 1987. Urban Growth Management and Groundwater Protection: Austin, Texas. *Planning for Groundwater Protection.* New York: Academic Press, Inc. pp. 261–287.

Callahan, M.A., Slimak, M.W., Gabel, N.W., May, I.P., Fowler, C.F., Freed, J.R., Jennings, P., Durfee, R.L., Whitmore, F.C., Maestri, B., Mabey, W.R., Holt, B.R., and Gould, C. 1979. *Water Related Environmental Fates of 129 Priority Pollutants.* U.S. Environmental Protection Agency, Monitoring and Data Support Division, EPA-4-79-029a and b. Washington, DC.

Carnes, R.A., and Lossin, R.D. 1970. An Investigation of the pH Characteristics of Compost. *Compost Science* 11(5):18–21.

Carr, D.J., Geinopolos, A., and Zanoni, A.E. 1983. *Characteristics and Treatability of Urban Runoff Residuals.* U.S. Environmental Protection Agency, Municipal Environmental Research Laboratory, EPA-600/S2-82-094, Cincinnati, OH.

Chambers, G.M., and Tottle, C.H. 1980. *Evaluation of Stormwater Impoundments in Winnipeg,* Report SCAT-1, Environment Canada, Ottawa.

Chaney, R.L., and Hundeman, P.T. 1979. Use of peat moss columns to remove cadmium from wastewaters. *Journal of the Water Pollution Control Federation* 51(1):17–21.

Chang, H., and Duke L.D. 2001. Pollutants in Runoff from Auto Dismantlers and Urban Watersheds. *5th International Conf.: Diffuse/Nonpoint Pollution and Watershed Management.* CD-ROM.

Chang, H.H. 1988. *Fluvial Processes in River Engineering.* John Wiley & Sons.

Chapelle, F.H. 2001. *Ground-Water Microbiology and Geochemistry.* New York: John Wiley and Sons, Inc.

Characklis, G.W., and Wiesner, M.R. 1997. Particles, Metals, and Water Quality in Runoff from Large Urban Watershed. *J. Environ. Eng.* 123(8):753.

Chase, William L. 1987. Reclaiming Wastewater in Phoenix, Arizona. In: *Irrigation Systems for the 21st Century—Proceedings of a Conference Sponsored by the Irrigation and Drainage Division of the American Society of Civil Engineers.* Portland, OR. July 28–30. ASCE, New York. pp. 336–343.

Chebbo, G., Gromaire, M-C., Garnaud, S., and Gonzalez A. 1999. The Experimental Urban Catchment "Le Marais" in Paris. In: *Proc. the Eighth International Conference on Urban Storm Drainage.* August 30–September 3, Sydney, Australia, ed. by I.B. Joliffe and J.E. Ball. The Institution of Engineers Australia, The International Association for Hydraulic Research, and The International Association on Water Quality. 1520.

Cheng, S. 1988. Detention storage optimization in least-cost sewer system design. *Second International Conference on Urban Storm Drainage,* Urbana, IL, June 1981.

Chow, Ven Te. 1959. *Open Channel Hydraulics.* McGraw-Hill.

Christoffersen, J., Christoffersen, M.R., Larsen, R., and Moller, I.J. 1991. Regeneration by surface-coating of bone char used for defluoridation of water. *Water Research* 25(2):227–229.

Chu, K.H., and Hashim, M.A. 2000. Adsorption of copper(II) and EDTA-chelated copper(II) onto granular activated carbons. *Journal of Chemical Technology and Biotechnology* 75:1054–1060.

Churchill, S.A., Walters, J.V., and Churchill, P.F. 1995. Sorption of heavy metals by prepared bacterial cell surfaces. *Journal of Environmental Engineering* 121(10):706–711.

City of Austin, Texas. 1988. *Design Guidelines for Water Quality Control Basins.* Environmental DCM, City of Austin Transportation and Public Services Department.

Clark, S. 2000. *Urban Stormwater Filtration: Optimization of Design Parameters and a Pilot-Scale Evaluation.* Ph.D. Dissertation. University of Alabama at Birmingham, Birmingham, AL. 450 pp.

Clark, S. 1996. *Evaluation of Filtration Media for Stormwater Runoff Treatment.* MSCE thesis prepared for the Department of Civil and Environmental Engineering, the University of Alabama at Birmingham, Birmingham, AL. 442 pp.

Clark, S., and Pitt, R. 1998. Stormwater Treatment at Critical Areas Volume 3: Evaluation of Filtration Media. EPA/600/X-98/XXX.

Clark, S., and Pitt, R. 1999. *Stormwater Treatment at Critical Areas: Evaluation of Filtration Media for Stormwater Treatment.* U.S. Environmental Protection Agency, Water Supply and Water Resources Division, National Risk Management Research Laboratory. Cincinnati, OH. 442 pp.

Clark, S. 1999. *Quality Assurance Project Plan: Natural Media Filtration.* U.S. Environmental Protection Agency, Urban Watershed Management Branch, National Risk Management Research Laboratory, Edison, NJ. 50 pp.

Clark, S., Pitt, R., and Brown, P. 2001a. Effect of anaerobiosis on filter media pollutant retention. 2001 Engineering Foundation Conference: *Stormwater BMP Effectiveness in Mitigating Urbanization Effects on Receiving Waters.* Snowmass, CO.

Clark, S., Rovansek, R., Wright, L., Heaney, J., Field, R., and Pitt, R. 2001b. Urban Wet Weather Flows 2000 Literature Review. *Water Environment Research* 73(5): Sept./Oct. (special CD-ROM issue).

Clark, S., Pitt, R., and Brown, P. 2001c: Pollution Retention by Filter Media under Aerobic versus Anaerobic Conditions. In: *WEFTEC 2001 Conf. Proc.* CD-ROM.

Clark, S., Pitt, R., and Field, R. 1995. Stormwater Treatment: Inlet Devices and Filtration. Presented at the *Stormwater NPDES Related Monitoring Needs* ASCE conference, Mt. Crested Butte, CO. Proceedings ed. by H. Torno. Published by ASCE, New York. pp. 641–650.

Clark, S., Brown, P., and Pitt, R. 2000. Wastewater Treatment using Low-Cost Adsorbents and Waste Materials. *2000 Water Environment Federation and Purdue University Industrial Wastes Technical Conference,* May. St. Louis, MO. Water Environment Federation, CD-ROM.

Clark, S., Pitt, R., Field, R., and Scholze, R. 1998a. Urban Runoff Filtration for Critical Source Areas. In: *Proc. Adv. in Urban Wet Weather Pollut. Reduction,* Cleveland, OH, WEF (CP3805). 177.

Clark, S., Pitt, R., Brown, P., and Field, R. 1998b. Treatment by Filtration of Stormwater Runoff Prior to Groundwater Recharge. In: *Proc. Water Environ. Fed. 71st Annu. Conf. Exposition,* Orlando, FL. 4:379.

Clark, S., Lawler, D.F., and Cushing, R.S. 1992. Contact filtration: particle size and ripening. *Journal of the American Water Works Association* 84(12):61–71.

Clark, S.E., 1996. *Evaluation of Filtration Media for Stormwater Runoff Treatment.* M.S.C.E. Thesis. University of Alabama at Birmingham, Birmingham, AL. 442 pp.

Clark, S., Pitt, R., Field, R., and Burian, S. 2002. Urban Wet Weather Flows—2001 Literature Review. *Water Environment Research* 74(5): Sept./Oct. (special CD-ROM issue).

Claytor, R. 1995. *Watershed Protection Techniques* 1(4): Center for Watershed Protection. Ellicott City, MD.

Claytor, R., and Schueler, T. 1996. *Design of Stormwater Filtering Systems* (DRAFT manual). Prepared for Chesapeake Bay Research Consortium in cooperation with U.S. EPA Region V Chesapeake Bay Research Consortium. Center for Watershed Protection, Endicott, MD. 230 pages.

Claytor, R.A. 1996. Habitat and biological monitoring reveals headwater stream impairment in Delaware's Piedmont. *Watershed Protection Techniques* 2(2):358–360.

Claytor, R.A. 1996a. An introduction to stormwater indicators: an urban runoff assessment tool. *Watershed Protection Techniques* 2(2):321–328.

Claytor, R.A. 1996b. Multiple indicators used to evaluate degrading conditions in Milwaukee County. *Watershed Protection Techniques* 2(2):348–351.

Claytor, R.A. 1998. An eight-step approach to stormwater retrofitting—How to get them implemented. Presented at the EPA Region V *Urban Stormwater Retrofit Conference.* U.S. Environmental Protection Agency. Chicago, IL.

Claytor, R.A., and Brown, W. 1996. *Environmental Indicators to Assess the Effectiveness of Municipal and Industrial Stormwater Control Programs.* Prepared for the U.S. EPA, Office of Wastewater Management. Center for Watershed Protection, Silver Spring, MD. 210 pp.

Clymo, R. 1963. Ion Exchange in Sphagnum and Its Relation to Bog Ecology. *Annals of Botany. New series* 27:309–24.

Cohen, A., Rollins, M., Zunic, W., and Durig, J. 1991. Effects of Chemical and Physical Differences in Peats on Their Ability to Extract Hydrocarbons from Water. *Water Research* 25(9):1047–60.

Collins, M.R., Eighmy, T.T., Fenstermacher, J.M., Jr., and Spanos, S.K. 1992. Removal of natural organic matter by conventional slow sand filtration. *Journal of the American Water Works Association* 84(5):80–90.

Colston, N.V., Jr. 1974. *Characterization and Treatment of Urban Land Runoff.* EPA-670/2-74-096, U.S. Environmental Protection Agency.

Commins, J., and Stribling, J. 1992. Wheaton Branch Retrofit Project: 1990–91 Biomonitoring Project. *ICPRB Report No. 92-1.* Interstate Commission on the Potomac River Basin, Rockville, MD.

Conradin, F. 1995. Zurich modifies collection system by building open brooks. *Water Environment & Technology* (Nov.):26–27.

Cook, W.L., Parrish, F., Satterfield, J.D., Nolan, W.G., and Gaffney, P.E. 1983. *Biological and Chemical Assessment of Nonpoint Source Pollution in Georgia: Ridge-Valley and Sea Island Streams.* Department of Biology, Georgia State University, Atlanta, GA.

Cooney, D.O. 1999. *Adsorption Design for Wastewater Treatment.* Boca Raton, FL: Lewis Publishers. 190 pp.

Cooper, D.C., Picardal, F., Rivera, J., and Talbot, C. 2000. Zinc immobilization and magnetite formation via ferric oxide reduction by *Shewanella putrefaciens* 200. *Environ. Sci. Technol.* 34:100–106.

Corsi, R., Greb, S.R., Bannerman, R.T., and Pitt, R. 1999. *Evaluation of the Multi-Chambered Treatment Train, a Retrofit Water-Quality Management Device.* U.S. Geological Survey Open-File Report 99-270. Middleton, WI.

Coupal, B., and Lalancette, J.M. 1976. The treatment of waste waters with peat moss. *Water Research* 10:1071–1076.

Cowan, W. 1956. Estimating Hydraulic Roughness Coefficients. *Agricultural Engineering* (July):473–475.

Cowgill, U.M. 1988. Sampling waters, the impact of sample variability on planning and confidence levels. In: *Principles of Environmental Sampling,* ed. by L.H. Keith. ACS Professional Reference Book. American Chemical Society. pp. 171–189.

Cowherd, C.J., Maxwell, C.M., and Nelson, D.W. 1977. *Quantification of Dust Entrainment from Paved Roadways.* EPA-450 3-77-027, U.S. Environmental Protection Agency, Research Triangle Park, NC.

Cox, M.B. and Palmer, V.J. 1948. Results of Tests on Vegetated Waterways and Method of Field Application. Stillwater, OK: Oklahoma Agricultural Experiment Station: 5–43.

Crites, Ronald W. 1985. Micropollutant Removal in Rapid Infiltration. In: *Artificial Recharge of Groundwater.* ed. by Takashi Asano. Boston: Butterworth Publishers. pp. 579–608.

Crittenden, J.C., Vaitheeswaran, K., Hand, D.W., Howe, E.W., Aieta, E.M., Tate, C.H., McGuire, M.J., and Davis, M.K. 1993. Removal of dissolved organic carbon using granular activated carbon. *Water Research* 27(4):715–721.

Croke, T. 2001. Accounting for bends in channel design. *Erosion Discussion* 6(1) (Summer). North American Green (available from their web page: www.nagreen.com).

Croke, T. 2000. Reliable channel design using turf reinforcement mats. *Erosion Discussion* 5(1) (June). North American Green (available from their web page: www.nagreen.com).

Crunkilton, R., and Kron, D. 1999. Assessing the Effectiveness of a Pilot-Scale Wet Detention Basin in Reducing Toxicity of Urban Runoff. In: *Comprehensive Stormwater & Aquatic Ecosystem Manage. Conf. Papers First South Pacific Conf.,* Auckland, New Zealand.

Crunkilton, R., Kleist, J., Ramcheck, J., DeVita, W., and Villeneueve, D. 1997. Assessment of the response of aquatic organisms to long-term *in-situ* exposures to urban runoff. Presented at the *Effects of Watershed Developments and Management on Aquatic Ecosystems* conference. Snowbird, UT, August 4–9, 1996, ed. by L.A. Roesner. ASCE, New York.

CSF Treatment Systems Inc, 1994. *Technical Memorandum: Three Year Performance Summary—185th Avenue.* Prepared by CSF Treatment Systems, Inc., Portland, OR.

CTA Inc. 1983. *Georgia Nonpoint Source Impact Assessment Study: Blue Ridge/Upland Georgia Cluster, Piedmont Cluster. and Gulf Coastal Plain Cluster.* Georgia Envir, Protection Division, Dept. of Natural Resources, Atlanta, GA.

da Silva, F.F., Wallach, R., and Chen, Y. 1993. Hydraulic properties of sphagnum peat moss and tuff (scoria) and their potential effects on water availability. *Plant and Soil* 154:119–126.

Dahab, M.F., Becker, H.L., and Riley, T.E. 1991. Treatment of a wood products superfund wastewater: a case study. *Canadian Journal of Civil Engineering* 18:654-662.

Daigle, J.Y. 1993. *Peat moss for Waste Water Treatment.* Open File Report 93-3. New Brunswick Natural Resources and Energy, Mineral Resources.

Dally, L.K., Lettenmaier, D.P., Burges, S.J., and Benjamin M.M. 1983. *Operation of Facilities for Urban Stream Quality Enhancement.* University of Washington, Department of Civil Engineering, Water Resources Series Technical Report No. 79, Seattle, WA.

Dalrymple, R.J., Hodd, S.L., and Morin, D.C. 1975. *Physical and Settling Characteristics of Particulates in Storm and Sanitary Wastewaters.* EPA-670/2-75-011. U.S. Environmental Protection Agency, Cincinnati, OH.

Darby, J., Lawler, D., and Wilshusen, T. 1991. Depth Filtration of Wastewater: Particle Size and Ripening. *Research Journal. Water Pollution Control Federation* 63(3):228–38.

Darnell, C.M., Lootah, H., and Al Maidoor, E.A. 2001. Stormwater Management Using Detention Ponds. In: *WEFTEC 2001 Conf. Proc.* CD-ROM.

Datta, D.K., and Subramanian, V. 1998. Distribution and Fractionation of Heavy Metals in the Surface Sediments of the Ganges-Brahmaputra-Meghna River System in the Bengal Basin. *Environ. Geol.* 36(1-2):93.

Davidova, E.G., and Kasparova, S.G. 1993a. Sorption of heavy metals by yeast cell walls. *Microbiology* 61(5):582–585.

Davidova, E.G., and Kasparova, S.G. 1993b. Adsorption of metals by yeast cell walls. *Microbiology* 61(6):716–719.

Davies, P.H. 1995. Factors in controlling nonpoint source impacts. Engineering Foundation/ASCE conference held in 1991 in Mt. Crested Butte, CO: *Stormwater Runoff and Receiving Systems: Impact, Monitoring and Assessment* (ed. by E.E. Herricks). Boca Raton, FL: Lewis/CRC Press. pp. 53–64.

Davis, A.P., and Burns, M. 1999. Evaluation of Lead Concentration in Runoff from Painted Structures. *Water Res.* (G.B.) 33(13):2949.

Davis, A.P., Shokouhian, M., and Ni, S.B. 2001. Loading Estimates of Lead, Copper, Cadmium, and Zinc in Urban Runoff from Specific Sources. *Chemosphere* 44:997.

Davis, A.P., Shokouhian, M., Sharma, H., Henderson, C., Winogradoff, D., and Coffman, L. 1998. Bioretention for Treatment of Urban Stormwater Runoff: Laboratory and Field Results. In: *Proc. Water Environ. Fed. 71st Annu. Conf. Exposition,* Orlando, FL 4:421.

Davis, R., Zhang, X., and Agarwala, J. 1989. Particle Classification for Dilute Suspensions Using an Inclined Settler. *Industrial and Engineering Chemistry Research* 28:785–93.

Davis, W.J., McCuen, R.H., and Kamedulski, G.E. 1978. The effect of storm water detention on water quality. *International Symposium on Urban Storm Water Management,* University of Kentucky, Lexington.

Day, G.E. 1980. Investigation of Concrete Grid Pavements. In: *Proceedings—National Conference on Urban Erosion and Sediment Control: Institutions and Technology,* EPA-905/9-80-002, U.S. Environmental Protection Agency, Chicago, IL.

De Vegt, A.L., Bayer, H.G., and Buisman, C.J. 1998. Biological sulfate removal and metal recovery from mine waters. *Mining Engineering* 50(11):67–70.

Debo, T.N., and Ruby, H. 1982. Detention basins—An urban experience. *Public Works* 113(1):42.

Debo, T.N., and Reese, A.J. 1995. Municipal Storm Water Management. Boca Raton, FL: Lewis Publishers.

Debo, T.N. 1982. Detention ordinances—Solving or causing problems? In: *Proceedings of the Conference on Stormwater Detention Facilities, Planning, Design, Operation, and Maintenance,* August, Henniker, NH, ed. by W. DeGroot. Published by the American Society of Civil Engineers, New York.

Deletic, A.B., and Maksimovic, C.T. 1998. Evaluation of water quality factors in storm runoff from paved areas. *Journal of Environmental Engineering* 124(9):869-879.

Dennison, M.S. 1996. Stormwater Discharges. Boca Raton, FL: Lewis Publishers.

Denver Regional Council of Governments. 1983. *Urban Runoff Quality in the Denver (Colorado) Region.* Prepared for the U.S. EPA. Washington, DC. PB85-101640.

DePinto, J.V., Young, T.C., and Martin, S.C. 1980. Aquatic sediments. *Journal of Water Pollution Control Federation* 52(6):1656–70.

Deram A., and Petit, D. 1997. Ecology of bioaccumulation in *Arrhenatherum elatius* L. (Poaceae) populations—applications of phytoremediation of zinc, lead and cadmium contaminated soils. *J. Exp. Bot.* 48(Suppl):98.

Dewberry and Davis, Inc. 1990. *Investigation of Potential Sediment Toxicity from BMP Ponds.* Prepared for the Northern Virginia Planning District Commission, the Occoquan Policy Board, and the Virginia State Water Control Board.

Diab, S., Kochba, M., and Avimelech, Y. 1993. Development of a biofilter for turbid and nitrogen-rich irrigation water; b. removal of phosphorus, algae and clay. *Bioresource Technology* 44(2):137–140.

Dierkes, C., and Geiger, W.F. 1999. Pollution Retention Capabilities of Roadside Soils. *Water Sci. Technol.* (G.B.) 39(2):201.

Diniz, E.V. 1993. Hydrologic and water quality comparisons of runoff from porous and conventional pavements. In: *Integrated Stormwater Management.* ed. by R. Field, M.L. O'Shea, and K.K. Chin. Boca Raton, FL: Lewis Publishers. pp. 149–162.

Diniz, E.V. August 1980. *Porous Pavement; Phase I, Design and Operational Criteria.* EPA-600/2-80-135. U.S. Environmental Protection Agency, Cincinnati, OH.

Doll, E.H. and Fredenhagan, V.B. 1954. Grassed Waterways. *Agricultural Engineering* (June):417–419.

Dombeck, G.D., Perry, M.W., and Anderson, M.B. 1999. Trace Metals Assimilation in Treatment Wetland Sediments. In: *Proc. Water Environ. Fed. 72nd Annual Conf. Exposition* [CD-ROM], New Orleans, LA.

Donigian, A.S., Jr., and Crawford, N.H. 1976. *Modeling Nonpoint Pollution from the Land Surface.* EPA-600/3-76-083, U.S. Environmental Protection Agency, Athens, GA.

Draper, N., and Smith, H. 1981. *Applied Regression Analysis,* second edition. New York: John Wiley & Sons, Inc. 700 pp.

Drapper, D., Tomlinson, R., and Williams, P. 2000. Pollutant Concentrations in Road Runoff: Southeast Queensland Case Study. *J. Environ. Eng.* 126–313.

Driscoll, E.D. 1986. Detention and retention controls for urban stormwater. In: *Engineering Foundation Conference: Urban Runoff Quality—Impact and Quality Enhancement Technology* June, Henniker, NH, ed. by B. Urbonas and L.A. Roesner. Published by the American Society of Civil Engineers, New York.

Driscoll, E.D. 1989. Long-term performance of water quality ponds. In: *Engineering Foundation Conference: Design of Urban Runoff Quality Controls.* Potosi, MO, July 10–15, 1988, ed. by L.A. Roesner, B. Urbonas and M.B. Sommen. Published by the American Society of Civil Engineers, New York, NY.

Driscoll, E.D., and Strecker. E.W. 1993. Assessment of BMPs being used in the U.S. and Canada. Proceedings of the *6th International Conference on Urban Storm Drainage,* ed. by J. Marsalek and H.C. Torno, September 12–17, Niagara Falls, ON. IAWQ, London. Published by Seapoint Publishers, Victoria, BC. pp. 945–950.

Droste, R.L., and Johnston, J.C. 1993. Urban snow dump quality and pollutant reduction in snowmelt by sedimentation. *Canadian Journal of Civil Engineering* 20:9–21.

Duchene, M., McBean, E., and Thomson, N. 1993. Infiltration characteristics associated with the use of trenches in stormwater management. In: Proceedings of the *6th International Conference on Urban Storm Drainage,* ed. by J. Marsalek and H.C. Torno, September 12–17, Niagara Falls, ON, IAWQ, London. Published by Seapoint Publishers, Victoria, BC. pp. 1115–1120.

Dull, L. 1984. *Survey of Oil/Water Separating System Technologies—Methods and Their Applications in Controlling Pollutants of Concern in Wastewater from Transit Bases.* Seattle Metro, Seattle, WA.

Dupre, B., Viers, J., Dandurand, J.L., Polve, M., Benezeth, P., Vervier, P., and Braun, J.J. 1999. Major and Trace Elements Associated With Colloids in Organic-Rich River Waters: Ultrafiltration of Natural and Spiked Solutions. *Chemical Geology* 160(1-2):63.

Duru, J.O. 1981. On-site detention: A stormwater management or mismanagement technique. *Second International Conference on Urban Storm Drainage,* June, Urbana, IL.

Durum, W.H. 1974. Occurrence of Some Trace Metals in Surface Waters and Groundwaters. In: *Proceeding of the Sixteenth Water Quality Conference,* Am. Water Works Assoc., et al. *Univ. of Illinois Bull.* 71(108).

Dushenkov, S., Vasudev, D., Kapulnik, Y., Gleba, D., Fleisher, D., Ting, K.C., and Ensley, B. 1997. Removal of Uranium from water using terrestrial plants. *Environ. Sci. Technol.* 31:3468–3474.

Dushenkov, V., Kumar, P.B.A.N., Motto, H., and Raskin, I. 1995. Rhizofiltration: The use of plants to remove heavy metals from aqueous streams. *Environ. Sci. Technol.* 29:1239–1245.

Dyer, S.D., and White, C.E. 1996. A watershed approach to assess mixture toxicity via integration of public and private databases. In: *Abstract Book: SETAC 17th Annual Meeting.* p. 96. Washington, DC.

Ebbert, J.C., Poole, J.E., and Payne, K.L. 1983. *Data Collected by the U.S. Geological Survey During a Study of Urban Runoff in Bellevue, Washington, 1979-82.* Preliminary U.S. Geological Survey Open-File Report, Tacoma, WA.

Ebbs S.D., and Kochian, L.V. 1998. Phytoextraction of zinc by oat (*Avena sativa*), barley (*Hordeum vulgare*),and Indian Mustard (*Brassica juncea*). *Environ. Sci. Technol.* 32:802–806.

Ebbs S.D., Stephen, D., and Kochian, L.V. 1997. Toxicity of zinc and copper to Brassica species: implications for phytoremediation. *J. Environ. Qual.* 26:776–781.

Edwards, D.R., Moore, P.A., Jr., Workman, S.R., and Bushee, E.L. 1999. Runoff of Metals from Alum-Treated Horse Manure and Municipal Sludge. *J. Am. Water Resour. Assoc.* 35(1):155.

Edwards, M., and Benjamin, M.M. 1989. Adsorptive filtration using coated sand: a new approach for treatment of metal bearing wastes. *Journal of the Water Pollution Control Federation* 61:1523–1533.

Ehrenfeld, J.G., and Schneider, J.P. 1983. *The Sensitivity of Cedar Swamps to the Effects of Non-Point Pollution Associated with Suburbanization in the New Jersey Pine Barrens.* PB8-4-136779, U.S. Environmental Protection Agency, Office of Water Policy, Washington, DC.

Ehrlich, G.G., Godsy, E.M., Pascale, C.A., and Vecchioli, John. 1979. Chemical Changes in an Industrial Waste Liquid During Post-Injection Movement in a Limestone Aquifer, Pensacola, Florida. *Groundwater* 17(6):562–573.

Eighmy, T.T., Collins, M.R., Spanos, S.K., and Fenstermacher, J. 1992. Microbial populations, activities and carbon metabolism in slow sand filters. *Water Research* 26(10):1319–1328.

Eiland, F. 1983. A simple method for quantitative determination of ATP in soil. *Soil Biol. Biochem.* 15:665–670.

Ellis, J.B. 1999a. Vegetative Systems for Highway Discharges: Design and Operation. In: *Comprehensive Stormwater & Aquatic Ecosystem Manage. Conf. Papers First South Pacific Conf.*, Auckland, New Zealand.

Ellis, J.B. 1999b. Design Considerations for the use of Vegetative Controls for the Treatment of Highway Discharges. *IAHS Publication (Int. Assoc. of Hydrol. Sci)* 259:357.

Ellis, J.B. 2000a. Risk Assessment Approaches for Ecosystem Responses to Transient Pollution Events in Urban Receiving Waters. *Chemosphere* 41:85.

Ellis, J.B. 2000b. Infiltration Systems: A Sustainable Source-Control Option for Urban Stormwater Quality Management? *J. Chartered Instit. Water Environ. Manage.* 14:27.

Ellis, J.B. 1993. Wetland BMP design for urban runoff pollution control in Europe and Australia. In: Proceedings of the *6th International Conference on Urban Storm Drainage,* ed. by J. Marsalek and H.C. Torno, In: September 12–17, Niagara Falls, ON. IAWQ, London. Published by Seapoint Publishers, Victoria, BC. pp. 957–962.

Ellis, J.B., Hamilton, R., and Roberts, A.H. 1981. Composition of Suspended Solids in Urban Stormwater. *Second International Conference on Urban Storm Drainage,* June, Urbana, IL.

Emmerling-DiNovo, C. 1995. Stormwater detention basins and residential locational decisions. *Water Resources Bulletin* 31(3):515–521.

Environment Canada. 1980. Rideau River Water Quality and Stormwater Monitoring Study. MS Report OR-29, Ontario Ministry of the Environment.

Environmental Science & Technology. 1996a. Toxicity of aquatic mixtures yielding to new theoretical approach. 30(4)(April):155a–156a.

Environmental Science & Technology. 1996b. News Briefs. 30(7)(July):290a.

Estèbe, A., Belhomme, G., Lecomte, S., Videau, V., Mouchel, J-M., and Thévenot, D.R. 1996. Urban Runoff Impacts on Particulate Metal Concentrations in River Seine: Suspended Solid and Sediment Transport. In: *Proc. 7th Int. Conf. on Urban Storm Drainage,* Hannover, Germany, IAHR/IAWQ Joint Committee on Urban Storm Drainage. 575.

Eyde, T.H. 1993. Zeolites. *Mining Engineering* 50.

Fan, C., Field, R., O'Connor, T., and Stinson, M. K. 1998. Urban Wet-Weather Flows Toxic Pollutants: Characterization and Enhanced Sedimentation Treatment. In: *Water Resources and the Urban Environment, Proceedings of the 25th Annual Conference on Water Resource Planning and Management,* June 7–10, Chicago, IL. 535.

Fan, C-Y., Field, R., Sullivan, D., and Lai, F-h. 2001. Toxic Pollutants in Urban Wet-Weather Flows: An Overview of the Multi-Media Transport, Impacts, and Control Measures. In: *Proc. ASCE EWRI Conf.—Bridging the Gap: Meeting the World's Water and Environmental Resources Challenges.* CD-ROM.

Fan, C-Y. Field, R., Heaney, J., Pitt, R., Clark, S., Wright, L., Rovansek, R., and Olivera, S. 2000. Urban Wet Weather Flows. 1999 Literature Review. *Water Environment Research* 72(5), Sept./Oct. (CD-ROM). 199 pp.

Farooq, S., and Al-Yousef, A-K. 1993. Slow sand filtration of secondary effluent. *Journal of Environmental Engineering* 119(4):615–630.

Fassman, E.A., and Yu, S.L. 2001. Comparison of Pollutant Removal Performance of Wetland Vegetation. In: *Proc. ASCE EWRI Conf.—Bridging the Gap: Meeting the World's Water and Environmental Resources Challenges.* CD-ROM.

Fatoki, O.S., and Mathabatha, S. 2001. An Assessment of Heavy Metal Pollution in the East London and Port Elizabeth Harbours. *Water SA* 27:233.

Ferarra, R.A. 1982. Stormwater detention and nonpoint source pollution control. *1982 International Symposium on Urban Hydrology, Hydraulics and Sediment Control,* July, University of Kentucky, Lexington.

Ferguson, Bruce K. 1990. Role of the Long-Term Water Balance in Management of Stormwater Infiltration. *Journal of Environmental Management* 30(3):221–233.

Field, R., and Tafuri, A.N. 1977. *Urban Runoff Pollution Control Technology Overview.* U.S. Environmental Protection Agency, EPA-600/2-77-047, Cincinnati, OH.

Field, R., and Pitt, R. 1990. Urban storm-induced discharge impacts: U.S. Environmental Protection Agency research program review. *Water Science and Technology* 22(10/11):1–7.

Field, R., and Pitt, R. 1994a. Combined sewer overflow control through in-receiving water storage: an efficiency evaluation. *Water Environment Federation Specialty Conference on A Global Perspective for Reducing CSOs: Balancing Technologies, Costs, and Water Quality,* July, Louisville, KY.

Field, R., and Pitt, R. 1994b. Optimization of CSO storage and treatment systems. *Water Environment Federation Specialty Conference on A Global Perspective for Reducing CSOs: Balancing Technologies, Costs, and Water Quality,* July, Louisville, KY.

Field, R., Brown, M.P., Vilkelis, W.V. 1994 a. *Stormwater Pollution Abatement Technologies.* U.S. Environmental Protection Agency, Office of Research and Development. Cincinnati, OH. EPA/600/R-94/129.

Field, R., Pitt, R., Lalor, M., Brown, M.P., Vilkelis, W.V., and Thackston, E. 1994b. Investigation of dry-weather pollutant entries into storm-drainage systems. *J. Envir. Eng.* 120(5):1044–1066.

Field, R., and Cibik, C. 1980. Urban runoff and combined sewer overflows. *Journal of Water Pollution Control Federation* 52(6):1290–1307.

Field, R., and Turkeltaub, R. 1981. Urban runoff receiving water impacts: program overview. *Journal of Environmental Engineering* 107:83-100.

Field, R., Struzeski, E.J., Jr., Masters, H.E., and Tafuri, A.N. 1973. *Water Pollution and Associated Effects from Street Salting*. EPA-R2-73-257, U.S. Environmental Protection Agency, Cincinnati, OH.

Field, R., O'Shea, M., and Chin, K., eds. 1993. *Integrated Stormwater Management.* Boca raton, FL: Lewis Publishers.

Field, R., Pitt, R., Hsu, K., Borst, M., DeGuida, R., Fan, C-Y. Heaney, J., Perdek, J., and Stinson, M. 1997. Urban Wet Weather Flow-1996 Literature Review. *Water Environment Research* 69(4):426–444.

Field, R., Pitt, R., Brown, M., and O'Connor, T. 1995. Combined sewer overflow control using storage in seawater. *Water Research* (Elsevier Science) 29(6):1505–1514.

Field, R., O'Connor, T., Fan, C-Y., Pitt, R., Clark, S., Ludwig, J., Hendrix, T. 1998. Urban Wet Weather Flows—1997 Literature Review. *Water Environment Research* 70(4).

Fifty, W.F., Brust, C.J., and Albert, J.L. 2001. Investigating the Impact of Wastewater and Storm Water Discharges on Coastal Receiving Waters. In: *WEFTEC Latin America 2001 Conf. Proc.* CD-ROM.

Figura, P., and McDuffle, B. 1979. Use of chelex resin for determination of labile trace metal fractions in aqueous ligand media and comparison of the method with anodic stripping voltammetry. *Anal. Chem.* 51:20–125.

Figura, P., and McDuffle, B. 1980. Determination of labilities of soluble trace metal species in aqueous environmental samples by anodic stripping voltammetry and chelex column and batch methods. *Anal. Chem.* 52:1433–1439.

Figura, P., and McDuffle, B. 1977. Characterization of the calcium form of Chelex-100 for trace metal studies. *Anal. Chem.* 49:1950–1953.

Florence, T.M. 1977. Trace metal species in fresh water. *Water Research* 11:681–687.

Florence, T.M., and Bately. G.E. 1980. Chemical speciation in natural waters. *CRC Crit. Rev. Anal. Chem.* 9:219–296.

Ford, T.E. 1993. Aquatic Microbiology: An Ecological Approach. Boston, MA: Blackwell Scientific Publications. pp. 463–466.

Forest Products Research Laboratory. 1995. *AgroFiber* (Fact Sheet).

Forndran, A., Field, R., Dunkers, K., and Moran, D. 1991. Balancing flow for CSO abatement. *Water Environment & Technology* (May), pp. 54–58.

Förster, J. 1999. Variability of Roof Runoff Quality. *Water Sci. and Tech.* 39(5):137.

Förster, J. 1996. Heavy Metal and Ion Pollution Patterns in Roof Runoff. In: *Proc. 7th Int. Conf. on Urban Storm Drainage,* Hannover, Germany, IAHR/IAWQ Joint Committee on Urban Storm Drainage. 241.

Förster, J., and Knoche, G. 1999. Quality of Roof Runoff From Green Roofs. In: *Proc. the Eighth International Conference on Urban Storm Drainage.* August 30–September 3, Sydney, Australia, ed. by I.B. Joliffe and J.E. Ball. The Institution of Engineers Australia, The International Association for Hydraulic Research, and The International Association on Water Quality. 1312.

Fortier, S., and Scobey, F.C. 1926. Permissible canal velocities. *Trans. ASCE* 89 (paper No. 1588):940–984.

Fredrickson, J.K., Zachara, J.M., Kennedy, D.W., Dong, H., Onstott, T.C., Hinman, N.W., and Li, S-W. 1998. Biogenic iron mineralization accompanying the dissimilatory reduction of hydrous ferric oxide by a groundwater bacterium. *Geochim. Cosmochim. Acta* 62:3239–3257.

Free, B.M., and Mulamoottil, G.G. 1983. The limnology of Lake Wabukayne; A storm-water impoundment. *Water Resources Bulletin* 19(5):821.

Friedman, M., and Waiss, A.C. 1972. Mercury uptake by selected agricultural products and by-products. *Environmental Science and Technology* 6:457–458.

Fuji, S., Somiya, I., and Takeda, K. 1987. Water quality improvement of secondary effluent by an oxidation pond with subsequent sand filtration treatment. *Water Science and Technology* 19(12):89–99.

Fujita, S. 1994. Infiltration structures in Tokyo. *Water Science and Technology* 30(1):33–41.

Gabriel, J., Baldrian, P., Hladikova, K., and Hakova, M. 2001. Copper sorption by native and modified pellets of wood-rotting *basidomycetes. Letters in Applied Microbiology* 32:194–198.

Galli, J. 1990. *Peat-Sand Filters: A Proposed Stormwater Management Practice for Urbanized Areas.* Prepared for the Coordinated Anacostia Retrofit Program and Office of Policy and Planning, DC. Department of Public Works.

Galli, J., and Schueler, T. 1992. Wheaton Branch Stream Restoration Project. In: *Watershed Restoration Sourcebook.* Metropolitan Washington Council of Governments, Washington, DC.

Galvin, David D., and Moore, Richard K. 1982. *Toxicants in Urban Runoff, METRO Toxicant Program Report #2.* U.S. Environmental Protection Agency Grant #P-000161-01, Lacey, WA.

Gambrell, R.P., and Patrick, W.H., Jr. 1977. Chemical and microbiological properties of anaerobic sorts and sediments. In: *Plant Life in Anaerobic Environments,* ed. by R.M.M. Crawford Ann Arbor, MI: Ann Arbor Science Publishers.

Gardea-Torresdey, J.L., Tang, L., and Salvador, J.M. 1996. Copper adsorption by esterified and unesterified fractions of Sphagnum peat moss and its different humic substances. *Journal of Hazardous Materials* 48:191–206.

Garie, H.L., and McIntosh, A. 1986. Distribution of benthic macroinvertebrates in a stream exposed to urban runoff. *Water Resources Bulletin* 22:447.

Garie, H.L., and McIntosh, A. 1990. Distribution of benthic macroinvertebrates in a stream exposed to urban runoff. *Water Science and Technology* 22 (10/11).

Garnaud, S., Mouchel, J-M., Chebbo, G., and Thevenot, D.R. 1999a. Heavy Metal Concentrations in Dry And Wet Atmospheric Deposits In: Paris District: Comparison With Urban Runoff. *Sco. Total Environ.* 235(1–3):235.

Garnaud, S., Mouchel, J-M., and Thevenot, D.R. 1999b. Mobility Evolution of Particulate Trace Metals in Urban Runoff: From Street Deposits and Road Runoff to Combined Sewer Deposits and Catchment Outlet. In: *Proc. the Eighth International Conference on Urban Storm Drainage.* August 30–September 3, Sydney, Australia, ed. by I.B. Joliffe and J.E. Ball. The Institution of Engineers Australia, The International Association for Hydraulic Research, and The International Association on Water Quality. 1511.

Garnham, G.W. 1997. The use of algae as metal biosorbents. In: *Biosorbents for metal ions* (ed. J. Wase and C. Forster), pp. 11-38. Taylor & Francis Ltd.

Gbondo-Tugbawa, S., and Driscoll, C.T. 1998. Application of the Regional Mercury Cycling Model (RMCM) to Predict the Fate and Remediation of Mercury in Onondaga Lake, New York. *Water, Air and Soil Poll.* (Neth.) 105(1-2):417.

Gburek, W.J., and Urban, J.B. 1983. Storm Water Detention and Groundwater Recharge Using Porous Asphalt—Initial Results. *1983 International Symposium on Urban Hydrology, Hydraulics and Sediment Control,* July, University of Kentucky, Lexington.

Geldof, G., Jacobsen, P., and Fujita, S. 1993. Urban stormwater infiltration perspectives. In: Proceedings of the *6th International Conference on Urban Storm Drainage,* ed. by J. Marsalek and H.C. Torno, September 12–17, Niagara Falls, ON. IAWQ, London. Published by Seapoint Publishers, Victoria, BC. pp. 963–968.

Geldof, G., Jacobsen, P., and Fujita, S. 1994. Urban stormwater infiltration perspectives. *Water Science and Technology* 29(1–2):245–254.

German, Edward R. 1989. *Quantity and Quality of Stormwater Runoff Recharged to the Floridan Aquifer System Through Two Drainage Wells in the Orlando, Florida Area.* U.S. Geological Survey—Water Supply Paper 2344. Prepared in cooperation with the Florida Department of Environmental Regulation. USGS, Denver, CO.

Gietz, R.J. 1981. *Stormwater Runoff Treatment by Impoundment: Barrhaven Pilot Study,* Report SCAT-8. Environment Canada, Ottawa.

Gietz, R.J. 1983. *Urban Runoff Treatment in the Kennedy-Burnett Settling Pond.* For the Rideau River Stormwater Management Study, Pollution Control Division, Works Department, Regional Municipality of Ottawa-Carleton, Ottawa, ON.

Gilbert, R.O. 1987. *Statistical Methods for Environmental Pollution Monitoring.* New York: John Wiley & Sons, Inc.

Glass, G.E., and Sorensen, J.A. 1999. Six-year trend (1990–1995) of Wet Mercury Deposition in the Upper Midwest USA. *Environ. Sci. Technol.* 33(19):3303.

Glazewski, R., and Morrison, G.M. 1996. Photochemistry of Copper in an Urban River. In: *Proc. 7th Int. Conf. on Urban Storm Drainage,* Hannover, Germany. IAHR/IAWQ Joint Committee on Urban Storm Drainage. 797.

Glenn, D.W., and Sansalone, J.J. 2002. Accretion and partitioning of heavy metals associated with snow exposed to urban traffic and winter storm maintenance activities. *Journal of Environmental Engineering* 128(2):167–185.

Glenn, D.W., Tribouillard, T. and Sansalone, J.J. 2001. Temporal Deposition of Anthropogenic Pollutants in Urban Snow as Generated by Traffic. In: *WEFTEC 2001 Conf. Proc.* CD-ROM.

Goforth, G.F., Diniz, E.V., and Rauhut, J.B. 1984. *Stormwater Hydrological Characteristics of Porous and Conventional Paving Systems.* EPA-600/2-83-106, U.S. Environmental Protection Agency, Cincinnati, OH. (Feb. 1984, also dated Oct. 1983.)

Goldshmid, J. 1974. Water-Quality Aspects of Ground-Water Recharge in Israel. *Journal of the American Water Works Association* 66(3):163–166.

Gore & Storrie Ltd./Proctor & Redfern Ltd. 1981. Executive Summary Report on Rideau River Stormwater Management Study, Phase 1, Rideau River Stormwater Management Study. Ottawa and the Ontario Ministry of the Environment, Kingston, ON.

Gosset, T., Trancart, J.L., and Thevenot, D.R. 1986. Batch metal removal by peat: kinetics and thermodynamics. *Water Research* 20(1):21–26.

Goulet, R.R., Pick, F.R., and Droste, R.L. 2001. Test of the First-Order Removal Model for Metal Retention in a Young Constructed Wetland. *Ecol. Eng.* 17:357.

Greb, S.R., Bannerman, R.T., Corsi, S.R., and Pitt R.E. 2000. Evaluation of the Multichambered Treatment Train, a Retrofit Water-Quality Management Practice. *Water Environ. Res.* 72:207.

Green, M.B., and Martin, J.R. 1996. Constructed reed beds clean up storm overflows on small wastewater treatment works. *Water Environment Research* 68(6):1054–1060.

Greenberg, A., Clesceri, L., and Eaton, A., eds. 1992. *Standard Methods for the Examination of Water and Wastewater, 18th ed.* Washington, DC: American Public Health Association, American Water Works Association, Water Environment Federation.

Griffin, M.L., Richards, T., and Millican, J.M. 2001. City of Atlanta CSO System Evaluation. In: *WEFTEC 2001 Conf. Proc.* CD-ROM.

Grizzard, T.J., Randall, C.W., Weand, B.L., and Ellis, K.L. 1986. Effectiveness of extended detention ponds. In: *Engineering Foundation Conference: Urban Runoff Quality—Impact and Quality Enhancement Technology,* June, Henniker, NH, ed. by B. Urbonas and L.A. Roesner. published by the American Society of Civil Engineers, New York.

Gromaire, M.C., Garnaud, S., Ahyerre, M., and Chebbo, G. 2000. The Quality of Street Cleaning Waters: Comparison with Dry and Wet Weather Flows in a Parisian Combined Sewer System. *Urban Water* 2:39.

Gromaire-Mertz, M.C., Garnaud, S., Gonzalez, A., and Chebbo, G. 1999. Characterization of Urban Runoff Pollution in Paris. *Water Sci. Technol.* (G.B.). 39(2):1.

Grout, H., Wiesner, M.R., and Bottero, J.Y. 1999. Analysis of Colloidal Phases in Urban Stormwater Runoff. *Environ. Sci. Technol.* (G.B.) 33(6):831.

Gum, D.D. 1998. Comparing the lead uptake of three species of grass. Senior Thesis, College of Health and Natural Sciences, Lake Superior State University, Michigan.

Gupta, M., Mason, D., Clark, M., Meinholz, T., Hansen, C., and Geinopolos, A. 1977. *Screening/Flotation Treatment of Combined Sewer Overflows Volume 1—Bench Scale and Pilot Plant Investigations.* EPA-600/2-77-069a. U.S. Environmental Protection Agency, Cincinnati, OH.

Gupta, M.K., Agnew, R.W., and Kobriger, N.P. 1981. *Constituents of Highway Runoff,* Volume one of a six volume series. Federal Highway Administration, U.S. Department of Transportation, FHWA/RD-81/042, Washington, DC.

Gupta, V. 1998. Equilibrium Uptake, Sorption Dynamics, Process Development, and Column Operations for the Removal of Copper and Nickel from Aqueous Solution and Wastewater using Activated Slag, a Low-Cost Adsorbent. *Industrial and Engineering Chemistry Research* 37:192–202.

Haggerty, G.M., and Bowman, R.S. 1994. Sorption of chromate and other inorganic anions by organo-zeolite. *Environmental Science & Technology* 28(3):452–458.

Hall, M.J. 1990. The design of flood storage ponds: catering for both quality and quantity of runoff. In: *Developments in Storm Drainage—A Symposium on Infiltration and Storage of Stormwater in New Developments,* January, Sheffield City. pp. 21–33.

Hall, W.S. 1996. Bioavailability of Chromium, Copper, and Lead in Estuarine Wetland Sediments. In: *Proc. WEFTEC'96, 69th Annu. Conf. & Expo.*, Dallas, TX. 8–52.

Hamid, A., Wallace, L., and Bell, D. 2001. City of Atlanta Combined and Separate Sewer Overflow Control Program Meets the Challenge of the 21st Century. In: *WEFTEC 2001 Conf. Proc.* CD-ROM.

Hampson, Paul S. 1986. *Effects of Detention on Water Quality of two Stormwater Detention Ponds Receiving Highway Surface Runoff in Jacksonville, Florida.* U.S. Geological Survey Water-Resources Investigations Report 86-4151. Prepared in cooperation with the Florida Department of Transportation. USGS, Denver, CO.

Harada, S., and Ichikawa, A. 1993. A study on the performance of the drainage infiltration strata. In: *Proceedings of the 6th International Conference on Urban Storm Drainage,* ed. by J. Marsalek and H.C. Torno, September 12–17, Niagara Falls, ON. IAWQ, London. Published by Seapoint Publishers, Victoria, BC. pp. 1104–1109.

Hares, R.J., and Ward, N.I. 1999. Comparison of the Heavy Metal Content of Motorway Stormwater Following Discharge Into Wet Biofiltration and Dry Detention Ponds Along the London Orbital (M25) Motorway. *Sco. Total Environ.* 235(1-3):169.

Harper, H.H., and Herr, J.L. 1996. Alum treatment of stormwater—The first ten years: What have we learned and where do we go from here? North American Lake Management Society (NALMS) Conference presentation, October, Minneapolis/St. Paul.

Harper, Harvey H. 1988. *Effects of Stormwater Management Systems on Groundwater Quality.* Final Report for DER Project WM190. Florida Department of Environmental Regulation.

Harrington, B.W. 1986. *Feasibility and Design of Wet Ponds to Achieve Water Quality Control.* Sediment and Stormwater Division, Maryland Water Resources Administration.

Harris, B. Fluoranthene 1982. Reportable Quantity (RQ) Ranking Based on Chronic Toxicity. U.S. Environmental Protection Agency. Contract No. 68-03-3112. Cincinnati, OH.

Hatfield, J.L., Jaynes, D.B., Burkart, M.R., Cambardella, C.A., Moorman, T.B., Prueger, J.H., and Smith, M.A. 1999. Water Quality in Walnut Creek Watershed: Setting and Farming Practices. *J. Environ. Qua.* 28(1):11.

Hawley, M.E., Bondelid, T.R., and McCuen, R.H. 1981. A planning method for evaluating downstream effects of detention basins. *Water Resources Bulletin* 17(5):806.

Hazen and Sawyer, Inc.and Center for Watershed Protection. 1997. *Final Site Plan for BMP 67, Sub-Basin BGC-5.* Kensico Watershed Stormwater Best Management Facilities, Westchester, NY.

Heal, K. 1999. Metals in Sediments of Sustainable Urban Drainage Structures in Scotland. *IAHS Publication (Int. Assoc. of Hydrol. Sci.)* 259:331.

Hebrard, C., and Delolme, C. 1999. Role of the Biotic Compartment in the Transfer of Zinc Through the Vadose Zone—Application to an Infiltration System. *Water Sci. Technol.* (G.B.) 39(2):209.

Heimbuch, R.J. 1981. Rock filled detention reservoirs. *1981 International Symposium on Urban Hydrology, Hydraulics, and Sediment Control,* July, University of Kentucky, Lexington.

Heinzmann, B. 1993. Coagulation and flocculation of stormwater from a separated storm sewer—A new possibility for enhanced treatment. In: *2nd International Conference on: Upgrading of Wastewater Treatment Plants.* Berlin, September 21–24. Sponsored by IAWQ and EWPCA. pp. 304–313.

Henderson, R.W., Lightsey, G.R., and Poonawala, N.A. 1977. Competitive adsorption of metal ions from solutions by low-cost organic materials. *Bulletin of Environmental Contamination and Toxicology* 18(3):340–344.

Herr, J.L., and Harper, H.H. 1996. The evaluation and design of an alum stormwater treatment system to improve water quality in Lake Maggiore, St. Petersburg, Florida. North American Lake Management Society (NALMS) Conference presentation, October, Minneapolis/St. Paul, MN.

Herr, J.L., and Harper, H.H. 2000. Reducing Nonpoint Source Pollutant Loads to Tampa Bay Using Chemical Treatment. In: *WEFTEC2000, 73rd Annual Conference and Exposition,* October, Anaheim, CA. Water Environment Federation, CD-ROM.

Herricks, E.E, Milne, I., and Johnson, I. 1996a. A protocol for wet weather discharge toxicity assessment. In: *WEFTEC'96: Proceedings of the 69th Annual Conference & Exposition,* Dallas, TX. Volume 4, pp. 13–24.

Herricks, E.E., Brent, R., Milne, I., and Johnson, I. 1996b. Assessing the response of aquatic organisms to short-term exposures to urban runoff. In: *Effects of Watershed Development & Management on Aquatic Ecosystems,* Engineering Foundation Conference, August, Snowbird, UT. ASCE, NY.

Herricks, E.E., ed. 1995. *Stormwater Runoff and Receiving Systems: Impact, Monitoring and Assessment.* Conference of the Engineering Foundation/ASCE held in 1991 in Mt. Crested Butte, Colorado. Boca Raton, FL: Lewis/CRC Press. 458 pp.

Hey, D.L., and Schaefer, G.C. 1983. *An Evaluation of the Water Quality Effects of Detention Storage and Source Control.* U.S. Environmental Protection Agency and the Northeastern Illinois Planning Commission, Chicago, IL.

Hird, J., Sansalone, J., and Tramonte, J. 1976. Treatment of Elevated Highway Runoff Using Clarification Technology. In: *WEFTEC 2001 Conf. Proc.* CD-ROM. 2001.

Hittman Assoc. 1976. *Methods to Control Fine-Grained Sediments Resulting from Construction Activity.* U.S. Environmental Protection Agency, Pb-279 092, Washington, DC.

Hogland, W. 1990. Pervious asphalt construction—an overview of the situation in Sweden and the U.S.A. In: *Developments in Storm Drainage—A Symposium on Infiltration and Storage of Stormwater in New Developments.* January, Sheffield City. pp. 122–135.

Hollander, M., and Wolfe, D.A. 1999. *Nonparametric Statistical Methods,* second edition. New York: John Wiley & Sons, Inc. 743 pp.

Horwath K., and Paul, E.A. 1994. Microbial biomass. Chapter 36 in *Methods of Soil Analysis. Part 2. Microbiological and Biochemical Properties*. Madison, WI: SSSA. pp. 753–774

House, L.B., Waschbusch, R.J., and Hughes, P.E. 1993. *Water Quality of an Urban Wet Detention Pond in Madison, Wisconsin, 1987–1988.* U.S. Geological Survey, in cooperation with the Wisconsin Department of Natural Resources. USGS Open File Report 93-172.

Huang J.W., Chen, J., Berti, W.R., and Cunningham, S.D. 1997. Phytoremediation of lead-contaminated soils: role of synthetic chelates in lead phytoextraction. *Environ. Sci. Technol.* 31:800–805.

Huber, W.C., and Heaney, J.P. 1981. The U.S. EPA Storm Water Management Model, SWMM: A Ten Year Perspective. *Second International Conference on Urban Storm Drainage*, June, Urbana, IL.

Hütter, U., and Remmler, F. 1996. Stormwater infiltration at a site with critical subsoil conditions: Investigations of soil, seepage water, and groundwater. *7th International Conference on Urban Storm Drainage.* Hannover, Germany, Sept. 9–13. Ed. by F. Sieker and H-R. Verworn. IAHR/IAWQ. Hannover, Germany: SuG Verlagsgesellschaft. pp. 713–718.

Hvitved-Jacobsen, T., Keiding, K, and Yousef, Y.A. 1987. Urban runoff pollution removal in wet detention ponds. In: *Proceedings of the 4th International Conference on Urban Storm Drainage.* August, Lausanne, Switzerland. Water Resources Publications. pp. 137–142.

Hvitved-Jacobsen, T., Johansen, N.B., and Yousef, Y.A. 1994. Treatment systems for urban and highway runoff in Denmark. *The Science of the Total Environment.* No. 146/147: 499–506.

International Association on Water Quality. 1998. *19th Biennial Int. Conf.*, Vancouver, Can. 8:95.

Igloria, R., Hathorn, W.E., and Yonge, D.R. 1997. NOM and Trace Metal Attenuation During Storm-Water Infiltration. *J. Hydraulic Eng.* 2(3):120.

Ives, K.J. 1990. Testing of filter media *Aqua* 39(3):144–151.

Ives, K.J. 1989. Filtration studied with endoscopes. *Water Research* 23(7):861–866.

Jacobsen, P., and Mikkelsen, P.S. 1996. Reduction of urban runoff pollutant loads using detention ponds and stormwater infiltration techniques. In: *Proceedings of the 7th Conference of Urban Storm Drainage.* Sept. 9–13, Hannover, Germany.

James, W., and Robinson, M.A. 1982. Continuous models essential for detention design. *Proceedings of the Conference on Stormwater Detention Facilities*, American Society of Civil Engineers. 163–175.

Jewell, T.K., Adrian, D.D., and Hosmer, D.W. 1980. Analysis of Stormwater Pollutant Washoff Estimation Techniques. *International Symposium on Urban Storm Runoff*, July, University of Kentucky, Lexington.

Jirik, A.W., Bay, S.M., Greenstein, D.J., Zellers, A., and Lau, S.-L. 1998. Application of TIEs in Studies of Urban Stormwater Impacts on Marine Organisms. In: *ASTM Special Tech. Public.—Proc. 1997 7th Symp. on Toxicology and Risk Assessment: Ultraviolet Radiation and the Environment,* St. Louis, MO. 284.

Johnson, C.R. 1990. A New Look at Sewer Separation for CSO Control. *WPCF Specialty Conference Series: Control of Combined Sewer Overflows*, April 8–11, Boston, MA.

Johnson, I., Herricks, E.E., and Milne, I. 1996. Application of a test battery for wet weather discharge toxicity analyses. In: *WEFTEC'96: Proceedings of the 69th Annual Conference & Exposition,* Dallas, TX. Volume 4, pp. 219–229.

Jones, J.E., and Jones, D.E. 1982. Interfacing considerations in urban detention ponding. *Proceedings of the Conference on Stormwater Detention Facilities, Planning, Design, Operation, and Maintenance*, August, Henniker, NH, ed. by W. DeGroot. Published by the American Society of Civil Engineers, New York.

Jones, J. 1995. Beware the sediment scare. *Civil Engineering* (July):56–57.

Jones, J., Anderson, S., Fognani, J., McGregor, F.R., and Axley, T. 1996. Stormwater best management practices: When is sediment considered hazardous? *Water Environment & Technology* 8(10):51–57.

Joshi, A., and Chaudhuri, M. 1996. The removal of arsenic from ground water by iron-oxide-coated sand. *Journal of Environmental Engineering* 122(8):769–771.

Jury, W.A., Spencer, W.F., and Farmer, W.J. 1983. Model for Assessing Behavior of Pesticides and other Trace Organics using Benchmark Properties. I. Description of Model. *J. Environmental Quality* (12):558–564.

Kamedulski, G.E. and McCuen, R.H. 1979. The effect of maintenance on storm water detention basin efficiency. *Water Resources Bulletin* 15(4):1146.

Kao, M.-M. 1993. The evaluation of sawdust swine waste compost on the soil ecosystem, pollution and vegetable production. *Water Science and Technology* 27(1):123–131.

Kapoor, A., and Viraraghavan, T. 1997. Fungi as biosorbents. In: *Biosorbents for metal ions,* ed. by J. Wase and C. Forster. Taylor & Francis Ltd. pp. 67–86.

Karamanev, D.G., Belanger, M-C., Chavarie, C., Chaouki, J., and Mayer, R. 1994. Hydrodynamic characteristics of a trickling bed of peat moss used for biofiltration of wastewater. *Canadian Journal of Chemical Engineering* 72:411–417.

Karouna, N. 1989. Cost Analysis of Urban Retrofits. Unpublished Report. Metropolitan Washington Council of Governments, Washington, DC.

Karouna-Renier, N.K., and Sparling, D.W. 2001. Relationships Between Ambient Geochemistry, Watershed Land-Use and Trace Metal Concentrations in Aquatic Invertebrates Living in Stormwater Treatment Ponds. *Environ. Pollut.* 112:183.

Kercher, W., Landon, J.C., and Massarelli, R. 1983. Grassy Swales Prove Cost Effective for Water Pollution Control. *Public Works* (April):53–54.

Khan, O.U., Shawley, G.W., Chen, C.L., and Chen, C.W. 1997. Castro Valley Water Quality Modeling. In: *Proc. 24th Water Resour. Plann. Manage. Conf. Aesthetics in the Constructed Environment*, Houston, TX. p. 303.

Klein, R.D. 1979. Urbanization and stream quality impairment. *Water Resources Bulletin* 15.

Kleinbaum, D.G., Kupper, L.L., Muller, K.E., and Nizam, A. 1998. *Applied Regression Analysis and Other Multivariable Methods*, 3rd Edition. Pacific Grove, CA: Duxbury Press. 798 pp.

Kloiber, S.M., and Brewonik, P.L. 1996. Effectiveness of continuous alum treatment for phosphorus removal from stormwater. North American Lake Management Society (NALMS) Conference presentation, October, Minneapolis/St. Paul, MN.

Ko, D., C.K., Porter, J.F., and McKay, G. 1996. Correlation-based approach to the optimization of fixed-bed sorption units. *Industrial and Engineering Chemistry Research* 38:4868–4877.

Kobriger, N.P., Meinholz, T.L., Gupta, M.K., and Agnew, R.W. 1981. *Constituents of Highway Runoff. Vol. 3. Predictive Procedure for Determining Pollution Characteristics in Highway Runoff.* FHWA/RD-81/044. Federal Highway Administration, Washington, DC.

Koeppe, D.E., comp. 1977. Vol. IV: Soil-Water-Air-Plant Studies. In: *Environmental Contamination by Lead and Other Heavy Metals*, ed. by G. L. Rolfe and K. A. Peinbold, Institute for Environmental Studies, Univ. of Illinois, Urbana-Champaign.

Koerlsman, W., Van Kerkhoven, M.B., and Verhoeven, J.T.A. 1993. Release of inorganic N, P, and K in peat soils: effect of temperature, water chemistry and water level. *Biogeochemistry* 20:63–81.

Kolka, R.K., Grigal, D.F., Verry, E.S., and Nater, E.A. 1999. Mercury and Organic Carbon Relationships in Streams Draining Forested Upland/Peatland Watersheds. *J. Environ. Qual.* 28:766.

Krein, A., and Schorer, M. 2000. Road runoff pollution by polycyclic aromatic hydrocarbons and its contribution to river sediments. *Water Res.* (G.B.) 34:4110.

Krejci, V., Haldimann, P., and Grottker, M. 1993. Administrative aspects of stormwater infiltration in Switzerland. In: Proceedings of the *6th International Conference on Urban Storm Drainage*. Ed. by J. Marsalek and H.C. Torno. September 12–17, Niagara Falls, ON. IAWQ, London. Published by Seapoint Publishers, Victoria, BC. pp. 999–1004.

Kronis, H. 1982. *Physical-Chemical Treatment and Disinfection of Stormwater*. Research Report No 88, Wastewater Treatment Section, Ontario Ministry of the Environment, Toronto.

Ku, Henry F.H., and Simmons, Dale L. 1986. *Effect of Urban Stormwater Runoff on Groundwater beneath Recharge Basins on Long Island, New York*. U.S. Geological Survey Water-Resources Investigations Report 85-4088. Prepared in cooperation with Long Island Regional Planning Board, Syosset, NY. USGS, Denver, CO.

Ku, Henry F.H., Vecchioli, John, and Ragone, Stephen E. 1975. Changes in Concentration of Certain Constituents of Treated Waste Water during Movement through the Magothy Aquifer, Bay Park, New York. *Journal Research U.S. Geological Survey* 3(1):89–92.

Ku, Henry F.H., Hagelin, Nathan W. and Buxton, Herbert T. 1992. Effects of Urban Storm-Runoff Control on Ground-Water Recharge in Nassau County, New York. *Groundwater* 30(4):507–514.

Lager, J., Smith, W., and Tchobanoglous, G. 1977. *Catchbasin Technology Overview and Assessment*. EPA-600/2-77-051, U.S. Environmental Protection Agency, Cincinnati, OH.

Lager, J.A., Smith, W.G., Lynard, W.G., Finn, R.M., Finnemore, J. 1977. *Urban Stormwater Management and Technology: Update and User's Guide*. U.S. EPA. EPA-600/8-77-014 c.2.

Lai, C.H., Lo, S.L., and Lin, C.F. 1994. Evaluating an iron-coated sand for removing copper from water. *Water Science and Technology* 30(9):175–182.

Lainé, S., Poujol, T., Dufay, S., Baron, J.R., and Robert, P. Treatment of rainwater by air flotation-filtration-UV disinfection to ensure that treated water is of bathing water quality. *Water Science and Technology* 38(10):99-105.

Lake Tahoe Regional Planning Agency. 1978. *Lake Tahoe Basin Water Quality Management Plan, Volume II, Handbook of Best Management Practices*. Lake Tahoe, California.

Lancelette, J.M. 1974. Removal and recovery of metals from polluted waters. *U.S. Patent 3,790,370.*

Land and Water. 1996. Storm pond turned into selling feature. (Nov./Dec.) 40(6):35.

Laplace, D., Bachoc, A., Sanchez, Y., and Dartus, D. 1992. Truck sewer clogging development—description and solutions. *Water Science and Technology* 25(8):91–100.

Larsen, V.J., and Schierup, H.H. 1981. The use of straw for removal of heavy metals from wastewater. *Journal of Environmental Quality* 10(2):188–193.

Lau, Y.L., Marsalek, J., and Rochfort, Q. 2000. Use of a Biofilter for Treatment of Heavy Metals in Highway Runoff. *Water Qual. Res. J. Can.* 35:563.

Lawler, Matusky, and Skelly Engineers, and Fanning and Phillips. 1982. Chapter 5: *The Long Island Segment of the Nationwide Urban Runoff Program*, Final Draft Report to the Long Island Regional Planning Board, (June) Long Island, NY.

Leahy, J.J., and Birkinshaw, C. 1992. The chemical and physical structure of peat bitumens. *Journal of Chemical Technology and Biotechnology* 53:57–65.

Lee, G.F., and Jones-Lee, A. 1993. Water quality impacts of stormwater-associated contaminants: focus on real problems. *Water Science and Technology* 28(3-5):231–240.

Lee, G.F., and Jones-Lee, A. 1995a. Deficiencies in stormwater quality monitoring. In: *Stormwater NPDES Related Monitoring Needs*, Engineering Foundation Conference, Mt. Crested Butte, CO. ASCE, NY. 651–662.

Lee, G.F. and Jones-Le, A. 1995b. Issues in managing urban stormwater runoff quality. *Water/ Engineering Management* 142(5):51–53.

Lee, G.F., and Jones-Lee, A. 1996. Evaluation of the water quality significance of the chemical constituents in aquatic sediments: Coupling sediment quality evaluation results to significant water quality impacts. *WEFTEC'96, Surface Water Quality and Ecology, parts 1 and 2*. 1996 Water Environment Federation Technical Exposition and Conference, Dallas, TX. Alexandria, VA: WEF.

Legrand, J., Maillot, H., Nougarède, F., and Defontaine, S. 1994. A device for stormwater treatment in the urban development zone of Annœullin (in French). *TSM* (11):639–643.

Legret, M., Colandini, V., and LeMarc, C. 1996. Effects of a Porous Pavement with Reservoir Structure on the Quality of Runoff Water and Soil. *Sci. of the Total Environ.* 102:335.

Legret, M., Nicollet, M., Miloda, P., Colandini, V., and Raimbault, G. 1999. Simulation of Heavy Metal Pollution from Stormwater Infiltration Through a Porous Pavement with Reservoir Structure. *Water Sci. Technol.* (G.B.) 39(2):119.

Lehmann, E.L., and D'Abrera. 1975. *Nonparametrics: Statistical Methods Based on Ranks*. San Francisco, CA: Holden-Day. 425 pp.

Lenet, D.R., Penrose, D.L., and Eagleson, K.W. 1981. Variable effects of sediment addition on stream benthos. *Hydrobiologia* 79:187.

Lenet, D., and Eagleson, K. 1981. *Ecological Effects of Urban Runoff on North Carolina Streams*, North Carolina Division of Environmental Management, Biological Series #104. North Carolina Dept. of Natural Resources and Community Development, Raleigh, NC.

Lenet, D.R., Penrose, D.L., and Eagleson, K. 1979. *Biological Evaluation of Non-Point Sources of Pollutants in North Carolina Streams and Rivers*, North Carolina Division of Environmental Management, Biological Series #102. North Carolina Dept. of Natural Resources and Community Development, Raleigh, NC.

Lenhart, J.H. 1997. Processed deciduous leaves for stormwater treatment. *Land and Water* 54.

Leppert, D. 1990. Heavy metal sorption with clinoptilolite zeolite: Alternatives for treating contaminated soil and water. *Mining Engineering* 42(6):604–608.

Lettenmaier, D.P., and Dally, L.K. 1983. *Stormwater Quality Evaluation for Metro Transit East Operating Base*. Department of Civil Engineering, University of Washington, Seattle, WA.

Levesque, L.M.J., and De Boer, D.H. 2000. Trace Element Chemistry of Surficial Fine-Grained Laminae in the South Saskatchewan River, Canada. In: *IAHS Pub. Symp. Role of Erosion and Sediment Transport in Nutrient and Contaminant Transfer*, July, Waterloo, ON, Can. 263:183.

Lewis, C.L., Sax, T.P., and Duke, L.D. 2000. Estimating Pollutant Loading from Industrial Activities in an Urban Watershed. In: *2000 Joint Conference on Water Resources Engineering and Water Resources Planning and Management*, July, Minneapolis, MN. American Society of Civil Engineers, CD-ROM.

Lind, B.B., and Karro, E. 1995. Stormwater infiltration and accumulation of heavy metals in roadside green areas in Goteborg, Sweden. *Ecological Engineering* 5(4):533–529.

Lindsay, W.L. 1979. *Chemical Equilibria in Soils*. New York: John Wiley and Sons.

Lindstrom, M., and Hakanson, L.A. 2001. Model to Calculate Heavy Metal Load to Lakes Dominated by Urban Runoff and Diffuse Inflow. *Ecological Modelling* 137:1.

Line, D.E., Arnold, J.A., Jennings, G.D., and Wu, J. 1996. Water Quality of Stormwater Runoff from Ten Industrial Sites. *Water Res. Bull.* 32(4)807.

Line, D.E., Wu, J., Arnold, J.A., Jennings, G.D., and Rubin, A.R. 1997. Water Quality of First Flush Runoff From 20 Industrial Sites. *Water Environ. Res.* 69(3):305.

Linsey, G., Roberts, L., and Page, W. 1992. Inspection and Maintenance of Infiltration Facilities. *Journal of Soil and Water Conservation* 47(6):481–486.

Liu, D., Sansalone, J.J., and Cartledge, F.K. 2001a. Divalent Heavy Metal Removal Using Manganese Oxide Coated Polymeric Media for Storm Water Treatment. In: *WEFTEC 2001 Conf. Proc.* CD-ROM.

Liu, D.F., Teng, Z., Sansalone, J.J., and Cartledge, F.K. 2001b. Surface Characteristics of Sorptive-Filtration Storm Water Media. I: Low-Density (Rho(S)< 1.0) Oxide-Coated Buoyant Media. *J. Environ. Eng.* 127:868.

Liu, D.F., Teng, Z., Sansalone, J.J., and Cartledge, F.K. 2001c. Surface Characteristics of Sorptive-Filtration Storm Water Media. II: Higher Specific Gravity (Rho(S)> 1.0) Oxide-Coated Fixed Media. *J. Environ. Eng.* 127:879.

Lloyd, J.W., Lerner, D.N., Rivett, M.O., and Ford, M. 1988. Quantity and Quality of Groundwater beneath an Industrial Conurbation—Birmingham, UK. In: *Proceedings of the Conference—Hydrological Processes and Water Management in Urban Areas.* April 24–29, Duisburg, Federal Republic of Germany. International Hydrological Programme, UNESCO. pp. 445–452.

Lo, K.W., Ooi, P.L., and Lee, S.-L. 1990. Dynamic replacement and mixing of organic soils with sand charges. *Journal of Geotechnical Engineering* 116(10):1463–1482.

Loague, K.M., and Freeze, R.A. 1985. A Comparison of Rainfall-Runoff Modeling Techniques on Small Upland Catchments. *Water Resources Research* 21(2):229–248.

Lohani, V.K., Chanat, J., Kibler, D.F., and Fedak, R. 1999. Model Choice and Basin Scale for Urbanizing Watersheds in Appalachian Headwater Areas. *ASCE 26th Annu. Conf. Water Resour. Plann. Manage. 1999 Annu. Conf. Environ. Eng.*, Tempe, AZ.

Long, E.R., MacDonald, D.D., Smith, S.L., and Calder, F.D. 1996. Incidence of adverse biological effects within ranges of chemical concentrations in marine and estuarine sediments. *Environmental Management* 19:81–97.

Loumbourdis, N.S., and Wray, D. 1998. Heavy-Metal Concentration in the Frog Rana Ridibunda from a Small River of Macedonia, Northern Greece. *Environ. Intl.* 24(4):427.

Lowengart, A., Diab, S., Kochba, M., and Avnimelech, Y. 1996. Development of a biofilter for turbid and nitrogen-rich irrigation water; a: organic carbon degradation and nitrogen removal processes. *Bioresource Technology* 44:131–135.

Luzkow, S.M., Scherger, D.A., and Davis, J.A. 1981. Effectiveness of two in-line urban storm water best management practices (BMP's). *1981 International Symposium on Urban Hydrology, Hydraulics, and Sediment Control*, University of Kentucky, Lexington.

Macaskie, L.E. and Basnakova, G. 1998. Microbially-enhanced chemisorption of heavy metals: a method for the bioremediation of solutions containing long-lived isotopes of neptunium and plutonium. *Environ. Sci. Technol.* 32:184–187.

Macdonald, K., Jefferies, C., and Dring, D. 1999. A Preliminary Assessment of the Efficiency of Stormwater Runoff Ponds in the Scottish Climate. In: *Proc. the Eighth International Conference on Urban Storm Drainage.* August 30–September 3, Sydney, Australia, ed. by I.B. Joliffe and J.E. Ball. The Institution of Engineers Australia, The International Association for Hydraulic Research, and The International Association on Water Quality. 882.

Magmedov, V.G., Zakharchenko, M.A., Yakovleva, L.I., and Ince, M.E. 1996. Use of constructed wetlands for the treatment of runoff and drainage waters: the UK and Ukraine experience. *Water Science and Technology* 33(4-5):315–323.

Magnuson, M.L., Kelty, C.A., and Kelty, K.C. 2001. Trace Metal Loading on Water-Borne Soil and Dust Particles Characterized Through the Use of Split-Flow Thin Cell Fractionation. *Analytical Chemistry* 73:3492.

Makepeace, D.K., Smith, D.W., and Stanley, S.J. 1995. Urban stormwater quality: Summary of contaminant data. *Critical reviews in Environmental Science and Technology* 25(2):93–139.

Malmquist, P-A., Larm, T., Bennerstedt, K., and Wranghede, A-K. 1999. Sources of Pollutants Discharging to Lake Trekanten, Stockholm. In: *Proc. the Eighth International Conference on Urban Storm Drainage*. August 30–September 3, Sydney, Australia, ed. by I.B. Joliffe and J.E. Ball. The Institution of Engineers Australia, The International Association for Hydraulic Research, and The International Association on Water Quality. 1736.

Malmquist, Per-Arne. 1978. Atmospheric Fallout and Street Cleaning—Effects on Urban Stream Water and Snow. *Prog. Wat. Tech.* 10(5/6):495–505. (Pergamon Press, Great Britain.)

Malterer, T., McCarthy, B., and Adams, R. 1996. Use of peat in waste treatment. *Mining Engineering* (48):53–56.

Mancini, J., and Plummer, A. 1986. Urban runoff and water quality criteria. In: *Urban Runoff Quality—Impact and Quality Enhancement Technology*. Ed. by B. Urbonas and L.A. Roesner. Engineering Foundation Conference, Henniker, NH. ASCE, NY. pp. 133–149.

Manning, M.J., Sullivan, R.H., and Kipp, T.M. 1976. *Nationwide Evaluation of Combined Sewer Overflows and Urban Stormwater Discharges, Vol. III: Characterization of Discharges*. U.S. Environmental Protection Agency, Cincinnati, OH.

Marcy, S., and Gerritsen, J. 1996. Developing deverse assessment endpoints to address multiple stressors in watershed ecological risk assessment. In: *Abstract Book: SETAC 17th Annual Meeting*, Nov. 17–21, Washington, DC. p. 96.

Markich, S.J., and Brown P.L. 1998. Relative Importance of Natural and Anthropogenic Influences on the Fresh Surface Water Chemistry of the Hawkesbury-Nepean River, South-Eastern Australia. *Sci. Total Environ*. 217(3):201.

Marques, M., and Hogland, W. 2001. Stormwater Run-off and Pollutant Transport Related to the Activities Carried Out in a Modern Waste Management Park. *Waste Man. Res*. 19:20.

Marsalek, J., Weatherbe, D., and Zukovs, G. 1982. Institutional aspects of stormwater detention. In: *Proceedings of the Conference on Stormwater Detention Facilities, Planning, Design, Operation, and Maintenance*, August, Henniker, NH, ed. by W. DeGroot. Published by the American Society of Civil Engineers, New York.

Marsalek, J., Watt, W.E., Anderson, B.C., and Jaskot, C. 1997. Physical and Chemical Characteristics of Sediments from a Stormwater Management Pond. *Water Qual. Res. J. of Canada* 32(1):89.

Marsalek, J., Rochfort, Q., Brownlee, B., Mayer, T., and Servos M. 1999. Exploratory Study of Urban Runoff Toxicity. *Water Science Technology* 39(12):33–39.

Martin, E. and Miller, R.A. 1987. Efficiency of an urban stormwater detention system. In: *Proceedings of the 4th International Conference on Urban Storm Drainage*. Lausanne, Switzerland. August, Water Resources Publications. pp. 143–148.

Martin, E.H. 1988. Mixing and residence times of stormwater runoff in a detention system. In: *Engineering Foundation Conference: Design of Urban Runoff Quality Controls*. Potosi, Missouri. July 10–15, ed. by L. A. Roesner, B. Urbonas and M.B. Sommen, June 1989. pp. 164–179. Published by the American Society of Civil Engineers, New York.

Marton, J., and Mohler, I. 1988. The Influence of Urbanization on the Quality of Groundwater. In: *Proceedings of the Conference—Hydrological Processes and Water Management in Urban Areas*. April 24–29, Duisburg, Federal Republic of Germany. International Hydrological Programme, UNESCO. pp. 453–460.

Mason, C.F., Norton, S.A., Fernandez, I.J., and Katz, L.E. 1999a. Deconstruction of the Chemical Effects of Road Salt on Stream Water Chemistry. *J. Environ. Qual.* 28:82.

Mason, J.M., Jr. 1982. On-site stormwater detention: An overview. *Public Works*. (February).

Mason, J.M., Jr. 1981. Multistaged outlet control for small on-site detention. *1981 International Symposium on Urban Hydrology, Hydraulics, and Sediment Control*, July, University of Kentucky, Lexington.

Mason, R.P., and Sullivan, K. A. 1998. Mercury and Methylmercury Transport Through an Urban Watershed. *Water Res.* (GB) 32(2):321.

Mason, R.P., Lawson, N.M., Lawrence, A.L., Leaner, J.J., Lee, J.G., and Sheu, G.R. 1999b. Mercury in the Chesapeake Bay. *Marine Chemistry* 65(1-2):77.

Mason, Y., Ammana, A.A., Ulrich, A., and Sigg, L. 1999c. Behavior of Heavy Metals, Nutrients, and Major Components during Roof Runoff Infiltration. *Environ. Sci. Technol.* 33(10):1588.

Maxted, J. and Shaver, E. 1997. The use of retention basins to mitigate stormwater impacts on aquatic life. *Engineering Foundation Conference: Effects of Watershed Development & Management on Aquatic Ecosystems*. August 4–9, 1996. Snowbird, Utah. Engineering Foundation, New York.

Mayer, T., Marsalek, J., and Delos Reyes, E. 1996. Nutrients and metal contaminants status of urban stormwater ponds. *Journal of Lake and Reservoir Management* 12(3):348–363.

McComas, S.R., and Sefton, D.F. 1985. Comparison of stormwater runoff impacts on sedimentation and sediment trace metals for two urban impoundments. In: *Lake and Reservoir Management: Practical Applications*, proceedings of the Fourth Annual Conference and International Symposium, North American Lake Management Society, October, 1984, McAfee, NJ.

McCuen, R.H. 1998. *Hydrologic Analysis and Design*, 2nd Edition. Prentice Hall.

McCuen, R.H. 1980. Water quality trap efficiency of storm water management basins. *Water Resources Bulletin* 16(1):15–21.

McCuen, R.H. 1986 Stormwater modeling into the 21st century. *Sediment and Stormwater Conference*, sponsored by the Maryland Sediment and Stormwater Division, Water Resources Administration, Salisbury.

McCuen, R.H., Walesh, S.G., and Rawis, W. 1984. *Control of Urban Stormwater Runoff by Detention and Retention*, Misc. Public. No. 1428. U.S. Dept. of Agriculture.

McKay, G. *Use of Adsorbents for the Removal of Pollutants from Wastewaters*. Boca Raton, FL: CRC Press. 186 pp.

Medeiros, C., and Coler, R.A. 1982. *A Laboratory/Field Investigation into the Biological Effects of Urban Runoff*. Water Resources Research Center, University of Massachusetts, Amherst, MA.

Medeiros, C., Coler, R.A., and Calabrese, E.J. 1984. A laboratory assessment of the toxicity of urban runoff on the fathead minnow (*Pimephales promelas*). *Journal of Environmental Science Health* A19:847.

Merkel, B., Grossman, J., and Udluft, P. 1988. Effect of Urbanization on a Shallow Quarternary Aquifer. In: *Proceedings of the Conference—Hydrological Processes and Water Management in Urban Areas*. April 24–29, Duisburg, Federal Republic of Germany. International Hydrological Programme, UNESCO. pp. 461–468.

Metcalf and Eddy, Inc. 1991. *Wastewater engineering: treatment, disposal, and reuse*. New York: McGraw-Hill Publishing Company.

Metropolitan Washington Council of Governments. 1992. *Analysis of Urban BMP Performance and Longevity*. Prince George's County, Department of Environmental Resources.

Mickelson, S., Shuman, R., and Munger, S. 1998. An Intensive Field Sampling Program in Support of a Numerical Toxicant Fate and Transport Model and Risk Assessment of the Duwamish River and Elliott Bay, Washington. In: *Proc. Adv. in Urban Wet Weather Pollut. Reduction*, Cleveland, OH, WEF (CP3805). 337.

Mikkelsen, P.S., Weyer, G., Berry, C., Walden, Y., Colandini,V., Paulsen, S., Grotehusmann, D., and Rohlfing, R. 1994. Pollution from urban stormwater infiltration. *Water Science and Technology* 29(1-2):293–302.

Mikkelsen, P.S., Jacobsen, M., and Boller, M. 1996a. Pollution of Soil and Groundwater from Infiltration of Highly Contaminated Stormwater—A Case Study. In: *Proc. 7th Int. Conf. On Urban Storm Drainage*, Hannover, Germany. IAHR/IAWQ Joint Committee on Urban Storm Drainage. 707.

Mikkelsen, P.S., Arngjerg-Nielsen, K., and Harremoës, P. 1996b. Consequences for established design practice from geographical variation of historical rainfall data. In: *Proceedings: 7th International Conference on Urban Storm Drainage*. Sept. 9–13, Hannover, Germany.

Mikkelsen, P.S., Häfliger, M., Ochs, M., Tjell, J.C., Jacobsen, M., and Boller, M. 1996c. Experimental assessment of soil and groundwater contamination from two old infiltration systems for road run-off in Switzerland. *Science of the Total Environment*.

Mitchell, G., and McDonald, A.T. 1992. Discolouration of water by peat following induced drought and rainfall simulation. *Water Research* 26(3):321–326.

Mitchell, G.F., Hunt, C.L., and Su., Y. 2001. Using a Wetland to Mitigate Highway Storm Water Runoff. In: *Proc. ASCE EWRI Conf.—Bridging the Gap: Meeting the World's Water and Environmental Resources Challenges*. CD-ROM.

Moffa, P.E. 1996. The Control and Treatment of Industrial and Municipal Stormwater. New York: Van Nostrand Reinhold.

Moffa, P.E. 1989. Control and Treatment of Combined Sewer Overflows. New York: Van Nostrand Reinhold.

Mohaupt, V., Sieber, U., Van Den Roovaart, J., Verstappen, C.G., Langenfeld, F., and Braun, M. 2001. Diffuse Sources of Heavy Metals in the Rhine Basin. *Water Sci. Tech.* 44:41.

Moore, J.C., Hansen, D.J., Garnas, R.L., and Goodman, L.R. 1985. A sand/granular carbon filtration treatment system for removing aqueous pesticide residues from a marine toxicology laboratory effluent. *Water Research* 19(12):1601–1604.

Moreno-Grau, S., Moreno, J., Bayo, J., Angosto, J.M., Elvira, B., and Moreno-Clavel, J. 2000. Heavy Metals in Atmospheric Aerosols, Rainfall, Soils and Ground Water in Industrial Cartegena, Spain. *2000 Annual Meeting and International Conference of the American Institute of Hydrology*, November, Research Triangle Park, NC. American Institute of Hydrology.

Morrisey, D.J., Roper, D.S., and Williamson, R.B. 1997. Biological effects of the build-up of contaminants in sediments in urban estuaries. Presented at the *Effects of Watershed Developments and Management on Aquatic Ecosystems* conference. August 4–9, 1996, Snowbird, UT, ed. by L.A. Roesner. ASCE, New York. pp. 1–20.

Morrisey, D.J., Williamson, R.B., Van Dam, L., and Lee, D.J. 2000. Stormwater Contamination of Urban Estuaries. 2. Testing a Predictive Model of the Build-Up of Heavy Metals in Sediments. *Estuaries* 23:67.

Moskus, P.E., Dilks, D.W., and Schecter, D.S. 2001. Watershed and Water Quality Modeling as Part of a Watershed Protection Plan for Cherokee County, Georgia. *WEFTEC 2001 Conf. Proc.* CD-ROM.

Mosley, L.M., and Peake, B.M. 2001. Partitioning of Metals (Fe, Pb, Cu, Zn) in Urban Run-Off From the Kaikorai Valley, Dunedin, New Zealand. *New Zealand J. Marine Freshwater Res.* 35:615.

Moss Tested to Remove Manganese From Mine Drainage. 1984. *Coal Age 89*: 28.

Mote Marine Laboratory. 1984. *Biological and Chemical Studies on the Impact of Stormwater Runoff upon the Biological Community of the Hillsborough River, Tampa, Florida*. Stormwater Management Division, Dept. of Public Works, Tampa, FL.

Mothersill, C.L., Anderson, B.C., Watt, W.E., and Marsalek, J. 2000, Biological Filtration of Stormwater: Field Operations and Maintenance Experiences. *Water Qual. Res. J. Can.* 35:541.

Munger, S., Strand, J., Stark, K., Silver, K., Laetz, C., Georgiana, T., McElhany, D., Li, K., and Mickelson, S. 1998. Use of Transplanted Mussels to Monitor Chemical Contaminants in the Duwamish River Estuary, Washington. In: *Proc. Watershed Manage—Moving from Theory to Implementation*, Denver, CO. WEF (CP3804). 241.

Murphy, W. 1975. *Roadway Particulate Losses*. American Public Works Assoc. Unpublished.

Musiake, K., Okamura, J., and Matsushita, J. 1990. The urban hydrological system with shallow infiltration facilities—a case study of an area in Tokyo Metropolis. In: *Proceedings of the 5th International Conference on Urban Storm Drainage*. July, Osaka, Japan. pp. 113–118.

Naghavi, B., and Malone, R.F. 1986. Algae removal by fine sand/silt filtration. *Water Research* 20(3):377–383.

Namasivayam, C. and Periasamy, K. 1993. Bicarbonate-treated peanut hull carbon for mercury (II) removal from aqueous solution. *Water Research* 27(11):1163–1168.

Natarajan, U., and Rajagopal, R. 1993. Surveying The Situation. *Environmental Testing and Analysis* (Jan-Feb.):40–50.

National Academy of Science, National Research Council. 1994. *Ground Water Recharge Using Waters of Impaired Quality*, by Committee on Ground Water Recharge, Water Science and Technology Board, Commission on Geosciences, Environment, and Resources, National Research Council. National Academy Press, Washington, DC.

Nawang, W.M., and Saad, S. 1993. Stormwater infiltration investigation using porous pavement In: *Proceedings of the 6th International Conference on Urban Storm Drainage*, ed. by J. Marsalek and H.C. Torno, September 12–17, Niagara Falls, ON. IAWQ, London. Published by Seapoint Publishers, Victoria, BC. pp. 1092–1097.

Neff, J.W., Foster, R.S., and Slowey, J.F. 1978. *Availability of Sediment-Adsorbed Heavy Metals to Benthos with Particular Emphasis on Deposit-Feeding Infauna*, Technical Report D-78-42. Office, Chief of Engineers, U.S. Army, Washington, DC.

Nellor, Margaret H., Baird, Rodger B., and Smyth, John R. 1985. Health Aspects of Groundwater Recharge. In: *Artificial Recharge of Groundwater*, ed. by Takashi Asano. Boston, MA: Butterworth Publishers. pp. 329–356.

Nelson, P.O., and Yang, M. 1995. Equilibrium adsorption of chlorophenols on granular activated carbon. *Water Environment Research* 67:892–898.

Neto, J.A.B., Smith, B.J., and McAllister, J.J. 2000. Heavy Metal Concentrations in Surface Sediments in a Nearshore Environment, Jurujuba Sound, Southeast Brazil. *Environ. Pollut.* 109(1).

Netzer, A., and Hughes, D.E. 1984. Adsorption of copper, lead, and cobalt by activated carbon. *Water Research* 18(8):927–933.

Netzer, A., Wilkinson, P., and Beszedits, S. 1974. Removal of trace metals from wastewater by treatment with lime and discarded automotive tires. *Water Research* 8:813–817.

Newberry, G.P., Yonge, D.R. 1996. The retardation of heavy metals in stormwater runoff by highway grass strips. Research Report WA-RD 404.1 Washington State Transportation Commision; U.S. Department of Transportation.

Niemczynowicz, J. 1990. Swedish stormwater detention practices. In: *Proceedings of the 5th International Conference on Urban Storm Drainage*. July, Osaka, Japan. pp. 197–202.

Niemczynowicz, J., and Hogland, W. 1987. Tests of porous pavement in Lund, Sweden. In: *Proceedings of the 4th International Conference on Urban Storm Drainage*. August 1987, Lausanne, Switzerland. Water Resources Publications. pp. 79–80.

Nightingale, H.I., Ayars, J.E., McCormick, R.L., and Cehrs, D.C. 1983. Leaky Acres Recharge Facility: A Ten-Year Evaluation. *Water Resources Bulletin* 19(3):429–437.

Nightingale, H.I., and Bianchi, W.C. 1977a. Ground-Water Chemical Quality Management by Artificial Recharge. *Groundwater* 15(1):15–22.

Nightingale, H.I., and Bianchi, W.C. 1977b. Ground-Water Turbidity Resulting from Artificial Recharge. *Groundwater* 15(2):146–152.

Nightingale, H.I. 1987a. Water Quality Beneath Urban Runoff Water Management Basins. *Water Resources Bulletin* 23(2):197–205.

Nightingale, Harry I. 1987b. Accumulation of As, Ni, Cu and Pb in Retention and Recharge Basins Soils from Urban Runoff. *Water Resources Bulletin* 23(4):663–672.

Nix, S.J., and Durrans, S.R. 1996. Off-line stormwater detention systems. *Water Resources Bulletin* 32(6):1329–1340.

Norberg-King, Teresa J., Durhan, Elizabeth J., Ankley, Gerald T., and Robert, Eric. 1991. Application of Toxicity Identification Evaluation Procedures to the Ambient Waters of the Colusa Basin Drain, California. *Environmental Toxicology and Chemistry* 10:891–900.

Novotny, V., and Chesters, G. 1981. *Handbook of Nonpoint Pollution*. New York: VanNostrand Reinhold.

Novotny, V., and Olem, H. 1994. *Water Quality: Prevention, Identification , and Management of Diffuse Pollution*. New York: Van Nostrand Reinhold.

Novotny, V., Muehring, D., Zitomer, D.H, Smith, D.W., and Facey, R. 1998a, Cyanide and Metal Pollution by Urban Snowmelt: Impact of De-icing Compounds. *IAWQ 19th Biennial Int. Conf.*, Vancouver, Can., 8:219.

Novotny, V., Smith, D., and Bartosová, A. 1998b. Winter Diffuse Pollution of Urban Watersheds Affected by Deicing Operations. In: *Proc. Water Environ. Fed. 71st Annu. Conf. Exposition*, Orlando, FL. 4:389.

Nowakowska-Blaszczyk, A., and Zakrzewski, J. 1996. The Sources and Phases of Increase of Pollution in Runoff Waters in Route to Receiving Waters. In: *Proc. 7th Int. Conf. on Urban Storm Drainage*, Hannover, Germany, IAHR/IAWQ Joint Committee on Urban Storm Drainage. 49.

O'Connor, T., Field, R., Fischer, D., Rovansek, R., Pitt, R. Clark, S., and Lama, M. 1999. Urban Wet Weather Flows—1998 Literature Review. *Water Environment Research* 70(4).

O'Shea, M.L., Borst, M., Liao, D., Yu, S.L., and Earles, T.A. 1999. Constructed Wetlands for Stormwater Management. *ASCE 26th Annu, Conf. Water Resour. Plann. Manage. 1999 Annu. Conf. Environ. Eng.*, Tempe, AZ.

Oberts, G. 1990. Design considerations for management of urban runoff in wintry conditions. In: *Proceedings: International Conference on Urban Hydrology under Wintry Conditions*. March 19–21, Narvik, Norway.

Oberts, G. 1994. Performance of stormwater ponds and wetlands in winter. Technical note 16. *Watershed Protection Techniques* 1(2):64–68.

Oliver, L.J., and Grigoropoulis, S.G. 1981. Control of storm-generated pollution using a small urban lake. *Journal Water Pollution Control Federation* 53(5).

Olivieri, A., Hall, T., and Bruinsma, D. 2000. TMDL as a Regulatory Process: Conducting the Copper and Nickel TMDL in South San Francisco Bay. *Watershed 2000 Management Conference,* July 2000, Vancouver, British Columbia. Water Environment Federation, CD-ROM.

Ontario (Province of). 1984. *Urban Drainage Design Guidelines* (Draft). Urban Drainage Policy Implementation Committee, Technical Sub-Committee No. 3.

Ontario Ministry of the Environment. 1986. *Humber River Water Quality Management Plan.* Toronto Area Watershed Management Strategy. Toronto, ON.

Ontario Ministry of the Environment. 1983. *Rideau River Stormwater Management Study. Kingston, Ontario, Operation of Facilities for Urban Stream Quality Enhancement.* University of Washington, Department of Civil Engineering, Water Resources Series Technical Report No. 79, Seattle, WA.

Oscanyan, P.C. 1975. Design of sediment basins for construction sites. *National Symposium on Urban Hydrology and Sediment Control*, University of Kentucky, Lexington.

Othmer, E.F., Jr., Friedman, G., Borroum, J.S., and Currier, B.K. 2001. Performance Evaluation of Structural BMPs: Drain Inlet Inserts (Fossil Filter and StreamGuard) and Oil/Water Separator. In: *Proc. ASCE EWRI Conf.—Bridging the Gap: Meeting the World's Water and Environmental Resources Challenges*. CD-ROM.

Painter, T.J. 1991. Preservation in Peat. *Chemistry and Industry:* 421–424.

Palmer, V.J. 1946. Retardance Coefficients for Low Flow in Channels Lined with Vegetation, *Transactions of the American Geophysical Union* 27(I):187–197.

Parker, J.T.C., Fossum, K.D., and Ingerssol, T.L. 2000. Chemical Characteristics of Urban Stormwater Sediments and Implications for Environmental Management, Maricopa County, AZ. *Environ. Manage.* 26:99.

Parmer, K., and Pitt, R. 1995. *Quality Assurance Project Plan; Effects, Sources, and Treatability of Stormwater Toxicants*. Contract No. CR819573. Cincinnati, OH: U.S. EPA.

Paschka, M.G., Ghosh, R.S., and Dzombak, D.A. 1999. Potential Water-quality Effects from Iron Cyanide Anticaking Agents in Road Salt. *Water Environ. Res.* 71(6):1235.

PEDCo-Environmenal, Inc. 1977. *Control of Re-entrained Dust from Paved Streets*. EPA-907/9-77-007, U.S. Environmental Protection Agency, Kansas City, MO.

Pedersen, E.R. 1981. *The Use of Benthic Invertebrate Data for Evaluating Impacts of Urban Stormwater Runoff.* Masters thesis submitted to the College of Engineering, University of Washington, Seattle.

Perez-Candela, M., Martin-Martinez, J.M., and Torregrosa-Macia, R. 1995. Chromium (VI) removal with activated carbons. *Water Research* 29(9):2174–2180.

Perkins, M.A. 1982. *An Evaluation of Instream Ecological Effects Associated with Urban Runoff to a Lowland Stream in Western Washington.* U.S. Environmental Protection Agency, Corvallis Environmental Research Laboratory, Corvallis, OR.

Persson, L.B.B., Morrison, G.M., Roux, M.S., Petterson, T.J.R., Ellis J., and Rauch, S. 1999. Improved Assessment of Metal Concentrations in Stormwater Runoff With a Passive Sampler. In: *Proc. the Eighth International Conference on Urban Storm Drainage*. August 30–September 3, Sydney, Australia, ed. by I.B. Joliffe and J.E. Ball. The Institution of Engineers Australia, The International Association for Hydraulic Research, and The International Association on Water Quality. 1505.

Petersen, C.R., Faarbaek, T., Jensen, G.H., Weyer, G., Fujita, S., Ishikawa, K., Geldof, G., Stenmark, C., and Pratt, C.J. 1993. Urban stormwater infiltration design practice and technology: state of the art assessment. In: *Proceedings of the 6th International Conference on Urban Storm Drainage*, ed. by J. Marsalek and H.C. Torno. September 12–17, Niagara Falls, ON. IAWQ, London. Published by Seapoint Publishers, Victoria, BC. pp. 969–974.

Peterson, J.G. 1999. The Aurora Project: Community Watershed Partnerships in Action. In: *Transactions of the Sixty-Fourth North America Wildlife and Natural Resource Conference*, ed. by R.E. McCabe, and S.E. Loos. 336.

Petterson, T.J.R. 1999. The Effects of Variations of Water Quality on the Partitioning of Heavy Metals in a Stormwater Pond. In: *Proc. the Eighth International Conference on Urban Storm Drainage*. August 30–September 3, Sydney, Australia, ed. by I.B. Joliffe and J.E. Ball. The Institution of Engineers Australia, The International Association for Hydraulic Research, and The International Association on Water Quality. p. 1943.

Phillips, G.R., and R.C. Russo. 1978. *Metal Bioaccumulation in Fishes and Aquatic Invertebrates: A Literature Review*. EPA-600-3-78-103, U.S. Environmental Protection Agency, Duluth, MN.

Pisano, W.C., and Brombach, H. 1996. Solids settling curves: Wastewater solids data can aid design of urban runoff controls. *Water Environment & Technology* 8(4):27–33.

Pitt, R., and McLean, J. 1983. *Toronto Area Watershed Management Strategy Study: Humber River Pilot Watershed Project*. Ontario Ministry of the Environment, Toronto, ON.

Pitt, R. 1986a. Runoff Controls in Wisconsin's Priority Watersheds. In: *Conference on Urban Runoff Impact and Quality Enhancement Technology*. June, Henniker, NH, ed. by B. Urbonas and L.A. Roesner, Proceedings published by the American Society of Civil Engineering, NY.

Pitt, R. 1986b. Runoff controls in Wisconsin's Priority Watersheds. In: *Advanced Topics in Urban Runoff Research* conference. Proceedings ed. by B. Urbonas and L.A. Roesner. Engineering Foundation and ASCE, NY. pp. 290–313.

Pitt, R. 1991a. Biofiltration Management Practices: Design Guidelines and Background Information for Designing Infiltration devices and Grass Swales for Stormwater Quality Control. University of Alabama at Birmingham.

Pitt, R. 1991b. Biological impacts associated with urban runoff. In: *Effects of Urban Runoff on Receiving Systems: An Interdisciplinary Analysis of Impact, Monitoring, and Management*, Engineering Foundation Conference, Mt. Crested Butte, CO. ASCE, NY.

Pitt, R. 1994. Water quality trends from stormwater controls. In: *Stormwater NPDES Related Monitoring Needs,* ed. by H.C. Torno. Engineering Foundation and ASCE. pp. 413–434.

Pitt, R, Clark, S., and Parmer, K. 1995b. *Potential Groundwater Contamination from Intentional and Non-Intentional Stormwater Infiltration.* Cooperative Agreement No. CR 819573, U.S. Environmental Protection Agency, Cincinnati, OH.

Pitt, R. 1995a. Biological Effects of Urban Runoff Discharges. In: *Stormwater Runoff and Receiving Systems: Impact, Monitoring, and Assessment*, ed. by E.E. Herricks. Engineering Foundation and ASCE. Boca Raton, FL: CRC/Lewis. pp. 127–162.

Pitt, R. 1995b. Effects of Urban Runoff on Aquatic Biota. In: *Handbook of Ecotoxicology,* ed. by D.J. Hoffman, B.A. Rattner, G.A. Burton, Jr., and J.Cairns, Jr. Boca Raton, FL: Lewis Publishers/CRC Press, pp. 609–630.

Pitt, R. 1995c. Preliminary Investigation of EquaFlow™ system for Town Lake, Austin, TX. Unpublished report prepared for Loomis and Associates and the City of Austin, TX.

Pitt, R. 1997. Unique Features of the Source Loading and Management Model (SLAMM). In: *Advances in Modeling the Management of Stormwater Impacts, Volume 6*, ed. by W. James. Computational Hydraulics International, Guelph, ON and Lewis Publishers/CRC Press. pp. 13–37.

Pitt, R., and Amy, G. 1973. *Toxic Materials Analysis of Street Surface Contaminants.* EPA-R2-73-283, U.S. Environmental Protection Agency, Washington, DC.

Pitt, R., and Shawley, G. 1982. A *Demonstration of Non-Point Source Pollution Management on Castro Valley Creek.* Alameda County Flood Control and Water Conservation District (Hayward, CA) for the Nationwide Urban Runoff Program. U.S. Environmental Protection Agency, Water Planning Division, Washington, DC.

Pitt, R., and McLean, J. 1986. *Toronto Area Watershed Management Strategy Study: Humber River Pilot Watershed Project.* Ontario Ministry of the Environment, Toronto, ON.

Pitt, R., and Voorhees, J. 1995. Source loading and management model (SLAMM). *Seminar Publication: National Conference on Urban Runoff Management: Enhancing Urban Watershed Management at the Local, County, and State Levels*, March 30–April 2, 1993. Center for Environmental Research Information, U.S. Environmental Protection Agency. EPA/625/R-95/003. Cincinnati. OH. pp. 225–243.

Pitt, R., and Voorhees, J. 2002. SLAMM, the Source Loading and Management Model. In: *Management of Wet-Weather Flow in the Watershed,* ed. by Dan Sullivan and Richard Field. Boca Raton, FL: CRC Press.

Pitt, R., and Dunkers, K. 1992. Combined treatment of stormwater and lakewater in Lake Rönningesjön, Sweden. In: *12th Annual International Symposium, North American Lake Management Society*, November. Cincinnati, OH.

Pitt, R., and Dunkers, K. 1993. Lake water quality improvements from treatment of stormwater using the Flow Balancing Method. *66th Annual Water Environment Federation Conference*. October. Anaheim, CA.

Pitt, R., and Bozeman, M. 1982. *Sources of Urban Runoff Pollution and Its Effects on an Urban Creek*, EPA-600/52-82-090. U.S. Environmental Protection Agency, Cincinnati, OH.

Pitt, R., and Bozeman, M. 1979. *Lake Merritt Management Plan*. City of Oakland Department of Public Works, Oakland, CA.

Pitt, R., and Bissonnette, P. 1984. *Bellevue Urban Runoff Program Summary Report*, PB84 237213. U.S. Environmental Protection Agency, Water Planning Division. Washington, DC. 173 pp.

Pitt, R. and Sutherland, R. 1982. *Washoe County Urban Stormwater Management Program; Volume 2: Street Particulate Data Collection and Analyses*. Washoe Council of Governments, Reno, NV.

Pitt, R. 1979. *Demonstration of Nonpoint Pollution Abatement Through Improved Street Cleaning Practices*. EPA-600/2-79-161. U.S. Environmental Protection Agency, Cincinnati, OH.

Pitt, R. 1993a. Detention pond design for water quality improvement. *National ASCE Hydraulic Conference*. July. San Francisco, CA.

Pitt, R. 1993b. The stormwater quality detention pond model (DETPOND). *26th Annual Water Resources Conference*, October. University of Minnesota.

Pitt, R. 1985. *Characterizing and Controlling Urban Runoff through Street and Sewerage Cleaning*. EPA/600/S2-85/038. PB 85-186500. U.S. Environmental Protection Agency, Storm and Combined Sewer Program, Risk Reduction Engineering Laboratory. Cincinnati, OH. 467 pp.

Pitt, R. 1987. *Small Storm Urban Flow and Particulate Washoff Contributions to Outfall Discharges*. Ph.D. dissertation, Dept. of Civil and Environmental Engineering, University of Wisconsin, Madison, WI. Listed in Dissertation Abstracts International, University Microfilms International 49(1):1988.

Pitt, R., and Field, R. 1998. An Evaluation of Storm Drainage Inlet Devices for Stormwater Quality Treatment. In: *Proc. Water Environ. Fed. 71st Annu. Conf. Exposition*, Orlando, FL. 6:827.

Pitt, R., and Lalor, M. 2001. The Role of Pollution Prevention in Stormwater Management. In: *Models and Applications to Urban Water Systems, Monograph 9*. p. 1.

Pitt, R., and Barron, P. 1990. *Assessment of Urban and Industrial Stormwater Runoff Toxicity and the Evaluation/Development of Treatment for Runoff Toxicity Abatement—Phase I*. U.S. Environmental Protection Agency, Office of Research and Development, Edison, NJ.

Pitt, R., and Clark, S. 1998. *Laboratory Safety Manual and Standard Operating Procedures: UAB Water Quality Research Laboratories*. Department of Civil and Environmental Engineering, University of Alabama at Birmingham. 340 pp.

Pitt, R., Field, R., Lalor, M., and Brown, M. 1995. Urban stormwater toxic pollutants: assessment, sources, and treatability. *Water Environment Research* 67(3):260–275.

Pitt, R., Clark, S., and Parmer, K. 1994. *Protection of Groundwater from Intentional and Non-intentional Stormwater Infiltration*. U.S. Environmental Protection Agency, EPA/600/SR-94/051. PB94-165354AS, Storm and Combined Sewer Program, Cincinnati, OH. 187 pp.

Pitt, R., Clark, S., Parmer, K., and Field, R. 1996. *Groundwater Contamination from Stormwater*. ISBN 1-57504-015-8. Chelsea, MI: Ann Arbor Press, Inc. 219 pp.

Pitt, R., Robertson, B., and Field, R. 1998a. Innovative Multi-Chambered Stormwater Control Device for Critical Source Areas. In: *Proc. Adv. in Urban Wet Weather Pollut. Reduction*, Cleveland, OH, WEF (CP3805). 141.

Pitt, R. Clark, S., Lantrip, J., and Day, J. 1998b. *Telecommunication Manhole Water and Sediment Study*. Bellcore, Inc. (now Telcordia), Special Report SR-3841. Morriston, NJ. With further support from NYNEX, BellSouth, Bell Atlantic, GTE, SNET, Pacific Bell, US West, Ameritech and AT&T. *Vol. 1: Evaluation of Field Test Kits* (483 pp.); *Vol. 2: Water and Sediment Characteristics* (1290 pp.); *Vol. 3: Discharge Evaluation Report* (218 pp.); *Vol. 4: Treatment of Pumped Water* (104 pp.).

Pitt, R., Robertson, B., Barron, P., Ayyoubi, A., and Clark, S. 1999a. *Stormwater Treatment at Critical Areas: The Multi-Chambered Treatment Train (MCTT)*. EPA Report No. EPA/600/R-99/017. U.S. Environmental Protection Agency, National Risk Management Laboratory, Cincinnati, OH. 505 pp.

Pitt, R., Robertson, B., Barron, P., Ayyoubi, A., and Clark, S. 1999b. *Stormwater Treatment at Critical Areas: The Multi-Chambered Treatment Train (MCTT)*. EPA/600/R-99/017. U.S. Environmental Protection Agency, Wet Weather Flow Management Program, National Risk Management Research Laboratory, Cincinnati, OH. 505 pp.

Pitt, R., Lantrip, J., Harrison, R., Henry, C., and Hue, D. 1999c. *Infiltration through Disturbed Urban Soils and Compost-Amended Soil Effects on Runoff Quality and Quantity*. EPA 600/R-00/016. U.S. Environmental Protection Agency, Water Supply and Water Resources Division, National Risk Management Research Laboratory, Cincinnati, OH. 231 pp.

Pitt, R., Lilburn, M., Nix, S., Durrans, S.R., Burian, S., Voorhees, J., and Martinson, J. 1999d. *Guidance Manual for Integrated Wet Weather Flow (WWF) Collection and Treatment Systems for Newly Urbanized Areas (New WWF Systems)*. Cooperative agreement #CX 824933-01-0. U.S. Environmental Protection Agency. 612 pp.

Pitt, R., Harrison, R., Lantrip, J., and Henry, C.L. 1999e, Infiltration Through Disturbed Urban Soils and Compost-Amended Soil Effects on Runoff Quality And Quantity. In: *Proc. Water Environ. Fed. 72nd Annu. Conf. Exposition*, [CD-ROM], New Orleans, LA.

Poertner, H.G. 1974. *Practices in Detention of Urban Stormwater Runoff.* OWRR Contract No. 14-31-0001-3722. American Public Works Association, Chicago, IL.

Pratt, C.J. 1998. Design Guidelines for Porous/Permeable Pavements. Sustaining Urban Water Resour. in the 21st Century. In: *Proc. Eng. Found. Conf.*, Malmo, Sweden, 1997 [CD ROM].

Pratt, C.J. 1996. Research and development in methods of soakaway design. *Journal of the Chartered Institution of Water and Environmental Management* 10(1):47–51.

Pratt, C.J. 1990. Permeable pavements; design and maintenance. In: *Developments in Storm Drainage—A Symposium on Infiltration and Storage of Stormwater in New Developments*, Jan., Sheffield City. pp. 136–151.

Pratt, C.J. 1993. An assessment of effectiveness of a BMP strategy in urban drainage in the U.K. In: *Proceedings of the 6th International Conference on Urban Storm Drainage.* Edited by J. Marsalek and H.C. Torno. Niagara Falls, Ontario. September 12-17, 1993. International Association on Water Quality, London. Published by Seapoint Publishers, Victoria, BC, pp. 951-956.

Pratt, C.J. and Powell, J.J.M. 1993. A new UK approach for the design of subsurface infiltration systems. In: *Proceedings of the 6th International Conference on Urban Storm Drainage*, ed. by J. Marsalek and H.C. Torno. September 12–17, Niagara Falls, ON. IAWQ, London. Published by Seapoint Publishers, Victoria, BC. pp. 987–992.

Pratt, C.J., Mantle, J.D.G., and Schoffield, P.A. 1989. Urban stormwater reduction and quality improvement through the use of permeable pavements. In: *Urban Discharges and Receiving Water Quality Impacts*. Oxford: Pergamon Press, pp. 123–132.

Pratt, C.J., Mantle, J.D.G., and Schoffield, P.A. 1996. UK research into the performance of permeable pavement, reservoir structures in controlling stormwater discharge quantity and quality. *International Conference NOVATECH 95*, Lyon, pp. 337–344.

Pratt, J.M., Coler, R.A., and Godfrey, P.J. 1981. Ecological effects of urban stormwater runoff on benthic macroinvertebrates inhabiting the Green River, Massachusetts. *Hydrobiologia* 83:29.

Pruitt, Janet B., Troutman, David A., and Irwin, G.A. 1985. *Reconnaissance of Selected Organic Contaminants in Effluent and Groundwater at Fifteen Municipal Wastewater Treatment Plants in Florida, 1983-84*. U.S. Geological Survey Water-Resources Investigation Report 85-4167.

Prych, E.A., and Ebbert, J.C. Undated. *Quantity and Quality of Storm Runoff from Three Urban Catchments in Bellevue, Washington.* Preliminary U.S. Geological Survey Water Resources Investigations Report, Tacoma, WA.

Pyzoha, D. 1994. *Implementing a Stormwater Management Program*. Boca Raton, FL: Lewis Publishers.

Quasebarth, T., Schroeder, D., Chappell, R., Churchwell, R., and Lippner, G. 2001. An Investigation of Factors Influencing Solids Transport and Deposition into Highway Drain Inlets. In: *Proc. ASCE EWRI Conf.—Bridging the Gap: Meeting the World's Water and Environmental Resources Challenges*. CD-ROM.

Rael, J., Shelton, S., and Dayaye, R. 1995. Permeable barriers to remove benzene: candidate media evaluation. *Journal of Environmental Engineering* 121(5):411–415.

Ragone, Stephen E. 1977. *Geochemical Effects of Recharging the Magothy Aquifer, Bay Park, New York, with Tertiary-Treated Sewage*. Geological Survey Professional Paper 751-D. Prepared in cooperation with the Nassau County Department of Public Works. USGS, Washington, DC.

Rainbow, P.S. 1996. Chapter 18: Heavy Metals in Aquatic Invertebrates. In: *Environmental Contaminants in Wildlife; Interpreting Tissue Concentrations*. Beyer, W.N., Heinz, G.H., and Redmon-Norwood, A.W. (Eds.), CRC/Lewis Press, Boca Raton, FL. 405 pgs.

RAM Services, Inc. *ENRETECH I* (Leaflet. RAM Services, Inc., Birmingham, AL. 1995.)

Randall, C.W. 1982. Stormwater detention ponds for water quality control. In: *Proceedings of the Conference on Stormwater Detention Facilities, Planning, Design, Operation, and Maintenance*, August, Henniker, NH, ed. by W. DeGroot. Published by the American Society of Civil Engineers, New York.

Randall, C.W., Ellis, K., Grizzard, T.J., and Knocke, W.R. 1982. Urban runoff pollutant removal by sedimentation. In: *Proceedings of the Conference on Stormwater Detention Facilities, Planning, Design, Operation, and Maintenance*, Henniker, NH, ed. by W. DeGroot. Published by the American Society of Civil Engineers, New York.

Raskin, I., Kumar, N., Dushenkov, S., and Salt, D. 1994. *Curr. Opin. Biotechnol.* 5:285–290.

Rate, A.W., Robertson, A.E., and Borg, A.T. 2000. Distribution of Heavy Metals in Near-Shore Sediments of the Swan River Estuary, Western Australia. *Water, Air, Soil Pollut.* 124:155.

Rauch, S., and Morrison, G.M. 1999. Platinum Uptake by the Freshwater Isopod *Asellus aquaticus* in Urban Rivers. *Sco. Total Environ.* 235(1-3):261.

Rauch, S., Morrison, G.M., Motelica-Heino, M., and Donard, O. 1999. Evaluation of Speciation, Transport and Ecological Risk of Palladium, Platinum and Rhodium in Urban Stormwater Systems. In: *Proc. the Eighth International Conference on Urban Storm Drainage*. August 30–September 3, Sydney, Australia, ed. by I.B. Joliffe and J.E. Ball. The Institution of Engineers Australia, The International Association for Hydraulic Research, and The International Association on Water Quality. 202.

Raudkivi, A. J., and Tan, S. K. 1984. Erosion of cohesive soils, *Journal of Hydraulic Research* 22(4):217–233.

Ray, S., and White, W. 1976. Selected aquatic plants as indicator species for heavy metal pollution. *Journal of Environmental Science and Health* A11:717.

Razack, M., Drogue, C. and Baitelem, M. 1988. Impact of an Urban Area on the Hydrochemistry of a Shallow Groundwater (Alluvial Reservoir) Town of Narbonne, France. In: *Proceeding of the Conference—Hydrological Processes and Water Management in Urban Areas*. International Hydrological Programme, UNESCO, April 24–29, Duisburg, Federal Republic of Germany. pp.487–494.

Redmon-Norwood, A.W., Eds., Boca Raton, FL: CRC/Lewis Press. 405 pp.

Ree, W.O. 1949. Hydraulic Characteristics of Vegetation for Vegetated Waterways. *Agricultural Engineering* (April):184–189.

Ree, W.O. 1958. Retardation Coefficients for Row Crops in Diversion Terraces. *Transactions of the ASAE*: 78–80.

Ree, W.O. 1960. Friction Factors for Vegetation Covered, Light Slope Waterways. Research Report No. 335, USDA. 1–34.

Ree, W.O. and Palmer, V.J. 1949. Flow in Channels Protected by Vegetative Linings. Technical Bulletin No. 967, USDA. 1–22, 44–48.

Reed, B.E., Robertson, J., and Jamil, M. 1995. Regeneration of granular activated carbon (GAC) columns used for removal of lead. *Journal of Environmental Engineering* 121(9):653–662.

Remcale, J. 1990. The cell wall and metal binding. In: *Biosorption of Heavy Metals*, ed. by B. Volesky. Boca Raton, FL: CRC Press, Inc. pp. 83–92.

Rhoads, B.L., and Cahill, R.A. 1999. Geomorphological Assessment of Sediment Contamination in an Urban Stream System. *Applied Geochemistry* 14(4):459.

Rice, R.G., and Robson, C.M. 1982. Biological Activated Carbon: Enhanced Aerobic Biological Activity in GAC Systems. Ann Arbor, MI: Ann Arbor Science.

Richards, T.S., and Kreutzberger, W.A. 2001. Development of Site Specific Metals Criteria for CSO Discharges to Headwater Streams. In: *WEFTEC 2001 Conf. Proc.* CD-ROM.

Richey, J.S. 1982. *Effects of Urbanization on a Lowland Stream in Western Washington*, Doctor of Philosophy dissertation, University of Washington, Seattle.

Richey, J.S., Perkins, M.A., and Malueg, K.W. 1981. The effects of urbanization and stormwater runoff on the food quality in two salmonid streams. *Verh. Internat. Werein. Limnol.* 21:812–818, Stuttgart.

Richman, M. 1997. Compost media capture pollutants from stormwater runoff. *Water Environment and Technology* (July):21–22.

Riley, S.J., and Banks, R.G. 1996. The Role of Phosphorus and Heavy Metals in the Spread of Weeds in Urban Bushlands; an Example from the Lane Cove Valley, NSW, Australia. *Sci. of the Total Environ.* 102(1-3):39.

Robertson, B. 1995. *Evaluation of a Multi-Chambered Treatment Train (MCTT) for Treatment of Stormwater Runoff from Critical Pollutant Source Areas.* MSCE Thesis, Department of Civil and Environmental Engineering. University of Alabama at Birmingham. 268 pp.

Robertson, B., Pitt, R., Ayyoubi, A., and Field, R. 1995. A Multi-Chambered Stormwater Treatment Train. In: *Stormwater NPDES Related Monitoring Needs,* ed. by H.C. Torno. Engineering Foundation and ASCE. pp. 631–640.

Robinson, S.M., Arnold, W.D., and Byers, C.H. 1994. Mass-transfer mechanisms for zeolite ion exchange in wastewater treatment. *AIChE Journal* 40(12):2045–2054.

Rochfort, Q., Grapentine, L., Marsalek, J., Brownlee, B., Reynoldson, T., Thompson, S., Milani, D., and Logan, C. 2000. Using Benthic Assessment Techniques to Determine Combined Sewer Overflow and Stormwater Impacts in the Aquatic Ecosystem. *Water Qual. Res. J. Can.* 35:365.

Rochfort, Q.J., Anderson, B.C., Crowder, A.A., Marsalek, J., and Watt, W.E. 1997. Field-scale Studies of Subsurface Flow Constructed Wetlands for Stormwater Quality Enhancement. *Water Qual. Res. J. Canada* 32(1):101.

Rodgers, J.H., Lehman, R.W., and Gladden, J.B. 2001. Wetlands for Industrial Wastewater Treatment at the Savannah River Site: a Case Study. In: *WEFTEC 2001 Conf. Proc.* CD-ROM.

Roesner, L.A., Urbonas, B.R., and Sonnen, M.B., eds. 1989. Design of Urban Runoff Quality Controls. In: *Proceedings of an Engineering Foundation Conference on Current Practice and Design Criteria for Urban Quality Control*, American Society of Civil Engineers, NY.

Rolfe, G.L., and Reinhold, K.A. July 1977a. *Vol. I: Introduction and Summary. Environmental Contamination by Lead and Other Heavy Metals.* Institute for Environmental Studies, University of Illinois, Champaign-Urbana.

Rolfe, G.L., Haney, A., and Reinhold, K.A. 1977b. *Environmental Contamination by Lead and Other Heavy Metals. Vol. II: Ecosystem Analysis*, Institute for Environmental Studies, University of Illinois, Urbana-Champaign.

Rose, S., Crean, M.S., Sheheen, D.K., and Ghazi, A.M. 1987. Comparative Zinc Dynamics in Atlanta Metropolitan Region Stream and Street Runoff. *Environ. Geol.* 40:983. 2001.

Rospond, F.J. 1976. Roof detention of rainfall to limit urban runoff. *National Symposium on Urban Hydrology, Hydraulics, and Sediment Control*, July, University of Kentucky, Lexington.

Rosted Petersen, C.O., Jacobsen, P., and Mikkelsen, P.S. 1994. Design of stormwater infiltration for reduction of combined sewer overflow (CSO). *Water Science and Technology* 30(1):53–61.

Rozan, T.F., and Benoit, G. 2001, Mass Balance of Heavy Metals in New Haven Harbor, Connecticut: Predominance of Nonpoint Sources. *Limnology Oceanography* 46:2032.

Rubin, A.J., ed. 1976. *Aqueous-Environmental Chemistry of Metals*. Ann Arbor, MI: Ann Arbor Science Publishers.

Rupperd, Y. 1993. A lamellar separator for urban street runoff treatment (in French). *Bulletin de Liaison des Laboratoires des Ponts et Chaussées* (183):85–90.

Rutherford, J.B. 1977. On-site multi-functional stormwater detention facilities; A case study. *International Symposium on Urban Hydrology, Hydraulics, and Sediment Control*, University of Kentucky, Lexington.

Ruthven, D.M. 1988. Zeolites as selective adsorbents. *Chemical Engineering Progress* 84(2):42–50.

Salbu, B., and Steinnes, E., eds. 1995. *Trace Elements in Natural Waters*. Boca Raton, FL: CRC Press. 295 pp.

Salo, John E., Harrison, Doug, and Archibald, Elaine M. 1986. Removing Contaminants by Groundwater Recharge Basins. *Journal of the American Water Works Association* 78(9):76–81.

Sansalone, J. 1996. Immobilization of Metals and Solids Transported in Urban Pavement Runoff. *North American Water and Environ. Congress '96*, Anaheim, CA. ASCE, NY.

Sansalone, J.J., and Buchberger S.G. 1997. Partitioning and First Flush of Metals in Urban Roadway Storm Water. *J. Environ. Eng*. 123(2):134.

Sansalone, J., Mishra, S.K., and Singh, V.P. 2001. The Role of Anthropogenic Hydrology on the Transport, Partitioning and Solid-Phase Distribution of Heavy Metals in Urban Storm Water—Implications for Treatment. In: *WEFTEC 2001 Conf. Proc*. CD-ROM.

Sansalone, J.J. 1999a. Adsorptive Infiltration of Metals in Urban—Media Characteristics. *Sco. Total Environ.* 235(1-3):179.

Sansalone, J.J. 1999b. In-situ performance of a passive treatment system for metal source control. *Water Science and Technology* 39(2):193-200.

Sansalone, J.J. and Buchberger, S.G. 1996. Characterization of Solid and Metal Element Distributions in Urban Highway Stormwater. In: *Proc. 7th Int. Conf. on Urban Storm Drainage*, Hannover, Germany. IAHR/IAWQ Joint Committee on Urban Storm Drainage. 253.

Sansalone, J.J., and Hird, J.P. 2000. Controlling Heavy Metals with Exfiltration. *Watershed & Wet Weather Tech. Bull.* Vol. 5(2).

Sansalone, J.J., and Hird, J.P. 1999. A passive source control prototype for heavy metals transported by urban stormwater. In: *Proceedings of the Water Environment Federation's Annual Conference and Exposition*, WEFTEC'99, October. New Orleans, LA. Water Environment Federation, Alexandria, VA. CD-ROM.

Sanudo-Wilhelmy, S.A., and Gill, G.A. 1999. Impact of the Clean Water Act on the Levels of Toxic Metals in Urban Estuaries: The Hudson River Estuary Revisited. *Environ. Sci. Technol.* 33(20):3477.

Sar, P., Kazy, S.K., and Singh, S.P. 2001. Intracellular nickel accumulation by *Pseudomonas aeruginosa* and its chemical nature. *Letters in Applied Microbiology* 32:257-261.

Sartor J., and Boyd, G. 1972. *Water Pollution Aspects of Street Surface Contaminants*. EPA-R2-72-081. U.S. Environmental Protection Agency.

Satpathy, J.K., and Chaudhuri, M. 1995. Treatment of cadmium-plating and chromium-plating wastes by iron oxide-coated sand. *Water Environment Research* 67(5):788–790.

Saygideger, S. 1998. Bioaccumulation and Toxicity of Zinc in Spirogyra Fluviatilis Hilse (*Chlorophyta*). *Water, Air, Soil Pollut.* 101(1-4):323.

Schaefer, J., Eckhardt, J-D., Berner, Z.A., and Stueben, D. 1999. Time-Dependent Increase of Traffic-Emitted Platinum-Group Elements (POE) in Different Environmental Compartments. *Environmental Science and Technology* 33(18):3166.

Schiffer, Donna M. 1989. *Effects of Three Highway-Runoff Detention Methods on Water Quality of the Surficial Aquifer System in Central Florida.* U.S. Geological Survey Water-Resources Investigations Report 88-4170. Prepared in cooperation with the Florida Department of Transportation. USGS, Denver, CO.

Schillinger, J.E., and Stuart, D.G. 1978. *Quantification of Non-Point Water Pollutants from Logging, Cattle Grazing, Mining, and Subdivision Activities.* U.S. Environmental Protection Agency. PB 80-174063.

Schimmenti, F.G. 1980. Stormwater detention basins must control more than runoff. *American City and County* (December).

Schneider, Brian J., Ku, Henry F.H., and Oaksford, Edward T. 1987. *Hydrologic Effects of Artificial-Recharge Experiments with Reclaimed Water at East Meadow, Long Island, New York.* U.S. Geological Survey Water-Resources Investigations Report 85-4323. Prepared in cooperation with the Nassau County Department of Public Works, Syosset, NY. USGS, Denver, CO.

Scholes, L., Shutes, R.B.E., Revitt, D.M., Forshaw, M., and Purchase, D. 1998. Treatment of Metals in Urban Runoff by Constructed Wetlands. *Sci. Total Environ.* 214(1-3):211.

Scholes, L.N.L., Shutes, R.B.E., Revitt, D.M., Purchase, D., and Forshaw, M. 1999. The Removal of Urban Pollutants by Constructed Wetlands During Wet Weather. *Water Sci. Technol.* (G.B.) 40(3)333.

Schueler, T., ed. 1996a. Ditches or biological filters? Classifying the pollutant removal performance of open channels *Watershed Protection Techniques* 2(2):379–383.

Schueler, T., ed. 1996b. Stream channel geometry used to assess land use impacts in the Pacific Northwest. *Watershed Protection Techniques* 2(2):345-348.

Schueler, T., ed. 1996c. Characterization of Heavy Metals in Santa Clara Valley. *Watershed Protect. Techniques* 2(2):353.

Schueler, T., ed. 1996d. Vegetated rock filter treats stormwater pollutants in Florida. *Watershed Protection Techniques* 2(2):372–374.

Schueler, T.R. 1986. *New Guidebook Review Paper No. 2, Wet Pond BMP Design*. Metropolitan Washington Council of Governments, Washington, DC.

Schueler, T.R., and Galli, J. 1992. The environmental impact of stormwater ponds. *Engineering Foundation Conference: Urban Runoff and Receiving Systems: An Interdisciplinary Analysis of Impact, Monitoring, and Management*. August 4–9, Mt. Crested Butte, Colorado. Published by the American Society of Civil Engineers, New York.

Schueler, T.R. 1992. Design of Stormwater Wetland Systems: Guidelines for Creating Diverse and Effective Stormwater Wetlands in Mid-Atlantic Region. Metropolitan Washington Council of Governments. Publication No. 92710.

Schueler, T.R. 1994a. Technical Note 31: Runoff and Groundwater Dynamics of Two Swales in Florida. *Watershed Protection Techniques* 1(3):120–121

Schueler, T.R. 1994b. Technical Note 32: Performance of grassed swales along east coast highways. *Watershed Protection Techniques* 1(3):122–123.

Schueler, T.R. 1994c. Technical Note 65: Pollutant Removal Pathways in Florida Swales. *Watershed Protection Techniques* 2(1):299–301.

Schueler, T. 1987. *Controlling Urban Runoff: A Practical Manual for Planning and Designing Urban Best Management Practices*. Metropolitan Washington Council of Governments. 280 pp.

Scott, J.B., Steward, C.R., and Stober, Q.J. 1982. *Impacts of Urban Runoff on Fish Populations in Kelsey Creek, Washington*. Contract No. R806387020, U.S. Environmental Protection Agency, Corvallis Environmental Research Laboratory, Corvallis, OR.

Seaburn, G.E., and Aronson, D.Q. 1974. *Influence of Recharge Basins on the Hydrology of Nassau and Suffolk Counties, Long Island, New York*. U.S. Geological Survey Water-Supply Paper 2031. USGS, Washington, DC.

Sedlak, D.L., Phinney, J.T. and Bedsworth, W.W. 1997. Strongly Complexed Cu and Ni in Wastewater Effluents and Surface Runoff. *Environ. Sci. Technol.* 31(10):3010.

Seidl, M., Huang, V., and Mouchel, J.M. 1998. Toxicity of combined sewer overflows on river phytoplankton; the role of heavy metals. *Environ. Pollution* 101(1)107–116.

Selenska-Pobell, S., Panak, P., Miteva,V., Boudakov, I., Bernhard, G., and Nitsche, H. 1999. Selective accumulation of heavy metals by three indigenous Bacillus strains, *B. cereus, B. megaterium*, and *B. sphaericus* from drain waters of a uranium waste pile. *FEMS Microbiology Ecology* 29:59–67.

Selim, M.I., and Wang, J. 1994. Fate of atrazine in biologically active granular activated carbon. *Environmental Toxicology and Chemistry* 13(1):3–8.

Setmire, J.G., and Bradford, W.L. 1980. Quality of Urban Runoff, Tecolote Creek Drainage Area, San Diego County, CA. U.S. Geological Survey. PB81-159451. Menlo Park, CA.

Shafer, M.M., Overdier, J.T., Phillips, H., Webb, D., Sullivan, J.R., and Armstrong, D.E. 1999. Trace Metal Levels and Partitioning in Wisconsin rivers. *Water, Air, Soil Pollut.* 110(3-4);273.

Shaheen, D.G. 1975. *Contributions of Urban Roadway Usage to Water Pollution*. 600/2-75-004. U.S. Environmental Protection Agency. Washington, DC.

Sharma, D.C., and Forster, C.F. 1993. Removal of hexavalent chromium using sphagnum moss peat. *Water Research* 27(7):1201–1208.

Shaver, E. 1994. Beach community adds sand filters to storm drains. *Water Environment and Technology* 5(5):18.

Shaver, Earl. 1991. Sand Filter Design for Water Quality Treatment. In: *Proceedings from an Engineering Foundation Specialty Conference*, Crested Butte, CO.

Shelley, P.E., and Gaboury, D.R. 1986. Estimation of Pollution from Highway Runoff—Initial Results. *Conference on Urban Runoff Quality—Impact and Quality Enhancement Technology*, June, Henniker, NH, ed. by B. Urbonas and L.A. Roesner. Proceedings published by the American Society of Civil Engineering, New York.

Shen, H.W. 1981. Some Basic Concepts on Sediment Transport in Urban Storm Drainage Systems. *Second International Conference on Urban Storm Drainage*, June, Urbana, IL.

Shikowshi, D.M., and Viraraghavan, T. 1993. Removal of chromium from wastewater by peat filters. *Journal of Environmental Science and Health* A28(5):967–981.

Shirmohammadi, A., and Knisel, W.G. 1989. Irrigated Agriculture and Water Quality in the South. *Journal of Irrigation and Drainage Engineering*, 115(5):791-806.

Shutes, R.B.E., Revitt, D.M., Mungur, A.S., and Scholes, L.N.L. 1997a. The Design of Wetland Systems for the Treatment of Urban Runoff. *Water Sci. Technol.* 35(5):19.

Shutes, R.B.E., Revitt, D.M., Munger, A.S., and Scholes, L.N.L. 1997b. Design of Wetland Systems for Treatment of Urban Runoff. *Water Qual. Int.* (March/April):35.

Sieker, F., and Verworn, H-R. eds. 1996. *Proc. 7th Int. Conf. On Urban Storm Drainage*, IAHR/IAWQ. Hannover, Germany: SuG-Verlagsgesellschaft.

Singer, M.J., and Blackard, J. 1978. Effect of Mulching on Sediment in Runoff from Simulated Rainfall. *Soil Sci. Soc. Am. J.* 42:481–486.

Singleton, R. 1993. The sulfate-reducing bacteria: An overview. In: The *Sulfate-Reducing Bacteria: Contemporary Perspectives*, ed. by J.M. Odom and R. Singleton. New York: Springer-Verlag. pp. 1–20.

Skinner, L., de Peyster, A., and Schiff, K. 1999. Developmental Effects of Urban Storm Water in Medaka (*Oryzias latipes*) and Inland Silverside (*Menidia beryllina*). *Arch. Environ. Contam.Toxicol.* 37(2):227.

Smith, E. 2001. Pollutant Concentrations of Stormwater and Captured Sediment in Flood Control Sumps Draining an Urban Watershed. *Water Res.* 35:3117.

Smith, D., Reale, D., and Coleman, M. 2000. Source Control: The First Step in Cleaning Up Commencement Bay. In: *WEFTEC2000, 73rd Annual Conference and Exposition*, October, Anaheim, CA. Water Environment Federation, CD-ROM.

Smith, E.B., and Swanger, L.K. 2000. Toxicity and Worldwide Environmental Regulation of Lead-Free Solders. *Trans. Instit. Metal Finish.* 78:B18.

Smith, W.G. 1982. Water Quality Enhancement through Stormwater Detention. In: *Proceedings of the Conference on Stormwater Detention Facilities, Planning, Design, Operation, and Maintenance*, August, Henniker, NH, ed. by W. DeGroot. Published by the American Society of Civil Engineers, NY.

Snodgrass, W.J., Kilgour, B.W., Leon, L., Eyles, N., Parish, J., and Barto, D.R. 1998. Applying ecological criteria for stream biota and an impact flow model for evaluation sustainable urban water resources in southern Ontario. In: *Sustaining Urban Water Resources in the 21st Century.* Proceedings of an Engineering Foundation Conference, ed. by A.C. Rowney, P. Stahre, and L.A. Roesner. Sept. 7–12, 1997, Malmo, Sweden. To be published by ACSE, NY.

Soil Conservation Service (now NRCS). 1982. *Ponds—Planning, Design, Construction.* Agriculture Handbook Number 590, U.S. Dept. of Agriculture

Soil Conservation Service (now NRCS). 1986. *Urban Hydrology for Small Watersheds.* Tech. Release No. 55, (June) U.S. Dept. of Agriculture.

Soil Conservation Service. 1954. *Handbook for Channel Design for Soil and Water Conservation*. SCS-TP-61.

Soil Conservation Service. 1977. *Design of Open Channels*. TR-25.

Solo-Gabriele, H., and Perkins, F. 1997. Watershed-Specific Model for Streamflow, Sediment, and Metal Transport. *J. Environ. Eng.* 123(1):61.

Solomon, R.L., and Natusch, D.F.S. 1977. Vol: III: Distribution and Characterization of Urban Dists. In: *Environmental Contamination by Lead and Other Heavy Metals*, ed. by G.L. Rolfe and K.G. Reinbold. Institute for Environmental Studies, Univ. of Illinois, Urbana-Champaign.

Southeast Michigan Council of Governments. 1981. *Detention Basin Maintenance, Program Needs and Financing Options*. Technical Bulletin No. 3, May. Detroit, MI.

Spehan, R.L., Anderson, R.L., and Fiandt, J.T. 1978. Toxicity and bioaccumulation of cadmium and lead in aquatic invertebrates. *Environmental Pollution* 15:195.

Spring, R.J., Howell, R.B., and Shirley, E. 1978. *Dustfall Analysis for the Pavement Storm Runoff Study* (I-405 Los Angeles). Office of Transportation Laboratory, California Dept. of Transportation, Sacramento.

Squires, Rodney C., and Johnston, William R. 1987. Selenium Removal—Can We Afford It? In: *Irrigation Systems for the 21st Century—Proceeding of a Conference*. Sponsored by the Irrigation and Drainage Division of the American Society of Civil Engineers. NY. pp. 455–466.

Squires, Rodney C., Groves, G. Raymond, and Johnston, William R. 1989. Economics of Selenium Removal from Drainage Water. *Journal of Irrigation and Drainage Engineering* 115(1):48–57.

Stack, W., and Belt, K. 1996. Protecting Baltimore's Waters from Toxic Substances in Urban Storm Drains. In: *Abstract Book: SETAC 17th Annu. Meeting*, Washington, DC, *Soc. Environ. Toxicol. and Chem.* 104.

Stahre, P. 1993. Assessment of BMP's being used in Scandinavia. In: *Proceedings of the 6th International Conference on Urban Storm Drainage*, ed. by J. Marsalek and H.C. Torno. September 12–17, Niagara Falls, ON. IAWQ, London. Published by Seapoint Publishers, Victoria, BC. pp. 939–944.

Stahre, P., and Urbonas, B. 1990. *Stormwater Detention for Drainage, Water Quality and CSO Management*. Englewood Cliffs, NJ: Prentice Hall. 340 pp.

Stanley, D.W. 1990. Pollutant removal by a stormwater dry detention pond. *Water Environment Research* 68(6):1076–1083. Sept./Oct. 1996.

Steenhuis, T.S., Staubitz, W., Andreini, M.S., Surface, J., Richard, T.L., Paulsen, R., Pickering, N.B., Hagerman, J.R., and Geohring, L.D. 1990. Preferential movement of pesticides and tracers in agricultural soils. *Journal of Irrigation and Drainage Engineering* 116(1):50–66.

Stenmark, C. 1990. Local infiltration of urban storm water in cold climate. In: *Proceedings of the 5th International Conference on Urban Storm Drainage*. July, Osaka, Japan. pp. 809–814.

Stenstrom, M.K., Lau, S.-L., Lee, H.-H., Ma, J.-S., Ha, H., Kim, L.-H., Khan, S., and Kayhanian, M. 2001. First Flush Stormwater Runoff from Highways. In: *Proc. ASCE EWRI Conf.—Bridging the Gap: Meeting the World's Water and Environmental Resources Challenges*. CD-ROM.

Stephenson, D. 1996. Evaluation of effects of urbanization on storm runoff. In: *7th International Conference on Urban Storm Drainage*. Sept. 9–13, Hannover, Germany, ed. by F. Sieker and H-R. Verworn. IAHR/IAWQ. Hannover, Germany: SuG-Verlagsgesellschaft. pp. 31–36.

Stephenson, J.B., Zhou, W.F., Beck, B.F., and Green, T.S. 1999. Highway Stormwater Runoff in Karst Areas—Preliminary Results of Baseline Monitoring and Design of a Treatment System for a Sinkhole in Knoxville, Tennessee. *Engineering Geology* 52(1-2):51.

Striegl, R.G. 1985. *Effects of Stormwater Runoff on an Urban Lake, Lake Ellyn at Glen Ellyn, Illinois*, USGS Open File Report 84-603. U.S. Geological Survey, Lakewood, CO.

Sutherland, R., and McCuen, R.H. 1978. Simulation of Urban Nonpoint Source Pollution. *Water Resources Bulletin* 14(2):409–428.

Sutherland, R.A., Tack, F.M.G., Tolosa, C.A., and Verloo, M.G. 2000. Operationally Defined Metal Fractions in Road Deposited Sediment, Honolulu, Hawaii. *J. Environ. Qual.* 29:1431.

Sutherland, R.C. 1996. Studies show sweeping has beneficial impact on stormwater quality. *APWA Reporter* pp. 8–23.

Sutherland, R.C., and Jelen, S.L. 1996. Sophisticated stormwater quality modeling is worth the effort. In: *Advances in Modeling the Management of Stormwater Impacts*, ed. by W. James. Guelph, ON: Computational Hydraulics International.

Sztruhar, D., and Wheater, H.S. 1993. Experimental and numerical study of stormwater infiltration through pervious parking lots. In: *Proceedings of the 6th International Conference on Urban Storm Drainage*, ed. by J. Marsalek and H.C. Torno. September 12–17, Niagara Falls, ON. IAWQ, London. Published by Seapoint Publishers, Victoria, BC. pp. 1098–1103.

Taylor, S.M., Hansen, L. and BeitiAa, C. 2001. Assessment of Costs and Benefits of Detention for Water Quality Enhancement. In: *Proc. ASCE EWRI Conf.—Bridging the Gap: Meeting the World's Water and Environmental Resources Challenges*. CD-ROM.

Terstriep, M.L., Bender, G.M., and Noel, D.C. 1982. *Final Report—NURP Project, Champaign, Illinois: Evaluation of the Effectiveness of Municipal Street Sweeping in the Control of Urban Storm Runoff Pollution.* State Water Survey Division, Illinois Dept. of Energy and Natural Resources, Champaign-Urbana.

Tetra Tech. 1991. *Detention and Retention Effects on Groundwater, A Literature Review.* Contract No. 68-C9-0013. U.S. Environmental Protection Agency. Watershed Management Unit. Water Division, Region V, Chicago, IL.

Thompson, J.B., and Beveridge, T.J. 1993. Interactions of metal ions with bacterial surfaces and the ensuing development of minerals. In: *Particulate Matter and Aquatic Contaminants*, ed. by S.S. Rao. Boca Raton, FL: Lewis Publishers. pp. 65–104.

Thompson, M.K., and James, W. 1995. Provision of parking lot pavements for surface water pollution control studies. *Modern Methods for Modeling the Management of Stormwater Impacts*, ed. by W. James. Guelph, ON: Computational Hydraulics International. pp. 381–398.

Thomson, N.R., McBean, E.A., Snodgrass, W., and Monstrenko, I.B. 1997a. Highway Stormwater Runoff Quality—Development of Surrogate Parameter Relationships. *Water, Air, Soil Pollut.* (Neth.) 94(3-4):307.

Thomson, N.R., McBean, E.A., Snodgrass, W., and Monstrenko, I.B. 1997b. Sample Size Needs for Characterizing Pollutant Concentrations in Highway Runoff. *J. Environ. Eng.* 123(10):1061.

Tobiason, J.E., Johnson, G.S., Westerhoff, P.K., and Vigneswaran, B. 1993. Particle size and chemical effects on contact filtration perfomance. *Journal of Environmental Engineering* 119(3):520–539.

Tobiason, S.A., and Logan L.R.J. 2000. Stormwater Whole Effluent Toxicity (WET) Testing and Source Tracing at Sea-Tac International Airport. In: *WEFTEC2000, 73rd Annual Conference and Exposition*, October, Anaheim, CA. Water Environment Federation, CD-ROM.

Totura, G. 1996. Innovative uses of specialty ion exchange resins provide new cost-effective options for metal removal. *Environmental Progress* 15(5), 208-212.

Tramier B. 1983. *Water Treatment Technology*—IP 84-011. Institute of Petroleum. London, England.

Trauth, R., and Xanthopoulos, C. 1996. Non-point pollution of groundwater in urban areas. In: *7th International Conference on Urban Storm Drainage*. Sept. 9–13. Hannover, Germany. Ed. by F. Sieker and H-R. Verworn. IAHR/IAWQ. Hannover, Germany: SuG-Verlagsgesellschaft. pp. 701–706.

Treweek, Gordon P. 1985. Pretreatment Processes for Groundwater Recharge. In: *Artificial Recharge of Groundwater*, ed. by Takashi Asano. Boston: Butterworth Publishers. pp. 205–248.

Tsai, P., Hansen, E., and Lee, K. 2001. Atmospheric Deposition of Trace Metals in San Francisco Bay Area. In: *WEFTEC 2001 Conf. Proc.* CD-ROM.

Tsihrintzis, V.A., and Hamid, R. 1997. Urban Stormwater Quantity/Quality Modeling Using the SCS Method and Empirical Equations. *Water Resour. Bull.* 33(1):163.

Turer, D., Maynard, J.B., and Sansalone, J.J. 2001. Heavy Metal Contamination in Soils of Urban Highways: Comparison Between Runoff and Soil Concentrations at Cincinnati, Ohio. *Water Air Soil Pollut.* 132:293.

U.S. Army Corps of Engineers (COE) 1975. Hydraulic Engineering Center. *Urban Storm Water Runoff: STORM. Generalized Computer Program.* 723-58-L2520. Davis, CA.

U.S. Army Corps of Engineers (COE). 1994. *Engineering and Design: Channel Stability Assessment for Flood Control Projects*. EM 1110-2-1418. U.S. Army Engineer Waterways Experiment Station, Vicksburg, MS.

U.S. Army Corps of Engineers (COE). undated. *Hydraulic Design of Flood Control Channels.* EM 1110-2-1601. U.S. Army Engineer Waterways Experiment Station, Vicksburg, MS.

U.S. Department of Agriculture (USDA). 1940. Hydraulic Studies of Conservation Structures at the Outdoor Hydraulics Laboratory, Spartanburg, South Carolina. USDA: 1–18.

U.S. Department of Agriculture (USDA). 1954. Handbook of Channel Design for Soil and Water Conservation. Technical Paper TP-61, USDA: 1–34.

U.S. Department of Agriculture (USDA). 1977. *Design of Open Channels*. Technical Release No. 25, Soil Conservation Service, Washington, DC.

U.S. Department of Agriculture (USDA). 1987. Stability Design of Grass-Lined Open Channels. Agricultural Handbook 667.

U.S. Environmental Protection Agency. EPA-600/8-77-014. U.S. Environmental Protection Agency. Cincinnati, OH. pp. 90–92.

U.S. Environmental Protection Agency. 1976. *Areawide Assessment Manual.* Three Volumes. Municipal Envir. Research Lab., Cincinnati, OH.

U.S. Environmental Protection Agency. 1978. *Environmental Monitoring Series: Mercury, Lead, Arsenic, and Cadmium in Biological Tissue,* EPA-600/4-78-051. Environmental Monitoring and Support Laboratory, Las Vegas, NV.

U.S. Environmental Protection Agency. 1982. *Handbook of Sampling and Sample Preservation of Water and Wastewater*. EPA 600/4-82-029 (including later additions under this report number).

U.S. Environmental Protection Agency. 1983. *Methods for Chemical Analysis of Water and Wastes*. EPA 600/4-79-020 (including later additions and revisions under this report number.) U.S. Environmental Protection Agency, Washington, DC. 1979; revised.

U.S. Environmental Protection Agency. 1983a. *Final Report for the Nationwide Urban Runoff Program.* Water Planning Division, PB 84-185552. Washington, DC.

U.S. Environmental Protection Agency. 1983b. *Methods for Chemical Analysis of Water and Wastes,* EPA-600/4-79-020. U.S. Environmental Protection Agency, Cincinnati, OH.

U.S. Environmental Protection Agency. 1986. *Methodology for Analysis of Detention Basins for Control of Urban Runoff Quality.* Water Planning Division, EPA 440/5-87-001, Washington, DC.

U.S. Environmental Protection Agency. 1986. *Quality Criteria for Water*. EPA 440/5-86-001. U.S. Environmental Protection Agency, Washington, DC.

U.S. Environmental Protection Agency. 1997 . *EPA Methods and Guidance for the Analysis of Water*. EPA 821/C-97-001. U.S. Environmental Protection Agency, Washington, DC.

U.S. Environmental Protection Agency. 1999a. *Stormwater Technology Fact Sheet: Vegetated Swales*. EPA Office of Water, Document 832-F-99-006.

U.S. Environmental Protection Agency. 1999b. *Post-construction Storm Water Management in New Development and Redevelopment: Grassed Swale*. EPA Office of Water, Storm Water Phase II BMP's. http://www.tetratech-test.com/bmpmanual/htmlfolder/post_24.htm

U.S. Environmental Protection Agency, 1999. EPA 600/R-00/016. Water Supply and Water Resources Division, National Risk Management Research Laboratory, Cincinnati, OH. 231 pp.

U.S. Public Health Service. 1981. *Second Annual Report on Carcinogens*. U.S. Dept. of Health and Human Services.

Uguccioni, M., and Zeiss, C. 1997. Improvement of leachate prediction through municipal solid waste layers. *Journal of the American Water Resources Association* 33(6):1265–1278.

Urban Runoff Toxicity. *Water Sci. and Tech.* (G.B.) 39(12):33.

Urbonas, B., and Stahre, P. 1993. Stormwater: Best Management Practices and Detention. Saddle River, NJ: Prentice Hall.

Urbonas, B. 1999. Design of a sand filter for stormwater quality enhancement. *Water Environment Research* 71(1):102–113.

Urrutia, M.M. 1997. General bacterial sorption processes. In *Biosorbents for metal ions* (ed. by J.D.A. Wase and C. Forster). Taylor and Francis. pp. 39–66.

Valiron, F. 1992. Usual techniques for stormwater pollutant removal in urban areas (in French). *Provisory report for the Seine-Normandie Water Agency*, February, 61 pp.

Van Benschoten, J.E., Reed, B.E., Matsumoto, M.R., and McGarvey, P.J. 1994. Metal removal by soil washing for an iron oxide coated sandy soil. *Water Environment Research* 66(2):168–174.

Van Buren, M.A., Watt, W.E., and Marsalek, J. 1996. Enhancing the removal of pollutants by an onstream pond. *Water Science and Technology* 33(4-5):325–332.

Van Der Tak, L.D., Bennett, M.S., Kiernan, J., and Pocci, F.J. 1996. Application of Clean Sampling Methods to Quantifying Trace Amounts of Mercury in CSO Discharges in New Jersey. In: *Proc. WEFTEC'96, 69th Annu. Conf. & Expo.*, Dallas, TX, Water Environ. Fed. 8–54.

van Geen, A., and Luoma, S.N. 1999. A Record of Estuarine Water Contamination from the Cd Content of Foraminiferal Tests in San Francisco Bay, California. *Marine Chemistry* 64(1-2):57.

Vaughn. 1988. Industrial applications of zeolites. *Chemical Engineering Progress* 84:32-41.

Vedagiri, U., and Ehrenfeld, J. 1991. Effects of sphagnum moss and urban runoff on bioavailability of lead and zinc from acidic wetlands of the New Jersey Pinelands. *Environmental Pollution* 72:317–330.

Vignoles M., and Herremans, L. 1995. Metal pollution of sediments contained in runoff water in the Toulouse city. (in French). In: *NOVATECH 95, 2nd International Conference on Innovative Technologies in Urban Storm Drainage*. May 30–June 1, Lyon, France. Organized by Eurydice 92 and Graie. pp. 611–614.

Viraraghavan, T., Henning, M.K., Mourits, F., and Ranganathan, R. 1989. Coalescence/Filtration of Water-In-Oil Emulsions. *43rd Purdue Industrial Waste Conference Proceedings*. Chelsea, MI: Lewis Publishers. 435–440.

Volesky, B. 1990. Removal and recovery of heavy metals by biosorption. In: *Biosorption of Heavy Metals*, ed. by B. Volesky. Boca Raton, FL: CRC Press, Inc. pp. 7–43.

W&H Pacific. 1992a. *Compost Storm Water Treatment System: Executive Summary*. W&H Pacific, Portland, OR.

W&H Pacific. 1992b. *Methods and Results Summary: Compost Storm Water Filter System*. W&H Pacific, Portland, OR.

Wada, Y., and Miura, H. 1990. Effect and evaluation of storm runoff control by permeable combined infiltration facilities for controlling storm runoff. In: *Proceedings of the 5th International Conference on Urban Storm Drainage*. July, Osaka, Japan. pp. 789–794.

Wada, Y., and Miura, H. 1999. Combined Sewer Overflow Control Using Large-Scale Storage Pipe for Flood Control. In: *Proc. the Eighth International Conference on Urban Storm Drainage*. August 30–September 3, Sydney, Australia, ed. by I.B. Joliffe and J.E. Ball. The Institution of Engineers Australia, The International Association for Hydraulic Research, and The International Association on Water Quality, 1663.

Walesh, S. 1982. Ad hoc session on quantity issues. In: *Proceedings of the Conference on Stormwater Detention Facilities, Planning, Design, Operation, and Maintenance*, August, Henniker, NH, ed. by W. DeGroot, published by the American Society of Civil Engineers, New York.

Walker, W.J., McNutt, R.P., and Maslanka, C.A.K. 1999. Potential Contribution of Urban Runoff to Surface Sediments of the Passaic River: Sources and Chemical Characteristics. *Chemosphere* 38(2):363.

Wallinder, I.O., and Leygraf, C. 2001. Seasonal Variations in Corrosion Rate and Runoff Rate of Copper Roofs in an Urban and a Rural Atmospheric Environment. *Corrosion Sci.* 43:2379.

Walpole, R.E., and Myers, R.H. 1993. *Probability and Statistics for Engineers and Scientists*, fifth edition. Saddle River, NJ: Prentice-Hall, Inc. 725 pp.

Walsh, P.M., Barrett, M.E., Malina, J., and Charbeneau, R. 1997. Use of Vegetative Controls for Treatment of Highway Runoff. University of Texas at Austin. CRWR Online Report 97-5.

Wang, T. 1981. Transport Deposition and Control of Heavy Metals in Highway Runoff. Washington State Dept. of Transportation. 102 pp.

Wang, T.S., Spyridakis, D.E., Mar, B.W., and Horner, R.R. 1980. *Transport Deposition and Control of Heavy Metals in Highway Runoff.* WA-RD-39.10, U.S. Dept. of Transportation.

Wanielista, M.P., Yousef, Y.A., and Harper, H.H. 1983. Hydrology/Hydraulics of Swales. *Open Channel Hydraulics Workshop*, by the University of Central Florida and American Society of Civil Engineers, Florida Section, Sept., Orlando.

Wanielista, Martin P., Charba, Julius, Dietz, John, Lott, Richard S., and Russell, Bryan. 1991. *Evaluation of the Stormwater Treatment Facilities at the Lake Angel Detention Pond, Orange County, Florida*. State of Florida Department of Transportation Environmental Research Paper FL-ER-49-91. National Technical Information Service, Springfield, VA.

Warick, R.M. 1997. A new method for detecting pollution effects on marine macrobenthic communities. *Marine Biology* 92:557–562. 1986.

Waschbusch, R.J. 1999. *Evaluation of the Effectiveness of an Urban Stormwater Treatment Unit in Madison, Wisconsin*, 1996-97. U.S. Geological Survey Open-File Report 99-4195. Middleton, WI.

Watts, R.J. 1998. *Hazardous Wastes: Sources, Pathways, Receptors*. New York: John Wiley & Sons, Inc. 760 pp.

Watzin, M.C., McIntosh, A.W., Brown, E.A., Lacey, R., Lester, D.C., Newbrough, K.L., and Williams, A.R. 1997. Assessing sediment quality in heterogeneous environments: A case study of a small urban harbor in Lake Champlain, Vermont, USA. *Environmental Toxicology and Chemistry* 16(10):2125–2135.

Webster, J.J., Hampton, G.J., and Leach, F.R. 1984. ATP in soil: A new extractant and extraction procedure. *Soil Biol. Biochem.* 16:335–342.

Webster, J.G., Brown, K.L., and Webster, K.S. 2000. Source and Transport of Trace Metals in the Hatea River Catchment and Estuary, Whangarei, New Zealand. *New Zealand J. Marine Freshwater Res.* 34:187.

Webster's New Universal Unabridged Dictionary. 1996. Barnes and Noble, Inc.

Weis, P., and Weis, J.S. 1996. Leaching from Chromated-copper-arsenic (CCA) Treated Wood and Effects in Chesapeake Bay. *Abstract Book: SETAC 17th Annu. Meeting*, Washington, DC, *Soc. Environ. Toxicol. and Chem.* 88.

Whipple, W., and Hunter, J.V. 1981. Settleability of urban runoff pollution. *Journal WPCF* 53(12):1726.

Wiegand, C., Schueler, T., Chittenden, W., and Jellick, D. 1986. Comparative costs and cost effectiveness of urban best management practices. In: *Engineering Foundation Conference: Urban Runoff Quality—Impact and Quality Enhancement Technology*, June, Henniker, NH, ed. by B. Urbonas and L. A. Roesner. Published by the American Society of Civil Engineers, New York.

Wiese, J., and Scmitt, T.G. 1999. An Evaluation of the Pollutant Discharge of an Entire River Catchment Area Caused by Stormwater Runoff. In: *Proc. the Eighth International Conference on Urban Storm Drainage*. August 30–September 3, Sydney, Australia, ed. by I.B. Joliffe and J.E. Ball. The Institution of Engineers Australia, The International Association for Hydraulic Research, and The International Association on Water Quality, 1335.

Wigginton, B., and Lenhart, J. 2000. Using iron-infused media and stormfilter technology for the removal of dissolved phosphorus from stormwater discharges. www.stormwatermgt.com

Wigington, P.J., Randall, C.W., and Grizzard, T.J. 1983. Accumulation of selected trace metals in soils of urban runoff detention basins. *Water Resources Bulletin* 19(5):709.

Wilber, W.G., and Hunter, J.V. 1980. *The Influence of Urbanization on the Transport of Heavy Metals in New Jersey Streams.* Water Resources Research Institute, Rutgers University. New Brunswick, NJ.

Wilde, F.D. 1994. *Geochemistry and Factors Affecting Ground-Water Quality at Three Storm-Water-Management Sites in Maryland: Report of Investigations No. 59*. Department of Natural Resources, Maryland Geological Survey, Baltimore, MD. Prepared in Cooperation with the U.S. Department of the Interior Geological Survey, The Maryland Department of the Environment, and The Governor's Commission on Chesapeake Bay Initiatives.

Williams, S.T., McNeilly, T., and Wellington, E.M.H. 1977. The decomposition of vegetation growing on metal mine waste. *Soil Biol. Biochem.* 9:271–275.

Williams, D.T., and Julien, P.Y. 1989. Examination of stage-discharge relationships of compound/composite channels, In: *Proceedings of the International Conference on Channel Flow and Catchment Runoff*, ed. by B.C. Yen, University of Virginia, Charlottesville, 22-26 May. pp. 478–488.

Williams, L.H. 1982. Effectiveness of stormwater detention. In: *Proceedings of the Conference on Stormwater Detention Facilities, Planning, Design, Operation, and Maintenance*, August, Henniker, NH, ed. by W. DeGroot. Published by the American Society of Civil Engineers, New York.

Williamson, R.B., and Morrisey, D.J. 2000. Stormwater Contamination of Urban Estuaries. 1. Predicting the Build-Up of Heavy Metals in Sediments. *Estuaries* 23:56.

Wilson, L.G. 1967. Sediment Removal from Flood Water. *Transactions American Society of Agricultural Engineers* 10(1):35–37.

Wilson, L.G., Osborn, M.D., Olson, K.L., Maida, S.M., and Katz, L.T. 1990. The Groundwater Recharge and Pollution Potential of Dry Wells in Pima County, Arizona. *Groundwater Monitoring Review* 10:114–121.

Wind, S. 1996. Three years after—failure or success. *Land and Water* 40(6):37.

Wingington, P.J., Randall, C., and Grizzard, T. 1986. Accumulation of Selected Trace Metals in Soils of Urban Runoff Swale Drains. *Water Resources Bulletin* 22(1):73–78.

Winogradoff, D.A., and Coffman, L.S. 1999. Bioretention: Water Quality Performance Data and Design Modifications. *ASCE 26th Annu Conf. Water Resour. Plann. Manage. 1999 Annu. Conf. Environ. Eng.*, Tempe, AZ.

Wittenberg, H., and Sivapalan, M. 1999. Watershed Groundwater Balance Estimation Using Streamflow Recession Analysis and Baseflow Separation. *J. Hydrol.* (Neth.) 219(1-2):20.

Wojtenko, I., Fan, C.-Y., and Field, R. 2001a. U.S. EPA CSO Capstone Report: Control and Treatment. 5th International Conf.: Diffuse/Nonpoint Pollution and Watershed Management. CD-ROM.

Wojtenko, I., Ray, A., and Field, R. 2001b. Treatment of Stormwater by Natural Organic Materials. In: *Proc. ASCE EWRI Conf.—Bridging the Gap: Meeting the World's Water and Environmental Resources Challenges*. CD-ROM.

Wolff, J., Ebeling, J., and Muller, A. 1988. Waste Water Irrigation Suited to the Environment as Shown by the Example of "Abwasserverband Wolfsburg." *Hydrological Processes and Water Management in Urban Areas—Proceedings of the Conference*, April 24–29, International Hydrological Programme, UNESCO, Duisburg, Federal Republic of Germany. pp. 599–607.

Wong, M.Y., Sauser, K.R., Chung, K.T., Wong, T.Y., and Liu, J.K. 2001. Response of the Ascorbate-Peroxidase of Selenastrum Capricornutum to Copper and Lead in Stormwaters. *Environ. Monitoring Assessment* 67:361.

Woodward, S., and Rock, C.A. 1995. Control of Residential Stormwater by Natural Buffer Strips. *Lake and Reservior Management* 11(1):37–45.

Woodward-Clyde Consultants. 1994. *Sunnyvale Detention Basin Demonstration Project*. Prepared for Santa Clara Valley Nonpoint Source Pollution Control Program.

Woodward-Clyde Consultants. 1990. *Evaluation of Existing Stormwater Pump Stations in Santa Clara County for Pollution Control Potential*. Prepared for Santa Clara Valley Water District.

Yamada, K. 1993. Infiltration pitch for individual houses to mitigate storm water impact. In: *Proceedings of the 6th International Conference on Urban Storm Drainage*, ed. by J. Marsalek and H.C. Torno. September 12–17, Niagara Falls, ON. IAWQ, London. Published by Seapoint Publishers, Victoria, BC. pp. 1110–1114.

Yang, O.B., Kim, J.C., Lee, J.S., and Kim, Y.G. 1993. Use of activated carbon fiber for direct removal of iodine from acetic acid solution. *Industrial Engineering and Chemistry Research* 32:1692–1697.

Yousef, Y.A., Wanielista, M.P., and Harper, H.H. 1986. Design and effectiveness of urban retention basins. In: *Engineering Foundation Conference: Urban Runoff Quality–Impact and Quality Enhancement Technology*, June, Henniker, NH, ed. by B. Urbonas and L. A. Roesner. Published by the American Society of Civil Engineers, New York.

Yousef, Y.A., and Lin, L.Y. 1990. Toxicity of bottom sediments in detention ponds. In: *Environmental Engineering: Proceedings of the 1990 Specialty Conference*. pp. 809–816. American Society of Civil Engineers, New York.

Yousef, Y.A., Harper, H.H., Wiseman. L.P., and Bateman, J.M. 1985. Consequential species of heavy metals in highway runoff. *Transportation Research Record*, #1017.

Yu, S.L., Kasnick, M.A., and Byearne, M.R. 1993. A level spreader/vegetative buffer strip system for urban stormwater management. In: *Integrated Stormwater Management*, ed. by R. Field, M.L. O'Shea, and K.K. Chin. Boca Raton, FL: Lewis Publishers. pp. 93–103.

Yu, S.L., Liao, S-L., Fitch, G.M., and Long, S. 1996. Constructed wetlands for stormwater management. In: *69th Annual Water Environment Federation Technical Exposition and Conference (WEFTEC)*. Dallas, TX. Water Environment Federation, Alexandria, VA. Vol. 7, parts I, II, and III. pp. 167–178.

Yuan, Y., Hall, K., and Oldham, C.A. 2001. Preliminary Model for Predicting Heavy Metal Contaminant Loading From an Urban Catchment. *Sci. Total Environ.* 266:299.

Zaman, S. 1999. Assessment of Stormwater Quality and Pollution Pathways Identification in Urbanised Catchments. In: *Proc. the Eighth International Conference on Urban Storm Drainage*. August 30–September 3, Sydney, Australia, ed. by I.B. Joliffe and J.E. Ball. The Institution of Engineers Australia, The International Association for Hydraulic Research, and The International Association on Water Quality. 1744.

Zamzow, M.J., and Murphy, J.E. 1992. Removal of metal cations from water using zeolites. *Separation Science and Technology* 27(14):1969–1984.

Zhou, P., Teng, Z., and Sansalone, J. 2001. Analysis of Geo-Spatial Distribution of Heavy Metals in Highway Surficial Soils Loaded by Urban Storm Water. In: *WEFTEC 2001 Conf. Proc.* CD-ROM.

Zobrist, J., Muller, S.R., Ammann, A., Bucheli, T.D., Mottier, V., Ochs, M., Schoenenberger, R., Eugster, J., and Boller, M. 2000. Quality of Roof Runoff for Groundwater Infiltration. *Water Res.* (G.B.) 34:1455.

WASTEWATER UTILITY

Alabama
Montgomery Water Works & Sanitary Sewer Board

Alaska
Anchorage Water & Wastewater Utility

Arizona
Gila Resources
Glendale, City of, Utilities Department
Mesa, City of
Phoenix Water Services Department
Pima County Wastewater Management

Arkansas
Little Rock Wastewater Utility

California
Calaveras County Water District
Central Contra Costa Sanitary District
Contra Costa Water District
Crestline Sanitation District
Delta Diablo Sanitation District
Dublin San Ramon Services District
East Bay Municipal Utility District
El Dorado Irrigation District
Escondido, City of
Fairfield-Suisun Sewer District
Irvine Ranch Water District
Las Virgenes Municipal Water District
Lodi, City of
Los Angeles, City of
Los Angeles County, Sanitation Districts of
Napa Sanitation District
Orange County Sanitation District
Palo Alto, City of
Riverside, City of
Sacramento Regional County Sanitation District
San Diego Metropolitan Wastewater Department, City of
San Francisco, City & County of
San Jose, City of
Santa Barbara, City of
Santa Rosa, City of
South Bayside System Authority
South Orange County Wastewater Authority
Union Sanitary District
West Valley Sanitation District

Colorado
Boulder, City of
Colorado Springs, City of
Littleton/Englewood Water Pollution Control Plant
Metro Wastewater Reclamation District, Denver

Connecticut
New Haven, City of, WPCA

District of Columbia
District of Columbia Water & Sewer Authority

Florida
Broward, County of
Fort Lauderdale, City of
Gainesville Regional Utilities
JEA
Kissimmee, City of, Department of Water Resources
Miami-Dade Water & Sewer Authority
Orange County Utilities Department
Orlando, City of
Reedy Creek Improvement District
Seminole County Environmental Services
St. Petersburg, City of
Stuart Public Utilities
Tallahassee, City of
Tampa, City of
West Palm Beach, City of

Georgia
Atlanta Department of Watershed Management
Augusta, City of
Clayton County Water Authority
Cobb County Water System
Columbus Water Works
Fulton County
Gwinnett County Department of Public Utilities
Macon Water Authority
Savannah, City of

Hawaii
Honolulu, City and County of

Idaho
Pocatello, City of

Illinois
American Bottoms Wastewater Treatment Plant
Dupage, County of
Greater Peoria Sanitary District
Kankakee River Metropolitan Agency
Metropolitan Water Reclamation District of Greater Chicago
North Shore Sanitary District
Wheaton Sanitary District

Indiana
Fort Wayne, City of

Iowa
Ames, City of
Des Moines Metro Wastewater Reclamation Authority

Kansas
Johnson County Unified Wastewater Districts
Unified Government of Wyandotte County/Kansas City, City of

Kentucky
Louisville & Jefferson County Metropolitan Sewer District

Louisiana
Sewerage & Water Board of New Orleans

Maine
Bangor, City of
Portland Water District

Maryland
Anne Arundel County Bureau of Utility Operations
Howard County Department of Public Works
Washington Suburban Sanitary Commission

Massachusetts
Boston Water & Sewer Commission
Upper Blackstone Water Pollution Abatement District

Michigan
Ann Arbor, City of
Detroit, City of
Grand Rapids, City of
Holland Board of Public Works
Lansing, City of
Owosso Mid-Shiawassee County WWTP
Saginaw, City of
Wayne County Department of Environment
Wyoming, City of

Minnesota
Rochester, City of

Missouri
Independence, City of
Kansas City Missouri Water Services Department
Little Blue Valley Sewer District
Metropolitan St. Louis Sewer District

Nebraska
Lincoln Wastewater System

Nevada
Henderson, City of

New Jersey
Passaic Valley Sewerage Commissioners

New York
New York City Department of Environmental Protection

North Carolina
Cary, Town of
Charlotte/Mecklenburg Utilities
Durham, City of
Metropolitan Sewerage District of Buncombe County
Orange Water & Sewer Authority

Ohio
Akron, City of
Butler County Department of Environmental Services
Columbus, City of
Metropolitan Sewer District of Greater Cincinnati
Northeast Ohio Regional Sewer District
Summit, County of

Oklahoma
Tulsa, City of

Oregon
Clean Water Services
Eugene/Springfield Water Pollution Control
Water Environment Services

Pennsylvania
Philadelphia, City of
University Area Joint Authority, State College

South Carolina
Charleston Commissioners of Public Works
Mount Pleasant Waterworks & Sewer Commission
Spartanburg Sanitary Sewer District

Tennessee
Cleveland, City of
Knoxville Utilities Board
Murfreesboro Water & Sewer Department
Nashville Metro Water Services

Texas
Amarillo, City of
Austin, City of
Dallas Water Utilities
Denton, City of
El Paso Water Utilities
Fort Worth, City of
Gulf Coast Waste Disposal Authority
Houston, City of
San Antonio Water System
Trinity River Authority

Utah
Salt Lake City Corporation

Virginia
Alexandria Sanitation Authority
Arlington, County of
Fairfax County Virginia
Hampton Roads Sanitation District
Henrico, County of
Hopewell Regional Wastewater Treatment Facility
Loudoun County Sanitation Authority
Lynchburg Regional WWTP
Prince William County Service Authority
Richmond, City of

Washington
Edmonds, City of
Everett, City of
King County Department of Natural Resources
Seattle Public Utilities
Yakima, City of

Wisconsin
Green Bay Metro Sewerage District
Madison Metropolitan Sewerage District
Milwaukee Metropolitan Sewerage District
Racine, City of
Sheboygan Regional Wastewater Treatment
Wausau Water Works

Australia
South Australian Water Corp.
Sydney Water Corp.
Water Corp. of Western Australia

Canada
Greater Vancouver Regional District
Toronto, City of, Ontario

Mexico
Servicios de Agua y Drenaje de Monterrey, I.P.D.

United Kingdom
Yorkshire Water Services Limited

STORMWATER UTLITY

California
Los Angeles, City of, Department of Public Works
Monterey, City of
Santa Rosa, City of

Colorado
Boulder, City of

Georgia
Griffin, City of

Kansas
Overland Park, City of

Kentucky
Louisville & Jefferson County Metropolitan Sewer District

Maine
Portland Water District

North Carolina
Cary, Town of
Charlotte, City of, Stormwater Services

Pennsylvania
Philadelphia, City of

STATE

Fresno Metropolitan Flood Control District, Calif.
Urban Drainage & Flood Control District, Colo.

CORPORATE

ADS Environmental Services
The ADVENT Group Inc.
Alan Plummer & Associates
Alden Research Laboratory
Alpine Technology Inc.
Aquateam–Norwegian Water Technology Centre A/S
BaySaver Inc.
BioVir Inc.
Black & Veatch
Boyle Engineering Corporation
BPR CSO
Brown & Caldwell
Burns & McDonnell
CABE Associates Inc.
The Cadmus Group
Camp Dresser & McKee Inc.
Carollo Engineers Inc.
Carpenter Environmental Associates Inc.
CDS Technologies Inc.
Chemtrac Systems Inc.
CH2M HILL
Clancy Environmental Consultants Inc.
Damon S. Williams Associates, LLC
David L. Sheridan, P.C.
Earth Tech Inc.
Ecolab Water Care Services
EMA Inc.
The Eshelman Company Inc.
Finkbeiner, Pettis, & Strout Inc.
Frontier Geosciences Inc.
ftn Associates Inc.
Gannett Fleming Inc.
GE Betz
Golder Associates Inc.
Greeley and Hansen LLC
The HACH Company
Hazen & Sawyer, P.C.
HDR Engineering Inc.
HNTB Corporation
HydroQual Inc.
Insituform Technologies Inc.
Institute for Environmental Technology & Industry, Korea
Jacobson Helgoth Consultants Inc.
Jason Consultants Inc.
Jordan, Jones, & Goulding Inc.
KCI Technologies Inc.
Kelly & Weaver, P.C.
Kennedy/Jenks Consultants
Komline Sanderson Engineering Corporation
Lawler, Matusky & Skelly Engineers, LLP
Limno-Tech Inc.
Lombardo Associates Inc.
Malcolm Pirnie Inc.
McKim & Creed
MEC Analytical Systems Inc.
MWH
New England Organics
Odor & Corrosion Technology Consultants Inc. (OCTC)
ONDEO Degremont Inc.
Oswald Green, LLC
PA Government Services Inc.
Parametrix Inc.
Parsons
Post, Buckley, Schuh & Jernigan
R&D Engineering/Conestoga Rover & Associates
The RETEC Group
R.M. Towill Corporation
Ross & Associates Ltd.
Royce Technologies
SAIC Maritime Technical Group
Sear-Brown
Sierra Environmental Services
SYNAGRO
Tetra Tech Inc.
Trojan Technologies Inc.
URS Corporation
USFilter NATC
Vortechnics Inc.
Wade-Trim Inc.
Weston Solutions Inc.
Woodard & Curran
WRc/D&B, LLC
WWETCO, LLC

INDUSTRY

American Electric Power
ChevronTexaco Energy Research & Technology Company
The Coca-Cola Company
Dow Chemical Company
DuPont Company
Eastman Kodak Company
Merck & Company Inc.
ONDEO Services
Procter & Gamble Company
PSEG Services Corp.
Severn Trent Services Inc.
Shell Global Solutions
Thames Water Plc
United Water Services LLC

Note: List as of 5/12/03

WERF Product Order Form

As a benefit of joining the Water Environment Research Foundation, subscribers are entitled to receive one complimentary copy of all final reports and other products.Additional copies are available at cost (usually $10).To order your complimentary copy of a report, please write "free" in the unit price column.WERF keeps track of all orders.If the charge differs from what is shown here, we will call to confirm the total before processing.

Name ______________________ Title ______________________

Organization ______________________

Address ______________________

City ________ State ________ Zip Code ________ Country ________

Phone ________ Fax ________ Email ________

Stock #	Product	Quantity	Unit Price	Total
			Postage & Handling	
			VA Residents Add 4.5% Sales Tax	
			Canadian Residents Add 7% GST	
			TOTAL	

Method of Payment: (All orders must be prepaid.)

❑ Check or Money Order Enclosed

❑ Visa ❑ Mastercard ❑ American Express

Account No. ________ Exp. Date ________

Signature ________

Shipping & Handling:

Amount of Order	United States	Canada & Mexico	All Others
Up to but not more than:	Add:	Add:	Add:
$20.00	$5.00*	$8.00	50% of amount
30.00	5.50	8.00	40% of amount
40.00	6.00	8.00	
50.00	6.50	14.00	
60.00	7.00	14.00	
80.00	8.00	14.00	
100.00	10.00	21.00	
150.00	12.50	28.00	
200.00	15.00	35.00	
More than $200.00	Add 20% of order	Add 20% of order	

*minimum amount for all orders

Note: Please make checks payable to the Water Environment Research Foundation.

To Order (Subscribers Only):

Log on to www.werf.org and click on the "Product Catalog."

Phone: (703) 684-2470
Fax: (703) 299-0742.

WERF
Attn: Subscriber Services
635 Slaters Lane
Alexandria, VA 22314-1177

To Order (Non-Subscribers):

Non-subscribers may be able to order WERF publications either through WEF (www.wef.org) or IWAP (www.iwapublishing.com).Visit WERF's website at www.werf.org for details.

IWA Publishing's authorised EU representative for General Product Safety Regulations is Diane D'Arras, 15 rue Duret, 75116 Paris, France, e-mail: safety@iwap.co.uk.

Printed and bound by CPI Group (UK) Ltd, Croydon, CR0 4YY
12/05/2026
02108876-0011